Research and Development of Modified Building Materials

Research and Development of Modified Building Materials

Editor

Maria Stefanidou

MDPI • Basel • Beijing • Wuhan • Barcelona • Belgrade • Manchester • Tokyo • Cluj • Tianjin

Editor
Maria Stefanidou
Aristotle University of Thessaloniki
Greece

Editorial Office
MDPI
St. Alban-Anlage 66
4052 Basel, Switzerland

This is a reprint of articles from the Special Issue published online in the open access journal *Materials* (ISSN 1996-1944) (available at: https://www.mdpi.com/journal/materials/special_issues/modified_building_materials).

For citation purposes, cite each article independently as indicated on the article page online and as indicated below:

LastName, A.A.; LastName, B.B.; LastName, C.C. Article Title. *Journal Name* **Year**, *Volume Number*, Page Range.

ISBN 978-3-0365-6742-6 (Hbk)
ISBN 978-3-0365-6743-3 (PDF)

Contents

About the Editor

Maria Stefanidou

Maria Stefanidou Professor, Aristotle University of Thessaloniki. Research interests related to materials for the consolidation and restoration of monuments and historical buildings and modified traditional and modern materials using new technologies.

Preface to "Research and Development of Modified Building Materials"

This book represented an effort to gather research works from scholars all over the world dealing with composites. In construction, composites have a key role in the performance and functionality of buildings, and as technology is continuously opening up new horizons, new, innovative and high-tech materials are being derived, covering different needs.

I want to thank all those contributing to the creation of this e-book. It requires devotion and hard work to manage to deliver something to the scientific community. I hope it will be of use to as many as possible.

For me, it was an exciting journey, and I learnt a lot!

Maria Stefanidou
Editor

materials

Article

Effect of Single-Walled Carbon Nanotubes on Strength Properties of Cement Composites

Jiayuan Kang, Salam Al-Sabah * and Roger Théo

School of Civil, Structural & Environmental Engineering, University College Dublin, D04 V1W8 Dublin 04, Ireland; jiayuan.kang@ucdconnect.ie (J.K.); theo.roger@ucdconnect.ie (R.T.)

* Correspondence: salam.al-sabah@ucd.ie

Received: 8 January 2020; Accepted: 12 March 2020; Published: 13 March 2020

Abstract: This study aimed to investigate the effects of single-walled carbon nanotubes (SWCNTs) on strength the properties of cement composites when surfactant (SAA) was applied as the dispersion method. TritonX-100 (TX10) was used as the SAA to pretreat SWCNTs, which has been proved to perform well in dispersing the agglomerates of SWCNTs. In this study, four different concentration of SWCNTs, namely 0.00 wt%, 0.02 wt%, 0.04 wt%, and 0.06 wt% by the mass of cement, were used to prepare cement composite specimens. The compressive strength and flexural strength of specimens were tested and recorded. The results show that the compressive and flexural strengths of cement composites decreased with the increase in the concentration of SWCNTs without the addition of TX10. However, when SWCNT suspensions were pretreated with TX10, the strength variation pattern changed; the compressive and flexural strengths of cement composites increased as a function of the concentration of SWCNTs, although there were reductions compared to non-TX10-treated specimens at all concentrations of SWCNTs. Furthermore, the relationship between the strength of cement composites and bulk density of specimens was considered.

Keywords: single wall carbon nanotubes; surfactant; TritonX-100; mass of specimens; compressive strength; flexural Strength; bulk density

1. Introduction

Concrete is known to have high compression strength but weak tensile capacity. Many scholars use steel fiber, polyvinyl alcohol (PVA) fiber, polypropylene (PP) fiber, carbon fiber, etc. as micro-reinforcement paste in concrete to prepare composite materials with better bearing capacity [1–8]. The results show that the mix of fiber can effectively enhance the tensile strength and toughness. The problem is that the strength of toughening fibers used in concrete are relatively low, and may cause larger porosity, imposing a negative effect on the mechanical properties of concrete. High strength fibers that are smaller than microfibers for applications of high-performance fiber reinforced concrete composite still has potential. Thus, the feasibility of applying carbon nanotubes (CNTs), as potential nanofibers, into cementitious composites has been widely studied due to CNTs' unique properties.

CNTs are known to have spectacular properties, such as high specific conductivity, high Young's modulus, high yield strength, good thermal properties, etc. [9–21]. CNTs' thermal conductivity is at least twice that of diamond [22], and its electricity conductivity is 1000 folds that of copper [12]. Specifically for mechanical properties, CNTs have an average Young's modulus around 1 TPa, which is approximately five times greater than steel, and the tensile strength is up to 63 GPa which is approximately 50 times greater than steel [12–17,19]. CNTs have great potential for reinforcing cementitious composites.

However, when comparing CNTs to traditional microfibers, it is evident that using CNTs in a cement matrix is challenging. One of the main obstacles is the method to disperse CNTs homogeneously in the cement matrix due to their strong van der Waals force (VDW), which originates from their polarizable extended π-electron systems [20], causing CNTs to have a tendency to agglomerate [23–25]. This behavior can cause negative effects to the mechanical properties of the cement composites [25]. A homogeneous morphology possesses an improved mechanical resistance, as mentioned by Gavallaro et al. [26] and Lisuzzo et al. [27]. According to Cota et al. [28], the mixing process using a Hobart mixer, commonly used to prepare mortar paste, cannot ensure proper dispersion of CNTs within the cementitious matrix, resulting in large CNT clusters within the hydrated paste. Since the CNTs are not well dispersed in the cement matrix, their spectacular physical properties cannot be fully revealed. The agglomerates and bundles could lead to a decrease in mechanical performance of cementitious materials [25]. Therefore, finding dispersion methods is an important aspect for further study.

Dispersion performance is a key for improving the interface bonding force among CNTs and cement matrix. In previous studies [24,29–34], it was found that surfactant (SAA), ultrasonic dispersion, chemical covalent modification, and electricity field induced method are the main methods of dispersing CNT, among which SAA and sonication are the most widely used methods. SAA, as the most widely used method, has many options, such as Sodium dodecyl benzene sulfonate (SDBS), sodium deoxycholate (NaDC), polyethylene glycol octylphenol ether (TX10), Gum Arabic (GA), cetyl trimethyl ammonium bromide (CTAB), and dodecyltrimethyl-ammonium bromide (DTAB), which can make carbon nanotubes disperse and stay stable for a long time.

Many scholars have investigated the possibility and feasibility of introducing CNTs into the field of civil engineering, and have found positive effects of CNTs on cementitious composites. Wang et al. [35] found that the addition of multi-walled carbon nanotubes (MWCNTs), with surfactant GA and sonication treatment, improved the mechanical properties of cement composites. The flexural toughness and fracture energy of cement-based composites were increased with 0.08 wt% addition of MWCNTs. Sindu et al. [36] also found that the combination of GA and sonication can improve the mechanical properties of cement composites. Wang et al. [35] found that, when the concentration of MWCNTs was increased to a certain amount, a negative effect would occur. Wu et al. [37] found that the addition of MWCNTs could significantly relieve the mechanical properties of cement-based composite materials. When the amount of MWCNTs reached its optimal value, i.e., 0.1%, the improvement of the mechanical properties of cement-based composite materials was the best. Fatemi and Foroutan [38] concluded that TX10 was the best at dispersing MWCNTs. Luo et al. [39] found that, by using SDBS and TX10 with sonication, the MWCNT–cement composites saw 29.10% and 20.8% increase in flexural and compressive strength, respectively. Mohamed et al. [40] found that, when sonication was applied to CNT/SAA solution to disperse MWCNTs, higher strength of specimens was obtained. Ana et al. [41] found that the addition of MWCNTs treated with superplasticizer (SP) and sonication improves the compressive strength of cement composites. Similarly, Li et al. [42] concluded that, by dispersing SWCNTs using sonication and SP, a 26.3% increase in strength of cement composites was obtained. Parveen et al. [43] concluded that the compressive strength and flexural strength of mortars saw 19% and 7%, respectively, increase when 1% SWCNTs were dispersed by sonication and Pluronic F-127, a novel dispersing agent. Based on the research presented above, it can be concluded that, when SAA and sonication are combined, a positive effect on the mechanical properties on cement composites can be obtained within a certain range of concentrations of CNTs.

However, some studies show a slight increase or even a reduction in mechanical properties of CNT–cement composites when SAA and sonication are used. Camacho et al. [44] concluded that the addition of CNT to Portland cement mortars shows little effect on bending strength and apparent density of mortars. Collins et al. [33] found the compressive strength was decreased to four times less compared to a plain cement composite. Similarly, Shao et al. [45] concluded that, when surfactants, namely SDS (Sodium dodecyl sulfate), TX10, and Polyethylene glycol sorbitan monolaurate (Tween-20), and sonication were combined to disperse CNTs in cement matrix, the compressive strength decreased compared to plain cement composites. Sobolkina et al. [46] also found that the use of SDS led to a drop in the strength of cement composites due to the formation of foam. Kim and Chung [47] found that less yield stress was obtained when SWCNTs were treated by applying sonication and SDS or Sodium deoxycholate (DOC). These results conflict with the conclusions obtained in the above studies.

It is worth mentioning the research of Bharj et al. [48], who found that, by using sonication only (without SAA) to disperse CNTs, it was possible to improve the mechanical properties of cement composite paste. Therefore, sonication probably always has a positive effect on strength for cement composites by dispersing CNTs in matrix, but it should be noted that ultrasonication can result in fragmentation of CNTs and poor stability of dispersion [49]. Therefore, SAA plays an important role in producing a stable CNT dispersion. Different types of SAA can produce either positive or negative effects on cement composites, regardless of their CNT dispersion performance. It has also been found that adding some types of surfactants into cement matrix can cause formation of foam or prevent hydration reaction among cement [45,46,50]. However, few studies have mentioned the formation of foam in sample preparation and the change of rheological behavior of slurry. According to Parveen et al. [43], the surfactant Pluronic F-127 works similarly to polycarboxylate-based superplasticizers, which can disperse cement particles and modify the fluidity of mortar [51]. It is known that plasticizers can keep the slump and cement dosage unchanged while significantly reducing water consumption of mixing additives. However, few studies have considered this and changed the water/cement ratio to keep the slump result constant. Moreover, there are few studies on the effect of using surfactant only to modify the cement composites.

In addition, most studies use MWCNTs as reinforcement materials in cement-based composites [52]. As mentioned above, CNTs have spectacular properties. However, SWCNTs and MWCNTs have many different properties such as Young's modulus [53,54], electrical conductivity, thermoelectric properties, and optical properties [55]. Table 1 shows the differences between SWCNTs and MWCNTs [53,54].

Table 1. Differences between SWCNTs and MWCNTs.

	SWCNT	MWCNT
Structure	Diameter: 0.5–3.0 nm Length: 100 nm–1 cm	Diameter: 5–100 nm Length: 100 nm–1 cm
Typical Young's modulus	1.3 TPa	1.8 TPa
Tensile strength	Up to 53 GPa	Up to 63 GPa
Electrical conductivity	10,000 S/cm	6000 S/cm
Heat conductivity	Max. 6000 W/m·k	Max. 3000 W/m·k

Table 1 shows that the mechanical properties of MWCNTs are generally better than SWCNTs. This is why most studies use MWCNTs to reinforce cementitious composites, as well as their relative lower cost. Most literature reports are related to mechanical properties. However, in recent years, the needs of different functions of cementitious composites are proposed. Heeyoung et al. [56] investigated the heating characteristics of CNT–cement mortars by using SWCNTs and MWCNTs. They concluded that cement mortars mixed with SWCNTs are more effective for modifying the heating characteristics compared to MWCNTs mixed mortars. However, Luigi et al. [57] found that MWCNT–cement composites showed better electrical properties, which can be applied for future work of monitoring the stress level of concrete element. Heeyooung et al. [58] found that adding CNTs to cement-based materials enhances their electrical and thermal characteristics, and they found that adding SWCNTs to concrete grout is more suitable than MWCNT for identifying voids in the duct through electrical resistance analysis. Jin et al. [59] concluded that extremely small amounts of SWCNTs can be used as optical strain sensors if the SWCNTs can be homogeneously dispersed in cement composites. These studies aimed to fully use the better electrical properties and thermal properties of SWCNTs; however, the mechanical properties of CNTs added to cement-based composites were not considered. For future application, on the premise of meeting the above cementitious composite functions, it is worth studying their mechanical properties.

This study investigated the effect of different concentrations of SWCNTs on the strength properties of cement composites when SAA dispersion method was used. TX10 was used as SAA because it has shown a higheer dispersing power than SDBS, CTAB, and DTAB for SWCNT suspensions [49]. In addition, the effect of TX10 on strength properties of cement composites was investigated and a comparison of strength of adding SWCNTs and MWCNTs to cement composites after 28 days of curing was also conducted.

2. Materials and Methods

2.1. Materials

2.1.1. Carbon Nanotubes

The SWCNTs and MWCNTs used in experiments were acquired from RheinChemie-Lanxess, Germany. They are water-based suspensions and can be directly used for production of formulations; their labels are Rhenofit® CNT-3 and Rhenofit® CNT-2, respectively. Table 2 shows the technical data of Rhenofit® CNT-3 and Rhenofit® CNT-2. Figure 1 shows their transmission electron microscope (TEM) images [60]. Agglomerates of CNTs can be observed in the pictures.

Table 2. Brief technical data sheet of Rhenofit® CNT-3/Rhenofit® CNT-2.

	Rhenofit® CNT-3	**Rhenofit® CNT-2**
Composition	0.2 wt% single wall carbon nanotubes	2 wt% Multi wall carbon nanotubes
	0.4 wt% dispersant agent in water	4 wt% dispersant agent in water
Appearance	stable black suspension	stable black suspension
Odor	no odor	no odor
Compatibility	can be mixed with liquid systems and some polar organic solvents	can be mixed with liquid systems and some polar organic solvents
pH Value	7.7	8–10

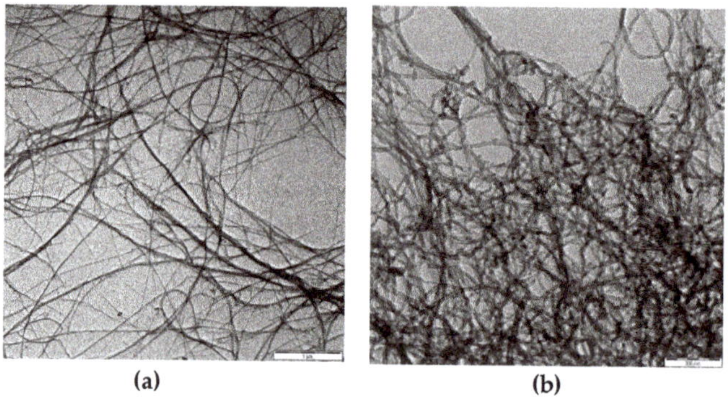

Figure 1. Transmission electron microscope images of: Rhenofit® CNT-3 (**a**); and Rhenofit® CNT-2 (**b**) from [60].

2.1.2. Cement and Sand

Ordinary Portland cement CEM I 42.5 R (rapid hardening) manufactured by Irish Cement (Ireland) was used in the following experiments. Rapid hardening Portland cement (RHPC) is a special purpose cement used in concrete to achieve a higher rate of early strength development, compared to using normal cement. Its main particle size is 5–30 µm, it has a relative density of 2.75–3.20 g/cm^3, and its apparent density is 0.9–1.5 g/cm^3.

Standard silica sand from Glenview (Ireland) was used in the experiments. Sand was dried by placing into oven for 24 h at a temperature of 105–110 °C. The particle size of the sand was variable. To prepare cement mortars, the sand was sieved through a 600-µm sieve. This was carried out to reduce the voids formed when preparing specimens.

2.1.3. Surfactant (SAA)

SAAs, such as TX10, SDS, GA, etc., are amphiphilic in nature and can be used as surface active agents. Structurally, they have a hydrophilic polar head and hydrophobic long tail [61]. They can be used to lower the surface tension between two liquids, between a gas and a liquid, or between a liquid and a solid.

TX10 from Carl Roth GmbH + Co KG (Karlsruhe, Germany) was used in experiments. TX10 is known for having good dispersing performance on SWCNTs [38,49]. Furthermore, it was found by Carriçot et al. [62] that, for different surfactants, the optimum mass ratio of CNTs to dispersant are different. Therefore, the optimum dosage of TX10 to treat SWCNTs should be researched. Wang et al. [63] measured that the saturation adsorption ratio of surfactant vs. SWNT is ca. 0.004 mol/g. With the increase of the concentration of TX10, no SWCNTs in suspension liquid will react with extra TX10.

TX10 has an average molecular mass of 625, a density of 1.065 g/cm^3, and a critical micelle concentration (CMC) of 0.22–0.24 × 10^{-3} moles/liter in water at 25 °C. One mole of TX10 is 625 g on average. The linear molecular formula is $(C_2H_4O)nC_{14}H_{22}O$. It is a nonionic surfactant.

Therefore, 1 g of SWCNTs needs the addition of 0.004 mol TX10 (2.5 g) to reach saturation dispersion [63]. In this experimental set, the dosage of TX10 should be determined. The maximum concentration of SWCNTs is 0.06 wt% by the mass of cement. In each mix group, the amount of cement was 1833 g; thus, the amount of added SWCNTs could be calculated as around 1.0998 g. Thus, the dosage of TX10 was around 2.75 g. Considering TX10 may negatively affect the cementitious matrix, 2.75 g TX10 was used for all concentrations of SWCNTs (0.02 wt%, 0.04 wt%, and 0.06 wt%), and, for all concentration, SWCNTs were well dispersed. In general, the mass ratio of TX10 was 0.15 wt% by the mass of cement.

2.2. Specimens Preparation

2.2.1. Mortar Mix Ratio

The Mix Ratio 1 proportion is shown in Table 3. Normal mortar specimens, consisting of cement, sand, and water, were prepared following Mix Ratio 1. However, for mortars treated by TX10, Mix Ratio 2 was adjusted, as shown in Table 4. This is discussed in Section 2.2.4.

Table 3. Mix Ratio 1 used for mortars (no TX10).

		1:3:0.7 Mix Ratio of Mass
Cement	**Sand**	**Water**
1	3	0.70

Table 4. Mix Ratio 2 used for mortars (TX10 added).

		1:3:0.55 Mix Ratio of Mass
Cement	**Sand**	**Water**
1	3	0.55

Slump tests were conducted to determine the workability of cement mortar mixed according to the mix ratios. The slump test was carried out as prescribed by ASTM C 143-10 and BS 1881:103:1993. According to Jensen et al. [64], surfactant TX100 acts as a plasticizer to some extent. In the following tests, the slump test was carried out to ensure that the mortar mix has similar workability. Mortar was divided into two groups: untreated plain mortars and TX10-treated mortars. For plain mortars, the slump test result was 52 mm.

According to BS EN 206-1:2000, Table 5 gives the European classification of slump test. It can be concluded that the workability of mortar with Mix Ratio 1 belongs to S2 classification.

Table 5. Classification of Workability and Magnitude of Slump.

Classification of Workability	Slump (mm)
S1	10–40
S2	50–90
S3	100–150
S4	≥160

2.2.2. Concentration of SWCNTs in Mortar Preparation

Ferro et al. [65] found that, when very small quantities of MWCNTs (i.e., 0.025–0.08% wt. cement) were added, the mechanical properties of cement composites materials were noticeably improved, and the quantity of CNTs generally matched with the mass of cement of mixing ratio. Therefore, in this study, three concentrations of SWCNT were used: 0.02 wt%, 0.04 wt%, and 0.06 wt% by the mass of cement. In addition, experiments studying the effect of MWCNTs on mechanical properties of CNTs were also carried out to see the difference of strength between specimens with SWCNTs and MWCNTs. The concentrations of MWCNTs were 0.00 wt%, 0.04 wt%, 0.08 wt%, and 0.12 wt% by the mass of cement.

2.2.3. Mortar Mix Design without TX10 Treatment

Table 6 shows the experimental mix design of SWCNT-added mortars without treatment of TX10 and the quantities of the materials. The mixes were named according to: SWCNTXN, where SWCNT represents the material SWCNT; X stands for the concentration of SWCNTs with 0, 2, 4, 6 representing 0.00 wt%, 0.02 wt%, 0.04 wt%, 0.06 wt% respectively; and N represents no SAA added.

Table 6. Mix design of SWCNT-added mortar without treatment of TX10.

Mix Name	Cement (g)	Sand (g)	Water (g)	SWCNT (%wt, cement)	Rhenofit® CNT-3 (g)	Total Mass of Solution (g)
SWCNT0N	1833	5499	1283.1	0.00	0.0	1283.1
SWCNT2N	1833	5499	1099.8	0.02	183.3	1283.1
SWCNT4N	1833	5499	916.4	0.04	366.7	1283.1
SWCNT6N	1833	5499	733.2	0.06	549.9	1283.1

Table 7 shows the experimental design of MWCNT-added mortars without treatment of TX10 and quantities of each materials.

Table 7. Mix design of mortar MWCNTs added without treatment of TX10.

Mix Name	Cement (g)	Sand (g)	Water (g)	MWCNT (%wt, cement)	Rhenofit® CNT-2 (g)	Total Mass of Solution (g)
MWCNT0N	1842	5529	1289.4	0.00	0.0	1289.4
MWCNT4N	1842	5529	1106.1	0.04	36.8	1289.4
MWCNT8N	1842	5529	922.8	0.08	73.6	1289.4
MWCNT12N	1842	5529	739.5	0.12	110.4	1289.4

Rhenofit® CNT-3 and Rhenofit® CNT-2 are water-based CNT suspensions; 99.4% and 96% of the solutions is water, respectively. Thus, during preparation of specimens, the water contained in the

Rhenofit® CNT-3 and Rhenofit® CNT-2 should be subtracted. Tables 6 and 7 show the mix proportion of the materials. Since Rhenofit® CNT-3 and Rhenofit® CNT-2 are water-based CNT suspensions, as the concentration of SWCNTs and MWCNTs increased, less water was needed. This was to keep the quantity of water in mortars uniform.

Mortars with no CNTs were set as control samples, to compare with CNT-containing mortars.

2.2.4. Mortar Mix Design with TX10 Treatment

When TX10 was added as an additive in preparing mortars, it was found that TX10 works as a plasticizer; the workability was greatly improved when Mix Ratio 1 (1:3:0.7) was used. Therefore, the slump test was carried out to determine a new mix ratio for TX10 contained samples. The results show that, when the water/cement ratio was 0.55, the slump test result was 54 mm, similar to that of plain mortar. Therefore, for TX10-treated mortar, the new Mix Ratio 2 of cement:sand:water was 1:3:0.55.

Following Mix Ratio 2, mix designs for TX10 added cement mortars were set. Table 8 shows the experiment groups of mortar with TX10 treatment, and the quantity of each materials. The quantity of TX10 was 2.75 g; as mentioned in Section 2.1.3, the mass of TX10 is 0.15 wt% by the mass of cement. The mixes were named according to: SWCNTXS, where SWCNT represents for the material SWCNT; X stands for the concentration of SWCNTs with 0, 2, 4, and 6 representing 0.00 wt%, 0.02 wt%, 0.04 wt%, and 0.06 wt%, respectively; and S means SAA was mixed. As mentioned above, less water was needed as the concentration of SWCNTs increased.

Table 8. Mix design of SWCNT-added mortars with treatment of TX10.

Mix Name	Cement (g)	Sand (g)	Water (g)	SWCNT (%wt, cement)	Rhenofit® CNT-3 (g)	TX10 (g)
SWCNT0S	1833	5499	1008.2	0.00	0.0	2.75
SWCNT2S	1833	5499	824.85	0.02	183.3	2.75
SWCNT4S	1833	5499	641.55	0.04	366.6	2.75
SWCNT6S	1833	5499	458.25	0.06	549.9	2.75

Table 9 shows the experimental mix design of MWCNT-added mortars without treatment of TX10 and the quantities of the materials. It should be noted that, for all mix designs, the quantity of each material was overprepared for the required three cubes and three prisms. As TX10 was 0.15 wt% by the mass of cement, 2.75 g of TX10 was needed for MWCNT-added mortar.

Table 9. Mix design of MWCNT-added mortar with treatment of TX10.

Mix Name	Cement (g)	Sand (g)	Water (g)	SWCNT (%wt, cement)	Rhenofit® CNT-3 (g)	TX10 (g)
MWCNT0S	1842	5529	1013.1	0.00	0.0	2.76
MWCNT2S	1842	5529	976.3	0.04	36.8	2.76
MWCNT4S	1842	5529	939.5	0.08	73.6	2.76
MWCNT6S	1842	5529	902.7	0.12	110.4	2.76

2.2.5. Mixing Process

For specimens without treatment of TX10, the mix process was as follows:

1. Certain quantities of water and Rhenofit® CNT-3/CNT-2 suspension were mixed as Solution A, and were stirred.
2. All sand and half of Solution A were poured into the stir machine and stirred.
3. All cement was poured into the stir machine and stirred.
4. The rest of Solution A was poured into the stir machine.

For specimens with treatment of TX10, the mix process was as follows:

1. Certain quantities of water, Rhenofit® CNT-3/CNT-2, and TX10 were mixed as Solution B, and were stirred.
2. All sand and half Solution B were poured into the stir machine and stirred.
3. All cement was poured into the stir machine.
4. The rest of Solution B was poured into the mixing machine.

Note that, when TX10 was added, foam was generated in the cement matrix, resulting in the increase of porosity of samples. The mass of specimens and bulk density decreased compared to the specimens with no TX10 added. This is discussed in Section 3.4.

After mixing of materials, the composites were filled into cube and prism molds. The size of the cube mold was 100 mm × 100 mm × 100 mm. The size of the prism mold was 40 mm × 40 mm × 160 mm. After 24 h, the mortar specimens were placed into curing container at a constant temperature of 23 ± 1 °C and relative humidity of 100% according to the ASTM C192. The curing times of SWCNT-added specimens were set as 3, 7, and 28 days. MWCNT-added cement-based specimens were set for 28 days.

2.3. Test Methods

2.3.1. Compressive Test

To test the compressive strength of concrete, mortar cubes were prepared. After curing, the cubes were taken out of their curing containers, and the surfaces were mopped dry. A cube was placed at the center of the loading area of the compressive strength test machine. The loading rate was 0.6 MPa/s.

Loading stopped automatically when cubes were broken, and the compressive strength and force could be directly recorded.

The compressive strength by knowing the maximum applied force was calculated as follows:

$$R_c = F_c/10000 \quad (1)$$

where R_c is the compressive strength in MPa; F_c is the maximum load at the time of fraction in N; and 10,000 is the area of the face of the cube (100 mm × 100 mm).

2.3.2. Flexural Test

For flexural strength test, symmetrical three-point loading was used. In experiments, the 40 mm × 40 mm × 160 mm prisms were tested using three-point loading test. Prisms were placed in the loading area for processing the three-point bending test. Prisms were subjected to the load at a rate of 50 ± 10 N/s. The loading device was turned off when failure occurred.

The flexural strength by knowing the maximum applied force was calculated as follows:

$$R_f = (1.5 \times l \times F_f)/b^3 \quad (2)$$

where R_f is the flexural strength in MPa; b is the side of the square section of the prism in mm; F_f is the peak load applied to center of the prism at fracture in N; and l is the distance between supports in mm. For the three-point loading test device, l equals 100 mm and b equals 40 mm. During loading, the force applied was recorded. By using the maximum force recorded, the flexural strength can be calculated through Equation (2).

3. Results and Discussion

3.1. Comparison of Strength Properties of SWCNTs and MWCNTs Added Specimens

It is known that MWCNTs show better mechanical properties than SWCNTs, and many studies have been conducted to study the effect of MWCNTs on mechanical properties of cement composites.

However, SWCNTs are more advantaged in enhancing electrical properties and thermal properties of cementitious materials, which give SWCNTs great potential for functional cement composites. This section compares the difference of strength properties of SWCNT- and MWCNT-added specimens when they were prepared by the same method. Table 10 shows the 28-day compressive strength and flexural strength of the two types of cement-based specimens with or without the treatment of TX10.

Table 10. Strength properties of SWCNTs and MWCNT-added specimens after 28 days of curing.

SWCNTs					MWCNTs				
SWCNTs Concentration	No TX10		TX10 Added		MWCNTs Concentration	No TX10		TX10 Added	
	Com	Fle	Com	Fle		Com	Fle	Com	Fle
0.00 wt%	30.15	7.12	18.77	4.73	**0.00 wt%**	30.57	7.08	18.12	4.59
0.02 wt%	28.21	6.66	17.64	4.26	**0.04 wt%**	29.26	6.96	18.46	4.79
0.04 wt%	25.79	6.20	18.94	4.95	**0.08 wt%**	27.46	6.94	21.71	4.91
0.06 wt%	25.58	5.64	22.71	5.31	**0.12 wt%**	24.86	6.62	21.65	5.06

Table 10 shows that, for both specimens with no addition of CNTs and TX10, their strengths were around 30.5 MPa, resulting from the same mix ratio. With the addition of CNTs, the strength of both SWCNT–cement composites and MWCNT–cement composites decreased with the increase of concentration of CNTs when CNTs were directly used. This could be attributed to the increase of agglomerates of CNTs when the concentration of CNTs was increased. It can be observed that the compressive strength and flexural strength of MWCNT–cement composites were higher than those with SWCNTs. For instance, when the concentration of CNTs was 0.04 wt%, the compressive strength of MWCNTs was 27.46 MPa and the flexural strength was 6.96 MPa, 6.5% and 12.3% higher than those of SWCNT–cement composites. Thus, MWCNTs show better mechanical properties than SWCNTs, but the difference is relatively small. However, when CNTs were dispersed by TX10, SWCNT–cement composites show slightly better strength properties. When the concentration of CNTs was 0.04 wt%, the compressive and flexural strengths were 18.94 and 4.95 MPa, respectively, 2.6% and 3.2% higher than MWCNT–cement composites. This might be due to TX10, which shows the highest dispersing power when dispersing SWCNTs [49]. However, for MWCNTs, according to Luo et al. [39], TX10 was not suitable for dispersing MWCNTs. The reduction of strength after treatment of TX10 is discussed in following sections. Generally, MWCNTs show better mechanical properties, but the increase of strength properties is relatively small. There is potential to study the effect of SWCNTs on mechanical properties of cement-based composites. The following sections mainly discuss the strength properties of SWCNT-added cement composites.

3.2. Compressive Strength Test Results

To get compressive strength of mortars, three specimens were prepared. According to BS 1881-119:1983 and ASTM C 116-90, the average of the three values was taken as the representative compressive strength of the concrete. Table 11 shows the failure compressive strength of all specimens. The differences of SWCNTXN and SWCNTXS groups (where X equals 0, 2, 4, and 6) are water/cement ratio and TX10 content. Although the two groups had similar slump results, which indicates similar workability, the values of compressive strength shows great difference. Non-TX10-treated samples showed higher compressive strength than SAA treated samples. The cause might be the decrease of mass of cubes and prisms when TX10 was added. TX10 results in the formation of foam during the stir process, which retards or prevents the hydration of cement. The relationship between strength and mass of specimens are discussed in Section 3.3. In this section, the effects of SWCNTs and TX10 on sample strength are discussed.

Table 11. Compressive strength test results of SWCNT-added specimens.

Mix Name	Compressive Strength (MPa)											
	3 Days			Mean	7 Days			Mean	28 Days			Mean
SWCNT0N	18.70	19.24	21.71	19.88	25.01	25.16	25.28	25.15	29.45	29.73	31.26	30.15
SWCNT2N	19.03	20.46	21.06	20.18	25.34	25.55	25.89	25.59	28.07	28.09	28.48	28.21
SWCNT4N	18.73	20.47	20.75	19.98	23.13	24.74	24.89	24.25	24.94	25.68	26.76	25.79
SWCNT6N	17.84	18.92	22.22	19.66	21.51	22.54	22.70	22.25	25.42	25.65	25.67	25.58
SWCNT0S	13.98	14.44	14.46	14.29	15.84	16.24	16.55	16.21	18.56	18.86	18.89	18.77
SWCNT2S	13.66	13.88	14.45	13.99	15.76	15.85	16.67	16.09	17.55	17.64	17.72	17.64
SWCNT4S	15.97	16.12	16.37	16.15	18.43	18.71	18.72	18.62	18.64	18.96	19.22	18.94
SWCNT6S	18.88	18.98	19.23	19.03	20.76	21.09	22.43	21.43	21.87	22.94	23.32	22.71

3.2.1. Specimens without Treatment of TX10

The compressive strength of specimens with no TX10 added is plotted against time in Figure 2. The effect of SWCNTs on these specimens is discussed in this section.

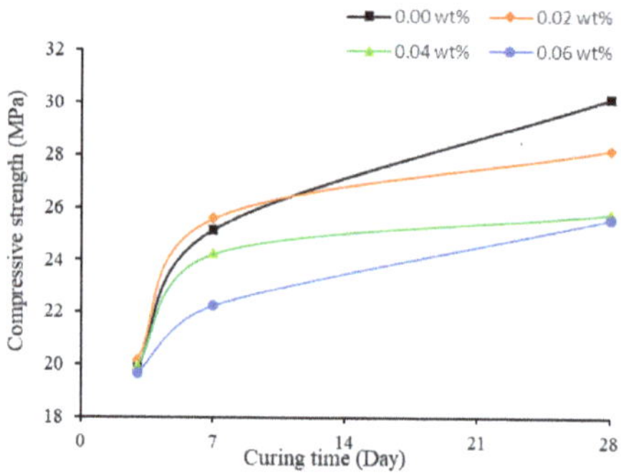

Figure 2. Compressive Strength of No TX10 contained mortar cubes.

Figure 2 shows variation of the compressive strength of specimens under four different SWCNT concentrations at different curing times. With no treatment of TX10, the higher was the concentration of the SWCNTs, the lower was the compressive strength of mortar cubes, which is similar to the results of Suprompituk et al. [66]. For plain cementitious mortar, the compressive strength could reach 30.15 Mpa after 28 days of curing, which decreased to 25.58 Mpa when the concentration of CNTs was increased to 0.06 wt%. The reason might be that, when SWCNTs were used directly, they easily tangled together, thus were not well dispersed in the cement matrix and formed agglomerates. This reduced the bond between the hydration products, and in turn the compressive strength of mortar was reduced.

Moreover, at early age, the differences in compressive strength were small. At three days of curing, they shared similar compressive strength, all around 20 MPa. Thus, at early age, SWCNTs showed little effect on the mechanical properties of mortars. In addition, t the compressive strength of SWCNT0N and SWCNT2N after three and seven days of curing were similar, although the strength of SWCNT2N was slightly higher. However, at 28 days of curing, the strength of SWCNT2N was 6.43% lower than SWCNT0N. The reason the strength of 0.02 wt% SWCNT-containing specimens was higher at early stage might be the positive effect of SWCNTs on early hydration reaction of cement. This phenomenon that CNTs may affect the early hydration progress of cement, producing higher hydration rates, was also found by Markar and Beaudoin [67] and Markar et al. [68]. In general, the reduction of 28-day strength with the increase of concentration of SWCNTs was related to the agglomerates of CNTs.

To analyze the early hydration rate, the evolution of the strength of mortars is plotted in Figure 3. Adding CNTs improved the evolution of strength of mortar in gaining strength more quickly. At three days of curing, plain mortar gained around 66% strength. With the increase of the quantity of SWCNTs, the hydration rate was improved. Obvious differences occurred at seven days of curing;

the compressive strength could reach around 90–94% of 28-day strength, which was higher than the 83.4% for SWCNT0N. However, when the concentration of CNTs reached 0.06 wt%, the percentage of gained strength at seven days of curing was around 87%, between plain mortars and SWCNT2N. This might be the negative effect of agglomerates. In general, adding SWCNTs can make cement composites gain strength more quickly at an early age. In addition, it was also found that, when the concentration of SWCNTs was 0.04 wt%, the evolution of compressive strength of samples was the highest.

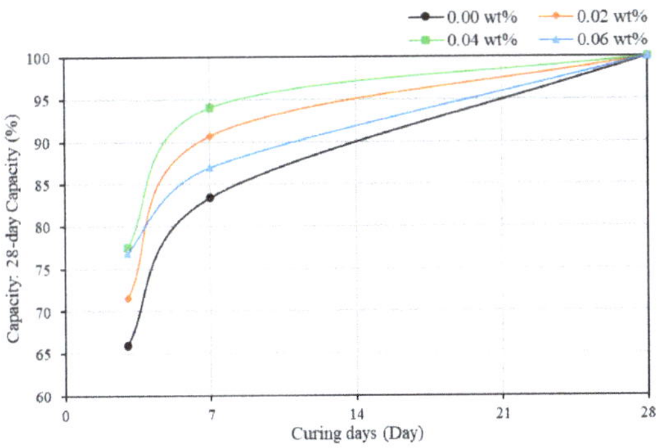

Figure 3. The evolution of the strength of mortars (no TX10) as a function of curing time.

Figure 4 shows the effect of concentration of SWCNTs on compressive strength at different curing times. It clearly shows that the higher was the concentration, the lower was the compressive strength. After 28 days of curing, specimens showed a higher compressive strength than seven- or three-day cured specimens. At three days of curing, the compressive strengths of specimens at the four concentrations showed similar values, around 20 MPa.

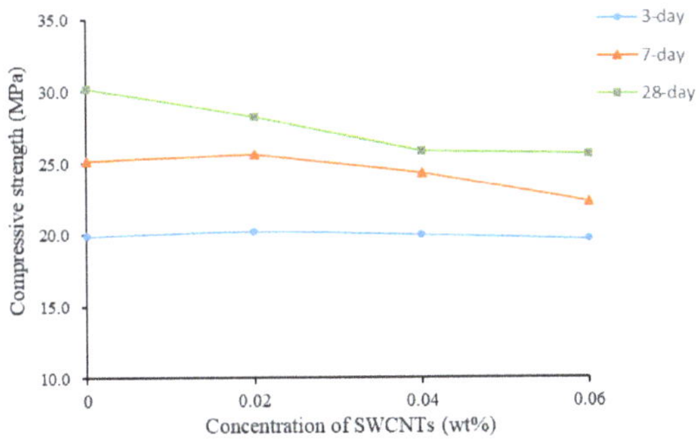

Figure 4. Compressive strength as a function of concentration of SWCNTs (No TX10).

3.2.2. Specimens with Treatment of TX10

When TX10 was added and mixed, the variation pattern of strength changed. The most obvious change shown in Table 11 is that the compressive strength of SWCNTXS (where X equals 0, 2, 4, and 6) was much lower than that of SWCNTXN (where X equals 0, 2, 4, and 6) at each curing time. The reasons might be the change of water/cement ratio and the effect of TX10 on the specimens. For samples with the addition of TX10, their masses were less than those of non-TX10-treated specimens because, when TX10 was added, much foam was generated in the cement paste, as mentioned in Section 3.2, causing high porosity in mortars, and in turn the decrease of the specimens' bulk density. Moreover, TX10 might have partly decreased the bond of cement reaction. Based on this situation, the effect of

SWCNTs and variation pattern on strength properties of mortars under the dispersion method of TX10 were studied.

Figure 5 shows that the compressive strength increased with the increase of concentration of SWCNTs; the higher was the concentration, the higher was the compressive strength. After 28 days of curing, the compressive strength of SWCNT6S increased 21% compared to SWCNT0S. The reason might be that TX10 changed the surface energy of SWCNT suspensions; getting SWCNTs well dispersed in water allowed the unique properties of CNTs to show up. Moreover, it should be noted that, at three days of curing, different from SWCNTXN (where X equals 0, 2, 4, and 6), which had no TX10 added and shared similar compressive strength of 20 MPa, the strength of TX10-treated mortars increased with the increase of concentration of CNTs, although the strength of SWCNT2S was slightly lower than that of SWCNT0S. Furthermore, Figure 6 shows that, with the addition of SWCNTs, the rapid hardening phenomenon of mortar was more obvious, even though CEM |, a type of rapid hardening cement, was used in experiments. At seven days of curing, when the concentration of SWCNTs was 0.04 wt%, the compressive strength could reach over 95% of the 28-day compressive strength. Compared to Figure 3, when the concentration of SWCNTs was 0.06 wt%, the percentage of strength gained could reach approximately 94%, which was higher than the 87% of non-TX10-treated specimens. Moreover, Figures 3 and 6 show that, for plain mortars with no CNTs added, TX10 decreased the hydration rate to some extent. The strength gained percentage decreased from 83.4% to 80% when TX10 was added.

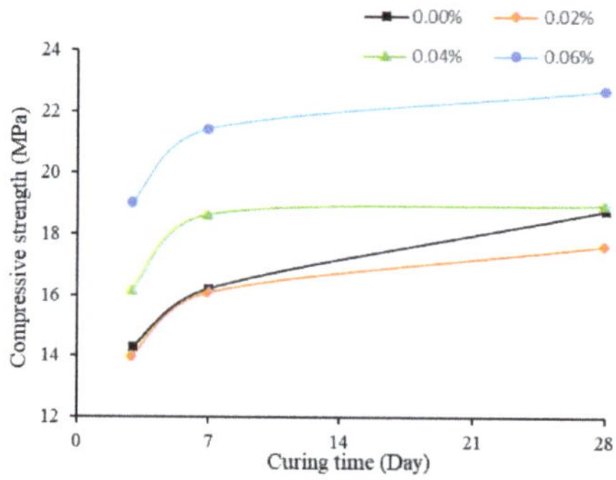

Figure 5. Compressive Strength of TX10 added mortar cubes.

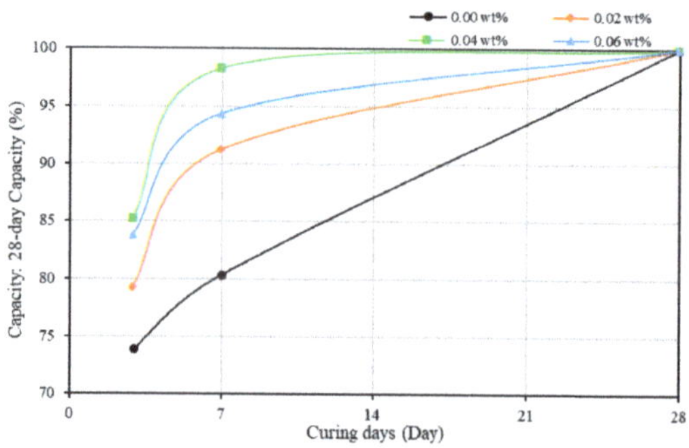

Figure 6. The evolution of the strength of mortars (SAA) as a function of curing time.

Figure 7 shows the compressive strength as a function of concentration of SWCNTs. The compressive strength was reduced initially, but then, with the increase of the concentration of SWCNTs, the compressive strength increased, reaching the peak when the concentration was 0.06 wt%. The reduction of strength might be due to the change of fluidity when CNTs were added, which requires more research. Thus, TX10 as a SAA worked well on dispersing SWCNTs in cement matrix, and SWCNTs helped enhance the interface bond force among aggregates and cement matrix. The agglomerates that might obstruct the cement pastes were reduced. Therefore, SAA is a useful dispersion method regardless of the negative effect on mechanical properties of cement composites. However, it should be noted that the addition of TX10 produced foam in the cement, resulting in the high porosity of mortars, causing a decrease in the bulk density of samples, and in turn the overall decrease of strength. Therefore, in future works, defoamer should be considered to study the possibility of improving the reduction of strength when TX10, or another SAA is added.

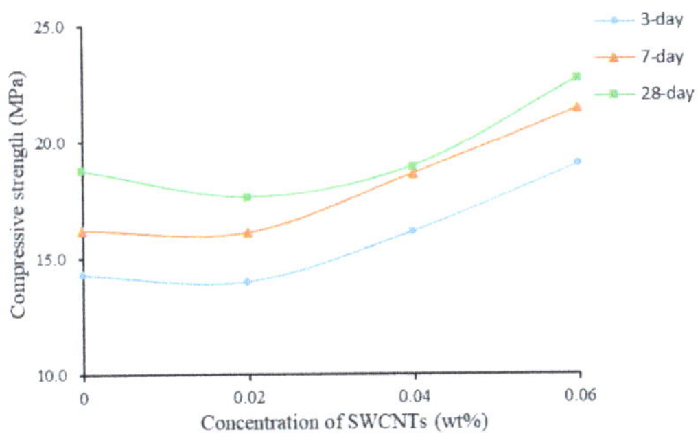

Figure 7. Compressive strength as a function of concentration of SWCNTs.

3.3. *Flexural Strength Test Results*

To obtain the flexural strength of concrete, three specimens were prepared. According to BS 1881-119:1983 and ASTM C 116-90, the average of the three values was taken as the representative flexural strength. The results of tested specimens are shown in Table 12.

Table 12. Flexural strength test results of SWCNT-added specimens.

	Flexural Strength (MPa)											
Mix Name	**3 Days**			**Mean**	**7 Days**			**Mean**	**28 Days**			**Mean**
SWCNT0N	5.02	5.34	5.35	5.24	6.47	6.70	6.84	6.67	6.76	7.11	7.52	7.12
SWCNT2N	5.12	5.25	5.35	5.24	6.46	6.71	6.73	6.63	6.08	6.60	7.15	6.66
SWCNT4N	5.13	5.41	5.46	5.34	5.94	6.12	6.27	6.11	6.12	6.24	6.25	6.20
SWCNT6N	4.55	4.63	5.10	4.76	5.19	5.57	5.79	5.52	5.29	5.48	6.14	5.64
SWCNT0S	4.11	4.25	4.32	4.23	4.31	4.40	4.66	4.46	4.69	4.71	4.78	4.73
SWCNT2S	4.00	4.02	4.03	4.02	3.99	4.25	4.31	4.18	4.20	4.27	4.32	4.26
SWCNT4S	4.27	4.32	4.40	4.33	4.54	4.64	4.79	4.66	4.68	4.97	5.21	4.95
SWCNT6S	4.16	4.58	4.63	4.46	4.66	4.88	5.50	5.01	5.21	5.29	5.42	5.31

3.3.1. Specimens without Treatment of TX10

Similar to the compressive strength of non-TX10-treated specimens, the flexural strength decreased with the increase of the concentration of SWWCNTs. Figure 8 shows the increase of flexural Strength of non-TX10-containing mortar prisms as a function of curing time. SWCNT-containing mortars could reach high flexural strength at early stage (seven days). After seven days of curing, the flexural strength reached stable value and continued curing gained little increase in flexural strength. The reason is that, with the addition of SWCNTs, the rapid hardening phenomenon of mortar was more obvious, even though early hardening cement was used. At 28 days of curing, SWCNT0S showed 6.75% increase in flexural strength compared to the seven-day strength, but, for SWCNTXS (where X equals 2, 4, and 6), the increase was little, only from 0.91% to 2.17%. This might be the crack-bridging effect in cement/CNTs composites produced by CNTs. CNTs have the ability to prevent the development of cracks [65,69]. TEM is needed for further works to analyze the crack-bridging effect of CNTs and to study the effect of CNTs on hydration products of cement/CNTs composites.

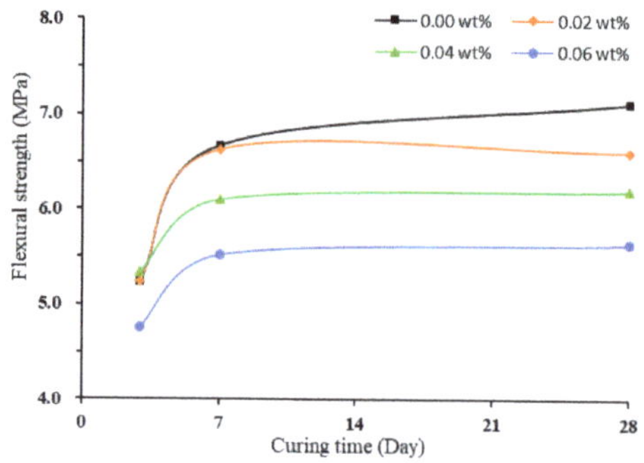

Figure 8. Flexural strength of non-TX10-added mortar prisms.

In addition, Figure 9 shows that the flexural strength decreased with the increase of the concentration of SWCNTs. It can also be observed that, when SWCNTs were added, the values of flexural strength after 7 and 28 days of curing were close for all concentrations of SWCNTs used.

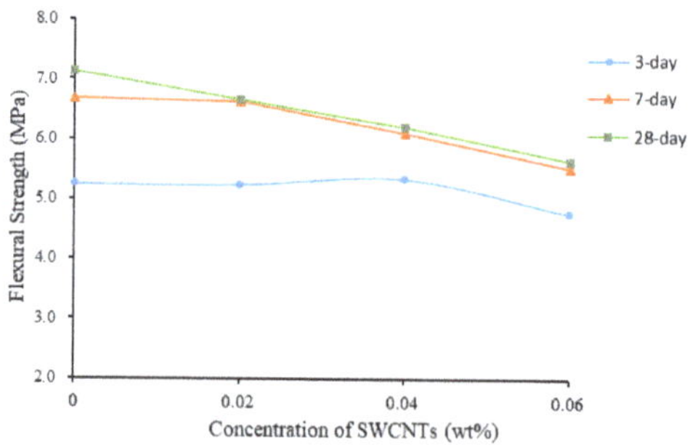

Figure 9. Flexural strength as a function of concentration of SWCNTs (No TX10).

3.3.2. Specimens with Treatment of TX10

When TX10 was added and mixed, the variation pattern of flexural strength changed. Figure 10 shows that the flexural strength was high initially, adding SWCNTs caused a slight reduction, and then the flexural strength increased with the increase of the concentration of SWCNTs. After 28 days of curing, flexural strength of SWCNT6S reached 5.31 MPa, 19.77% and 10.92% higher compared to flexural strength of SWCNT2S and SWCNT0S at 28 days of curing, respectively. This variation pattern was different to that when TX10 was not added, as shown in Figures 9 and 10. The dispersed CNTs showed positive effect on flexural strength of mortars.

Figure 11 shows that the flexural strength increased with the increase of concentration of SWCNTs, although the value of flexural strength of SWCNT0S was between those of SWCNT2S and SWCNT4S. By comparing Figures 8 and 11, it can be observed that, with the addition of TX10, the flexural strength of mortar with 0.06 wt% CNTs was the highest, resulting from the positive effect of CNTs. This finding was opposite to that of non-TX10-added mortars, where plain mortar presented the highest level.

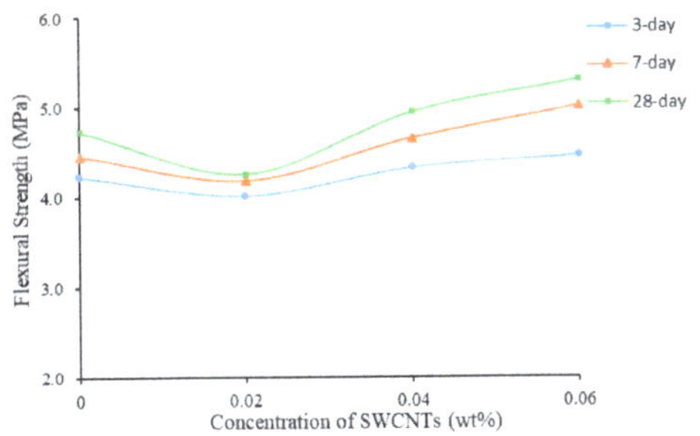

Figure 10. Flexural strength as a function of concentration of SWCNTs (TX10).

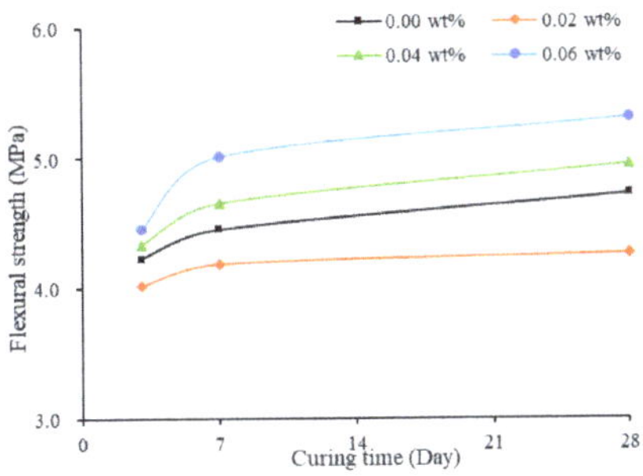

Figure 11. Flexural strength of TX10-added mortar prisms.

3.4. Relationship between Strength and Bulk Density of Cement Based Specimens

In the experiments, the mass of each tested cube and prism was recorded. According to the size of mortars and prisms, the bulk density of each specimens could be calculated. Table 13 shows the average bulk density of each set of mortars.

Table 13. Average bulk density of mortar specimens.

Types	Curing Days	0.00 wt% Cubes	0.00 wt% Prisms	0.02 wt% Cubes	0.02 wt% Prisms	0.04 wt% Cubes	0.04 wt% Prisms	0.06 wt% Cubes	0.06 wt% Prisms
Non-TX10-treated	3	2.094	2.164	2.122	2.164	2.115	2.082	2.107	2.129
	7	2.106	2.164	2.114	2.082	2.109	2.152	2.115	2.156
	28	2.105	2.074	2.114	2.176	2.108	2.121	2.105	2.078
	Mean	2.102	2.134	2.117	2.141	2.111	2.118	2.109	2.121
TX10-treated	3	1.882	1.891	1.902	1.871	1.939	1.867	1.999	2.000
	7	1.887	1.875	1.878	1.902	1.943	1.918	2.024	2.031
	28	1.934	1.871	1.873	1.855	1.893	1.852	1.963	1.965
	Mean	1.901	1.879	1.884	1.876	1.925	1.879	1.995	1.998

Table 13 shows that for non-TX10-added mortars, the bulk density was around 2.102–2.141 g/cm^3, higher than those of TX10-treated mortars (1.884–1.998 g/cm^3). This result proves that the porosity of mortars was increased due to the addition of TX10, which caused the reduction of strength of mortars. In addition, the strength of mortars was related to the bulk density of mortars. For instance, Figure 12 shows the variation of 28-day compressive strength and bulk density of TX10-treated specimens as a function of the concentration of SWCNTs. The average bulk density of plain cube mortars (no SWCNTs added) was 1.934 g/cm^3, which was reduced to 1.873 g/cm^3 with the increase of the concentration of CNTs to 0.02 wt%, and then increased to 1.893 and 1.963 g/cm^3 when the concentration of CNTs was increased to 0.04 wt% and 0.06 wt%, respectively. The variation pattern of 28-day compressive strength was similar to the variation of bulk density of specimens, i.e., the higher was the bulk density, the higher was the compressive strength.

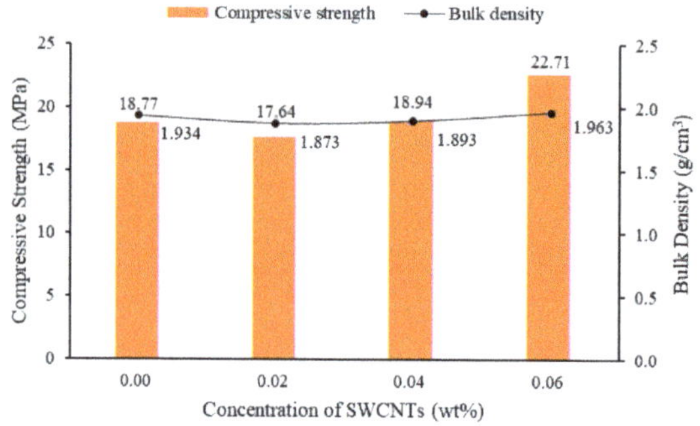

Figure 12. Variation of 28-day compressive strength and bulk density of TX10-treated specimens as a function concentration of SWCNTs.

Therefore, the strength (MPa)/mass of specimens (g) ratio was considered to standardize the law. It was found that the compressive and flexural strengths were related to the mass of the mortar cube/prism. Table 14 shows some of the data: average compressive and flexural strengths after seven days of curing are listed as illustration.

Table 14. Compressive (flexural) strength/mass ratio of cubes (prisms) at seven days of curing.

Mix Name	Cube Mass (g)	Comp (MPa)	Prism Mass (g)	Flex (MPa)	Comp/ Mass	Flex/ Mass	Difference
SWCNT0N	2106	25.15	554	6.67	0.01194	0.01205	0.921%
SWCNT2N	2114	25.59	533	6.63	0.01211	0.01244	2.725%
SWCNT4N	2109	24.25	551	6.11	0.01149	0.01108	3.568%
SWCNT6N	2115	22.25	552	5.52	0.01052	0.01000	4.943%
SWCNT0S	1887	16.21	480	4.46	0.00859	0.00928	8.032%
SWCNT2S	1878	16.09	487	4.18	0.00857	0.00859	0.233%
SWCNT4S	1943	18.62	491	4.66	0.00949	0.00949	0.000%
SWCNT6S	2024	21.43	498	5.01	0.01059	0.01006	5.005%

Comp represents compressive strength, and Flex represents flexural strength.

Table 14 shows that both the compressive strength and the flexural strength was related to the mass of the specimens. When SWCNTs were directly used, the general trend of strength/mass ratio was downward, which was same trend as for strength. When TX10 was added to pretreat SWCNTs, the ratio saw a slight increase when the concentration of SWCNTs increased, also similar to the variation pattern of strength.

The value of comp/mass of cubes ratio was close to the flex/mass ratio of prisms. The differences between the two ratios were generally less than 5%. Therefore, this relationship can be used to estimate the compressive strength (flexural strength) when the mass of the specimens and flexural strength (compressive strength) are known. However, in some cases, the difference reached 8–15%, which could be observed from results of SWCNTXS (where X equals 0, 2, 4, or 6) at three days of curing (Table 15). Its cause might be the errors in the process of preparing the specimens

Generally, it can be concluded that, when SWCNTs and MWCNTs were directly used to prepare mortars, specimens' strength decreased with the increase of concentration of CNTs due to the increase of CNTs' agglomerates. It was found that MWCNT-added specimens showed better strength properties than SWCNT-added specimens due to MWCNTs having better mechanical properties, although the differences of the strength of the cement composites were small. The effect of SWCNTs on strength properties of cement composites was mainly studied. When TX10, as a surfactant, was applied, the variation pattern was changed. The compressive and flexural strengths generally increased with the increase of concentration of SWCNTs, although TX10 produced a negative effect on cement composites by increasing porosity, causing a decrease in the bulk densities of mortar specimens. It was found that SWCNTs could help cement composites gain high strength more quickly at the early stage. In addition, it was found that the strength was related to the mass of the specimens.

Table 15. Compressive (flexural) strength/mass ratio of cubes (prisms) at three days of curing.

Mix Name	Cube Mass (g)	Comp (MPa)	Prism Mass (g)	Flex (MPa)	Comp/ Mass	Flex/ Mass	Difference
SWCNT0S	1882	14.29	484	4.23	0.00759	0.00874	15.152%
SWCNT2S	1902	14.00	479	4.02	0.00736	0.00839	13.995%
SWCNT4S	1939	16.15	478	4.33	0.00833	0.00906	8.764%
SWCNT6S	1999	19.03	512	4.46	0.00952	0.00871	8.508%

Where the comp represents compressive strength, and the flex represents flexural strength.

4. Conclusions

This paper presents the effect of SWCNTs (Rhenofit® CNT-3) on the strength properties of cement composites. In addition, the effect of the addition of TX10 (SAA) on the strength properties of cement composites is also discussed. The main results of this investigation are summarized as follows:

1. Directly using Rhenofit® CNT-3/CNT-2 (without addition of TX10) in cement matrix reduced the compressive strength of cement composites, resulting from the agglomeration of CNTs. The compressive and flexural strengths decreased as a function of the concentration of CNTs. Plain mortars showed the highest strength at this stage.

2. When TX10 was not added, MWCNTs (Rhenofit® CNT-2) added to cement-based specimens showed better strength properties than SWCNTs, resulting from the higher mechanical properties of MWCNTs. When TX10 was applied, the strength properties of SWCNT–cement composites were higher than those of MWCNT–cement composites, because TX10 was not suitable in dispersing MWCNTs.

3. When TX10 was applied, the compressive and flexural strengths of cement composites were greatly reduced compared to the case where no TX10 was used. In addition, the bulk density and mass of specimens were also reduced, from 2.102–2.141 to 1.884–1.998 g/cm^3. The reduction of strength and bulk density could be attributed to the formation of foam during the specimen preparation process and the retardation of hydration caused by TX10.

4. Even though strength reduction of cement composites occurred when TX10 was used to disperse SWCNTs, when the effect of SWCNTs on strength properties of cement composites was studied at this stage, it was found that the general variation trend of compressive and flexural strengths was upward; the strength increased with the increase of the concentration of SWCNTs. Conversely to when no TX10 was added to specimens, the maximum compressive and flexural strengths occurred when the concentration of SWCNTs was 0.06 wt%.

5. SWCNTs affected the early hydration progress of cement. SWCNTs could produce high hydration rates at early age: at seven days of curing, the compressive and flexural strengths could reach over 90% of the 28-day strength.

6. The variation of compressive and flexural strengths was related to the mass (bulk density) of mortar specimens. They shared similar variation patterns: the strength of specimens increased or decreased as a function of the mass of specimens.

7. The flexural strength/mass of cubes ratio was close to compressive strength/mass of prisms ratio, with difference less than 10%. This relation could be used to estimate the strength of specimens.

For future work, TEM should be applied to study the failure modes of specimens and the effect mechanism of CNTs on hydration products of cement/CNTs composites.

Author Contributions: Conceptualization, J.K. and S.A.-S.; methodology, J.K. and S.A.-S.; software, J.K. and R.T.; validation, J.K. and S.A.-S.; formal analysis, J.K. and R.T.; investigation, J.K.; resources, S.A.-S.; data curation, J.K.; writing—original draft preparation, J.K.; writing—review and editing, J.K. and S.A.-S.; supervision, S.A.-S.; project administration, S.A.-S. All authors have read and agreed to the published version of the manuscript.

Funding: This research received no external funding.

Acknowledgments: I thank Salam Al-Sabah for all his support throughout this project. I would also like to thank Derek Holmes for his assistance and for providing materials needed in the concrete laboratory in UCD. I must also mention Roger Théo who worked with me and helped me greatly in the laboratory.

Conflicts of Interest: The authors declare no conflict of interest.

References

1. Jose, N.-C.; Ariel, C.; Irene, G.-T.; Manuel, C.; Antonios, K. Effect of Metallic Waste Addition on the Physical and Mechanical Properties of Cement-Based Mortars. *Appl. Sci.* **2018**, *8*, 929.
2. Jang, J.G.; Kim, H.K.; Kim, T.S.; Min, B.J.; Lee, H.K. Improved flexural fatigue resistance of pva fiber-reinforced concrete subjected to freezing and thawing cycles. *Constr. Build. Mater.* **2014**, *59*, 129–135. [CrossRef]
3. Liang, N.H.; Liu, X.R.; Sun, J. Experimental study of compression for multi-scale polypropylene fiber concrete. *Appl. Mech. Mater.* **2012**, *174–177*, 1584–1588. [CrossRef]
4. Nuruddin, M.F.; Khan, S.U.; Shafiq, N.; Ayub, T. Strength development of high-strength ductile concrete incorporating metakaolin and pva fibers. *Sci. World J.* **2014**, *2014*, 387259. [CrossRef]
5. Shi, Z.Q.; Chung, D.D.L. Carbon fiber-reinforced concrete for traffic monitoring and weighing in motion. *Cem. Concr. Res.* **1999**, *29*, 435–439. [CrossRef]
6. de Sena Cruz, J.M.; Oliveira de Barros, J.A. Bond between near-surface mounted carbon-fiber-reinforced polymer laminate strips and concrete. *J. Compos. Constr.* **2004**, *8*, 519–527. [CrossRef]
7. Nelson, P.K.; Li, V.C.; Kamada, T. Fracture toughness of microfiber reinforced cement composites. *J. Mater. Civ. Eng.* **2002**, *14*, 384–391. [CrossRef]
8. Tesárek, P.; Holub, P.; Havrda, J.; Fládr, J.; Proek, Z.; Trejbal, J.; Plachy, T. Mechanical properties of cement composites reinforced by carbon microfibers: Compressive and bending strength. *Key Eng. Mater.* **2016**, *722*, 351–356. [CrossRef]
9. Arash, B.; Wang, Q.; Varadan, V.K. Mechanical properties of carbon nanotube/polymer composites. *Sci. Rep.* **2014**, *4*, 6479. [CrossRef]
10. Sinha, S.; Barjami, S.; Iannacchione, G.; Schwab, A.; Muench, G. Off-axis Thermal Properties of Carbon Nanotube Films. *J. Nanopart. Res.* **2005**, *7*, 651–657. [CrossRef]
11. Mahar, B.; Laslan, C.; Yip, R.; Sun, Y. Development of carbon nanotube-based sensors—A review. *IEEE Sens. J.* **2007**, *7*, 266–284. [CrossRef]
12. Rashad, A.M. Effect of carbon nanotubes (CNTs) on the properties of traditional cementitious materials. *Constr. Build. Mater.* **2017**, *153*, 81–101. [CrossRef]
13. Gopalakrishnan, K.; Birgisson, B.; Taylor, P.; Attoh-Okine, N. *Nano-Technology in Civil. Infrastructure*; Springer: Berlin/Heidelberg, Germany, 2011; pp. 1–1272.
14. Konsta-Gdoutos, M.S.; Metaxa, Z.S.; Shah, S.P. Highly dispersed carbon nanotube reinforced cement based materials. *Cem. Concr. Res.* **2010**, *40*, 1052–1059. [CrossRef]
15. O'Connell, M. *Carbon Nanotubes: Properties and Applications*; CRC Press: Boca Raton, FL, USA, 2006.
16. Salvetat-Delmotte, J.; Rubio, A. Mechanical properties of carbon nanotubes: A fiber digest for beginners. *Carbon* **2002**, *40*, 1729–1734. [CrossRef]
17. Saether, E.; Frankland, S.J.V.; Pipes, R.B. Transverse mechanical properties of single-walled carbon nanotube crystals. Part I: Determination of elastic moduli. *Compos. Sci. Technol.* **2003**, *63*, 1543–1550. [CrossRef]
18. Wang, X.; Li, Q.; Xie, J.; Jin, Z.; Wang, J.; Li, Y.; Jiang, K.; Fan, S. Fabrication of ultralong and electrically uniform single-walled carbon nanotubes on clean substrates. *Nano Lett.* **2009**, *9*, 3137–3141. [CrossRef]

19. Yu, M.F.; Lourie, O.; Dyer, M.J.; Moloni, K.; Kelly, T.F.; Ruoff, R.S. Strength and breaking mechanism of multiwalled carbon nanotubes under tensile loading. *Science* **2000**, *287*, 637–640. [CrossRef]
20. Lisuzzo, L.; Cavallaro, G.; Lazzara, G.; Milioto, S.; Parisi, F.; Stetsyshyn, Y. Stability of Halloysite, Imogolite, and Boron Nitride Nanotubes in Solvent Media. *Appl. Sci.* **2018**, *8*, 1068. [CrossRef]
21. Oscar, A.M.R.; Romildo, D.T.F. A review on the chemical, mechanical and microstructural characterization of carbon nanotubes-cement based composites. *Constr. Build. Mater.* **2017**, *154*, 697–710.
22. Khare, R.; Bose, S. Carbon nanotube based composites—A review. *J. Miner. Mater. Charact. Eng.* **2005**, *4*, 31–46. [CrossRef]
23. Liew, K.M.; Kai, M.F.; Zhang, L.W. Carbon nanotube reinforced cement composites: An overview. *Compos. Part A* **2016**, *91*, 301–323. [CrossRef]
24. Lourie, O.; Cox, D.; Wagner, H. Buckling and collapse of embedded carbon nanotubes. *Phys. Rev. Lett.* **1998**, *81*, 1638. [CrossRef]
25. Saez de Ibarra, Y.; Gaitero, J.; Erkizia, E.; Campillo, I. Atomic force microscopy and nanoindentation of cement pastes with nanotube dispersions. *Phys. Status Solidi (a)* **2006**, *203*, 1076–1081. [CrossRef]
26. Cavallaro, G.; Lazzara, G.; Lisuzzo, L.; Milioto, S.; Parisi, F. Filling of Mater-Bi with Nanoclays to Enhance the Biofilm Rigidity. *J. Funct. Biomater.* **2018**, *9*, 60. [CrossRef]
27. Lisuzzo, L.; Cavallaro, G.; Milioto, S.; Lazzara, G. Effects of halloysite content on the thermo-mechanical performances of composite bioplastics. *Appl. Clay Sci.* **2020**, *185*, 105416. [CrossRef]
28. Cota, F.P.; Panzera, T.H.; Schiavon, M.A.; Christoforo, A.L.; Borges, P.H.R.; Bowen, C.; Scarpa, F. Full factorial design analysis of carbon nanotube polymer-cement composites. *Mater. Res.* **2012**, *15*, 573–580. [CrossRef]
29. Gopalakrishnan, K.; Birgisson, B.; Taylor, P.; Attoh Okine, N.O. *Nanotechnology in Civil. Infrastructure a Paradigm Shift*; Springer: Heidelberg, Germany, 2011.
30. Yazdanbakhsh, A.; Grasley, Z.; Tyson, B.; Abu Al Rub, R.K. Distribution of carbon nanofibers and nanotubes in cement composites. *Transport. Res. Rec.* **2010**, *2142*, 89–95. [CrossRef]
31. Parveen, S.; Rana, S.; Fangueiro, R. A review on nanomaterial dispersion, microstructure, and mechanical properties of carbon nanotube and nanofiber reinforced cement composites. *J. Nanomater.* **2013**, *2013*, 710175. [CrossRef]
32. Li, G.Y.; Wang, P.M.; Zhao, X. Pressure-sensitive properties and microstructure of carbon nanotube reinforced cement composites. *Cem. Concr. Compos.* **2007**, *29*, 377–382. [CrossRef]
33. Collins, F.; Lambert, J.; Duan, W.H. The influences of admixtures on the dispersion, workability, and strength of carbon nanotube–OPC paste mixtures. *Cem. Concr. Compos.* **2012**, *34*, 201–207. [CrossRef]
34. Cwirzen, A.; Habermehl-Cwirzen, K.; Penttala, V. Surface decoration of carbon nanotubes and mechanical properties of cement/carbon nanotube composites. *Adv. Cem. Res.* **2008**, *20*, 65–73. [CrossRef]
35. Wang, B.; Han, Y.; Liu, S. Effect of highly dispersed carbon nanotubes on the flexural toughness of cement-based composites. *Constr. Build. Mater.* **2013**, *46*, 8–12. [CrossRef]
36. Sindu, B.S.; Sasmal, S. Properties of carbon nanotube reinforced cement composite synthesized using different types of surfactants. *Constr. Build. Mater.* **2017**, *155*, 389–399. [CrossRef]
37. Wu, X.; Dai, L. Carbon nano-tubes in improving the mechanical property of cement-based composite materials. *Frat. Integritá Strutt.* **2017**, *11*, 388–395. [CrossRef]
38. Fatemi, S.M.; Foroutan, M. Study of dispersion of carbon nanotubes by Triton X-100 surfactant using molecular dynamics simulation. *J. Iran. Chem. Soc.* **2015**, *12*, 1905–1913. [CrossRef]
39. Luo, J.; Duan, Z.; Li, H. The influence of surfactants on the processing of multi-walled carbon nanotubes in reinforced cement matrix composites. *Phys. Status Solidi* **2009**, *206*, 8. [CrossRef]
40. Mohsen, M.O.; Taha, R.; Abu Taqa, A.; Al-Nuaimi, N.; Al-Rub, R.A.; Bani-Hani, K.A. Effect of Nanotube Geometry on the Strength and Dispersion of CNT-Cement Composites. *J. Nanomater.* **2017**, *2017*, 6927416. [CrossRef]
41. Evangelista, A.C.J.; de Morais, J.F.; Tam, V.; Soomro, M.; Di Gregorio, L.T.; Haddad, A.N. Evaluation of Carbon Nanotube Incorporation in Cementitious composites Materials. *Materials* **2019**, *12*, 1504. [CrossRef]
42. Li, X.; Wei, W.; Qin, H.; Hu, Y.H. Co-effects of graphene oxide sheets and single wall carbon nanotubes on mechanical properties of cement. *J. Phys. Chem. Solids* **2015**, *85*, 39–43. [CrossRef]
43. Parveen, S.; Rana, S.; Fangueiro, R.; Paiva, M.C. Microstructure and mechanical properties of carbon nanotube reinforced cementitious composites developed using a novel dispersion technique. *Cem. Concr. Res.* **2015**, *73*, 215–227. [CrossRef]

44. del Carmen Camacho, M.; Galao, O.; Baeza, F.J.; Zornoza, E.; Garcés, P. Mechanical properties and durability of cnt cement composites. *Materials* **2014**, *7*, 1640–1651. [CrossRef] [PubMed]
45. Shao, H.; Chen, B.; Li, B.; Tang, S.; Li, Z. Influence of dispersants on the properties of CNTs reinforced cement-based materials. *Constr. Build. Mater.* **2017**, *131*, 186–194. [CrossRef]
46. Sobolkina, A.; Mechtcherine, V.; Khavrus, V.; Maier, D.; Mende, M.; Ritschel, M.; Leonhardt, A. Dispersion of carbon nanotubes and its influence on the mechanical properties of the cement matrix. *Cem. Concr. Compos.* **2012**, *34*, 1104–1113. [CrossRef]
47. Kim, J.-H.; Chung, C.-W. Rheological Properties of Cement Paste Mixed with Aqueously Dispersed Single-Walled Carbon Nanotubes. *J. Korea Inst. Build. Constr.* **2019**, *19*, 113–121.
48. Bharj, J.; Singh, S.; Chander, S.; Rabinder, S. Experimental study on compressive strength of cement-CNT composite paste. *Indian J. Pure Appl. Phys.* **2014**, *52*, 35–38.
49. Rastogi, R.; Kaushal, R.; Tripathi, S.K.; Sharma, A.L.; Kaur, I.; Bharadwaj, L.M. Comparative study of carbon nanotube dispersion using surfactants. *J. Colloid Interface Sci.* **2018**, *328*, 421–428. [CrossRef]
50. Nadiv, R.; Vasilyev, G.; Shtein, M.; Peled, A.; Zussman, E.; Regev, O. The multiple roles of a dispersant in nanocomposite systems. *Compos. Sci. Technol.* **2016**, *133*, 192–199. [CrossRef]
51. Yamada, K.; Takahashi, T.; Hanehara, S.; Matsuhisa, M. Effects of the chemical structure on the properties of polycarboxylate-type superplasticizer. *Cem. Concr. Res.* **2000**, *30*, 197–207. [CrossRef]
52. Han, B.; Sun, S.; Ding, S.; Zhang, L.; Yu, X.; Ou, J. Review of nanocarbon-engineered multifunctional cementitious composites. *Compos. Part A Appl. Sci. Manuf.* **2015**, *70*, 69–81. [CrossRef]
53. Salvetat, J.P.; Bonard, J.M.; Thomson, N.H.; Kulik, A.J.; Forró, L.; Benoit, W.; Zuppiroli, L. Mechanical properties of carbon nanotubes. *Appl. Phys. A* **1999**, *69*, 255–260. [CrossRef]
54. Lee, J.I.; Jung, H.T. Technical status of carbon nanotubes composites. *Korean Chem. Eng. Res.* **2008**, *46*, 7–14.
55. Bandaru, P.R. Electrical Properties and Applications of Carbon Nanotube Structures. *J. Nanosci. Nanotechnol.* **2007**, *7*, 1239–1267. [CrossRef] [PubMed]
56. Lee, H.; Kang, D.; Song, Y.M.; Chung, W. Heating Experiment of CNT Cementitious Composites with Single-Walled and Multiwalled Carbon Nanotubes. *J. Nanomater.* **2017**, *2017*, 3691509. [CrossRef]
57. Luigi, C.; Alessandra, B.; Fabio, C. Electrical properties of carbon nanotubes cement composites for monitoring stress conditions in concrete structures. *Appl. Mech. Mater.* **2011**, *82*, 118–123.
58. Lee, H.; Kang, D.; Kim, J.; Choi, K.; Chung, W. Void detection of cementitious grout composite using single-walled and multi-walled carbon nanotubes. *Cem. Concr. Compos.* **2019**, *95*, 237–246. [CrossRef]
59. Kim, J.H.; Rhee, I.; Jung, Y.C.; Ha, S.; Kim, Y.A. Optical sensitivity of DNA-dispersed single-walled carbon nanotubes within cement composites under mechanical load. *Carbon Lett.* **2017**, *24*, 90–96.
60. *Image of Rhenofit ®CNT-3/CNT-2 from Transmission Electron Microscope*, 16th ed.; RheinChemie Additives: Mannheim, Germany, November 2018. Available online: http://cnt.lanxess.com (accessed on 28 February 2020).
61. D'Souza, O.J.; Mascarenhas, R.J.; Thomas, T.; Basavaraja, B.M.; Saxena, A.K.; Mukhopadhyay, K.; Roy, D. Platinum decorated multi-walled carbon nanotubes/Triton X-100 modified carbon paste electrode for the sensitive amperometric determination of Paracetamol. *J. Electroanal. Chem.* **2015**, *739*, 49–57. [CrossRef]
62. Carriço, A.; Bogas, J.A.; Hawreen, A.; Guedes, M. Durability of multi-walled carbon nanotube reinforced concrete. *Constr. Build. Mater.* **2018**, *164*, 121–133. [CrossRef]
63. Wang, H.; Zhou, W.; Ho, D.L.; Winey, K.I.; Fischer, J.E.; Glinka, C.J.; Hobbie, E.K. Dispersing single-walled carbon nanotubes with surfactants: A small angle neutron scattering study. *Nano Lett.* **2004**, *4*, 1789–1793. [CrossRef]
64. Jensen, R.E.; O'Brien, E.; Wang, J.; Bryant, J.; Lewis, D.A. Characterization of epoxy-surfactant interactions. *J. Polym. Sci. Part B Polym. Phys.* **1998**, *36*, 2781–2792. [CrossRef]
65. Ferro, G.; Tulliani, J.M.; Musso, S. Carbon nanotubes cement composites. *Frat. Integrità Strutt.* **2011**, *18*, 34–44. [CrossRef]
66. Suprompituk, W.; Jaroenapibal, P. Improvement of Early Compressive Strength in Belite Cement by Incorporating Silica Coated Single-Walled Carbon Nanotubes. *Key Eng. Mater.* **2016**, *718*, 157–162. [CrossRef]
67. Markar, J.M.; Beaudoin, J.J. Carbon nanotubes and their applications in the construction industry. *Spec. Publ. Roy. Soc. Chem.* **2004**, 331–341.

68. Markar, J.M.; Margeson, J.C.; Luh, J. Carbon nanotube/cement composites early results and potential applications. In Proceedings of the NRC Publications Record, 3rd International Conference on Construction Materails: Performaing Innovations and Structural Implications, Vancouver, BC, Canada, 22–24 August 2005; pp. 1–10.
69. Raki, L.; Beaudoin, J.; Aali, R.A.; Makar, J. Cement and concrete nanoscience and nanotechnology. *Materials* **2010**, 3, 918–942. [CrossRef]

materials

MDPI

Article

Properties of Concrete Made with Low-Emission Cements CEM II/C-M and CEM VI

Anna Król [1,*], Zbigniew Giergiczny [2] and Justyna Kuterasińska-Warwas [3]

1 Faculty of Mechanical Engineering, Opole University of Technology, Prószkowska Str. 76, 45-758 Opole, Poland
2 Faculty of Civil Engineering, Silesian University of Technology, Akademicka Str. 5, 44-100 Gliwice, Poland; zbigniew.giergiczny@polsl.pl
3 Institute of Ceramics and Building Materials, Oświęcimska Str. 21, 45-641 Opole, Poland; j.kuterasinska@icimb.pl
* Correspondence: a.krol@po.edu.pl

Received: 21 April 2020; Accepted: 11 May 2020; Published: 14 May 2020

Abstract: The paper presents the composition and properties of low-emission ternary cements: Portland multicomponent cement CEM II/C-M and multicomponent cement CEM VI. In the ternary cements, Portland clinker was replaced at the levels of 40% and 55% with a mixture of the main components such as limestone (LL), granulated blast furnace slag (S) and siliceous fly ash (V). Portland multicomponent cements CEM II/C-M and CEM VI are low-emission binders with CO_2 emissions ranging from 340 (CEM VI) kg to 453 (CEM II/C-M) kg per Mg of cement. The results obtained indicate the possibility of a wider use of ground limestone (LL) in cement composition. This is important in the case of limited market availability of fly ash and granulated blast furnace slag. The tests conducted on concrete have shown that the necessary condition for obtaining a high strength class and durability of concrete from CEM II/C-M and CEM VI ternary cements is low water–cement ratio. Durability characteristics of concrete (carbonation susceptibility, chloride ion permeation, frost resistance) made of CEM II/C-M and CEM VI cements were determined after 90 days of hardening. This period of curing reflects the performance properties of the concrete in a more effective way.

Keywords: ternary cements; limestone; siliceous fly ash; granulated blast furnace slag; concrete properties; concrete durability; CO_2 emission

1. Introduction

In 2017, the global production of cement, the base component of concrete, amounted to almost 4.65 billion Mg [1]. For the production of 1 Mg of Portland cement clinker, about 1.7 Mg of natural resources are used, mainly carbonate raw materials such as limestone and marl. Thus, as a result of the clinker firing process, huge amounts of CO_2 are released into the atmosphere, the source of which is the thermal dissociation of carbonates in the raw material bulk (60%) and the emission of CO_2 from the combustion of technological fuel (40%) [2,3]. It is considered that cement production is responsible for about 7.4% of the world carbon dioxide emission (2.9 Mg in 2016) [4]. Therefore, the world cement industry has to meet the constantly growing environmental requirements, which mainly concern the reduction of dust and greenhouse gas emissions [5]. Unfortunately, the production of the basic component of cement, i.e., Portland clinker, is associated with the emission of CO_2, which is about 825–890 kg of CO_2 per Mg of clinker [6]. The world average is about 840 kg of CO_2 but the carbon dioxide emission level should be lower than 400 kg per Mg of cement. It is suggested that the emission levels reach around 350–410 kg per Mg of cement [4].

The possibilities of emission reduction include two solutions in the cement production process [2,7]:

- production of multicomponent cements CEM II-CEMV according to EN 197-1 [8] using significant quantities of main ingredients other than Portland clinker;
- modification of the production process of cement clinker by modification of the raw material set (belite clinkers, belite-sulphate-aluminate clinkers, etc.) and use of alternative (non-fossil) biomass-rich fuels.

In the case of production of CEM II-CEM V multicomponent cements, the main components are usually by-products of industrial processes such as siliceous fly ash (V), calcareous fly ash from coal dust combustion in the power industry or granulated blast furnace slag from iron metallurgy (S) [2,9,10]. The cements containing significant amounts of fly ashes and slags are characterized by low hydration heat (a feature important in the implementation of massive concrete structures), higher strength after longer curing periods and higher resistance to chemical aggression [2,9–11]. To ensure appropriate durability of concrete made of cement with lower clinker content in the assumed construction environment, the concrete composition (type and amount of cement, w/c ratio, type of admixtures and amount of concrete additives) should be properly designed, so that the concrete is characterized by a tight matrix. Determining the concrete tightness, e.g., by limiting the amount of water in the concrete mix or using cement with mineral additives, results in limiting the capillary porosity of the hardened cement slurry [2,12–14]. On the "macro" scale, it directly affects the depth of penetration of aggressive media and the size of capillary pull, whereas on the "micro" scale, it results in impeding the diffusion of aggressive ions into the cement matrix. However, the availability of fly ash and granulated blast furnace slag, with increasing cement production, is limited [15]; therefore, limestone (LL) is used increasingly often in cement composition. The main advantage of this component is its widespread availability and the fact that it can be obtained from the cement plants own raw material resources [16–18].

Calcium carbonate, the main component of limestone, reacts with calcium aluminates to form hydrated calcium carboaluminates. The presence of hydrated calcium carboaluminates inhibits the transition of ettringite to monosulfate, thus, in hydration products the amount of monosulfate decreases or disappears while the amount of ettringite increases [19]. The fact that calcite reacts with C_3A to form carboaluminates means that $CaCO_3$ may play, to a limited extent, the role of a regulator of setting time. This results in the reduction of the amount of gypsum, which is necessary to regulate the setting time [20].

In addition to the reaction with calcium aluminate, the addition of limestone to the cement may accelerate the C_3S phase reaction. This effect is explained by the nucleation effect, in which $CaCO_3$ grains act as additional crystallization germs for cement hydration products [19–21]. Limestone is a very soft component in comparison to Portland clinker. After the milling process, it has a much higher specific surface area and, as a micro-filler, influences the properties of cement composites, e.g., by reducing porosity, increasing strength in the initial period of hardening and improving workability, reducing water consumption and reducing water draining from the concrete mixture (so-called "bleeding") [19–24]. Bearing these facts in mind, the European Committee for Standardization CEN has undertaken standardization works aimed at extending the range of cements containing cement components other than Portland clinker in its composition. It is proposed to implement the non-harmonized standard prEN 197-5 [25], which extends the range of Portland multicomponent cements (the possibility of using several main components in the composition of cement) by a group of Portland multicomponent cements CEM II/C-M with a minimum content of Portland clinker of 50% and a newly created group of multicomponent cements CEM VI, in which the share of non-clinker components may be a maximum 65%.

This paper presents the results of research on Portland multicomponent cement CEM II/C-M with 40% of non-clinker main components and multicomponent cement CEM VI with 55% of these components. Ground granulated blast furnace slag (S), siliceous fly ash (V) and ground limestone (LL) were used as non-clinker main components. Concrete tests were performed for the analyzed cements CEM II/C-M and CEM VI. The basic properties of concrete mixture and hardened concrete were determined with a view to future use of cements in construction practice. The level of CO_2

emissions originating in the composition of CEM II/C-M and CEM VI cement was also calculated, as well as the level of CO_2 emissions from the production of concrete with the use of tested cements.

2. Materials and Methods

2.1. Characteristics of Components and Composition of Tested Cements

Three types of non-clinker ingredient were used in the study: granulated blast furnace slag from iron metallurgy, siliceous fly ash from the combustion of coal in power plants, and natural limestone.

The chemical composition of the cement components and selected physical properties are given in Table 1. Figures 1 and 2 show diffractograms of ground granulated blast furnace slag (S) and fly ash (V).

In the slag phase composition (S), the dominant component is the vitreous phase, the quantitative content of which (determined microscopically) is 98%. In fly ash, next to the vitreous phase, the main crystalline components identified are quartz, mullite, hematite and magnetite. The Portland cement CEM I is a semi-finished product with an increased SO_3 content (5.0%) in order to obtain a normal SO_3 content (max. 3.5%) when mixed with the other main components of the cement. Therefore, this cement is a semi-finished product in the process of manufacturing multicomponent cements. The clinker content in the cement was 90%.

Table 1. Chemical composition and physical properties of the main components of cement.

Type of Raw Material	Content of Component, (wt. %)									Specific Gravity (g/cm^3)	Specific Area ACC. Blaine (cm^2/g)
	SiO_2	Al_2O_3	Fe_2O_3	CaO	MgO	SO_3	Na_2O	K_2O	Cl^-		
Limestone (LL) (1)	5.4	1.3	1.1	49.7	1.8	0.03	<0.1	<0.1	0.005	2.70	6150
Granulated blast furnace slag (S)	40.5	7.4	1.26	43.7	5.0	0.14	0.77	0.45	0.046	2.92	3800
Siliceous fly ash (V) (2)	52.3	27.5	5.80	3.6	2.6	0.29	0.94	3.15	0.008	2.14	2750
Portland cement CEM I	20.65	5.1	2.57	62.94	1.4	5.0	0.15	0.63	0.07	3.16	4500

(1) $CaCO_3$ content calculated on the basis of the amount of CaO is 89 (wt.%), total organic carbon (TOC)-0.04 (wt.%), clay content-0.4 g/100 g, (2) LOI of siliceous fly ash V-2.43 (wt.%) (Category A according to EN 197-1:2012 [8]).

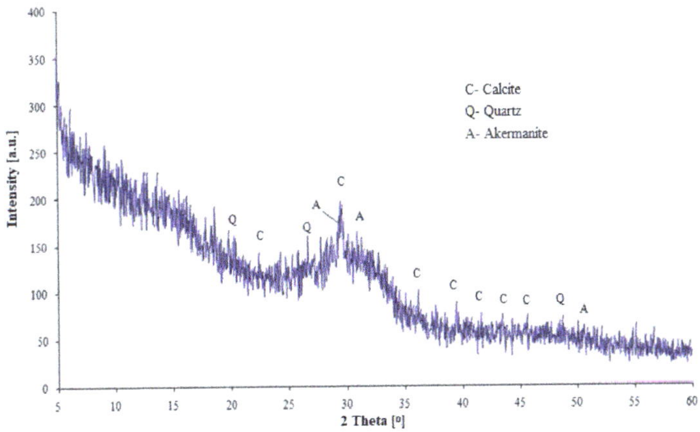

Figure 1. Diffractogram of granulated blast furnace slag.

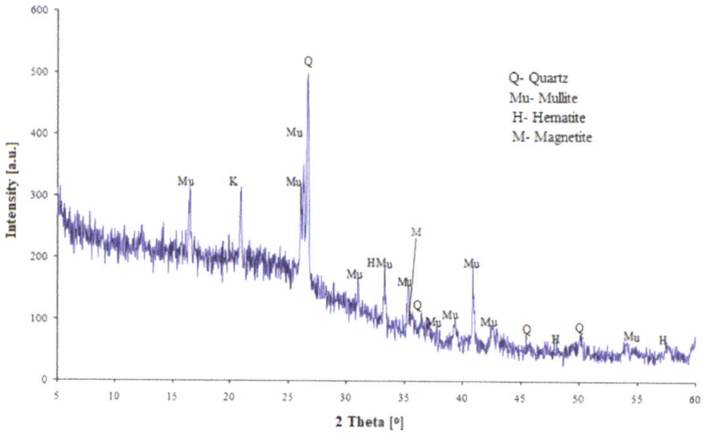

Figure 2. Diffractogram of siliceous fly ash.

When analyzing the properties of the main components of cement used, attention should be paid to the high specific surface area of limestone of 6150 cm^2/g (Table 1). Obtaining such a high specific surface area is relatively easy due to the very good granularity of the limestone. The granulometric composition of non-clinker cement components is shown in Figure 3.

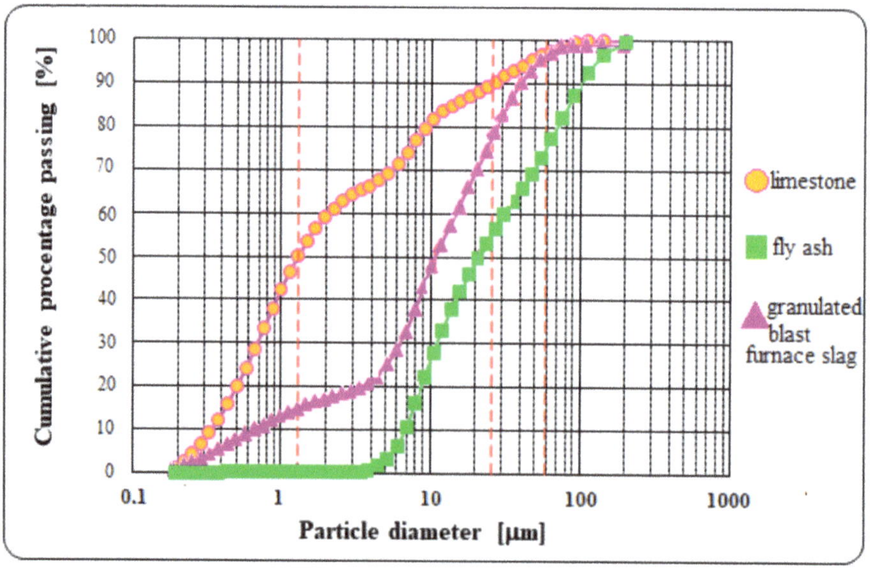

Figure 3. Particle size distribution of supplementary cementitious materials.

Two CEM II/C-M cements with a Portland clinker content of 54% and two CEM VI cements with a Portland clinker content of 40.5% were prepared for testing the ternary cements. The composition of the tested cements and CO_2 emission levels are given in Table 2. The CO_2 emission level from 1 Mg of cement was calculated assuming the average CO_2 emission level of production of 1 Mg of clinker at the level of 840 kg [4] and the clinker content in the composition of the tested ternary cements CEM II/C-M and CEM VI (Table 2). In the calculations, the CO_2 emission level related to the transport and grinding of the components into cement was omitted. The obtained CO_2 emission levels from 1 Mg of tested cements at the level of 340.2–453.6 kg allows the inclusion of CEM II/C-M and CEM VI cements into low-emission cements.

In order to evaluate the synergy effects of cement components, comparative cements containing one non-clinker component—slag cement C(55S), fly ash cement C(40V) and limestone cement C(40LL) were also tested (Table 2).

Table 2. Composition of tested cements and CO_2 emission level.

Cement Marking	Type of Cement	Content of Component, (wt. %)				CO_2 Emission Level from Mg of Cement, (kg)
		Cement CEM I	Slag S	Limestone LL	Fly Ash V	
C(30S-10LL)	CEM II/C-M (30S-10LL)	60	30	10	-	453.6
C(30V-10LL)	CEM II/C-M (30V-10LL)	60	-	10	30	453.6
C(35S-20LL)	CEM VI (35S-20LL)	45	35	20	-	340.2
C(35S-20V)	CEM VI (35S-20V)	45	35	-	20	340.2
C(40V) (1)	CEM (40V)	60	-	-	40	453.6
C(55S) (1)	CEM (55S)	45	55	-	-	340.2
C(40LL) (1)	CEM (40LL)	60	-	40	-	453.6

(1) comparative cements.

The properties of cements were determined according to the procedures of EN 196 and the density according to the standard on the properties of aggregates EN 1097-7 (Table 3).

Table 3. Procedures used to determine the properties of cement.

Property	Standard Test Method
Constancy of volume	EN 196-3:2016-11 [26]
Initial setting time	EN 196-3:2016-11 [26]
Specific surface area	EN 196-6:2011 [27]
Density	EN 1097-7:2008 [28]
Compressive strength	EN 196-1:2016-05 [29]

2.2. *Concrete Composition and Properties' Test Methods*

Based on Portland multicomponent cements CEM II/C-M and multicomponent cements CEM VI, concrete mixes with the following composition, given in Table 4, were designed.

Table 4. Mix proportions of concrete mixtures.

Type	Concrete Designation	w/c	Cement Content (kg/m^3)	Water (kg/m^3)	PCE Admixture (kg/m^3)	Coarse Aggregate (kg/m^3)		Sand 0–2 mm (kg/m^3)
						8–16 mm	2–8 mm	
I	C(30S-10LL)	0.60	300	180	-	680	530	680
	C(30V-10LL)		300	180	-	680	530	680
	C(35S-20LL)		300	180	-	680	530	680
	C(35S-20V)		300	180	-	680	530	680
II	C(30S-10LL)	0.35	340	120	10.2	725	565	725
	C(30V-10LL)		340	120	10.2	725	565	725
	C(35S-20LL)		340	120	10.2	725	565	725
	C(35S-20V)		340	120	10.2	725	565	725

Two types of concrete were prepared—type I containing 300 kg of cement in 1 m^3 of concrete mix at a ratio w/c = 0.60 and type II, containing 340 kg of cement in 1 m^3 of concrete mix at a ratio w/c = 0.35. Natural gravel aggregate with a fraction up to 16 mm and sand 0–2 mm as fine aggregate were used in the concrete mixture. A superplasticizer (PCE) based on polycarboxylate ether was used in the composition of concrete with reduced water–cement ratio (w/c = 0.35).

The properties of concrete mixtures and hardened concrete were tested according to the methodology included in the standards, which are presented in Table 5.

Table 5. Procedures used to determine the properties of the concrete mixture and hardened concrete.

Property	Standard Test Method (Procedures)
Consistency (fall cone method)	EN 12350-2:2011 [30]
Density of concrete mixture	EN 12350-6 [29,31]
Air content	EN 12350-7:2011 [32]
Compressive strength	EN 12390-3:2011 [33]
Absorption	PN-B-06250:1988 [34]
Depth of water penetration under pressure	EN 12390-8:2011 [35]
Depth of carbonation	prCEN/TS 12390-12:2010 [36]
Permeation of chloride ions	ASTM C 1202-05 [37]
Ordinary frost resistance	PN-B-06265:2018-10 [38]
De-icing salts frost resistance (surface scaling)	CEN/TS 12390-9:2007 [39]

3. Results and Discussion

3.1. Properties of CEM II/C-M and CEM VI Cements

The properties of the cements are presented in Table 6. The density of the cements was lowest for those containing siliceous fly ash (Table 6). The specific surface area of ternary cements ranged from 4350 to 4750 cm^2/g. Cement slurries with ternary cements did not show volume changes due to swelling (Table 6). The setting time of cements and other properties are closely related to their composition and the amount of mineral additives introduced (Table 6).

Table 6. Physical and mechanical properties of cements.

No.	Cement Designation	Constancy of Volume (mm)	Density (g/cm^3)	Initial Setting Time (h, min)	Specific Surface Area (cm^2/g)	Compressive Strength (MPa)				
						2 Days	7 Days	28 Days	90 Days	360 Days
1.	C(30S-10LL)	0.5	3.00	2:15	4750	19.3	36.4	60.1	69.2	74.9
2.	C(30V-10LL)	0	2.70	2:40	4700	18.4	28.5	40.9	54.9	64.1
3.	C(35S-20LL)	0	2.96	2:45	4600	12.1	25.0	46.8	56.2	61.0
4.	C(35S-20V)	0	2.82	3:25	4350	10.7	22.8	45.4	59.1	64.8
5.	C(40V)	0	2.64	3:00	4300	15.5	26.4	36.4	49.4	61.3
6.	C(55S)	0	2.98	3:10	4330	10.5	25.7	51.5	64.2	71.5
7.	C(40LL)	1	2.92	1:55	5765	16.5	27.5	33.2	37.0	39.0

In most cases, the highest increases in strength of standard cement mortars can be observed between 7 and 28 days of curing (Figure 4). During this period, the strength increases quite significantly in the case of cements containing blast furnace slag (the slag is hydraulically active and begins to react with water a long time before the ash pozzolanic reaction begins). The highest strength increase in this period is observed in comparative slag cement C(55S) on Figure 4. Partial replacement of blast furnace slag, both with limestone LL and fly ash V, slightly reduces the strength increase between the 7th and 28th days of hardening. Omitting the small influence of limestone on the increase in early strength, it is a rather chemically inert component in the cement system, and therefore its addition to slag causes a decrease in later strength. Replacement of slag with fly ash also slows down the dynamics of strength growth. This can be explained by the fact that the pozzolanic reaction of fly ash begins intensify only after 28 days of curing (when the amount of $Ca(OH)_2$ from cement hydration increases). The final strength (after 360 days) of multicomponent cement C(30S-10LL) is similar to that of comparative cement C(55S), while in the case of cement C(35S-20V) there is a slight decrease in the final strength, compared to cement C(55S) (Figure 4). This decrease can be explained by the lower activity of fly ash in relation to ground granulated blast furnace slag, which is a component with latent hydraulic activity (with a CaO content of approximately 40–44% and after heat treatment in a blast furnace under conditions similar to those in a rotary kiln for Portland clinker production).

The partial replacement of fly ash by limestone C(30V-10LL) in cement composition, allows a cement to be made with higher early strength (after 2 days) and higher final strength than the reference cements C(40V) and C(40LL). The higher early strength in the presence of limestone may result from the caulking effect as well as the small amount of carboaluminates formed. On the basis of the results obtained, it can also be observed that the addition of LL limestone decreases the strength of cement mortars to a greater extent in combination with siliceous fly ash (C(30V-10LL)), while the use of limestone (even up to 20%) in combination with ground granulated blast furnace slag (cements: C(30S-10LL), C(35S-20LL)) gives a much smaller decrease, slightly less than in the case of using the S slag composition with fly ash V-cement C(35S-20V).

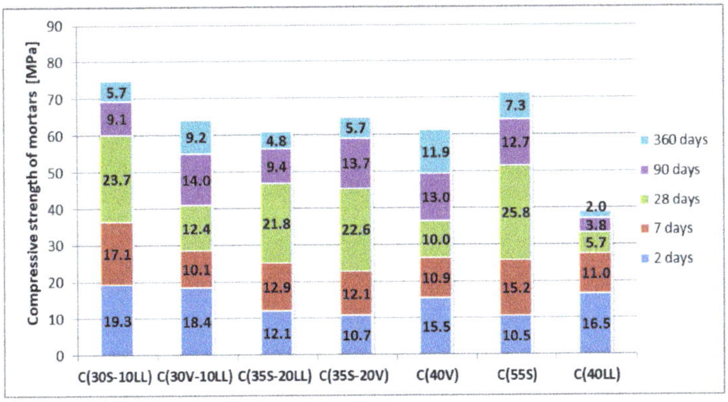

Figure 4. Increase in compressive strength depending on the curing time of cement mortars.

3.2. *Concrete Properties with CEM II/C-M and CEM VI Low-Emission Cements*

3.2.1. Properties of Concrete Mixture

The properties of concrete mixture are presented in Table 7. The consistency (fall cone method) of concrete mixtures at w/c = 0.60 corresponded to class S3 for concrete mixture with cement C(30S-10LL) and classes S2 for other cements. Reduction of the water content to the level w/c = 0.35 caused a decrease in consistency to class S1.

The density of all concrete mixtures with the ratio w/c = 0.60 was similar and ranged from 2350 to 2360 kg/m^3. The air content was low and ranged from 0.6% to 1.0%. Decreasing the water-cement ratio to the level w/c = 0.35 resulted in an increase in the density of concrete mixtures by about 70–80 kg/m^3 and an increase in air content to 1.6–2%. The increase in aeration of the concrete mixture is typical of using liquefying admixture.

Table 7. Properties of concrete mixtures.

Property	Ratio w/c	Concrete Designation—Corresponding to the Composition of the Cement			
		C(30S-10LL)	C(30V-10LL)	C(35S-20LL)	C(35S-20V)
Consistency, (mm)	0.60	110 S3 *	60 S2 *	50 S2 *	50 S2 *
	0.35	30 S1 *	20 S1 *	30 S1 *	30 S1 *
Density, (kg/m^3)	0.60 0.35	2360 2430	2350 2420	2350 2430	2350 2420
Air content, (vol.%)	0.60 0.35	0.8 1.6	0.8 1.8	1.0 2.0	0.6 2.0

* Consistency class acc. EN 12350-2:2011 [30].

3.2.2. Properties of Hardened Concrete

- Compressive strength

The compressive strength was determined, after 2, 7, 28 and 90 days of curing, on 10 cm cubic samples. For concretes with w/c = 0.35, strength tests were also performed after 1 day of curing. The results are presented in Figures 5 and 6.

The highest compressive strength, at w/c = 0.35 and 0.60, was obtained by concrete made of Portland multicomponent cement C(30S-10LL). The lowest compressive strength was achieved with C(30V-10LL) on Portland ash and lime cement, despite the fact that it contained more Portland clinker compared to the other two CEM VI cement concretes. Limestone and siliceous fly ash show a synergistic effect only with early strength (2 days) and only with 10% addition of limestone, which confirms the results obtained by De Werdt et al. [40–42]. The results obtained from strength tests of concretes made of slag-calcareous cements provide a reason for the prospective wider use of limestone in cement composition.

Decreasing the water-cement ratio (w/c) from 0.60 to 0.35 resulted in a significant increase in the compressive strength of concretes made of all cements tested (Figure 7). This increase is particularly visible in the initial period of concrete hardening, i.e., until the 7th day. It can be noted that the lowest results were obtained for concrete using cement C(30V-10LL), however, the compressive strength after 28 days was nearly 70 Mpa and was almost twice as high as the strength obtained at w/c = 0.60. To sum up, it should be stated that a low w/c ratio is a very effective factor in shaping the strength characteristics of concrete made of cements with low Portland clinker content (Figure 7).

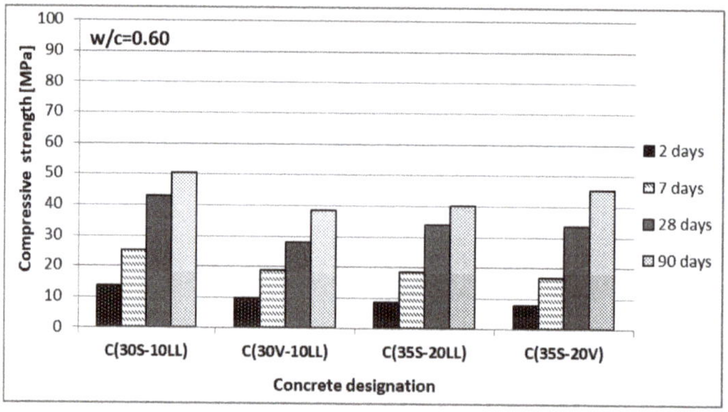

Figure 5. Compressive strength of concrete (w/c = 0.60).

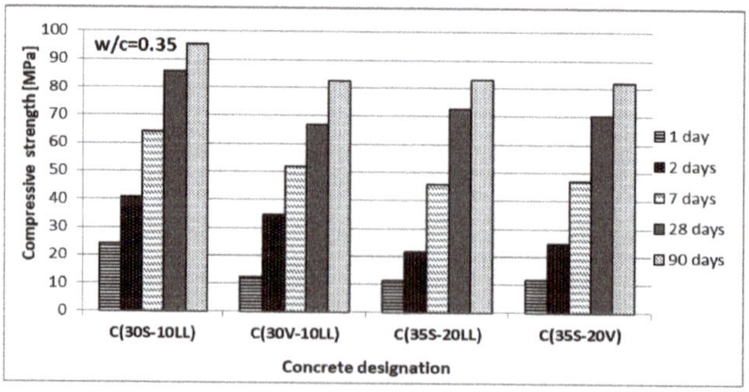

Figure 6. Compressive strength of concrete (w/c = 0.35).

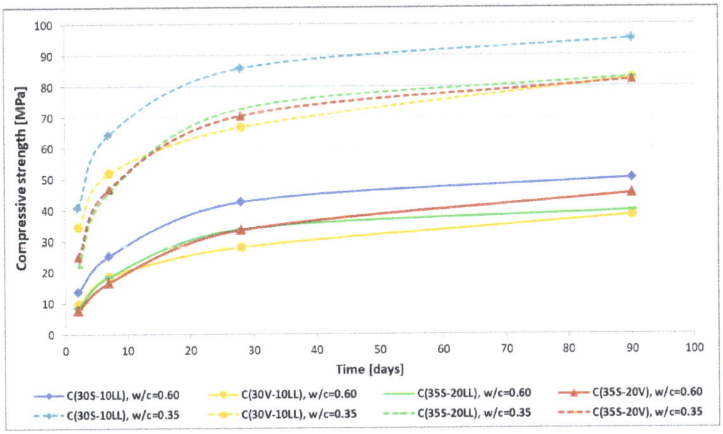

Figure 7. The impact of reduced w/c ratio on the compressive strength of concretes made of tested ternary cements.

- Water absorption and water penetration under pressure

All tested concretes, at the same w/c, show similar water absorption. For concretes with w/c = 0.60 the absorption varies between 6.6% to 7.0%, while for concretes with reduced w/c = 0.35 it is much lower and ranges from 3.3% to 4.1% (Figure 8). Extension of the curing time to 90 days resulted in a slight decrease in absorption of the tested concretes, most noticeable for concrete made with C(35S-20V) cement.

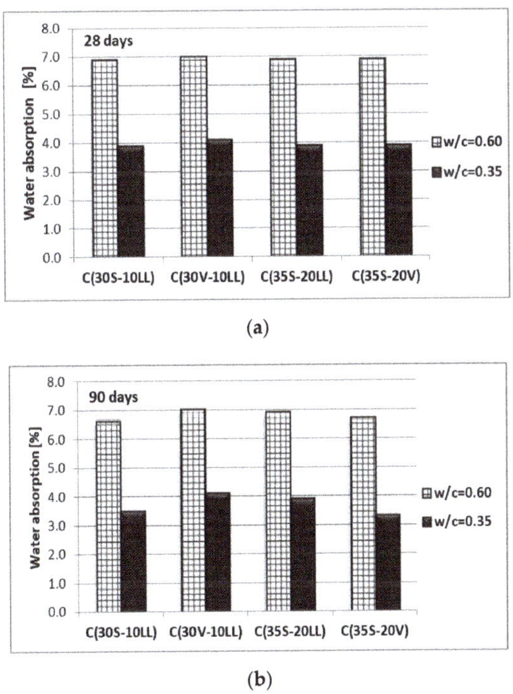

Figure 8. Water absorption of concrete after: (**a**) 28 days of curing, (**b**) 90 days of curing.

The results of the study on the depth of water penetration under pressure (Figure 9) show that concretes with w/c = 0.35 are characterized by very high tightness, especially after 90 days of curing. The depth of water penetration under pressure at w/c = 0.35 was maximum 15 mm for concrete with cement C(35S-20V) after 28 days of curing. Concretes with w/c = 0.60 showed water penetration depth after 28 days of hardening at the level from 15.3 mm to 43.7 mm and from 7.7 mm to 14.7 mm for concrete curing for 90 days (Figure 9).

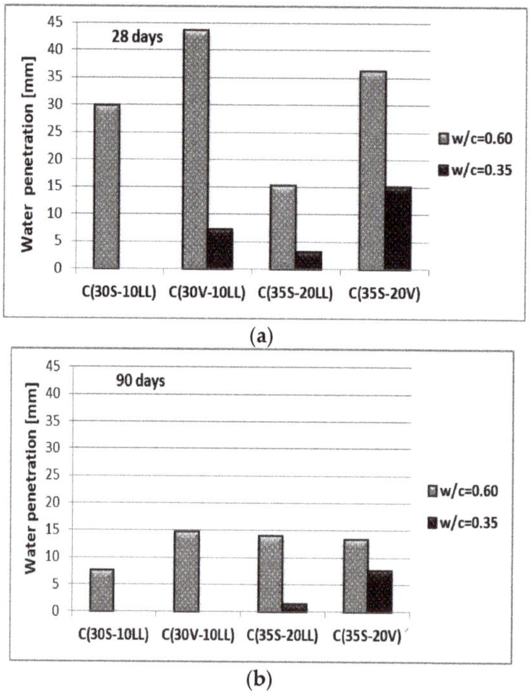

Figure 9. Water penetration depth of concrete after: (**a**) 28 days of curing, (**b**) 90 days of curing

Omitting the influence of the w/c ratio, the differences in the depth of water penetration inside the concrete matrices primarily result from the different activity of the main components of cements used. The most active component, apart from Portland clinker, is ground granulated blast furnace slag, whereas fly ash is a component with pozzolanic activity (ability to react in the presence of moisture, with $Ca(OH)_2$ from the hydration of silicate phases of Portland clinker). The impact of this reaction on the properties of mortar (concrete) is earliest visible about 28 days and later (Figure 9b). The addition of limestone improves the porosity of the cement-ash/slag system. Limestone, as a soft component, is ground into very fine grains, which fill the voids between cement and ash/slag grains. It results in increased early strength (after 2 days) in relation to the cement included only fly ash. After a longer period of time (28 days and later), cements containing granular blast furnace slag in the composition with ash (S, V) or limestone (S, LL) have higher strength and tightness.

- Carbonation susceptibility

The type of cement used was assessed for its susceptibility to carbonation (Figure 10). The test was carried out using an elevated CO_2 concentration 4%, the test duration was 70 days (accelerated method). Analyzing the results obtained for concretes at w/c = 0.60, it can be seen that the highest depth of carbonation is characterized by concrete with Portland multicomponent cement C(30V-10LL), after 28 days of hardening the depth of carbonation reaches 29.7 mm, and after 90 days it is 18.6 mm. Reducing the water–cement ratio to the level w/c = 0.35 very effectively lowered the depth of concrete carbonation (Figure 10). A significant decrease in the depth of carbonation linked with the extension of curing period should be associated with the activity of the cement components used, mainly ground granulated blast furnace slag (hydraulic activity) and fly ash (pozzolanic activity). Additional amounts of products formed later (after 90 days of curing) from the course of reaction between cement hydration products and active mineral additives, settle in the pores of hardening cement slurry and hinder the permeation and penetration of aggressive ions [21,23].

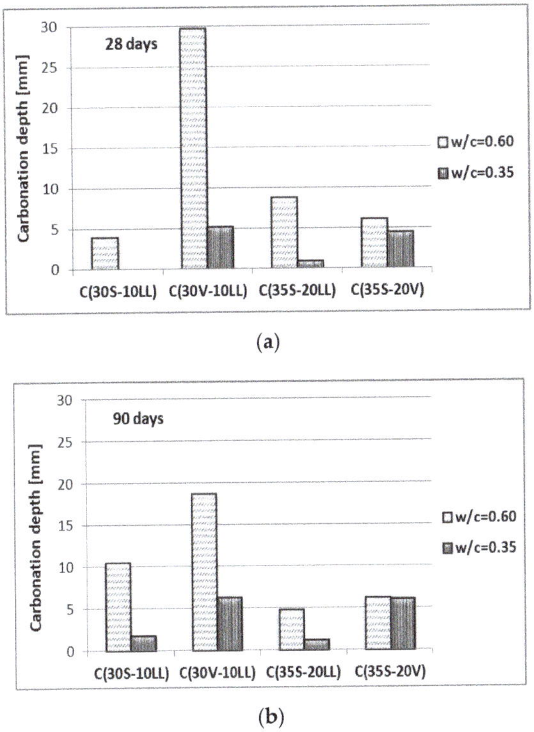

Figure 10. Carbonation depth of concrete after: (**a**) 28 days of curing, (**b**) 90 days of curing.

- Chloride ions permeation

The permeability limitation of the concrete matrix is confirmed by the results of chloride ion permeation (Figure 11). Extending the curing period to 90 days or decreasing the water and cement ratio to w/c = 0.35 results in a significant decrease in the permeation of chloride ions corresponding to low or very low permeation class according to ASTM C 1202-05 [37] for both test dates. The differences in chloride ion permeability between concrete samples can be explained in the same way as was described in the water penetration analysis, as this feature is strictly related to tightness of concrete.

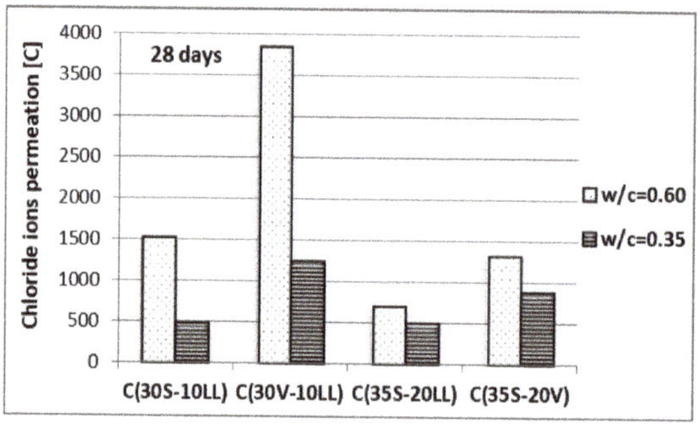

(a)

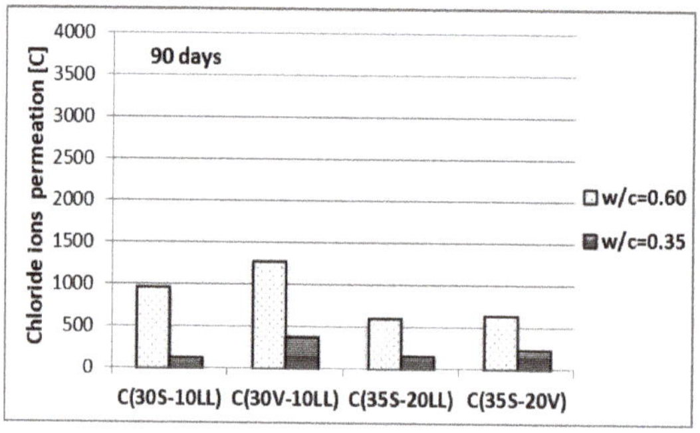

(b)

Figure 11. Chloride ions permeation of concrete after: (**a**) 28 days of curing, (**b**) 90 days of curing.

- Freeze–thaw resistance

An important feature of concrete, used in areas with minus temperatures, is its resistance to such environmental impact. Resistance of concrete to cyclic freezing and unfreezing was determined by the ordinary method (150 cycles of freezing at −18 °C and unfreezing of concrete at 18 °C, duration time of 1 cycle was 6 h (3 h of freezing and 3 h of thawing)) for concretes with coefficient w/c = 0.35. The test was performed after 28 and 90 days of concrete curing. The result of the test is positive if the decrease of compressive strength is less than 20% and the loss of mass is not more than 5% of weight. The test of concrete surface resistance to frost (56 cycles) in the presence of NaCl de-icing salt was also performed. Concrete was evaluated after 28 days of curing. The test results are presented in Table 8 and Figures 12 and 13.

Table 8. Results of frost-resistance tests.

Concrete Age (Days)	Tested Properties	Concrete Designation, w/c = 0.35			
		C(30S-10LL)	C(30V-10LL)	C(35S-20LL)	C(35S-20V)
28	Strength of samples after freeze-thaw cycles, (MPa)	86.7	63.4	77.2	76.4
	Strength of reference samples, (MPa)	93.8	78.5	83.7	82.7
	Decrease in strength, (%)	7.6	19.2	7.7	7.7
	Loss of sample weight after test, (wt. %)	0.0	0.1	0.1	0.1
	Scaling of the material after 56 freeze-thaw cycles in the presence of NaCl salt, (kg/m^2)	0.12	2.77	0.89	0.55
90	Strength of samples after freeze-thaw cycles, (MPa)	94.8	71.6	84.6	85.7
	Strength of reference samples, (MPa)	95.0	84.3	85.9	84.9
	Decrease in strength, (%)	0.2	15.1	1.5	1.0
	Loss of sample weight after test, (wt. %)	0.0	0.0	0.1	0.1

The type of applied cement affects the durability of concrete under cyclic freezing and unfreezing conditions, especially when de-icing agents are used. The worst results were obtained for concrete made of Portland multicomponent cement C(30V-10LL) (Figures 12 and 13; Table 8). The decrease in compressive strength after the frost resistance test using the normal method reached 19.2% for concrete subjected to alternating temperatures after 28 days of curing and 15.1% for concrete subjected to the test after 90 days of curing. For comparison, concretes made of other cements were characterized by strength decreases at a much lower level of 7.6–7.7% for 28-day samples and 0.2–1.5% for 90-day samples. Concrete samples after the frost-resistance test did not show significant changes in mass.

When analyzing the results of the surface resistance of concrete to frost in the presence of de-icing salt (Table 8, Figure 13), it is clear that only concrete made of Portland multicomponent cement C(30S-10LL) can be classified as resistant. However, remaining concretes, especially concrete made of Portland multicomponent cement C(30V-10LL), show considerable scaling.

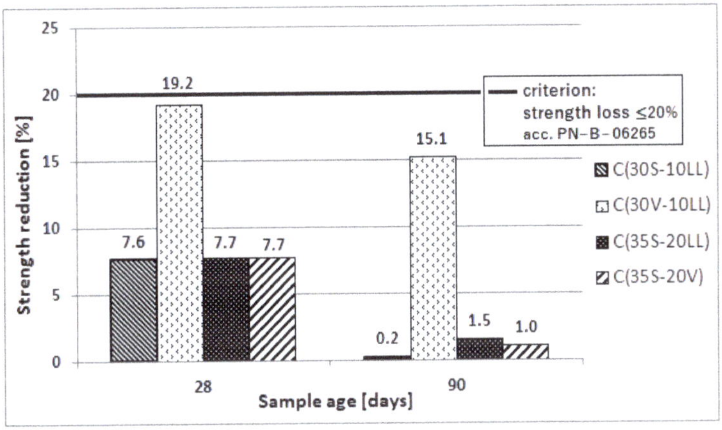

Figure 12. Frost resistance of concretes tested by the standard method [38].

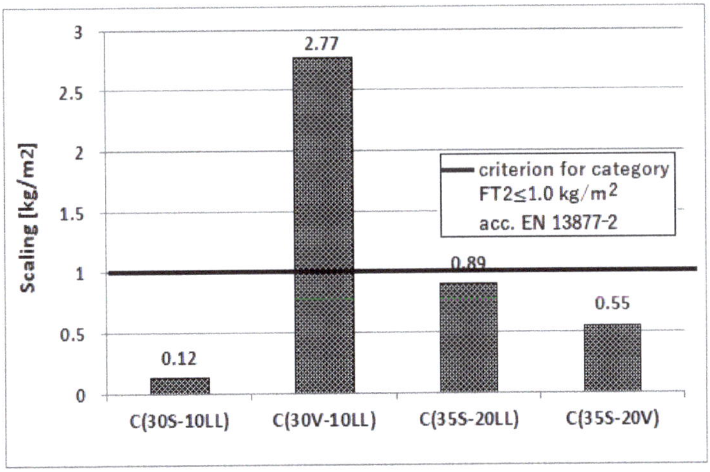

Figure 13. Frost resistance of concretes in the presence of de-icing salt in relation to resistance category according to EN 13877-2:2013-08 [43].

3.2.3. CO_2 Emissions from Concrete Made of CEM II/C-M and CEM VI Cement

The concrete emission level was calculated based on the level of CO_2 emissions from 1 Mg of cement (Table 2). The results are presented in Table 9. The obtained CO_2 emission levels per 1 m^3 of concrete are very low (Table 2). When converting the level of CO_2 emission into 1 MPa of compressive strength, it can be observed that an important factor is the w/c ratio in concrete. With a lower w/c ratio (0.35), the strength levels are much higher and the difference in the level of the obtained compressive strength depending on the composition of the cement starts to fade. The CO_2 emission level per 1 MPa is significantly reduced also, e.g., in the case of CEM II/C-M (30V-10LL) cement from 4.9 kg (w/c = 0.60) to 2.3 kg (w/c = 0.35). Considering the compressive strength after 90 days, CO_2 emissions are lower, on average by about 0.7 kg at w/c = 0.60 and about 0.2 kg at w/c = 0.35. This is due to the fact that cements with a high content of the main components other than Portland clinker (mainly granulated blast furnace slag and/or fly ash) have a significant strength increment between 28 and 90 days of hardening.

Table 9. CO_2 emissions from 1 m^3 concrete and converted to 1 MPa of 28-day and 90-day concrete compressive strength.

Concrete Designation	Cement Content (kg/m^3 Concrete)	w/c Ratio	CO_2 Emission (kg/m^3 Concrete)	28-Day Concrete Compressive Strength $f_{cm,cube\ 28}$ (MPa)	CO_2 Emission (kg) Converted into 1 MPa $f_{cm,cube\ 28}$	90-Day Concrete Compressive Strength $f_{cm,cube\ 90}$ (MPa)	CO_2 Emission (kg) Converted into 1 MPa $f_{cm,cube\ 90}$
C(30S-10LL)	300.0	0.60	136.1	42.6	3.2	50.3	2.7
C(30V-10LL)	300.0		136.1	27.9	4.9	38.3	3.6
C(35S-20LL)	300.0		102.1	33.8	3.0	39.8	2.6
C(35S-20V)	300.0		102.1	33.5	3.0	45.3	2.3
C(30S-10LL)	340.0	0.35	154.2	85.8	1.8	95.3	1.6
C(30V-10LL)	340.0		154.2	66.7	2.3	82.6	1.9
C(35S-20LL)	340.0		115.7	72.6	1.6	83.0	1.4
C(35S-20V)	340.0		115.7	70.4	1.6	82.0	1.4

4. Conclusions

Based on the results of the research carried out, the following conclusions were drawn:

1. The results of the research as well as ecological aspects (mainly reduction of CO_2 emissions) confirmed the advisability of further development of the assortment of general-use cements with the following ternary cements: Portland multicomponent cement CEM II/C-M and multicomponent cement CEM VI. The properties of these cements and the concretes made with them are the result of the properties of the main components and the synergistic interaction of the composition: granulated blast furnace slag S-limestone LL, siliceous fly ash V-limestone LL, granulated blast furnace slag S-siliceous fly ash V.
2. When evaluating the synergy effect of additives in the composition of cements, it can be observed that much better strength effects are obtained in the composition of ground granulated blast furnace slag (S) with limestone (LL) than fly ash (V) with limestone (LL). This is due to the higher activity of the ground slag compared to fly ash (V). The results obtained also confirm a very good synergy effect, known from construction literature and practice, between slag (S) and fly ash (V). A characteristic feature of CEM II/C-M and CEM VI ternary cements is a low level of early compressive strength (up to 7 days) and a significant increase in compressive strength at later stages (especially between 28 and 90 days of hardening).
3. A typical attribute of CEM II/C-M and CEM VI ternary cements is the low level of early compressive strength (up to 7 days) and a significant increase in compressive strength at a later date (especially between 28 and 90 days of hardening).
4. The results obtained indicate the possibility of wider use of ground limestone LL in cement (concrete) composition. This is important in terms of the limited market availability of fly ash and granulated blast furnace slag.

5. The results obtained from tests on concrete showed that a low w/c ratio is a prerequisite for obtaining high strength class and durability of concrete made of CEM II/C-M and CEM VI ternary cements. This can be achieved by using the latest generation of liquefying admixtures (superplasticizers).
6. The durability characteristics of concrete (carbonation susceptibility, chloride ion permeation, frost resistance) from CEM II/C-M and CEM VI cements should be determined after 90 days of hardening. This period of curing better reflects the performance properties of the concrete.
7. Portland multicomponent cements CEM II/C-M and CEM VI are low-emission binders with CO_2 emissions ranging from 340 (CEM VI) kg to 453 (CEM II/C-M) kg per Mg of cement. Because of a high content of non-clinker main components, these cements characterize a much lower degree of hydration after 28 days of hardening than ordinary Portland cement CEM I. The presence of slag (S) and fly ash (V) results in a significant increase in the strength of low-emission cements between 28 and 90 days of hardening, and hence a better effect in terms of reducing emissions per 1 MPa is achieved. In some countries, Poland among others, for this reason the quality of concrete with low-emission cements is assessed after 90 days.

Author Contributions: Conceptualization, A.K. and Z.G.; Investigation, J.K.-W.; methodology: Z.G.; resources: A.K. and J.K.-W.; supervision, A.K.; writing—original draft, A.K., Z.G. and J.K.-W.; writing—review & editing, A.K., Z.G. and J.K.-W. All authors have read and agreed to the published version of the manuscript.

Funding: This research received no external funding.

Conflicts of Interest: The author declares no conflict of interest.

References

1. The European Cement Association. *Activity Report*, 1st ed.; CEMBUREAU: Brussels, Belgium, 2017; pp. 1–42.
2. Giergiczny, Z. Fly ash and slag. *Cem. Concr. Res.* **2019**, *124*, 487–495. [CrossRef]
3. World Business Council for Sustainable Development. *Cement Sustainability Initiative: Getting the Numbers Right*; Project Emissions Report; WBCSD: Geneva, Switzerland, 2016.
4. Sanjuán, M.Á.; Andrade, C.; Mora, P.; Zaragoza, A. Carbon dioxide uptake by cement-based materials: A Spanish case study. *Appl. Sci.* **2020**, *10*, 339. [CrossRef]
5. Bjegovic, D.; Strimer, N.; Serdar, M. Ecological aspects of concrete production. In Proceedings of the Second International Conference on Sustainable Construction Materials and Technologies, Ancona, Italy, 28–30 June 2010; pp. 1483–1492.
6. World Business Council for Sustainable Development (WBCSD). *Cement Industry Energy and CO_2 Performance. Getting the Numbers Right (GNR) Project*, 1st ed.; World Business Council for Sustainable Development: Geneva, Switzerland, 2018.
7. Baran, T.; Garbacik, A.; Ostrowski, M.; Radelczuk, H. Methods of production of low emission Portland clinker. In Proceedings of the Dni Betonu Conference, Wisła, Poland, 1–2 July 2016; pp. 439–448.
8. EN 197-1:2011. *Cement-Part 1: Composition, Specifications and Conformity Criteria for Common Cements*; British Standards Institution: London, UK, 2011.
9. Król, A. The role of the granulated blast furnace slag in sustainable cement production and waste management. *Econ. Environ. Stud.* **2017**, *17*, 613–624. [CrossRef]
10. Król, A. The role of the silica fly ash in sustainable waste management. In Proceedings of the E3S Web of Conferences, 1st International Conference on the Sustainable Energy and Environment Development (SEED), Cracow, Poland, 17–19 May 2016; Volume 10.
11. Khan, M.I.; Lynsdale, C.J. Strength, permeation, and carbonation of high-performance concrete. *Cem. Concr. Res.* **2002**, *32*, 123–131. [CrossRef]
12. Zhang, Y.; Kong, X. Influences of superplasticizer, polymer latexes and asphalt emulsions on the pore structure and impermeation of hardened cementitious materials. *Constr. Build. Mater.* **2014**, *53*, 392–402. [CrossRef]
13. Giergiczny, Z. *Fly Ash in Cement and Concrete Composition*; Silesian University of Technology Publishing House: Gliwice, Poland, 2013.

14. Sathyan, D.; Anand, K.B. Influence of superplasticizer family on the durability characteristics of fly ash incorporated cement concrete. *Constr. Build. Mater.* **2019**, *204*, 864–874. [CrossRef]
15. Scrivener, K.L.; Gartner, E.M. Eco-efficient cements: Potential economically viable solutions for a low-CO_2 cement-based materials industry. *Cem. Concr. Res.* **2018**, *114*, 2–26. [CrossRef]
16. Gholampour, A.; Ozbakkaloglu, T. Performance of sustainable concretes containing very high volume class-F fly ash and ground granulated blast furnace slag. *J. Clean. Prod.* **2017**, *162*, 1401–1417. [CrossRef]
17. Gursel, A.P.; Maryman, H.; Ostertag, C. A life-cycle approach to environmental, mechanical, and durability properties of "green" concrete mixes with rice husk ash. *J. Clean. Prod.* **2016**, *112*, 823–836. [CrossRef]
18. Proske, T.; Rezvani, M.; Palm, S.; Muller, C.; Graubner, C.A. Concretes made of efficient multi-composite cements with slag and limestone. *Cem. Concr. Compos.* **2018**, *89*, 107–119. [CrossRef]
19. Zajac, M.; Rossberg, A.; Le Saout, G.; Lothenbach, B. Influence of limestone and anhydrite on the hydration of Portland cements. *Cem. Concr. Compos.* **2014**, *46*, 99–108. [CrossRef]
20. Matschei, T.; Lothenbach, B.; Glasser, F.P. The role of calcium carbonate in cement hydration. *Cem. Concr. Res.* **2007**, *37*, 551–558. [CrossRef]
21. Hea, H.; Courardb, L.; Pirardc, E. Particle packing density and limestone fillers for more sustainable cement. *Key Eng. Mater.* **2012**, *517*, 331–337. [CrossRef]
22. Githachuri, K.; Alexander, M.G. Durability performance potential and strength of blended Portland limestone cement concrete. *Cem. Concr. Compos.* **2013**, *39*, 115–121. [CrossRef]
23. Chłądzyński, S.; Garbacik, A. *Multicomponent Cements in Construction Industry*; Association of Cement Producers: Cracow, Poland, 2008.
24. Mirvalad, S.; Nokken, M. Minimum SCM requirements in mixtures containing limestone cement to control thaumasite sulfate attack. *Constr. Build. Mater.* **2015**, *84*, 19–29. [CrossRef]
25. CEN/TC 51/WG prEN 197-5-Working Document. Available online: https://www.cen.eu/news/brochures/brochures/cen-cenelec_wp_2019.pdf (accessed on 1 April 2020).
26. EN 196-3:2016-11. *Methods of Testing Cement—Part 3: Determination of Setting Times and Soundness*; British Standards Institution: London, UK, 2016.
27. EN 196-6:2011. *Methods of Testing Cement—Part 6: Determination of Fineness*; British Standards Institution: London, UK, 2011.
28. EN 1097-7:2008. *Tests for Mechanical and Physical Properties of Aggregates—Part 7: Determination of the Particle Density of Filler-Pyknometer Method*; British Standards Institution: London, UK, 2008.
29. EN 196-1:2016-05. *Methods of Testing Cement—Part 1: Determination of Strength*; British Standards Institution: London, UK, 2016.
30. EN 12350-2:2011. *Testing Fresh Concrete—Part 2: Slump-Test*; British Standards Institution: London, UK, 2011.
31. EN 12350-6:2011. *Testing Fresh Concrete—Part 6: Density*; British Standards Institution: London, UK, 2011.
32. EN 12350-7:2011. *Testing Fresh Concrete—Part 7: Air Content. Pressure Methods*; British Standards Institution: London, UK, 2011.
33. EN 12390-3:2011. *Testing Hardened Concrete—Part 3: Compressive Strength of Test Specimens*; British Standards Institution: London, UK, 2011.
34. PN-EN 206-1:2003. *Concrete. Part 1: Requirements, Properties, Production and Compliance*; PN-B-06250:1988 standard withdrawn and replaced; British Standards Institution: London, UK, 2003.
35. EN 12390-8:2011. *Testing Hardened Concrete—Part 8: Depth of Penetration of Water Under Pressure*; British Standards Institution: London, UK, 2011.
36. CEN/TS 12390-12:2010. *Testing Hardened Concrete—Part 12: Determination of The Potential Carbonation Resistance of Concrete: Accelerated Carbonation Method*; British Standards Institution: London, UK, 2010.
37. C 1202-05. *Standard Test Method for Electrical Indication of Concrete's Ability to Resist Chloride Ion Penetration*; American Society for Testing and Materials: West Conshohocken, PA, USA, 2012.
38. WCA. *Concrete-Requirements, Properties, Production and Conformity-National Supplementation PN-B-06265:2018-10*; Association of Cement Producers: London, UK, 2018.
39. PKN-CEN/TS 12390-9:2007. *Testing Hardened Concrete—Part 9: Freeze-Thaw Resistance-Scaling*; European Committee for Standardization: Brussels, Belgium, 2007.
40. Giergiczny, Z.; Piechówka-Mielnik, M. Limestone-the main component of multi-component Portland cements CEM II A/B-M. *Bud. Tech. Archit.* **2011**, *2*, 72–76.

41. De Weerdt, K.; Kjellsen, K.O.; Sellevold, E.; Justnes, H. Synergy between fly ash and limestone powder in ternary cements. *Cem. Concr. Compos.* **2011**, *33*, 30–38. [CrossRef]
42. De Weerdt, K.; Ben Haha, M.; Le Saout, G.; Kjellsen, K.O.; Justnes, E.H.; Lothenbach, B. Hydration mechanisms of ternary Portland cements containing limestone powder and fly ash. *Cem. Concr. Res.* **2011**, *41*, 279–291. [CrossRef]
43. EN 13877-2:2013-04. *Concrete Pavements—Part 2: Functional Requirements for Concrete Pavements*; British Standards Institution: London, UK, 2013.

materials

MDPI

Article

Effect of Ultralight Filler on the Properties of Hydrated Lime Injection Grout for the Consolidation of Detached Historic Decorative Plasters

Andreja Padovnik * and Violeta Bokan-Bosiljkov

Faculty of Civil and Geodetic Engineering, University of Ljubljana, Jamova 2, SI-1000 Ljubljana, Slovenia; violeta.bokan-bosiljkov@fgg.uni-lj.si

* Correspondence: andreja.padovnik@fgg.uni-lj.si

Received: 9 June 2020; Accepted: 27 July 2020; Published: 29 July 2020

Abstract: Injection-grout density is an important parameter when its additional weight leads to consolidated decorative plasters becoming damaged. This is especially evident in larger detached areas. In this study, thin-walled soda–lime–borosilicate glass microspheres were used as a density-reducing constituent in hydrated lime grout mixtures. The normal density grout composition—one volume part hydrated lime and three volume parts inert limestone filler with 0.5% of the polycarboxylate ether-based superplasticiser—was modified with partial substitution of the limestone filler with lightweight glass microspheres. The following volumetric proportions between limestone filler and glass microspheres were used: 100%:0%, 67%:33%, 50%:50%, 33%:67%, and 0%:100%. With the increase of the glass microspheres' volume, the density of the grout is gradually reduced. Furthermore, there is a decrease in the stability and injectability of fresh grout. In its hardened state, the grout's strength again reduces gradually, but there is no significant change in the grout's water absorption and water-vapour resistance. The resistance of the grout to freezing–thawing and heating–cooling cycles using distilled water or salt solution is highly improved when the microspheres are present.

Keywords: detached decorative plasters; injection grout; glass microspheres; reduced density; stability; strength; durability

---

1. Introduction

Consolidation of detached plasters with historic value is an irreversible conservation treatment for re-establishing adhesion between delaminated decorative layers. Injection grouts, used for stabilisation of decorative plasters, have to be compatible with original historical materials. Furthermore, they should provide adequate flowability, injectability, and stability in a fresh state. Requirements regarding the grouts' hardened state are very often given in relation to the properties of detached historical plaster; they should exhibit similar dry porosity, water-vapour permeability, capillary water absorption, and mechanical strength. An example of such requirements summarised by Padovnik et al. [1] is given in Table 1.

Additionally, specific properties, such as a low density of the grout, its increased durability, etc., are sometimes required. For larger detachments, a high quantity of grout is used to re-establish adhesion between the delaminated layers; low density of the grout is a key property that must be prescribed to prevent the formation of new damage and/or falling of the plaster from wall or ceiling, during or after consolidation by grout injection. Said density can be considerably reduced by the incorporation of lightweight filler to the grout mixture. The types of fillers used for such purpose are mineral materials with high porosity and water absorption, such as pumice, perlite and expanded glass or clay aggregate [2,3], and glass microspheres [4,5]. Very often, however, a composite filler—a mixture of pumice and glass microspheres—is used in conservation practice [6–8].

The glass microspheres—also known as glass bubbles—are nonabsorbent and have a significant advantage over porous fillers, since they can preserve extremely low wet and dry densities. Their spherical morphology, regular surface texture and extremely small particle size can help achieve and maintain suitable viscosity and stability, as well as improved injectability of the grout [4,7]. Zajadacz and Simon [4] studied the grout composition with Scotchlite Bubbles and Silcosil (fine-ground silica). They attributed the improved injectability to the small particle size of the glass microspheres and silica filler. The glass microspheres were well dispersed inside the grout mixture and tended not to segregate from other constituents, resulting in improved stability. The viscosity of this grout, however, was higher than usual. Rickerby et al. [7] used a high proportion of glass microspheres in a grout based on mud and pumice. They found that the spherical morphology of the microspheres contributed to poor packing ability and reduced internal cohesion of the grout mixture. Furthermore, Rousset et al. [9] reported that injection grout, prepared using a hydrated lime binder and glass microspheres as filler, possessed lower shrinkage and good adhesion. In Pasian et al. [8], microstructural analyses of the grout, based on slaked lime, pumice powder, and glass microspheres, showed the presence of hydraulic components on the microspheres' surface; a result of the pozzolanic reaction. Additionally, it was evident from the backscattered electron images of the cross sections and broken sections that some glass microspheres in the studied grouts were broken.

Table 1. Requirements for nonstructural lime-based grout [1].

Property	Requirement
Consistency	Fluid enough to inject [4]
Bleeding	≤2% (EN 447:1996)
Shrinkage	As small as possible and <4% [10]
Wet and dry density	As low as possible
Capillary water absorption coefficient	50–100% of substrate mortar [4]
Water-vapour resistance factor (μ)	50–100% of substrate mortar [4]
Compressive strength	Lower than that of substrate (<60%) [4] 3.0–8.0 MPa [10] 0.4–2.5 MPa [11]
Splitting tensile strength	0.3–1.2 MPa [10] Lower than that of substrate [5]
Pull-off-strength	≥0.1 MPa [4] 0.1–0.3 MPa or cohesive rupture [12]

When using the glass microspheres as a weight-reducing filler in a hydrated lime grout composition, the viscosity of the paste must be high enough to prevent the segregation of solid particles. To provide both an increased viscosity and adequate injectability of the hydrated lime grout, superplasticiser (SP; polymeric admixture) can be incorporated into the grout mixture [13,14]. Subsequently, much lower water content is needed for adequate workability and injectability. Among three SP groups frequently used in the lime-based grouts—poly-naphthalene sulfonates (PNS), lignosulfonates (LS) and polycarboxylate ethers (PCE)—the PCE products seem to be the most efficient solution [1,13].

The objective of this study was to develop a low-density hydrated lime injection grout, possessing adequate properties in both its fresh and hardened state in order to be used for the consolidation of detached lime plasters with historical value. Additionally, the grout was to be resistant to freezing–thawing and heating–cooling cycles, if possible, also in the presence of salts. The reduced density of the grout was obtained by using glass microspheres in the composite filler, which also contained fine-ground limestone particles. The required viscosity of the lime paste—to prevent segregation of filler particles and provide adequate injectability of the grout—was achieved using a combination of reduced water content and PCE superplasticiser. A parametric study was carried out as a means of determining the optimal composition of the lightweight grout, which would, in turn, ensure its improved durability.

2. Materials and Methods

Commercially available, dry hydrated lime of the class CL 90-S (standard EN 459-1 [15]) was used as a binder, with its density being 2.22 g/cm^3. Finely ground limestone (hereafter limestone filler or LS) composed of 95.3% calcite and 4.7% dolomite was selected as the main filler. The limestone filler particles had a density of 2.76 g/cm^3; their maximum size was 100 μm, with 10%, 20%, 50%, and 90% of particles smaller or equal to 3 μm, 9 μm, 15 μm, and 40 μm, respectively. The water absorption of LS was 0.5%. The chemical compositions of the binder and limestone filler, determined by the X-ray fluorescence analysis (Bruker S8 TIGER, Anhovo, Slovenia) according to the EN 196-2:2013 standard [16], are given in Table 2.

Table 2. Chemical composition of the used hydrated lime (CL 90-S) and limestone filler (LS).

Compound	CL 90-S	LS
SiO_2	2.14	<0.01
Al_2O_3	0.58	0.15
Fe_2O_3	0.20	0.01
CaO	71.01	55.38
MgO	3.05	0.76
K_2O	0.05	0.01
Na_2O	0.02	<0.01
SO_3	0.14	0.01
LOI	23.38	44.02

Thin-walled soda–lime–borosilicate glass microspheres (3M Glass Bubbles K1) were used as a density-reducing constituent of the grout, with their typical density being 0.125 g/cm^3 and respective minimum and maximum densities being 0.10 g/cm^3 and 0.14 g/cm^3. The maximum size of the microspheres was 120 μm, with 10%, 50%, and 90% of particles smaller or equal to 30 μm, 65 μm, and 115 μm, respectively.

In order to obtain the adequate viscosity and injectability of the grout in a fresh state, a polycarboxylate ether-based superplasticiser (PCE-SP) with a relative density of 1.05 g/cm^3 ± 0.02 g/cm^3 and a pH value of 5.5 ± 1.0 was used as a highly efficient water-reducing agent.

All materials were stored in a room at a controlled temperature of 20 °C ± 1 °C and relative humidity of 60 ± 5%. The grout mixtures were prepared using tap water at a temperature of 20 ± 1 °C.

In previous research [1], the grout mixture based on 1 volume part hydrated lime and 3 volume parts limestone filler, with the addition of the PCE-SP chemical admixture, showed the best behaviour in the fresh and hardened states. This grout composition was selected as the normal density reference mixture to which properties of compositions with reduced densities were compared. Grouts with reduced densities were designed in such a way that the gradually increasing part of the limestone filler in the reference mixture was replaced with the same volume of glass microspheres. Consequently, five different grout compositions were obtained (Table 3), with the following volumetric proportions between limestone filler and glass microspheres: 100%:0%; 67%:33%, 50%:50%, 33%:67%, and 0%:100%.

Two parameters of the reference grout mixture remained unchanged throughout different grout compositions: the hydrated lime content and the dosage of PCE-SP, calculated as a percentage of the total solid materials mass, i.e., the binder and the fillers, which was 0.5% [1]. The water content of the mixtures was adjusted to obtain adequate workability of each particular injection grout. The workability was evaluated by conservator via injection of the grout through a 10 mL syringe by applying minimum pressure, using the procedure described in [1].

Table 3. Composition of tested injection grout mixtures.

Mixture Formulation	LS_{100}	LS_{67}-GM_{33}	LS_{50}-GM_{50}	LS_{33}-GM_{67}	GM_{100}
Binder: filler (limestone filler and glass microspheres) volume ratio	1:3	1:3	1:3	1:3	1:3
Binder/filler (limestone filler and glass microspheres) mass ratio	0.28	0.40	0.54	0.85	6.19
Limestone filler: glass microspheres volume ratio	3:0	2:1	1.5:1.5	1:2	0:3
Water/binder mass ratio	1.86	1.76	1.59	1.52	1.48
Water/(binder, limestone filler and glass microspheres) mass ratio	0.41	0.50	0.56	0.70	1.28
PCE-SP (%)	0.5	0.5	0.5	0.5	0.5

Note: Sample formulation is indicated by the following symbols: LS = limestone filler; GM = glass microspheres. The volume proportion (limestone filler and glass microspheres) is indicated by subscript numbers.

Mixture Preparation and Testing Methods

The grout mixtures were prepared with a simple handheld kitchen mixer, in order to simulate the preparation of injection grouts in the field. The mixer had a power of 300 W and five different mixing velocities. The metal whisk used was 8.5 cm long with a diameter of 4.6 cm. For compositions with glass microspheres, the microspheres were first mixed with 50% of the water content into a slurry. The binder and the limestone filler were then dry-mixed for 15 s at the low speed of 540 rpm. During the next 45 s (at 540 rpm), the microspheres slurry and 20% of the water content were added and mixing at low speed proceeded for a further 45 s. During the last 15 s of mixing at low speed, the PCE-SP and the remaining 30% of the water content were added to the mixture. After that, the mixing was stopped, and the sides of the mixer bowl were scraped; the mixer was then turned on again and mixing proceeded for 3 min at medium speed (1200 rpm).

For nonstructural grouts, several adaptations of commonly available standard test methods were needed. We followed the testing procedures proposed by Biçer-Şimşir and Rainer [17] or Padovnik et al. [1]. A brief description of the testing methods is given in the continuation. The tests were carried out in the laboratory with controlled temperature (20 ± 1 °C) and relative humidity of the air (60 ± 5% RH). First, the methods to evaluate fresh grouts' properties are given, followed by the tests carried out on hardened grouts.

The wet density of each grout was determined according to an adapted EN 1015-6 [18] standard procedure. The volume of the mixture was reduced from 1000 mL to 100 mL, using a metal cylindrical vessel. The filling of the measuring vessel was carried out in the same manner as in the case of soft mortar [18]. The wet density was calculated as a quotient of measured mass of the grout and the 100 mL volume.

The mini slump flow test [19] was used to determine the consistency of the grouts. A truncated cone-shaped mould (according to EN 459-2 [20]) placed at the centre of a smooth plate was filled with fresh grout. The average spread of the grout after lifting the mould was measured [1].

A bleeding test was carried out according to the adapted ASTM C940 [21] standard procedure. The volume of grout used in the test was reduced from 800 ± 10 mL to 80 ± 1 mL. Apart from this, the standard procedure was followed. A graduated cylinder of 100 mL was filled with 80 mL of the grout. Change in the accumulation rate of bleed water on the surface of the grout was observed over a period of time. The bleeding was calculated as a quotient between the volume of final bleed water and the initial volume of the grout.

The water-retaining ability of fresh grout was evaluated by the standard procedure prEN 1015-8:1999 [22,23]. The fresh grout was subjected to suction provided by filter papers, resulting in a loss of water. The mass of water remaining in the grout was expressed as a percentage of the grout's initial water content and reported as the water retentivity.

The injectability test was carried out according to the adapted standard procedure EN 1771:2004 [24]. Crushed lime mortar was used as a granular material, with a water absorption coefficient of 1.4 kg/(m^2·s$^{\frac{1}{2}}$) after 10 min, and total and capillary porosities equal to 27% and 26%, respectively. The cumulative amount of the granular material passing through 1 mm, 2 mm, 3 mm, and 4 mm sieves was 5%, 12%, 42%, and 99%, respectively [1]. This test determined the ability of fresh grout to fill a capillary network of granular material in a dry or prewetted state.

The drying shrinkage test with mortar cups was used to determine the reduction in grout volume after drying. The procedure proposed in [17] was applied. The mortar for the mortar cups was prepared with 1 volume part lime putty and 2 volume parts limestone sand (0/1 mm). It had a water absorption coefficient of 1.8 kg/(m^2·s$^{\frac{1}{2}}$) after 10 min, and the total and capillary porosities equal to 32% and 25%, respectively. Each cup had an outer diameter of 75 mm and a height of 30 mm. Dry and prewetted mortar cups were filled with grout mixtures; dimensional changes, as well as the crack-pattern development, were observed as the grout dried at 20 ± 1 °C T and 60 ± 5% RH.

The adapted settlement column segregation test [25] was used in order to assess the stability of fresh grout mixtures. A cylindrical plastic tube with an internal diameter of 22 mm and a height of 375 mm with three holes to facilitate the collection of a sample from the top, middle, and bottom levels (Figure 1) was filled with fresh grout and covered. After an hour, a sample from the top of the column was transferred into the first glass container. The same procedure was applied for the middle and bottom levels. For each of the three levels, the wet density of the collected sample was determined according to the adapted standard procedure EN 1015-6 [18], previously described. The stability of fresh grout is of utmost importance for the compositions where a combination of glass microspheres and superplasticiser is used, since such mixtures are particularly sensitive to segregation.

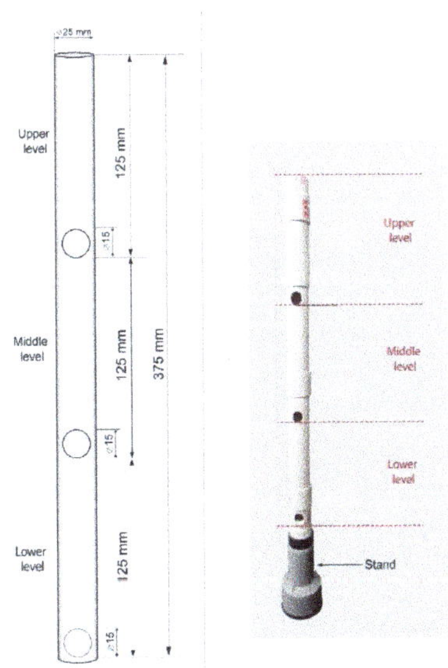

Figure 1. Cylindrical plastic tube dimensions.

In order to determine the properties of the grouts in their hardened state, cylindrical moulds were used to cast the specimens (at least three samples for each property and age). They were left in the moulds for 48 h and subsequently cured under controlled laboratory conditions (relative humidity 60 ± 5% and temperature 20 ± 1 °C) until testing. The hardened properties were determined at the grouts' age of 90 days, with compressive strength being measured again at 365 days.

The dry density was determined according to EN 1015-10 [26], on 50/50 mm cylinders.

The total and capillary porosities of the 50/50 mm cylinders were evaluated according to the Swiss standard SIA 262/1:2003, Appendix A [27]. Each specimen was subjected to different intensities of water saturation. Total and capillary porosities were calculated from the test results.

Water absorption by capillarity was measured following RILEM test No. II.6 [28]. Dry 50/50 mm cylinders were placed 2 mm deep in water and weighed at the prescribed intervals. The weight change of the specimen was used to calculate the amount of water absorbed after a predetermined time; subsequently, water absorption coefficients after 10 min and 24 h were calculated.

Water-vapour resistance coefficient (μ) was determined according to standard EN ISO 12572:2001 [29], using the dry cup method. A cylindrical specimen, with a diameter of 100 mm and height of 20 mm, was put on top of a vessel (cup) containing a desiccant (calcium chloride, $CaCl_2$), sealed to the vessel's rim, and placed in a humidity-controlled chamber at 23 ± 0.5 °C and 60 ± 3% RH. The rate of water-vapour transmission through the specimen from the controlled atmosphere to the inside of the cup was determined by periodic weighing of the cup with the desiccant and the specimen.

The compressive test was carried out according to the EN 1015–11/A1 procedure [30], on 50/50 mm cylinders.

The splitting tensile test was carried out according to the ASTM C496/C496M-11 procedure [31], on 50/50 mm cylinders.

The accelerated ageing was carried out on 50/50 mm cylinders, in order to assess the grouts' durability. The specimens were subjected to fourteen freezing–thawing and heating–cooling cycles, following the protocol in Figure 2. Before each cycle, the specimens were subjected to capillary absorption of 3% NaCl solution or distilled water for 30 min, following the RILEM test No. II.6 [28], described earlier. Results of the capillarity water absorption test showed that the amount of water absorbed after 30 min was close to the value measured at 24 h.

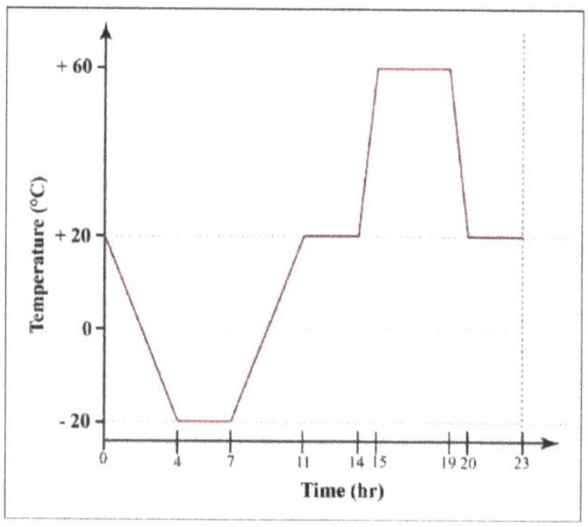

Figure 2. The protocol of the freezing–thawing and heating–cooling cycle used in the study.

Finally, to evaluate the re-attachment ability of the grout, a pull-off test—according to the standard EN 1015-12 [32]—was performed on panel sandwich models [1]. The models were prepared to simulate a smaller (2 mm) and a larger (5 mm) detachment of fine plaster (1:3 lime putty: fine sand 0/1 mm lime mortar) from the rough plaster (1:3 lime putty: coarse sand 0/4 mm lime mortar). At the age of one year, the simulated detachments were filled by the grout using a syringe.

3. Results and Discussion

3.1. Fresh State Properties

Fresh grout properties for different limestone filler/glass microsphere ratios are presented in Table 4.

Table 4. Fresh grout properties.

Mixture Formulation	Wet Density of Fresh Grout (g/cm^3)	Mini Slump Flow (mm)	Bleeding (%)	Water-Retention Capacity (%)
LS_{100}	1.73 ± 0.05	236 ± 5	1.5 ± 0.03	83 ± 1
LS_{67}-GM_{33}	1.51 ± 0.04	279 ± 6	1.7 ± 0.03	80 ± 2
LS_{50}-GM_{50}	1.31 ± 0.03	238 ± 16	1.7 ± 0.03	81 ± 3
LS_{33}-GM_{67}	1.14 ± 0.06	236 ± 10	0.6 ± 0.02	81 ± 1
GM_{100}	0.82 ± 0.02	217 ± 8	0.1 ± 0.00	82 ± 2

The wet densities of fresh grouts confirm that the glass microspheres are an efficient constituent used to reduce the grout's density. As expected, the lowest wet density was obtained for the grout GM_{100} (0.82 g/cm^3), where only glass microspheres were used as filler material, and the highest value was obtained for the grout LS_{100} (1.73 g/cm^3), where the limestone filler was the only mineral admixture.

The wet densities of grouts with composed limestone–glass microspheres filler lie between these two limits, their average values being that of 1.51 g/cm^3, 1.31 g/cm^3, and 1.14 g/cm^3 for the LS_{67}-GM_{33}, LS_{50}-GM_{50}, and LS_{33}-GM_{67} grout, respectively. The replacement of the limestone filler of a relatively high density (2.75 g/cm^3) with the same volume of glass microspheres of an extremely low typical density (0.125 g/cm^3) is the main parameter that governs the grout's weight reduction. The water content, which decreases with the increase in the glass microspheres' volume (Table 3), is one additional parameter that influences the volume of prepared grout. Another is the packing density of solid particles in the suspension; grain size distribution of the glass microspheres is much coarser compared to the limestone filler. When considering the reduced weight of the grout, the two additional parameters—water content and packing density—need to be considered as well. Based on obtained results, it is possible to conclude that the reduction of the grout's wet density with the incorporation of a relatively high volume of glass microspheres could be an effective method for reducing the weight of the grout, when the re-attachment of large plaster detachments needs to be carried out.

The mini-slump-flow value of the fresh grout evaluates its flowability under the action of self-weight. It is a measure for fresh grout consistency, which is often related to the grout's workability. In this study, the workability was evaluated via injection of the grout through a 10-mL syringe while applying minimum pressure [1], since the test method was set to reflect the conditions on the conservation site. The water content of the grout was reduced with the increasing volume of glass microspheres (Table 3), to obtain the same workability of the grouts. It is highly likely that with this, yield stress and viscosity of the lime paste (lime + water + PCE-SP) were increased as well. Comparing the consistency of the LS_{67}-GM_{33} grout to that of the LS_{100} grout, a considerable increase in the slump-flow value could be observed. Slump-flow values of the LS_{50}-GM_{50} and LS_{33}-GM_{67} grouts were the same as that of the LS_{100} grout; these two compositions show the same workability as the LS_{100} grout and, consequently, also the same slump-flow value. Mini-slump-flow value is often related to the paste's yield stress τ_0; paste is a generic name for the mixture of binder, filler particles smaller than 0.1 mm and water, and can also contain a chemical admixture. The value of τ_0 increases with an increase in paste density and decreases with the mini-slump-flow value increase. If we assume that only the paste's own weight is controlling the phenomena, the equation proposed in [33] is the following: $\tau_0 = C{\cdot}\rho/SF^5$. In this equation, C represents a constant that includes gravity and volume of the paste, ρ is the paste's density, and SF is the mini-slump-flow value. The equation can be used to explain the influence of the glass microspheres on the rheological properties of the grouts. It is clear that by replacing the fine limestone filler with coarser glass microspheres, the grout's yield stress (τ_0) decreases, despite the increase in yield stress and viscosity (μ) of the lime paste in the grout. For the constant SF value, the yield stress decrease is higher for grouts with a higher content of glass microspheres. The effect of glass microspheres on the rheological properties of the grout seems to be similar to that of air bubbles that are produced using an air-entraining agent (chemical admixtures) in cement paste. For the LS_{50}-GM_{50} grout, the obtained standard deviation of the test results was relatively high. Visual observation of the grout spreads revealed segregation between solid particles during the test; heavier particles settled to the bottom and lighter particles (the glass microspheres) accumulated on the fresh mixture's surface. A poor packing density of solid particles in the LS_{50}-GM_{50} grout could be responsible for the observed behaviour. The lowest slum-flow value was measured for the GM_{100} composition. A high water-content reduction of 20% in the GM_{100} composition was needed to obtain the required workability of the grout; it appears that complete elimination of the limestone filler particles significantly changed the rheology of the lightweight grout. From these results, we can conclude that there is no clear relationship between the workability test and the mini-slump-flow test results.

The grouts with limestone-filler content representing 50% or more of the total filler content (LS_{100}, LS_{67}-GM_{33}, LS_{50}-GM_{50}) showed a higher level of final bleeding, which ranged between 1.5% and 1.7%. In the mixtures where the prevailing part of the filler was composed of glass microspheres (LS_{33}-GM_{67} and GM_{100}), the final bleeding was between 0.1% and 0.6%. In all tested grouts, the final bleeding was lower than the standard limit value of 2% (EN 447 [34]; Table 1). These final bleeding values alone, however, are not enough when assessing the stability of the lightweight grout; important information can be provided by visual inspection of the sample appearance, as was the case for the GM_{100} grout (Figure 3), where it's lowest final bleeding of 0.1% was due to the fact that a big part of the bleed water was trapped between two layers of the tested sample. The trapped water was not considered when calculating the bleeding value; such behaviour of the hydrated lime grout was observed for the first time. It appears that local internal segregation of glass microspheres, bleed water, and (possibly) lime particles happened in the test sample. Internal segregation of particles was difficult to prove due to the same white colour of the lime, limestone filler, and glass microspheres.

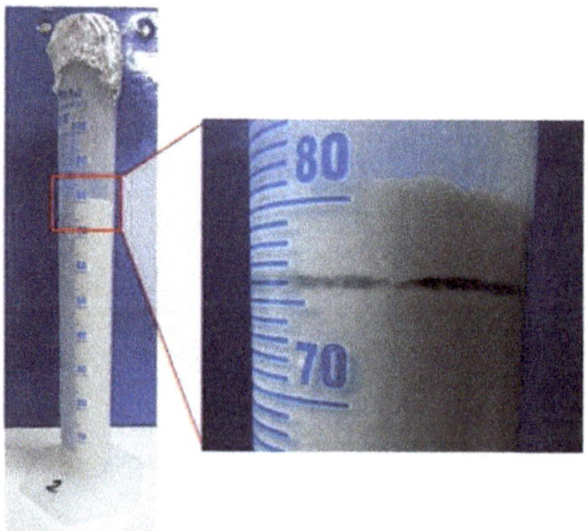

Figure 3. Water trapped between the two layers of the grout sample for the GM_{100} mixture.

The results of water-retention capacity range between 78% and 84% for all tested grouts (Table 4). Although the reference grout mixture (LS_{100}) seems to possess the highest water-retention capacity, the incorporation of glass microspheres did not significantly reduce this fresh grout property. Due to its high water retention, the grout resists releasing water into the highly porous media with high absorption capacity through which it flows. Consequently, the plugging of the grout inside the plasters can be prevented, and its drying shrinkage can be efficiently reduced [5]. The highest water retention was measured in the LS_{100} grout, which lacked the glass microspheres; this could be due to a lower content of free water, although this composition was prepared with the highest water content. The limestone filler particles are much finer than glass microspheres, and their shape is the same as the shape of crushed limestone aggregate grains. Thus, a significant reduction of free water content inside the LS_{100} grout can be attributed to a much higher surface at the same volume of particles (a spherical shape results in the lowest surface at a particular volume) and a higher ability of the limestone particles to capture water by adsorption and absorption. Another influencing parameter is the ability of the filler to increase the packing density of the grout's solid particles, reducing the free water content. Ince et al. [35] showed that the filler with an appropriate granular composition could optimally fill the voids within the grout matrix. As a result, less free water would be available in the mixture during the suction action provided by porous plaster; the free water could be easily removed from the grout.

The results of the stability test are presented in Figure 4. For the grouts LS_{100}, LS_{67}-GM_{33}, LS_{33}-GM_{67}, and GM_{100} the differences in the grout's wet density between the bottom and the top level of the testing column are low and equal to 0.01 or 0.02 g/cm^3. All measured densities are also in agreement with density values given in Table 4. Therefore, these grouts can be evaluated as stable. When observing the grout LS_{50}-GM_{50}, segregation of particles was noted. The highest density was present in the bottom third of the column (1.42 g/cm^3), while the lowest was in the top third of the column (1.35 g/cm^3). The interparticle forces in this composition were not strong enough to maintain a homogenous suspension of particles along the column height. Therefore, a higher percentage of the limestone filler particles settled towards the bottom of the column, while a larger amount of the light glass microspheres was able to rise towards the surface. The same behaviour was also observed during the mini-slump-flow test of the LS_{50}-GM_{50} grout. According to Rickerby et al. [7], the spherical morphology of glass microspheres and their coarser grain-size distribution may worsen the packing density of the composite filler. It seems that this was the case for the LS_{50}-GM_{50} composition. Injection grouts have to possess sufficient stability/homogeneity after mixing, during the whole injection process, and while setting is taking place. If the mixture segregates during the process of injecting or setting, the consolidation of air pockets cannot be successful. In their study, Miltiadou-Fezans and Tassios [36] concluded that, for each grout, the critical water-to-solids ratio resulting in segregation depends on an acceptable degree of instability, the specific surface of solids, and the percentage of superplasticiser used. Based on the trapped water detected following the bleeding test (Figure 3), high stability of the fresh GM_{100} grout is an unexpected result. One possible explanation for the fresh properties measured in the GM_{100} composition is a distributed segregation of solid particles; along the entire column, there can be a local settlement of lime-binder particles, as well as flowing of the glass microspheres towards the internal-bleed water surface.

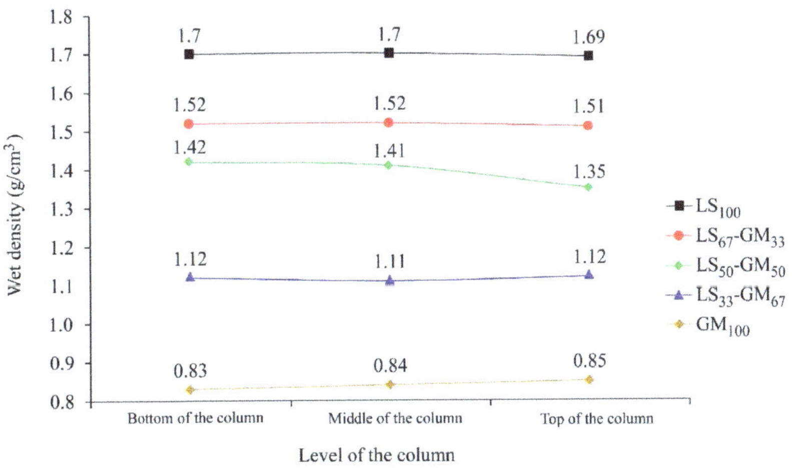

Figure 4. Results of the stability tests: measured grout's wet density from the bottom, middle, and top level of the testing column.

Figure 5 and Table 5 show the results of the drying shrinkage test inside the dry or prewetted mortar cup. From the results, it is clear that the resistance of the grout to drying shrinkage and, thus, to the formation of the separation ring and cracks inside of the grout, highly depends on the filler composition used. The lowest cracking was observed for grouts LS_{100} and GM_{100}, where a separation ring with a thickness of only 0.5 mm was formed in the dry cups; in the prewetted cups, the 0.5 mm separation ring was only formed in the LS_{100} composition. On the other hand, compositions with composed limestone particles and glass microspheres filler showed a weaker resistance to drying shrinkage; this was also expressed through the formation of cracks inside of the grout, observed in the LS_{67}-GM_{33} (dry and prewetted mortar cup) and LS_{33}-GM_{67} (dry mortar cup) compositions. It can be concluded that the combination of limestone particles with high density and modulus of elasticity and glass microspheres with extremely low density and modulus of elasticity, induces additional differential deformations in the grout that result in reduced resistance to the formation of cracks. On the other hand, a reduction of the water-to-binder ratio through the increase of the glass microspheres content decreases the sensitivity of the grout to shrinkage. These two influencing parameters with opposite effects are responsible for the observed response to drying shrinkage in each particular grout. Additionally, prewetting the mortar cups seems to be more efficient for compositions where the glass microspheres content in the filler is 50% or higher.

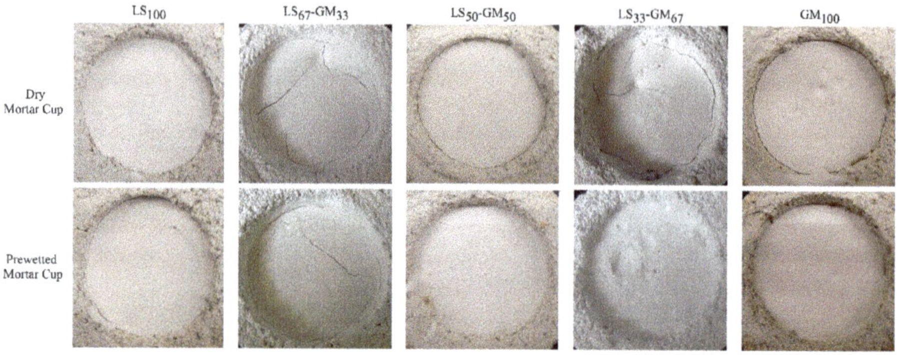

Figure 5. Grout mixture after drying in the dry and prewetted mortar cups.

Table 5. Drying shrinkage in dry and prewetted mortar cups.

Mixture Formulation	Dry Mortar Cup		Prewetted Mortar Cup	
	Separation Size (mm)	Crack Size (mm)	Separation Size (mm)	Crack Size (mm)
LS_{100}	0.5	0	0.5	0
LS_{67}-GM_{33}	0.2	0.5	0.2	0.5
LS_{50}-GM_{50}	1.0	0	0.1	0
LS_{33}-GM_{67}	1.0	0.5	0.1	0
GM_{100}	0.5	0	0	0

The separation ring between the mortar cup and the grout and/or the cracks in the grout might indicate an excessive water content in the mixture, which could weaken the bond between the grout and the plaster layers and reduce the grout strength [17].

The grout mixtures GM_{100} and LS_{50}-GM_{50} did not meet the requirements set for fresh grout properties. Due to their resistance to segregation not being high enough, we did not determine the injectability and hardened properties for these two mixtures.

The injectability curves of the grout mixtures LS_{100}, LS_{67}-GM_{33}, and LS_{33}-GM_{67} are given in Figure 6 for the prewetted and dry crushed lime mortar columns. From these curves, it can be noted that the glass microspheres have an essential influence on the ability of the grout to be injected into detached plaster; the increase in the volume of the microspheres decreases the injectability of the grout. The results obtained are not in line with the results of studies carried out by Zajadacz and Simon [4] and Rickerby et al. [7], where glass microspheres improved the injectability of tested grouts. The authors concluded that the improvement is due to the spherical morphology and small particle size of the glass microspheres. However, in [4] there is no information regarding detailed grout composition and mixing procedure, and the composition of earthen grout used in [7] is not comparable with the hydrated lime grout used in our study.

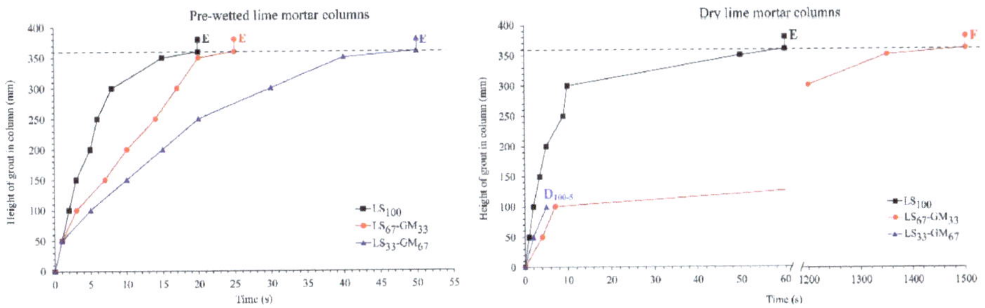

Figure 6. Injectability curves of the grout mixtures LS_{100}, LS_{67}-GM_{33}, and LS_{33}-GM_{67} for prewetted (left) and dry (right) lime mortar columns.

The results also demonstrated that prewetting of the crushed lime mortar improved the injectability of all three tested grouts. The ability of the grouts to be injected was classified following the proposal of Biçer-Şimşir and Rainer [37]. The mixtures LS_{100}, LS_{67}-GM_{33}, and LS_{33}-GM_{67} were classified as E (easy) when prewetted crushed mortar column was used. Additionally, the LS_{100} and LS_{67}-GM_{33} mixtures were classified as E (easy) and F (feasible), respectively, and the LS_{33}-GM_{67} mixture as D (difficult), when dry mortar column was used.

Lower bleeding and higher water-retention capacity of the LS_{100} mixture (Table 4) are the main influencing parameters responsible for better injectability of the grout not containing the glass microspheres.

3.2. *Properties in Hardened State*

Physical properties of 90-day-old hardened grouts are given in Table 6, in the form of an average value and associated standard deviation. Measured density was the highest for the LS_{100} grout (average value of 1.45 g/cm^3), which contained no glass microspheres, and the lowest for the LS_{33}-GM_{67} grout (average value of 0.85 g/cm^3) with microspheres occupying 2/3 of the filler volume. This shows that reducing the grout's weight by up to (approximately) 40%, in relation to the reference LS_{100} grout, can be achieved by replacing part of the limestone filler content with the same volume of glass microspheres. When comparing the fresh and hardened state densities of a particular grout, it is obvious that the drying of the grout is responsible for the reduction in density; this reduction is equal to 0.28 g/cm^3, 0.31 g/cm^3, and 0.29 g/cm^3 for the LS_{100}, LS_{67}-GM_{33}, and LS_{33}-GM_{67} grout, respectively.

Table 6. Physical properties of the LS_{100}, LS_{67}-GM_{33}, and LS_{33}-GM_{67} grouts in the hardened state: density, total, and capillary porosity, water absorption coefficient after 24 hr (W_{24}) and 10 min (W_{10}), and water-vapour resistance factor (μ).

Mixture Formulation	Density in the Hardened State (g/cm^3)	Total Porosity (%)	Capillary Porosity (%)	W_{24} (kg/(m$^2\sqrt{\text{min}}$))	W_{10} (kg/(m$^2\sqrt{\text{min}}$))	μ (–)
LS_{100}	1.45 ± 0.01	40 ± 1.0%	38 ± 1.0%	0.46 ± 0.03	2.11 ± 0.15	16 ± 1.0
LS_{67}-GM_{33}	1.20 ± 0.02	42 ± 2.0%	40 ± 0.4%	0.44 ± 0.02	2.20 ± 0.10	15 ± 0.9
LS_{33}-GM_{67}	0.85 ± 0.00	48 ± 1.0%	38 ± 0.2%	0.42 ± 0.01	2.14 ± 0.06	12 ± 0.2

The reductions are in good correlation with the capillary porosities of the grouts (Table 6), which would be expected due to the evaporable water being held in the capillary pores. The average capillary porosities of the three grouts are in a narrow range between 38% and 40%, despite the relatively large differences in their binder-to-water ratios. Said ratios are equal to 1.86, 1.76, and 1.52 for the LS_{100}, LS_{67}-GM_{33}, and LS_{33}-GM_{67} grout, respectively. The water absorption ability of the two filler materials needs to be addressed to explain these apparent inconsistencies of properties. The water absorption of the limestone filler is equal to 0.5%, and the water content of the product is less than 0.2%. Glass microspheres are nonporous; thus, they do not absorb water. The highest part of the added water was, therefore, absorbed by the filler particles in the LS_{100} grout and the lowest by the LS_{33}-GM_{67} grout particles; as a result, the narrow interval of the capillary porosities was obtained. Total porosity is the sum of the capillary pores and air pores. In the grout compositions containing glass microspheres, the spheres with broken glass walls can contribute to the measured air pores. The contribution of the glass spheres to the measured air pores' content only appeared to be significant in the LS_{33}-GM_{67} grout, which contained a high amount of glass microspheres. It seems that, during the mixing and/or the test execution, some glass microspheres may have become damaged, which is in line with the backscattered electron images in [8].

The amount of capillary water absorbed by the mixtures at the end of the test (after 24 h; W_{24}) is approximately the same for the three grout compositions, resulting in the W_{24} coefficient average values between 0.42 and 0.46 kg/(m$^2\sqrt{\text{min}}$). Obtained values are considerably lower than values given by Veiga [11] for the hydrated lime: sand (1:3) historic mortars, where W_{24} is in the range between 1.1 and 1.6 kg/(m$^2\sqrt{\text{min}}$). However, considering the requirement that the capillary water absorption of the grout must lie between 50% and 100% of the substrate mortar W_{24} [4], the obtained results are not far from meeting the required values. Another essential property of the grout is the initial water absorption, presented by the coefficient of capillary water absorption after 10 min [1,38]. From the results in Table 6, it is evident that the average initial water absorption of the three grouts (W_{10}) is approximately the same and ranging between 2.11 and 2.20 kg/(m$^2\sqrt{\text{min}}$). These coefficients are within the W_{10} interval for the fine and coarse lime mortars prepared using Slovenian hydrated limes and limestone sands, where values range between 1.10 and 2.60 kg/(m$^2\sqrt{\text{min}}$) [1,39].

The average value of the grouts' water-vapour resistance is lower or equal to 16 (Table 6), which is in line with the results obtained for lime-based mortars by Jornet et al. [38]. The grouts LS_{100} and LS_{67}-GM_{33}, with the highest contents of limestone filler, showed a slightly increased water-vapour resistance (16 and 15) compared to the grout LS_{33}-GM_{67} (μ = 12). Broken glass microspheres may be responsible for the obtained result.

Compressive and splitting tensile strengths are related to the total porosity; higher total porosity results in lower mechanical strength. That said, the total porosity values of the LS_{67}-GM_{33} and LS_{33}-GM_{67} grouts in Table 6 are underestimated due to the test method applied, which was unable to measure actual hollow volume inside of the glass microspheres. A higher actual total porosity than the one measured is evident from the densities of the grouts in the hardened state (Table 6).

The average values for mechanical strengths are presented in Table 7, along with the corresponding standard deviation. As expected, the glass microspheres decreased the compressive and splitting tensile strengths of the grouts considerably, compared to the reference LS_{100} composition. At the ages of 90 and 365 days, the average compressive strengths of the LS_{100}, LS_{67}-GM_{33}, and LS_{33}-GM_{66} grouts were 3.5 and 3.8 MPa, 1.8 and 2.3 MPa, and 1.4 and 1.4 MPa, respectively. This means that a reduction in compressive strength between 40% and 50% can be expected when replacing a third of the limestone filler volume with glass microspheres. When the replacement is increased to two-thirds, the same reduction goes up to about 60%. While the reference grout LS_{100} complies with the proposed range of compressive strengths given by Ferragni et al. [10] for hydraulic lime grouts (Table 1), the two compositions with the glass microspheres fulfil the requirements for repair lime-based mortars given by Veiga [11], where compressive strengths in the range of 0.4–2.5 MPa are proposed.

Moreover, Pasian et al. [8] studied grouts with reduced water content; they were prepared using slaked lime, pumice powder, quartz sand, and soda–lime–borosilicate glass microspheres. At 150 days, these grouts achieved an average compressive strength ranging from 1.15 MPa to 3.08 MPa. These values are in line with the LS_{67}-GM_{33} and LS_{33}-GM_{66} compressive strengths in Table 7.

Table 7. Compressive strength and splitting tensile strength of the grout mixtures.

Mixture Formulation	Average Compressive Strength 90 Days (MPa)	Average Compressive Strength 365 Days (MPa)	Average Splitting Tensile Strength 90 Days (MPa)
LS_{100}	3.5 ± 0.3	3.8 ± 0.4	0.16 ± 0.04
LS_{67}-GM_{33}	1.8 ± 0.2	2.1 ± 0.6	0.11 ± 0.07
LS_{33}-GM_{67}	1.4 ± 0.1	1.4 ± 0.1	0.08 ± 0.02

The injection grouts for stabilisation of detached plaster layers are expected to fail predominantly due to tensile stresses [17]. Their tensile strength should be lower than the tensile strength of the original plaster in order to prevent the occurrence of damage to the original material [5]. The average splitting tensile strength of tested grouts at the age of 90 days was between 0.08 and 0.16 MPa (Table 7). These values are well below the 0.3–1.2 MPa range proposed by Ferragni et al. [10]. On the other hand, they fulfil the requirement given in [5] and are close to values reported by Pasian et al. [8] for the nonstructural slaked lime grout and Veiga [11] for the rendering and plastering repair mortar for historic buildings.

The main influencing parameter governing the strength properties is the volume of the grout's solid constituents that can transfer stresses inside of the material; this is reflected in the grout's density and porosity. There are, however, additional parameters that contribute to the strength increase. The results show that grouts with higher limestone-filler content possess higher strength, due to their lower total porosity and better interlocking between the lime binder and the filler particles. The limestone filler is a compact carbonate with sharply cornered grains and a rough surface, which can absorb up to 0.5% of the water from the fresh grout. With water, some lime particles can also be absorbed, making the bond strength between the limestone filler and lime binder considerably higher compared to that between nonabsorbent glass microspheres and lime binder. This finding is supported by the study conducted by Lanas and Alvarez [40], where they concluded that the shape of grains, particle size distribution, and chemical and mineralogical composition of the filler influence the strength of grouts.

The mechanical strength and stiffness of the injection grout and historic lime plaster or render should be approximately equal in order to ensure adequate ductility and durability of the system. From the results in Table 7, it is evident that the glass microspheres are an efficient filler that can be used to adapt mechanical properties of the grout to the mechanical properties of historic plaster or render.

In addition to physical and mechanical properties of the hardened grouts, the durability of the grout mixtures needs to be addressed as well. Besides being an efficient weight-reducing filler, glass microspheres can be seen as a means to introduce stable micro air bubbles to the lime grout. These bubbles can increase the grout's resistance to extreme temperature fluctuations, such as freezing and thawing during the winter and heating and cooling during the summer. The comparison of average compressive strengths for mixtures LS_{100}, LS_{67}-GM_{33}, and LS_{33}-GM_{67}, at the age of 90 days and after the accelerated ageing using distilled water or de-icing salt (3% NaCl), is given in Figure 7.

The accelerated ageing of samples in the presence of distilled water shows that the glass microspheres increased the grout's resistance to the freezing–thawing and heating–cooling cycles. While the LS_{100} grout was damaged during the accelerated ageing and, as a result, the compressive strength was decreased from the reference value of 3.5 MPa to 2.7 MPa, compositions LS_{67}-GM_{33} and LS_{33}-GM_{67}—which contained the glass microspheres—were not damaged. Following ageing, the average compressive strength of the LS_{67}-GM_{33} grout increased from the reference value of 1.8 MPa to 2.1 MPa, while that of the LS_{33}-GM_{67} grout increased from 1.4 MPa to 1.5 MPa. Accelerated carbonation of the lime binder, due to wetting and drying, is most probably responsible for the observed strength increase. Similar behaviour was observed by Uranjek and Bokan-Bosiljkov [41] for lime mortar exposed to freezing and thawing cycles.

When the de-icing salt solution was used for the accelerated ageing, grout LS_{100} fell apart due to the combined effect of water freezing and salt crystallisation (Figure 8). Specimens made from the LS_{67}-GM_{33} and LS_{33}-GM_{67} grouts, on the other hand, had retained their shape but were damaged. Dusting, swelling, scaling, and formation of cracks appeared in the lower part of the LS_{67}-GM_{33} and LS_{33}-GM_{67} specimens after the sixth cycle of freezing–thawing and heating–cooling (Figure 8). The compressive strength of the LS_{67}-GM_{33} grout was reduced to 1.4 MPa (22% reduction), while that of the LS_{33}-GM_{67} grout was reduced to 0.9 MPa (36% reduction).

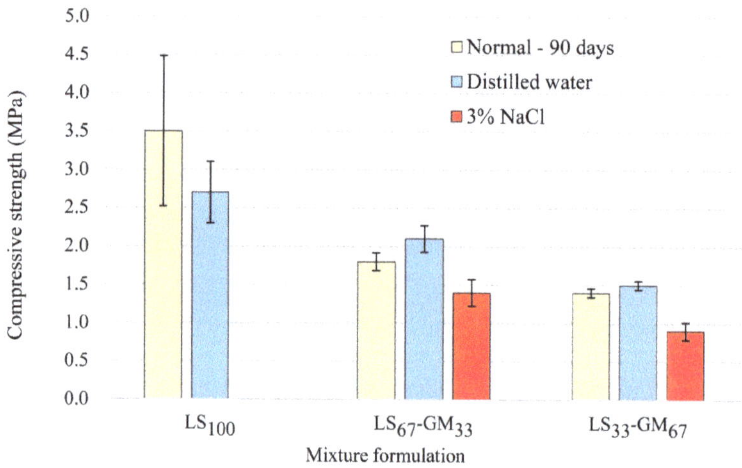

Figure 7. Comparison of compressive strengths of injection grouts at the age of 90 days (normal) and after the accelerated ageing using distilled water or the salt solution.

From the obtained results, it is evident that the durability of tested lime grouts is much higher when their ageing takes place in the presence of pure water, compared to using de-icing salt solution; the specimens aged using distilled water did not show visible damages (Figure 8). Moreover, the compressive strengths of the LS_{67}-GM_{33} and LS_{33}-GM_{67} grouts improved after ageing. We can conclude that the tested grouts are highly durable solutions that can consolidate detached plasters or renders if salt-induced problems are not present. The combination of ice formation and salt crystallisation is highly detrimental to the three grouts. However, by incorporating air bubbles in the lime grout using glass microspheres, high enough durability can still be obtained for applications where salts are present in the masonry walls. The highest durability in the presence of salts was obtained for the LS_{67}-GM_{33} grout, which shows that adequate balance of loadbearing capacity and micro air bubbles' volume is needed to provide adequate durability of the lime grout in an environment containing salts. We can conclude that glass microspheres have the same function as air bubbles in aerated cement mortars; they efficiently reduce the stresses arising from water freezing inside the hardened lime grout and, thus, prevent extensive damage to the grout.

The pull-off strengths, with information about the location of failure in the panel sandwich test, are presented in Table 8. The measured pull-off strength of each particular grout is smaller than its splitting tensile strength (Table 7). The pull-off strength of the LS_{100} grout in the 2 mm air pocket achieved the value of 0.1 MPa, which is lower than the cohesive strength of lime plaster (0.15 MPa). The failure was predominantly within the grout and partly along the interface between the grout and the fine plaster. The LS_{100} grout fulfilled the pull-off strength requirement given in Table 1.

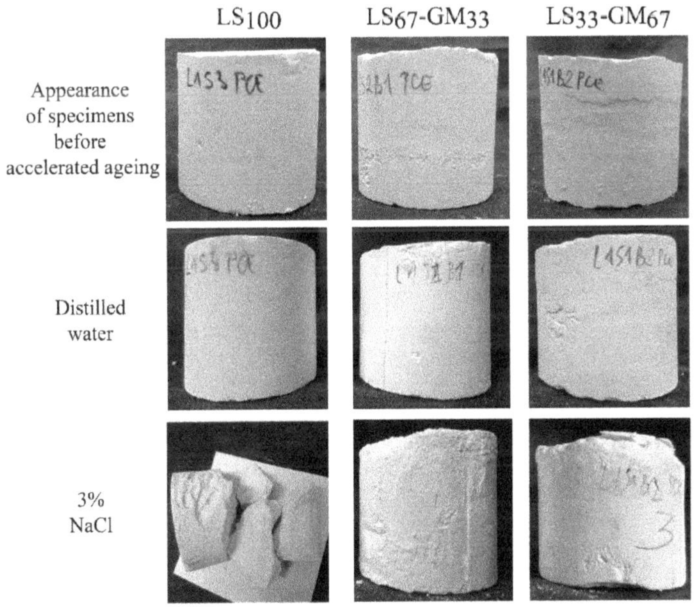

Figure 8. Specimens of mixtures LS_{100}, LS_{67}-GM_{33}, and LS_{33}-GM_{67}, saturated in the distillate water or the de-icing agent (3% NaCl), before and after freezing–thawing and heating–cooling cycles.

Table 8. The pull-off strengths of LS_{100}, LS_{67}-GM_{33}, and LS_{33}-GM_{67} grouts injected into the simulated air pockets with thicknesses of 2 and 5 mm.

Mixture Formulation	2 mm (MPa)	Location of Failure	5 mm (MPa)	Location of Failure
LS_{100}	0.10 ± 0.02	60% within the grout 40% along the grout–fine plaster interface	0.05 ± 0.03	20% within the grout 80% along the base–rough plaster interface
LS_{67}-GM_{33}	0.08 ± 0.02	15% within the grout 80% along the grout–rough plaster interface 5% within the rough plaster	0.05 ± 0.01	35% within the grout 60% along the base–rough plaster interface 5% within the rough plaster
LS_{33}-GM_{67}	0.07 ± 0.01	95% along the grout–rough plaster interface 5% within the rough plaster	-	
Area without the grout	0.15 ± 0.02			

The LS_{67}-GM_{33} and LS_{33}-GM_{67} grouts injected in the 2 mm air pocket showed a lower average pull-off strength of 0.08 and 0.07 MPa, respectively; the requirement given in Table 1 was subsequently not met. For the two grouts, failure was predominantly along the interface between the grout and the lower rough layer of the lime plaster. This shows that the bond between the lime plaster and the grout is the weakest link in the sandwich specimen consolidated using grouts with glass microspheres.

Pull-off tests carried out on panel sandwiches with thicker air pockets (5 mm) resulted in considerably reduced pull-off strengths; in the case of the LS_{100} and LS_{67}-GM_{33} grouts, they were equal to 0.05 MPa. The failure was predominantly along the interface between the base and the rough plaster. The LS_{33}-GM_{67} grout sandwich specimens already failed during the test disc installation. This suggests that the drilling of the specimens may have damaged the contacts between different layers of the sandwich panels. Subsequently, measured pull-off values can highly underestimate the actual bond strength between the grout and the plaster.

When comparing results in Table 8 with the pull-off strengths reported in comparable studies [6,12], all the values measured in this study were higher. In Pasian et al. [6] the pull-off strength was in the range of 0.032–0.041 MPa at 150 days, and, in Azeiteiro et al. [12], the maximum pull-off strength after 60 and 90 days was 0.015 MPa and 0.04 MPa, respectively.

4. Conclusions

This study aimed to develop lightweight hydrated lime injection grouts to be used for consolidation of larger decorative plaster detachments with historical value. A reduction of the grout's self-weight was performed by using glass microspheres as one of its constituents. In the 1:3 (volume ratio) hydrated lime: mineral filler grout, five volume ratios of limestone filler (LS) to glass microspheres (GM) were studied: 100%:0%; 67%:33%; 50%:50%; 33%:67%, and 0%:100%. Adequate workability was evaluated by a conservator—via injection of the grout through a syringe—and was set as the requirement for the grouts in the fresh state. Based on the test results obtained, the following main conclusions were drawn:

1. The glass microspheres are an efficient solution when a lower self-weight of the grout is required. However, in order to maintain the required workability and stability of fresh grout, the added water needs to be reduced when increasing the volume of glass microspheres. The effect of the glass microspheres on the rheological properties of fresh grout seems to be similar to that of air bubbles produced using air-entraining chemical admixture; they decrease yield stress and plastic viscosity of the grout.
2. The water-retention capacity of the tested grouts did not differ much between compositions, but the highest values were observed for the LS_{100} and GM_{100} grouts, which contained only one filler. This observation appears to be important when prewetting of the detached plaster is not feasible; namely, drying shrinkage of the grouts in dry mortar cups resulted in a thicker separation ring and/or crack formation in the LS_{67}-GM_{33}, LS_{50}-GM_{50}, and LS_{33}-GM_{67} compositions, compared to the LS_{100} and GM_{100} grouts. On the other hand, induced differential deformations, as a result of the very different moduli of elasticity of the limestone filler and glass particles, may have also influenced the damage pattern observed.
3. The packing density of solid particles (lime, limestone filler, and glass microspheres) in the grout seems to be an essential property controlling fresh-grout stability. The LS_{50}-GM_{50} and GM_{100} grouts were not stable enough to be applied in practice. On the other hand, compositions LS_{67}-GM_{33} and LS_{33}-GM_{67} fulfilled the stability requirement, as did the reference LS_{100} grout. As a result, further tests were performed only on these three grout compositions.
4. When the volume of glass microspheres in the grout was increased, the grout's injectability deteriorated. The reduction of injectability was not an issue for the prewetted lime mortar column in the testing assembly; all three compositions were classified as easily injectable (E). Problems arose when the dry lime mortar column was applied; in this version of the test, only the reference LS_{100} grout was classified as E. The LS_{67}-GM_{33} grout was classified as feasibly injectable (F), and LS_{33}-GM_{67} as difficultly injectable (D).
5. The average dry density of the hardened grouts was reduced from 1.45 g/cm^3 (LS_{100}) to 1.20 g/cm^3 (LS_{67}-GM_{33}) and 0.85 g/cm^3 (LS_{33}-GM_{67}). With the reduction of the grout's self-weight, its strength properties decreased as well. For example, the one-year average compressive strength decreased from 3.8 MPa (LS_{100}) to 2.1 MPa (LS_{67}-GM_{33}) and 1.4 MPa (LS_{33}-GM_{67}). We can conclude that by replacing part of the limestone filler with the glass microspheres, an adaptation of the grout's mechanical properties to the same properties of a particular historic plaster or render is possible. Disregarding the LS_{100} compressive strengths, which reach values typical of hydraulic lime grouts, all other measured strengths seem to fulfil requirements set for compatibility with historic lime plasters and renders.

6. The requirement of equal workability of the fresh grouts resulted in approximately the same capillary porosity and capillary water absorption properties of the LS_{100}, LS_{67}-GM_{33}, and LS_{33}-GM_{67} grouts.
7. When hydrated lime grout is exposed to extreme temperature fluctuations, such as freezing and thawing during the winter, and heating and cooling during the summer, glass microspheres inside the grout act similarly to the air bubbles (produced using an air-entraining agent) inside cement-based materials. The glass microspheres prevent entirely the damage resulting from the freezing of water inside of the grout. Moreover, they can accumulate part of the salts that crystallise within the grout and, thus, considerably reduce the damage due to salt crystallisation. It can be concluded that the glass microspheres are not only efficient in reducing the grout's density but can also significantly improve the grout's durability in extreme environments.

Author Contributions: Conceptualisation, A.P. and V.B.-B.; methodology, A.P. and V.B.-B.; validation, V.B.-B.; formal analysis, A.P. and V.B.-B.; investigation, A.P.; resources, V.B.-B.; writing—original draft preparation, A.P.; writing—review and editing, V.B.-B.; visualisation, A.P.; supervision, V.B.-B.; project administration, A.P.; funding acquisition, A.P. and V.B.-B. All authors have read and agreed to the published version of the manuscript.

Funding: This research was undertaken in a framework of the project C3330-17-529030 "Raziskovalci-2.0-UL-FGG-529030". Ministry of Education, Science and Sport of the Republic of Slovenia has approved the project. The investment is cofinanced by the Republic of Slovenia and the European Union under the European Regional Development Fund. The authors also acknowledge the financial support from the Slovenian Research Agency through the research core funding No. P2-0185.

Conflicts of Interest: The authors declare no conflict of interest.

References

1. Padovnik, A.; Piqué, F.; Jornet, A.; Bokan-Bosiljkov, V. Injection Grouts for the Re-Attachment of Architectural Surfaces with Historic Value—Measures to Improve the Properties of Hydrated Lime Grouts in Slovenia. *Int. J. Archit. Herit.* **2016**, *10*, 993–1007. [CrossRef]
2. Pachta, V.; Papadopoulos, F.; Stefanidou, M. Development and testing of grouts based on perlite by-products and lime. *Constr. Build. Mater.* **2019**, *207*, 338–344. [CrossRef]
3. Stefanidou, M. Cement-based renders with insulating properties. *Constr. Build. Mater.* **2014**, *65*, 427–431. [CrossRef]
4. Zajadacz, K.; Simoon, S. Grouting of architectural surfaces–The challenge of testing. In Proceedings of the International Seminar Theory and Practice in Conservation, Lisbon, Portugal, 4-5 May 2006; pp. 509–516.
5. Biçer-Simsir, B.; Griffin, I.; Palazzo-Bertholon, B.; Rainer, L. Lime-based injection grouts for the conservation of architectural surfaces. *Rev. Conserv.* **2009**, *10*, 3–17.
6. Pasian, C.; Pique, F.; Jornet, A.; Cather, S.A. Sandwich Specimen Preparation and Testing Procedure for the Evaluation of Non-Structural Injection Grouts for the Re-Adhesion of Historic Plasters. *Int. J. Archit. Herit.* **2019**, *6*, 1–2. [CrossRef]
7. Rickerby, S.; Shekede, L.; Zaixuan, F.; Wei, T.; Hai, Q.; Jinjian, Y.; Pique, F. Development and Testing of the Grouting and Soluble-Salts Reduction Treatments of Cave 85 Wall Paintings. In *Conservation of Ancient Sites on the Silk, Proceedings of the Second International Conference on the Conservation of Grotto Sites, Mogao Grottoes, Dunhuang, China, 28 June–3 July 2004*; Agnew, N., Ed.; Getty Conservation Institute: Los Angeles, CA, USA, 2010; pp. 471–479.
8. Pasian, C.; Secco, M.; Pique, F.; Artioli, G.; Rickerby, S.; Cather, S. Lime-based injection grouts with reduced water content: An assessment of the effects of the water-reducing agents ovalbumin and ethanol on the mineralogical evolution and properties of grouts. *J. Cult. Herit.* **2018**, *30*, 70–80. [CrossRef]
9. Rousset, B.; Gentile, S.; James, J.; Pozzi, B. Injection grouts for molasse sandstones: Preliminary assessments. In Proceedings of the RILEM Workshop Repair Mortars for Historic Masonry, Delft, The Netherlands, 26–28 January 2005.
10. Ferragni, D.; Forti, M.; Malliet, J.; Mora, P.; Teutonico, J.-M.; Torraca, G. Injection grouting of mural paintings and mosaics, in Adhesives and Consolidants. In Proceedings of the Preprints of the Contributions to the Paris Congress, London, UK, 2–8 September 1984; pp. 110–116.

11. Veiga, R. Conservation of historic renders and plasters: From laboratory to site. In Proceedings of the 2nd Conference on Historic Mortars-HMC 2010 and RILEM TC 203-RHM Final workshop, Prague, Czech Republic, 22–24 September 2010; RILEM Publications S.A.R.L.: Bagneux, France; pp. 1247–1256.
12. Azeiteiro, L.C.; Velosa, A.; Paiva, H.; Mantas, P.Q.; Ferreira, V.M.; Veiga, R. Development of grouts for consolidation of old renders. Construct. *Build. Mater.* **2014**, *50*, 352–360. [CrossRef]
13. Duran, A.; González-Sánchez, J.F.; Fernández, J.M.; Sirera, R.; Navarro-Blasco, Í.; Alvarez, J.I. Influence of two polymer-based superplasticisers (Poly-naphthalene Sulfonate, PNS, and Lignosulfonate, LS) on compressive and flexural strength, freeze-thaw, and sulphate attack resistance of lime-metakaolin grouts. *Polymers (Basel)* **2018**, *10*, 824. [CrossRef] [PubMed]
14. Gonzalez-Sanchez, J.F.; Taşcı, B.; Fernandez-Alvarez, J.M.; Navarro-Blasco, I.; Alvarez-Galindo, J.I. Combination of polymeric superplasticisers, water repellents and pozzolanic agents to improve air lime-based grouts for historic masonry repair. *Polymers (Basel)* **2020**, *12*, 887. [CrossRef] [PubMed]
15. EN 459-1. *Building Lime, Part 1: Definitions, Specifications and Conformity Criteria*; CEN: Brussels, Belgium, 2010.
16. EN 196-2. *Method of Testing Cement, Part 2: Chemical Analysis of Cement*; CEN: Brussels, Belgium, 2013.
17. Biçer-Şimşir, B.; Rainer, L. *Evaluation of Lime-Based Hydraulic Injection Grouts for the Conservation of Architectural Surfaces: A manual of Laboratory and Field Test Methods*; The Getty Conservation Institute: Los Angeles, CA, USA, 2013.
18. EN 1015-6. *Methods of Test for Mortar for Masonry-Part 6: Determination of Bulk Density of Fresh Mortar*; CEN: Brussels, Belgium, 1999.
19. Domone, P.; Hsi-Wen, C. Testing of binders for high performance concrete. *Cement. Concrete. Res.* **1997**, *27*, 1141–1147. [CrossRef]
20. EN 459-2. *Building Lime, Part 2: Test Methods*; CEN: Brussels, Belgium, 2010.
21. ASTM C940-16. *Standard Test Method for Expansion and Bleeding of Freshly Mixed Grouts for Preplaced-Aggregate Concrete in the Laboratory*; ASTM International: Harrisburg, PA, USA, 2016.
22. Padovnik, A.; Bokan-Bosiljkov, V. Adaptation of test methods used to evaluate the properties of non-structural grouts. In Proceedings of the 4th Historic Mortars Conference-HMC 2016, Santorini, Greece, 10–12 October 2016; Laboratory of Building Materials, Department of Civil Engineering, Aristotle University of Thessaloniki: Thessaloniki, Greece, 2016; pp. 572–579.
23. prEN 1015-8. *Methods of Test for Mortar for Masonry—Part 8: Determination of Water Retentivity of Fresh Mortar*; CEN: Brussels, Belgium, 1999.
24. EN 1771. *Products and Systems for the Protection and Repair of Concrete Structures, Test Methods—Determination of Injectability and Splitting Test*; CEN: Brussels, Belgium, 2004.
25. RILEM TC 145-WSM. Settlement column segregation test. In *The Workability and Rheology of Fresh Concrete: Compendium of Tests—Report of RILEM TC 145-WSM*; RILEM Publications S.A.R.L: Cachan Cedex, France, 2002; pp. 76–78.
26. EN 1015-10. *Methods of Test for Mortar for Masonry, Part 10: Determination of Dry Bulk Density of Hardened Mortar*; CEN: Brussels, Belgium, 1999.
27. SIA 262/1. *Construction En Béton-Spécifications Complémentaires, Appendix A*; SIA: Zurich, Switzerland, 2003.
28. RILEM TC 25-PEM. Test No. II. 6, Water absorption coefficient. *Mater. Struct.* **1980**, *13*, 173–253.
29. ISO 12572. *Hygrothermal Performance of Building Materials and Products—Determination of Water Vapour Transmission Properties—Cup Method*; ISO: Geneva, Switzerland, 2016.
30. EN 1015-11. *Methods of Test for Mortar for Masonry—Part 11: Determination of Flexural and Compressive Strength of Hardened Mortar*; CEN: Brussels, Belgium, 1999.
31. ASTM C496/C496M-1. *Standard Test Method for Splitting Tensile Strength of Cylindrical Concrete Specimens*; ASTM International: Harrisburg, PA, USA, 2004.
32. EN 1015-12. *Methods of Test for Mortar for Masonry—Part 12: Determination of Adhesive Strength of Hardened Rendering and Plastering Mortars on Substrates*; CEN: Brussels, Belgium, 2001.
33. Bouvet, A.; Ghorbel, E.; Bennacer, R. The mini-conical slump flow test: Analysis and numerical study. *Cem. Concr. Res.* **2010**, *40*, 1517–1523. [CrossRef]
34. EN 447. *Grout for Prestressing Tendons. Basic Requirement*; CEN: Brussels, Belgium, 2008.
35. Ince, C.; Ozturk, Y.; Carter, M.; Wilson, M. The influence of supplementary cementing materials on water retaining characteristics of hydrated lime and cement mortars in masonry construction. *Mater. Struct.* **2014**, *47*, 493. [CrossRef]

36. Miltiadou-Fezans, A.; Tassios, T.P. Stability of hydraulic grouts for masonry strengthening. *Mater. Struct.* **2013**, *46*, 1631–1652. [CrossRef]
37. Biçer-Şimşir, B.; Rainer, L. Field test methods for comparative evaluation of lime-based hydraulic injection grouts for the conservation of architectural surfaces. In Proceedings of the ICOM-CC 17th Triennial Conference, Melbourne, Australia, 15–19 September 2014.
38. Jornet, A.; Mosca, C.; Cavallo, G.; Corredig, G. Comparison Between Traditional Lime Based and Industrial Dry Mortars. In Proceedings of the 2nd Conference on Historic Mortars-HMC 2010 and RILEM TC 203-RHM Final Workshop, Prague, Czech Republic, 22–24 September 2010; RILEM Publications S.A.R.L.: Bagneux, France; pp. 227–237.
39. Padovnik, A. Consolidation of Detached Plaster Layers of Mural Paintings with Non-Structural Grouting. Ph.D. Thesis, Faculty of Civil and Geodetic Engineering, University of Ljubljana, Ljubljana, Slovenia, 26 May 2016.
40. Lanas, J.; Alvarez-Galindo, J.I. Masonry repair lime-based mortars: Factors affecting the mechanical behavior. *Cem. Concr. Res.* **2003**, *33*, 1867–1876. [CrossRef]
41. Uranjek, M.; Bokan-Bosiljkov, V. Influence of freeze–thaw cycles on mechanical properties of historical brick masonry. *Constr. Build. Mater.* **2015**, *84*, 416–428. [CrossRef]

materials

MDPI

Article

Long-Term Behavior and Durability of Alkali-Activated Clay Mortars

Aspasia Karozou [1,*], Stavroula Konopisi [1], Eleni Pavlidou [2] and Maria Stefanidou [1]

1 School of Civil Engineering, Dept. of Civil Engineering, Aristotle University of Thessaloniki, University Campus, 56124 Thessaloniki, Greece; konopisi@civil.auth.gr (S.K.); stefan@civil.auth.gr (M.S.)

2 Solid State Section, Physics Department, Aristotle University of Thessaloniki, Agiou Dimitriou, 54636 Thessaloniki, Greece; elpavlid@auth.gr

* Correspondence: akarozou@gmail.com

Received: 22 July 2020; Accepted: 25 August 2020; Published: 27 August 2020

Abstract: The need to increase the durability of clay-based materials, due to their inherent low strength and vulnerability in contact with water, led researchers to examine different options. In this paper, clay mortars were produced using four different activating solutions. Alkali hydroxides, alkali carbonates, and alkali silicates activating solutions were used. Interest is given to long term properties while their behavior to wetting–drying and freeze–thaw cycles is recorded. In total, the results of the experiments indicated the positive effect of the potassium metasilicate on mechanical characteristics presenting, however, low performance at wetting–drying. The combination of sodium metasilicate with sodium hydroxide solution has also presented a positive effect on both mechanical and physical properties. In contrast, sodium carbonate acted better in enhancing physical properties and granting water-resistant abilities. Moreover, the performance of the specimens mixed with water–glass addition presented excellent volume stability and low mass loss in durability tests.

Keywords: clay mortars; activating solutions; wetting–drying; freeze–thaw cycles; microstructure

1. Introduction

Low-cost materials that result from binders using activating solutions represent a sustainable option in the construction sector, not only for compatible and durable repair works but also as alternative solutions for new constructions. The "soft" nature of earthen materials is well-known, and up till now, various methods have been used to achieve stabilization, such as adding different binders, including Portland cement, fly ash, and lime [1,2]. Various issues emerge, however, such as the reduction of cement consumption in total and the reinforcement of earthen structures without the use of cement, that need to be examined to move towards sustainable building practices. The mechanism of alkaline activation of clays and soils as precursors, is under investigation in the last period, to enhance the physical-mechanical properties which are essential in determining the long-term durability of mortars [3]. As abundant materials, the development of alternative products could lead to many environmental, social, and economic advantages [4,5].

Chemical modification of earthen materials is likely to be achieved through inorganic polymerization. Specifically, alkali-activated products occur by dissolving through thermal treatment of the aluminosilicate network contained in a solid material which is called the precursor by using an alkali solution called the activator [6]. Nevertheless, the suitability of these aluminosilicate solid materials as precursors for the alkali activation process is a subject that should, in all cases, be examined before application. Usually, the preferable heating temperatures to achieve the polymerization of an aluminosilicate precursor such as clay are between 60 °C and 90 °C [7]. However, scientific studies prove that the alkali activation reaction of calcium–aluminosilicate systems can also take place in ambient conditions [6,8,9].

Moreover, different activators can be used according to the precursor in hand, thus, more suitable for earthen materials are considered the solutions of $Ca(OH)_2$, NaOH, and KOH. These activators are chosen for dissolving the clay minerals efficiently while modifying the binding networks [6,10,11]. In general, the nature of the precursors and activators determine the treating conditions applied to avoid efflorescence and achieve higher stability [12–14].

As for alkali-activated materials, capillary absorption tests have proven that their pore networks are sufficiently helical, leading to a low capillarity [15,16]. Other researchers have found that their permeability to pure water or solutions of different ions is higher or similar than that of cement concrete [17,18].

Nevertheless, the long-term behavior and the durability of alkali-activated clay mortars in wet–dry and freeze–thaw cycles are of interest since the longevity of these materials is crucial for their utilization both in old and in modern structures and requires further studying. Additionally, the lack of applicable regulations constitutes their use more difficult. Most studies concerning such durability tests deal with alkali-activated mortars and concretes, with admixtures of industrial waste binders, such as fly ash and slag [3,19]. Moreover, the durability tests usually follow up the standards for Portland cement and concern resistance to acid attack, high temperatures, fire, and freeze–thaw [19]. The performance of such alkali-activated materials under freeze–thaw durability experiments has been proven beneficiary, presenting high resistance [19].

The frost resistance of cement depends mainly on physicomechanical rather than chemical factors, such as the porous structure [20,21]. There is a variety of reports of the performance of alkaline activated materials in the freeze–thaw process, which in some cases perform better than comparable cement concrete exposed under the same conditions [22,23].

However, there is a little reference for wet–dry cycling in alkali-activated materials, since such a test is proven not to cause a significant deterioration in cement products [3]. Furthermore, different testing methods have been reported for wetting and drying, each one being adapted to simulate different environmental conditions [3,24]. Although the specimens were exposed to different environments (temperature, relative humidity) at different ages, dimensional and/or mass measurements did not show a significant effect on materials performance [25–27]. However, some damage may be attributed to calcium and alkali leaching of the material with efflorescence effects [28].

This study focuses on examining the long-term effect of various alkali activators in earthen mortars. The mortars were submitted in low thermal treatment (40 °C) and ambient conditions. Since state-of-the-art research is dealing with alkali-activated materials, their long-term behavior after exposure at durability cycles is of interest. For this reason, the mechanical and physical properties of the mortars are presented.

2. Materials and Methods

2.1. Materials and Mortars Design

In total, five different compositions were cast, including the reference one (recorded as A) with no activators, used for comparison reasons. The manufactured specimens were rectangular with dimensions (40 mm × 40 mm × 160 mm) that were used in every test conducted in this study besides open porosity. The specimens for the open porosity test were approximately 40 mm × 40 mm × 50 mm). The soil used was extracted from the island of Crete and has been characterized using XRD analysis, particle size distribution, chemical analysis, and Atterberg limits. The XRD analysis indicated that the soil particles consisted of quartz, calcium aluminum hydroxide, calcite, cancrinite, and a small percentage of muscovite. Moreover, the high content of calcium oxide of 25% was reported by the results of the chemical analysis using atomic absorption technique [29]. The specific gravity of the dry material was 1.96 g/cm^3 (ASTM-C188-95) [30], and the color characterization by Munsell charts is 5Y 7/1 light grey.

Moreover, the particle size distribution (Malvern 2000, Mastersizer, Thessaloniki, Greece) depicted that the particles of 2–500 μm size prevailed in the binder's mass (Figure 1). The above analysis indicated the high volumetric percentage of silt contained in the soil (2–63 μm) of approximately 56%, with a 31% being fine sand (63–250 μm) and 13% clay particles (<2 μm). Atterberg limits were estimated (ASTM-D4318-00), while the Plasticity Index of the soil was calculated at 23.2. The liquid limit was calculated at 47.2% and the plastic limit at 24%. Thus, using the plasticity chart by Casagrande (ASTM-D2487-17) [31], the soil was classified as lean clay (CL).

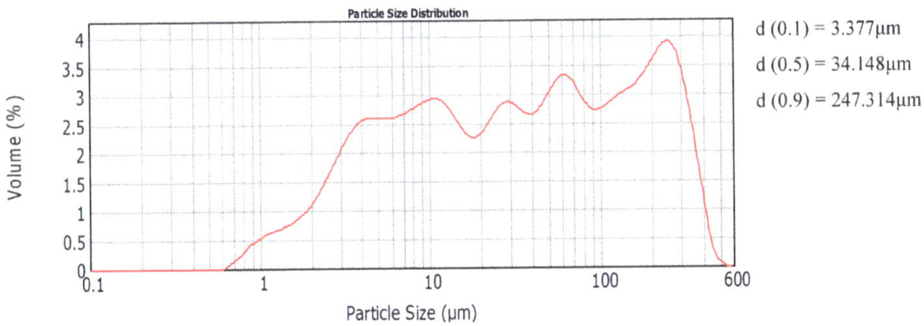

Figure 1. The particle size of Cretan clay used.

The details of the mortars' manufacture, as well as the activators used, can be found in a previous publication [29]. To clarify the acronyms PO stands for potassium metasilicate, which was used as such, SC for sodium carbonate solution, in a ratio of 70:30 water: sodium carbonate, WGS for sodium metasilicate, which was used as such, and WGN for sodium hydroxide solution mixed with water–glass at a 1:1 ratio. The water–glass used was of sodium silicate nature, with a molar ratio SiO_2/AlK_2O of 3.45, a pH concentration at 20 °C of 11.1, and a viscosity of 68 m·Pas at 20 °C. The activating solutions used in this work are of low viscosity (liquid form), and they are commercial products. These activators are primarily used for activation in slag, fly ash, bentonite, and pure kaolinitic clay systems [3,32,33]. However, they have not been thoroughly tested in such an impure clay system. The mortars' composition can be seen in Table 1, while the pH values, the details of the mortar mixtures, and curing conditions can be found in a previously published research paper [29]. The workability achieved was recorded by the flow table, as mentioned in EN 1015-3.

Table 1. Workability of the fresh mortars.

Mortar	Binder/Aggregate	Activator	L/S [1]	Workability [mm]
A	1:2.5	no activator	0.65	155
PO	1:2.5	potassium metasilicate	0.62	150
SC	1:2.5	sodium carbonate solution	0.71	150
WCS	1:2.5	water glass	0.95	155
WGN	1:2.5	water–glass with sodium hydroxide solution	0.69	155

[1] L/S = liquid/solid.

2.2. Methods

In all cases, various tests were performed in both non-alkaline (A) and alkaline activated mortars (PO, SC, WGS, and WGN) of different ages (28, 90, 180, and 365 days) after manufacture. The behavior of the mortars against water penetration was examined by the conduction of capillary absorption (UNI EN15801:2010), Karsten tube penetration (EN 16302:2013), and porosity (RILEM CPC11.3) tests. Capillary absorption tests were performed at the ages of 180 days and 365 days, while water absorption through Karsten tubes was performed at 90 days and 365 days. Moreover, the open porosity tests were conducted at all ages previously mentioned, with the use of heptane instead of water [34], due to the sensitive nature of the material against direct water contact. In this study, however, the results of the 180 days and 365 days are reported.

The capillary coefficient was calculated, while immediately after the completion of the capillary test, the drying test was carried out (EN 16322:2013). The drying index (ID) was then calculated, while the results are presented as a function of time expressed in hours. The Mi-t chart represents the first drying phase of the samples, while the drying index is calculated through the integral under this curve (EN 16322:2013). Thus, the equation for the calculation of the drying index is the following (Equation (1)):

$$ID = \int_0^{t_f} \frac{M_i dt}{M_{max} * t_f} \text{ where : } M_i = \frac{m_i - m_f}{A} \quad (1)$$

The symbol m_i (kg) stands for the mass of the sample at a given time t_i (h). Thus, m_f stands for the final mass of the sample recorded at the final time t_f (h). Consequently, the residual amount of water of the sample at a given time t_i per unit area in kg/m^2 is symbolized as M_i. Furthermore, M_{max} is the maximum mass difference of the sample that occurs at the beginning of the test at time t_0 (kg/m^2) [35,36]. The drying index can then be calculated by Equation (1) using the simplified Equation (2) according to the European standard 16322:2013 [36]:

$$ID = \frac{\sum_{i=1}^{i=n}\left[(t_i - t_{i-1}) * \left(\frac{M_{i-1}+M_i}{2}\right)\right]}{M_{max} * t_f} \quad (2)$$

Furthermore, through the Karsten tube test, the average water penetration values are calculated. The conduction of this experiment was at the ages of 90 days and 365 days. These ages were decided to have representative values both at an early age and in the long term. At the age of 28 days, the experiment was not conducted due to the vulnerable structure of the samples.

In the ages of 180 days and 365 days, the compressive and flexural strength of the mortars was examined (EN1015-11). The results recorded at the early age of 28 days and 90 days are presented in a previously published study [29]. Concerning the volume stability of the mortars, the linear shrinkage (DIN 18947:2013-08) and volume change were measured. For the latter experiment, the specimens were cured in a chamber with specific temperature and humidity conditions (23 ± 2 °C, 50% ± 5% relative humidity). The change in the dimensions and weight of the specimens was recorded daily until stabilization.

Durability tests were also carried out when the mortars reached the age of 90 days. This certain age was decided in order to allow the mortars of aerial nature (such as clay mortars) to gain strength and mass stability. Additionally, the freeze–thaw and wet–dry cycles applied were designed, taking into consideration the vulnerable nature of clay mortars and the realistic scenarios of exposing the mortars to deteriorating agents. These tests included freeze–thaw and wet–dry cycles, where the final mass loss, compressive strength values, porosity, and surface alteration through stereoscopic observation were recorded. The stereoscopic observation was conducted through a LEICA WILD M10 (Thessaloniki Greece) microscope for all mortars at the ages of 90 days and 365 days. Any surface modifications, including cracking and color alterations, as well as the roughness of the mortars, using qualitative, comparatively images under the microscope were recorded.

Furthermore, to define the modification of the inner structure, a microscopic examination by SEM (JEOL840A JSM, Thessaloniki, Greece) equipped with an EDS device was performed. Thus, the molar ratios of SiO_2/Al2O_3, CaO/SiO_2, and M_2O/Al_2O_3 (M = Na or K) at an early age and after the completion of one year were estimated indicatively.

3. Experimental Results

3.1. Physical Properties of the Mortars

The physical properties of the clay mortars were recorded and presented here after the age of 180 days to allow the carbonation mechanism to harden the soft clay structure. This decision was made after the experience recorded in previous tests [29]. The capillary absorption was measured at the 180 days and 365 days, with the capillary coefficient indicating the water absorption trend of the specimens. In all cases, the capillary coefficient was decreased through time (Table 2). The time intervals used for measuring the weight values were 0, 5, 10, 15, 30, 60, 90, 120, and 1440 min, respectively.

Table 2. Physical properties of the mortars.

Mortar	Capillary Coefficient [g/cm^2·min$^{1/2}$]		Drying Index (ID)		Porosity [%]	
	180 days	365 days	180 days	365 days	180 days	365 days
A	0.120	0.082	–	–	16.63	16.47
PO	0.075	0.065	0.095	0.075	16.04	15.14
SC	0.021	0.015	0.090	0.072	19.74	19.51
WGS	0.244	0.194	0.087	0.056	24.19	23.23
WGN	0.163	0.059	–	0.123	15.83	15.51

In general, WGS mortar presented the highest absorption rate due to capillary in the ages tested, without presenting any material loss (Figure 2). The decrease of the capillary coefficient by 20.7% at the age of 365 days indicates a more stable structure (Table 2). However, the results differ in the case of WGN mortar, since after 24 h in contact with water, the specimens suffered material loss without being able to complete the experiment at the age of 180 days. Nevertheless, the annual results showed a more stable structure, with WGN showing no material loss. Despite presenting a fast-initial absorption rate at the age of 180 days, when tested again at 365 days, the rate of absorption was reduced significantly, as can be noted by the significant difference of 63.6% between the two values of capillary coefficient (Table 2). Overall, the mortars that were activated with sodium metasilicate and sodium hydroxide solution (WGS, WGN) presented the highest absorption rate values at the age of 180 days. However, results differ at the age of 365 days for both mortars, since WGS showed the highest absorption rate, while the WGN mortar, as mentioned, had a significantly lower absorption rate. This fact is probably justified by the density of the geopolymer gel, being in the case of WGN less dense, and in the case of WGN, much denser [37], a fact justified by the porosity values as well. By the SEM analysis, in the case of WGN mortar, the loss of sodium through time (leaching effect) could have resulted in a less absorbent structure [37]. PO mortar presented low values of capillary coefficient at both ages tested, with higher final absorption value at 365 days (Figure 2). The low porosity values, as seen in Table 2 for both ages, indicate a dense formation that resulted in lower water uptake [12]. In both cases, SC mortar presented the lowest absorption rate through time, showing a 28% decrease in capillary coefficient values. Moreover, it is observed that the untreated mortar A was unable to complete the test until at all ages examined. Overall, the results come to an agreement with literature for alkali-activated metakaolin or natural pozzolan-based binders that are porous and present high capillary suction [3].

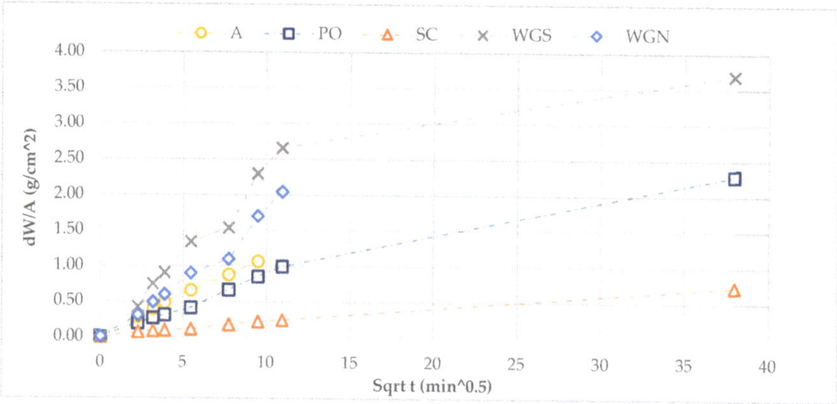

(**a**)

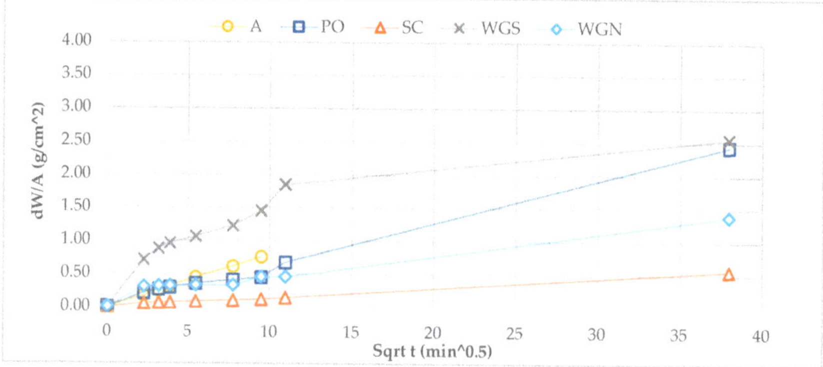

(**b**)

Figure 2. Capillary absorption of the mortars at (**a**) 180 days and (**b**) 365 days.

The conduction of the drying test started immediately after the completion of the capillary absorption test, as a reverse capillary test. In the case of the reference mortar A, this test was not able to be conducted since, at both ages, the samples were destroyed before completing the capillary absorption test. All other specimens were weighed using the same time intervals as the capillary absorption test and after that daily up to 960 h when all the samples have reached equilibrium with the environmental conditions (stable measurement). Weight stabilization of the specimens occurred at different times for each mortar during the total duration of the experiment. The determination of the drying curve was done after calculating the residual amount of water present in the specimen per unit area referred to as Mi (kg/m^2). Since the drying index describes the resistance of the material to drying, it can be claimed that a low value of ID reflects an overall easier drying behavior [35,38].

In total, ID values were decreasing for all the samples tested through time, while the highest ID value was recorded for the WGN samples in the long term.

Figure 3a,b depict the drying curves of the mortars at later ages. A higher slope of the curve to the horizontal axis reflects materials with high liquid conductivity (porous materials) [36]. The final time of the drying test at 180 days was approximately the same for all the samples tested (Figure 3). Moreover, it can be observed that WGS mortars have a higher liquid conductivity compared to the other two mortars, a fact that is also justified by the high porosity values measured at both ages (Table 2, Figure 3). Presenting the lowest values of drying index at both ages tested, WGS mortars have the fastest drying behavior comparatively, with a generally distinct and long first drying phase, a fact that agrees with their high porosity values. It is also noted that WGN mortars showed the highest resistance to drying at 365 days compared to all the treated mortars tested (Table 2). Despite presenting similarly low porosity and capillary coefficient values with PO mortar at the age of 365 days, the drying behavior of the WGN mortar is significantly different, exhibiting low liquid conductivity.

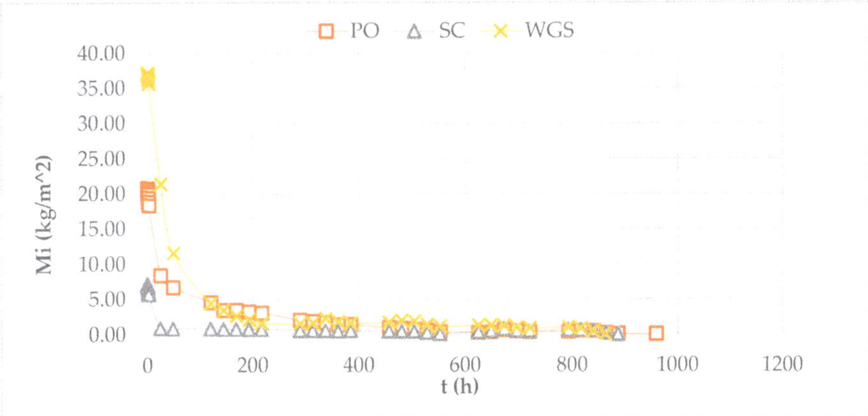

(**a**)

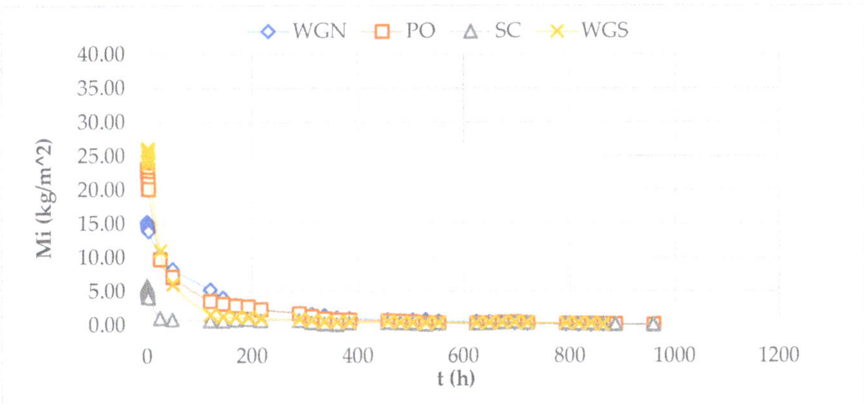

(**b**)

Figure 3. Drying curves of the mortars at (**a**) 180 days and (**b**) 365 days.

Moreover, during the conduction of the experiment, efflorescence was observed on the surface of the WGN mortars. Efflorescence indicates an excess amount of unreacted sodium oxide in the pore structure that is transferred to the surface of the sample, with the presence of water through capillary. Then, the transferred alkalis react with the atmosphere, thus causing carbonation known as efflorescence [6,39]. This phenomenon that also occurred in WGN mortar shows a low exchangeability, while it can lead to a further deterioration of the system.

The PO mortars presented a low resistance to drying, with a comparably high liquid conductivity, an interesting fact considering their low porosity values (Table 2). Additionally, the second most porous mortar SC also showed a fast-drying behavior with low values of ID and a shorter first drying phase.

In Figure 3, the final drying time of the mortars can be distinguished. The mortars PO and SC presented a more extended drying period, while all mortars previously tested showed improved drying behavior with lower ID values. Moreover, despite the reduction in porosity values through time, the ID index was not negatively affected, since the decrease of the annual values for all samples, indicates a faster drying behavior meaning a quicker elimination of moisture (Figure 3, Table 2).

The porosity results signify the porous structure of the WGS mortars since the porosity values were the highest recorded compared to the other mortars at all ages (Table 2, Figure 4). For PO and WGN mortars, it is noted that the porosity values remained relatively low, with the annual results being close to the values of the untreated mortar A. These values indicate the compact structure of these specimens. The high porosity values of WGS mortars, agree with the high absorption rate through capillary, while the values of the SC mortars reveal a porous structure. The high porosity values justify the low values of drying index at all ages for mortars SC and WGS.

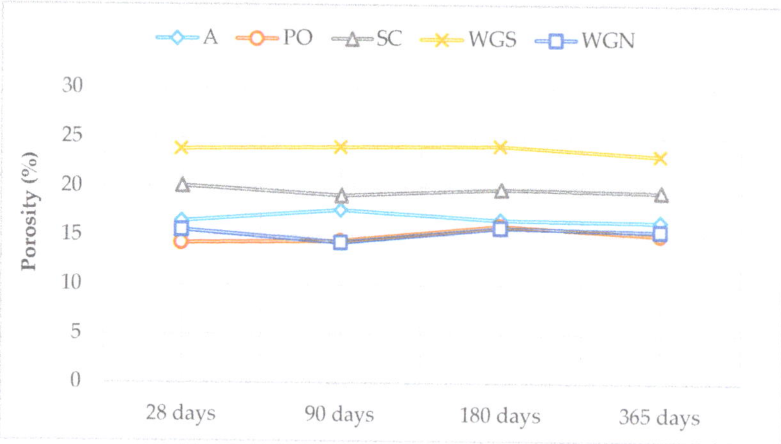

Figure 4. The porosity of the mortars at all ages tested.

The results of water penetration through Karsten tubes indicate the increased water absorption through time, of the most porous mortars SC and WGS. In general, it is observed that all treated mortars, besides SC, showed a higher absorption rate compared to the untreated mortar A, at all ages tested. The high tendency to water absorption of WGS and WGN mortars remains unchanged, presenting, however, a reverse behavior through time. The water absorption of WGS was increased from 90 days to 365 days by 93.2%, while the WGN mortars presented a decrease in water absorption by 53.9% (Table 3). It is noted that SC mortars showed a low water intake compared with the other treated mortars in every water absorption test conducted. PO mortars showed an average water penetration during the Karsten tube test, yet still higher than the untreated mortar A that had an overall low water intake.

Table 3. Water penetration and shrinkage of the mortars.

Mortar	Water Penetration [mL/min·cm^2]		Linear Shrinkage [%]		Volume Loss [%]	
	90 days	365 days	180 days	365 days	180 days	365 days
A	0.053	0.035	1.48	1.92	10.82	12.93
PO	0.139	0.076	0.60	0.64	2.31	2.61
SC	0.014	0.023	1.70	1.97	20.14	21.40
WGS	0.377	0.728	1.14	1.21	10.22	11.11
WGN	0.897	0.413	1.05	1.10	4.05	4.64

Linear shrinkage and volume loss of the mortars were recorded, up until 365 days after manufacture. In Table 3, both values at 180 days and 365 days are noted as to present their progress through time (Table 3). The long-term measurements were decided to test the probable instability of the mortars through time regarding volume loss and shrinkage. According to DIN 18947, the linear shrinkage should not be more than 2% [40]. Despite the reference mortar A and mortar SC, that have barely satisfied this requirement in the long term, all the other mortars meet the standard's requirements concerning linear shrinkage. Mortars treated with potassium metasilicate (PO) present an overall stable structure. In Figure 5, it can be detected that the percentage of volume loss through time is the lowest recorded. In general, mortars SC, WGS, and WGN presented higher values of linear and volume shrinkage, with the first showing a significant volume loss percentage in relation to the untreated mortar A, especially with the completion of one year. The total volume loss of SC mortars was 66.1% greater compared to the reference one, while PO mortars presented a 79.7% decrease in volume loss. WGS mortars presented a similar shrinkage behavior with the reference samples, showing an improvement in volume loss of 13.7%. WGN mortars presented overall good stability, with around 64% decrease in volume loss compared to A.

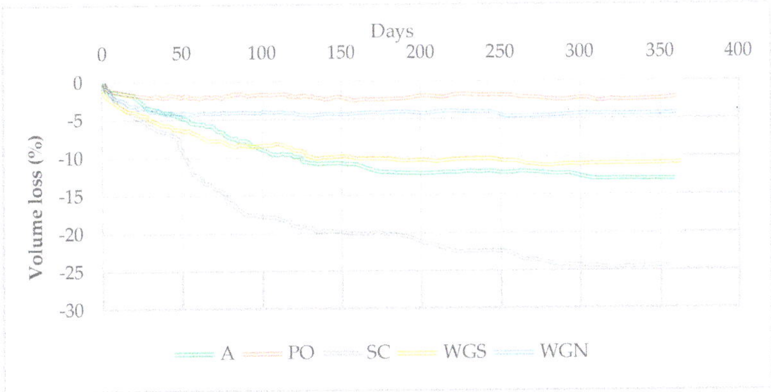

Figure 5. Volume loss (%) of the mortars.

Overall, the mortars with the lower porosity and liquid/solid ratios presented the most stable structure also in terms of volume loss and linear shrinkage. The use of potassium metasilicate and water–glass with sodium hydroxide solution as activators have been proved beneficial in making the structure of the samples more stable.

3.2. Mechanical Properties of Mortars

In Table 4, the mechanical properties of the mortars are presented. The values are the average of six samples in each case. Overall, both the PO and WGN mortars presented increased values of compressive strength. An impressive 302.83% increase in one year for PO and 195.93% for WGN compared to the reference A indicate the strengthening of the structure (Table 4). In total, PO mortar presented the best performance in terms of mechanical characteristics, granting the specific activator compatible with the precursor used in this study, matching the results that derive from SEM analysis. For clay-based materials, such values of compressive strength are considered exceptionally high, since the values that are usually expected of such systems are around 1–2 MPa [41,42].

Table 4. Mechanical properties of the mortars.

Mortar	Compressive Strength [MPa]				Flexural Strength [MPa]			
	180 days		365 days		180 days		365 days	
–	–	st.dev	–	st.dev	–	st.dev		st.dev
A	1.28	0.074	1.52	0.118	0.7	0.081	0.9	0.026
PO	5.12	0.690	6.13	0.548	1.84	0.611	2.55	0.390
SC	1.06	0.050	1.08	0.055	0.49	0.375	0.5	0.041
WGS	0.85	0.036	1.43	0.154	0.38	0.007	0.59	0.042
WGN	1.66	0.723	4.5	0.444	1.06	0.462	1.26	0.219

Moreover, concerning the mortar SC, it is also established in this study that sodium carbonate presents a slow strength development [32]. Even after one year, the compressive and flexural strength is considerably low, granting it as the one with the weakest behavior in terms of mechanical characteristics. The porosity values of SC also agree with the results of mechanical testing. Perhaps the slow strength development could be overcome with curing at a temperature higher than 40 °C [32] or even for an extended period, allowing a more stable and less porous structure.

The compressive test results seemed to agree with the porosity values also in the case of the WGN mortars. Notable is the fact that the WGN mortar showed the most significant compressive strength development from 180 days to 365 days by 171.19%. This fact could imply a lower strength development, especially when compared with the strong activator of potassium metasilicate. At the same time, it could also be linked to the loss of unreacted sodium through efflorescence, with an alteration of the Si/Al ratio in the structure as SEM indicates (see below). The high porosity values of WGS mortars indicate a weak performance in mechanical properties. This fact stands true for both ages, with the results at the age of 180 days being comparatively very low, while the annual values of compressive strength are comparable to the reference.

Flexural strength results presented a similar pattern, with PO and WGN mortars showing the most notable values, especially at the age of one year, with an impressive 183.33% and 39.71% increase, respectively (Table 4). In general, the mechanical characteristics are bond to the microstructure of the mortars. Through SEM analysis, it is noted that the differences in the values of Si/Al, Si/Ca, and Na, K/Al explain whether the formation of an inorganic polymeric network of alkali aluminosilicates was realized [3,43,44].

3.3. Durability Properties and Microscopic and Stereoscopic Observation

After reaching the age of 90 days, durability tests were carried out. These tests included freeze–thaw and wet–dry cycles, where the final percentage of mass loss, compressive strength values, porosity, and surface alteration through stereoscopic observation were recorded. The values given in Table 5 are averaged from three specimens. A full cycle in freeze–thaw durability tests consists of four hours in a chamber of −18 ± 2 °C, 10%RH, and the rest 20 h in ambient conditions (20 ± 2 °C, 65%RH). Moreover, a full wet–dry cycle includes wetting of the mortars by spraying approximately 4–5 mL of water per sample (4 cm × 4 cm × 16 cm prism) and letting them dry both in ambient conditions for three hours and then exposed them for 21 h at 40 °C. These conditions were decided based on experience due to the lack of regulations on clay-based materials. The conduction of the durability tests was until the completion of 50 cycles or until the destruction of the samples. The addition of potassium silicate in earth-based mortars proved disadvantageous concerning the wet–dry cycles since it has suffered the most significant dissolution, leading to a high amount of mass loss. Despite this deterioration, however, notable is the fact that the compressive strength value for wet–dry cycles was the highest one recorded. It is observed that both PO and WGN mortars suffered more significant mass loss than the reference mortar A regarding the wet–dry cycles, while SC mortars proved the most efficient. It is worth noting that the PO samples experienced deterioration after the completion of the 21st cycle. Thus, in Table 5, the mass change recorded is referred to as the mass of the samples after the cycle. In total, the freeze–thaw cycles showed a low mass-change effect, while the compressive strength results indicated an inadequate response to compressive strain, apart from the PO and WGN mortars. Generally, significant is the fact that despite the high deterioration of PO and WGN samples, their load-bearing capacity after the completion of the cycles was efficient.

Table 5. Mass change and compressive strength of the mortars after durability tests.

Mortar	Mass Change [%]		Compressive Strength [MPa]		Porosity [%]	
	Freeze–Thaw	Wet–Dry	Freeze–Thaw	Wet–Dry	Freeze–Thaw	Wet–Dry
A	−1.16	−2.56	0.97	1.38	15.42	14.82
PO	1.34	−10.41	5.46	4.88	15.38	15.19
SC	0.11	−0.01	0.62	1.10	19.17	19.33
WGS	0.66	−0.31	0.43	0.71	23.37	24.58
WGN	−0.27	−3.27	2.25	2.41	15.24	15.13

The morphological characteristics of the mortars are presented at the age of one year, and after the durability cycles. The mortars examined at the age of one year showed similar characteristics with the reference sample A. Shrinkage cracks were detected, and a porous structure was evident in all surfaces observed (Figure 6). The reference mortar showed a rough, porous surface, while PO mortar presented the most compact structure and smoother surface, with few shrinkage cracks and with pores of a mean diameter 300 µm. Efflorescence was also observed microscopically on the surface of the WGN samples and inside the mass of PO mortar. Moreover, surface shrinkage cracks were detected in all mortars, with WGS mortar presenting the more significant amount. Their width was ranging from 20–110 µm, while a darker color was detected.

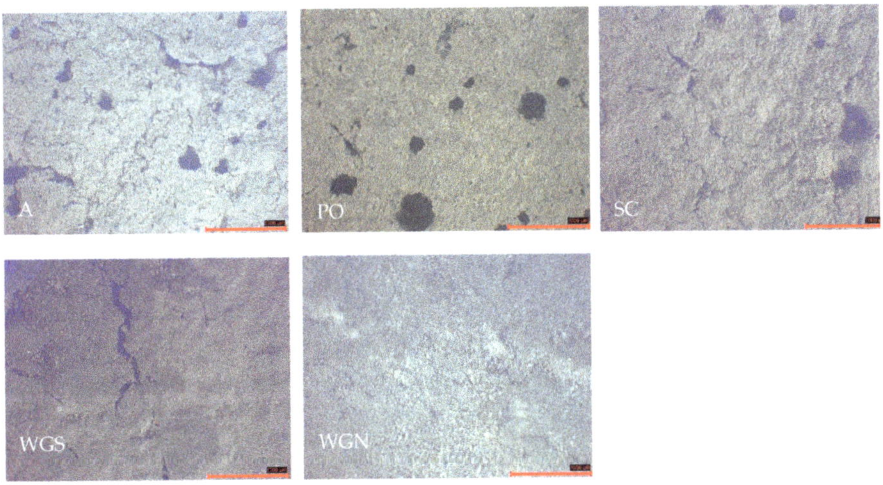

Figure 6. Stereoscopic observation of the mortars at the age of one year (scale 1000 µm).

After the conduction of freeze–thaw cycles, all the mortars presented a more cracked and rougher surface (Figure 7a). Additionally, the surface of the mortars after the conduction of the freeze–thaw cycles showed cracking and scaling that is explained by the expansion of the inner pore water. Thus, the mass increase of most of the mortars tested is justified, since the PO, SC, and WGS presented mostly cracks and not significant scaling. For WGS mortar, the width of the cracks was between 50–160 µm after the freeze–thaw cycles. Furthermore, for all the other mortars, the range of the cracks was 40–60 µm. A disruption between the binder and the aggregates led to the mass loss of the samples WGN and A [45].

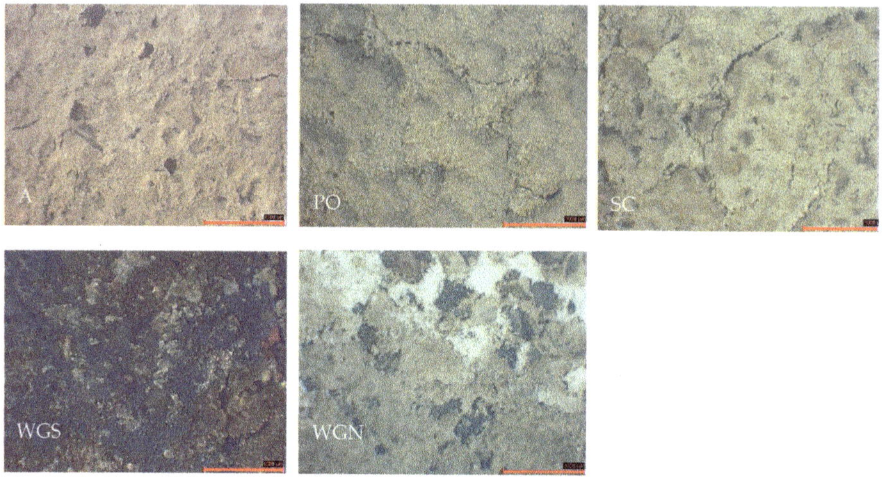

(**a**)

(**b**)

Figure 7. Stereoscopic observation of the mortars after the (**a**) freeze–thaw and (**b**) wet–dry cycles (scale: 1000 μm).

Moreover, the color alteration was apparent in WGS mortar, while in SC, some coloring spots were detected (Figure 7a). Both WGS and WGN mortars developed tarnishes and white agglomeration spots. The least porous mortar PO presented an excellent behavior in freeze–thaw cycles without showing any significant reduction in compressive strength after the completion of the cycles. The low porosity indicates a more stable mass, justifying the high values of compressive strength. The low porosity of the WGN mortar, however, is not consistent with higher strength development, presenting moderate mechanical characteristics after the freeze–thaw cycling.

The loose cohesion of the aggregates and the increase of cracks were evident after the completion of wet–dry cycles (Figure 7b). Tarnishes were again developed on the surface of the WGN mortar, while the SC mortar presented an overall good behavior against weathering. The compressive strength and porosity of the SC mortar after the wet–dry cycles are very close to the values of the annual results for the same mortar. That fact indicates the stability of the mortar against wet–dry cycles, presenting a water-resistant behavior, with almost no mass loss (Table 5). WGS mortar also presented low mass loss, yet the mechanical strength and porosity values were not improved. Significant is the fact that despite suffering mass loss, the compressive strength of the PO mortar was not significantly reduced. The results after the completion of the cycles are compared to the equivalent values of 180 days. Concerning WGN mortar, the results indicate a deterioration in mechanical strength results after the conduction of both weathering cycles. The WGN mortar suffered scaling and moderate mass loss after the wet–dry cycles, presenting, however, a very slightly improved behavior in mechanical characteristics compared to the freeze–thaw results.

To determine the nature of the efflorescence of WGN mortars, differential thermal, and thermogravimetric analysis (DTA/TG) was performed through a TA Instruments SDT 2960 analyzer (Thessaloniki Greece) (Figure 8). The results indicated the presence of sodium hydrogen carbonate ($NaHCO_3$) and sodium carbonate (Na_2CO_3) [46].

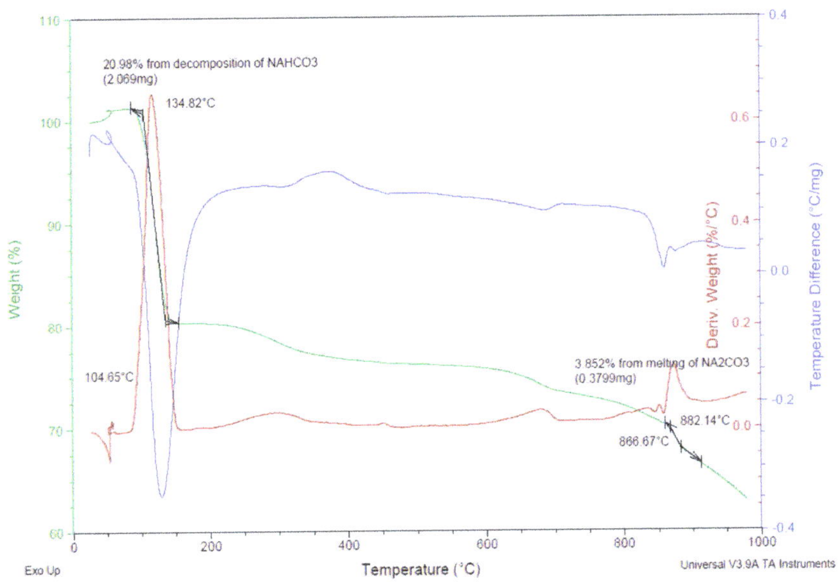

Figure 8. TG-DTA analysis of the efflorescence appearing on WGN mortars.

When observing the produced mortars under SEM (Figure 9a), the reference mortar A showed a loose crystal structure at an early age, while a smoother surface was observed through time (Figure 9b). In PO mortar, rod-like crystals were detected of a potassium-based compound, with a noticeable decrease of potassium in later age (Table 6). In the case of both PO and WGN mortars, a continuous structure with small pores and few cracks was observed, also showing excellent structure cohesion. As to compare the differences in the inner structure, SEM analysis was performed at an early age (28 days) and after one year. A rougher surface with formation of leaf-like crystals was detected through SEM for WGN mortar at an early age, however, in time, a decrease of sodium content by 94% was remarked (Table 6). The presented spectrums are the average of many images, where a whole area was analyzed.

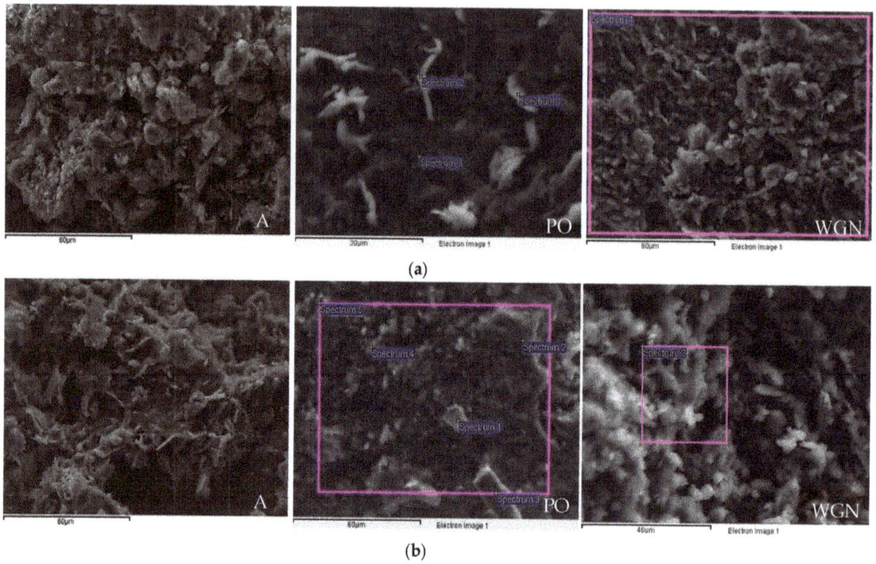

Figure 9. SEM images of the mortars (**a**) at an early age and (**b**) at 365 days (scale: 60 μm for A, 60 and 30 μm for PO, and 60 and 40 μm for WGN mortars).

Table 6. EDS spectrum analysis of (**1**) mortar PO and (**2**) mortar WGN (all results in atomic %).

(1) PO (a)							
Spectrum (Early Age)	Na	Al	Si	K	Ca	Fe	O
Spectrum 1	1.93	6.77	18.90	35.14	−1.19	1.08	36.52
Spectrum 2	4.14	7.63	22.01	23.46	2.49	0.58	39.52
Spectrum 3	6.08	10.45	32.51	1.48	0.26	0.68	48.74
PO (b)							
Spectrum (365 days)	Na	Al	Si	K	Ca	Fe	O
Spectrum 1	–	10.27	19.84	5.2	–	3.14	56.26
Spectrum 2	7.15	9.81	30.01	–	1.58	–	51.46
Spectrum 3	–	5.23	25.21	–	4.88	3.72	51.24
Spectrum 4	7.34	7.43	27.28	1.26	–	–	56.68
Spectrum 5	5.05	8.33	23.43	3.96	1.91	–	57.31
(2) WGN (a) and (b)							
Spectrum	Na	Al	Si	K	Ca	Fe	O
Spectrum 1 (early age)	9.01	4.28	19.57	1.50	4.00	1.66	59.10
Spectrum 3 (365 days)	1.54	5.57	19.49	0.83	3.70	2.25	60.83

The annual results of the SEM analysis are reported, together with the results of the 28 days for comparison reasons. From Table 6, the indicative atomic ratios of the modified compositions can be calculated. The atomic ratios calculated at 28 days were Si/Al = 3.06, Si/Ca = 20.32, and K/Al = 0.73, while at 365 days the ratios were Si/Al = 4.54, Si/Ca = 14.55, and K/Al = 0.50. The increase in the mechanical properties of the PO mortar can be explained by the increase of the Si/Al ratio through time [47,48]. For the WGN synthesis, the atomic ratios at 28 days were Si/Al = 4.57, Si/Ca = 4.89, and Na/Al = 2.10, while at 365 days, the ratios were Si/Al = 3.51, Si/Ca = 5.29, and Na/Al = 0.28. These results indicate the decrease of Si/Al ratio through time by 23.16%, as well as the higher decrease of Na/Al ratio by 95.4%, facts that may explain the lower compressive strength development of the

WGN mortar through time. The unbound quantity of sodium by the clay minerals in the structure probably justifies this fact. Thus, the efflorescence of WGN mortar can be explained.

4. Conclusions

The activation of earth materials rich in calcium was performed using different activators. Physical, mechanical, and durability properties were examined after one year. Overall, the most notable result is the high performance of potassium metasilicate (PO) in terms of mechanical properties, even after the conduction of durability tests. The achieved mechanical properties after durability tests at the age of 365 days (6.13 MPa in compressive and 3.28 MPa in flexural strength) constitute these mortars capable of different construction needs (keeping in mind their low mechanical properties as neat materials). The shrinkage trend of PO mortars was limited, judging by the low values of linear and volume shrinkage, up to the age of 365 days.

Furthermore, the improvement of their drying behavior without suffering material disintegration indicated the development of a more compact structure. The inadequate behavior of PO mortar at wet–dry cycles is a concern that should be further examined. However, the maintenance of its load-bearing capacity even after the completion of durability tests grand this mortar as a suitable building material and the specific activator very promising in order to allow further development of new advanced materials.

The water intake of sodium carbonate mortars (SC) in terms of capillary and water absorption by Karsten tubes was proven beneficial. Therefore, this water-resistant behavior of sodium carbonate shows great potential for special applications. While presenting a water-resistant behavior, the high values of porosity and, consequently, the low values of compressive strength diminish the value of sodium carbonate as an activator. Moreover, the high percentage of volume shrinkage of SC indicates an unstable structure, yet its high resistance against durability tests, especially in terms of wet–dry cycles, grand it as a promising water-resistant agent. Further study is required for the sodium carbonate as an activator that presented slow strength development and low values of compressive strength compared to the other mortars. This fact could be managed with the use of a combination of different activators or higher curing temperatures, as also literature suggests [38].

Concerning the behavior of the specimens mixed with water–glass (WGS), the results of mechanical and physical properties tested at different ages revealed a low performance. However, the fact that the specimens presented excellent stability in volume and linear shrinkage and low mass loss in durability tests, marks water–glass as a promising treating agent regarding mass stabilization.

The mechanical characteristics of the mortar with water–glass and NaOH solution (WGN), have been proven exceptional even in the long term. Therefore, this solution stands as another promising activator for clay mortars. Despite the efflorescence that was forming during the drying stage, it displayed a compact structure with low shrinkage tendency and satisfactory physical properties. Different methods of curing and perhaps a reduction in the dose of the activator are factors that could control efflorescence and should be further studied. The high tendency of these mortars to absorb water was evident during the capillary absorption test that caused the disintegration of the samples tested at a certain age.

In total, the combination of these activators with the specific earthen material to create advanced mortars proved satisfactory. Each agent has proven to act differently and enhance different properties. Further research is proposed to investigate the matter of the application of those agents at various earth-based materials and different percentages thoroughly. XRD analysis should be performed to establish the mineralogical composition of the newly formed materials. Moreover, further research on the reaction mechanism, the appropriate activators, and treatment should be withheld, to gain a better understanding of the conditions necessary to reach the optimum results.

Author Contributions: Conceptualization: M.S.; Methodology, Supervision, Review: S.K. and E.P.; Validation and Analysis: A.K.; Investigation, Writing, Draft Preparation, Resources: All authors have read and agreed to the published version of the manuscript.

Funding: The General Secretariat for Research and Technology (GSRT) and the Hellenic Foundation for Research and Innovation (HFRI), with grand number 347, have funded the research through the scholarship funding program for Ph.D. candidates of the author Karozou Aspasia.

Conflicts of Interest: The authors declare no conflict of interest.

References

1. Galán-Marín, C.; Rivera-Gómez, C.; Petric, J. Clay-based composite stabilized with natural polymer and fibre. *Constr. Build. Mater.* **2010**, *24*, 1462–1468. [CrossRef]
2. Bui, Q.B.; Morel, J.C.; Hans, S.; Meunier, N. Compression behaviour of non-industrial materials in civil engineering by three scale experiments: The case of rammed earth. *Mater. Struct./Mater. Et Constr.* **2009**, *42*, 1101–1116. [CrossRef]
3. Provis, J.L.; van Deventer, J.S. (Eds.) *Alkali Activated Materials, State-of-the-Art Report, RILEM TC 224-AAM*; Springer: Dordrecht, The Netherlands, 2014; ISBN 9789400776715.
4. Sameh, S.H. Promoting earth architecture as a sustainable construction technique in Egypt. *J. Clean. Prod.* **2014**, *65*, 362–373. [CrossRef]
5. Faqir, N.M.; Shawabkeh, R.; Al-Harthi, M.; Wahhab, H.A. Fabrication of Geopolymers from Untreated Kaolin Clay for Construction Purposes. *Geotech. Geol. Eng.* **2019**, *37*, 129–137. [CrossRef]
6. Provis, J.L. Alkali-activated materials. *Cem. Concr. Res.* **2017**, *114*, 40–48. [CrossRef]
7. Liew, Y.M.; Kamarudin, H.; Mustafa Al Bakri, A.M.; Bnhussain, M.; Luqman, M.; Khairul Nizar, I.; Ruzaidi, C.M.; Heah, C.Y. Optimization of solids-to-liquid and alkali activator ratios of calcined kaolin geopolymeric powder. *Constr. Build. Mater.* **2012**, *37*, 440–451. [CrossRef]
8. Papayianni, I.; Kesikidou, F.; Konopisi, S. Alkali Activation of HCFA Mixtures with Aluminosilicate Additives-Mechanical Characteristics. *Key Eng. Mater.* **2018**, *761*, 96–99. [CrossRef]
9. Papayianni, I.; Konopissi, S. Study on Geopolymerization of Highlime Fly Ashes, Concrete in the Low Carbon Era. In Proceedings of the International Conferenceheld, Scotland, UK, 9–11 July 2012.
10. Giannopoulou, I.; Panias, D. Structure, design and applications of geopolymeric materials. In Proceedings of the 3rd International Conference on Deformation Processing and Structure of Materials, Belgrade, Serbia, 20–22 September 2007; pp. 5–15.
11. Provis, J.L.; Palomo, A.; Shi, C. Advances in understanding alkali-activated materials. *Cem. Concr. Res.* **2015**, *78*, 110–125. [CrossRef]
12. Luukkonen, T.; Abdollahnejad, Z.; Yliniemi, J.; Kinnunen, P.; Illikainen, M. One-part alkali-activated materials: A review. *Cem. Concr. Res.* **2017**, *103*, 21–34. [CrossRef]
13. Duxson, P.; Mallicoat, S.W.; Lukey, G.C.; Kriven, W.M.; van Deventer, J.S.J. The effect of alkali and Si/Al ratio on the development of mechanical properties of metakaolin-based geopolymers. *Colloids Surf. A Physicochem. Eng. Asp.* **2007**, *292*, 8–20. [CrossRef]
14. Davidovits, J. Geopolymers – Inorganic polymeric new materials. *J. Therm. Anal.* **1991**, *37*, 1633–1656. [CrossRef]
15. Rodríguez, E.; Bernal, S.; Mejía De Gutiérrez, R.; Puertas, F. Alternative concrete based on alkali-activated slag. *Mater. Constr.* **2008**, *58*, 53–67. [CrossRef]
16. Bernal, S.A.; Mejía De Gutiérrez, R.; Pedraza, A.L.; Provis, J.L.; Rodriguez, E.D.; Delvasto, S. Effect of binder content on the performance of alkali-activated slag concretes. *Cem. Concr. Res.* **2011**, *41*, 1–8. [CrossRef]
17. Lloyd, R.R.; Provis, J.L.; Van Deventer, J.S.J. Pore solution composition and alkali diffusion in inorganic polymer cement. *Cem. Concr. Res.* **2010**, *40*, 1386–1392. [CrossRef]
18. Wongpa, J.; Kiattikomol, K.; Jaturapitakkul, C.; Chindaprasirt, P. Compressive strength, modulus of elasticity, and water permeability of inorganic polymer concrete. *Mater. Des.* **2010**, *31*, 4748–4754. [CrossRef]
19. Pacheco-Torgal, F.; Abdollahnejad, Z.; Camões, A.F.; Jamshidi, M.; Ding, Y. Durability of alkali-activated binders: A clear advantage over Portland cement or an unproven issue? *Constr. Build. Mater.* **2012**, *30*, 400–405. [CrossRef]
20. Janssen, D.J.; Snyder, M.B. *Resistance of Concrete to Freezing and Thawing*; National Sand and Gravel Association: Washington, DC, USA, 1994; ISBN 0309057736.
21. Valenza, J.J.; Scherer, G.W. Mechanism for salt scaling. *J. Am. Ceram. Soc.* **2006**, *89*, 1161–1179. [CrossRef]

22. Shi, C.; Krivenko, P.V.; Roy, D.M. *Alkali-Activated Cements and Concretes*; Taylor & Francis: Abingdon-on-Thames, UK, 2006.
23. Gifford, P.M.; Gillott, J.E. Freeze–thaw durability of activated blast furnace slag cement concrete. *Mater. J.* **1996**, *93*, 242–245.
24. Guo, J.J.; Wang, K.; Guo, T.; Yang, Z.Y.; Zhang, P. Effect of dry-wet ratio on properties of concrete under sulfate attack. *Materials* **2019**, *12*. [CrossRef]
25. Puertas, F.; Amat, T.; Fernández-Jiménez, A.; Vázquez, T. Mechanical and durable behaviour of alkaline cement mortars reinforced with polypropylene fibres. *Cem. Concr. Res.* **2003**, *33*, 2031–2036. [CrossRef]
26. Zhang, Z.; Yao, X.; Zhu, H. Potential application of geopolymers as protection coatings for marine concrete I. Basic properties. *Appl. Clay Sci.* **2010**, *49*, 1–6. [CrossRef]
27. Hakkinen, T. The influence of slag content on the microstructure, permeability and mechanical properties of concrete, Part 2 Technical properties and theoretical examinations. *Cem. Concr. Res.* **1993**, *23*, 518–530. [CrossRef]
28. Garrabrants, A.C.; Sanchez, F.; Kosson, D.S. Leaching model for a cement mortar exposed to intermittent wetting and drying. *Aiche J.* **2003**, *49*, 1317–1333. [CrossRef]
29. Karozou, A.; Konopisi, S.; Paulidou, E.; Stefanidou, M. Alkali activated clay mortars with different activators. *Constr. Build. Mater.* **2019**, *212*, 85–91. [CrossRef]
30. ASTM International. *ASTM C188-95, Standard Test Method for Density of Hydraulic Cement*; ASTM International: West Conshohocken, PA, USA, 2016; pp. 1–2. [CrossRef]
31. ASTM International. *ASTM D248—11 Standard Practice for Classification of Soils for Engineering Purposes (Unified Soil Classification System)*; ASTM International: West Conshohocken, PA, USA, 2006; pp. 1–5. [CrossRef]
32. Nicolas, R.S.; Provis, J.L.; Bernal, S.A.; van Deventer, J.S.J. Alkali-activated slag cements produced with a blended sodium carbonate/sodium silicate activator. *Adv. Cem. Res.* **2015**, 1–12. [CrossRef]
33. Steudel, A.; Mehl, D.; Emmerich, K. Simultaneous thermal analysis of different bentonite–sodium carbonate systems: An attempt to distinguish alkali-activated bentonites from raw materials. *Clay Miner.* **2013**, *48*, 117–128. [CrossRef]
34. Stefanidou, M. Methods for porosity measurement in lime-based mortars. *Constr. Build. Mater.* **2010**, *24*, 2572–2578. [CrossRef]
35. Liliana Sofia, P.N. Drying Index as an In-Service Parameter of Renders Applied on Exterior Facades Supervisor: Prof a Inês dos Santos Flores Barbosa Colen, Co-supervisor: Eng° LuísMiguel Cardoso da Silva October 2012 2. Performance of Renders Facing the Water. 2012. Available online: https://fenix.tecnico.ulisboa.pt/downloadFile/395144992891/extended%20abstract.pdf (accessed on 22 July 2020).
36. *EVS EN 16322-13 Conservation of Cultural Heritage—Test Methods—Determination of Drying Properties*; European Standard: Brussels, Belgium, 2013.
37. Ismail, I.; Bernal, S.A.; Provis, J.L.; San Nicolas, R.; Brice, D.G.; Kilcullen, A.R.; Hamdan, S.; Van Deventer, J.S.J. Influence of fly ash on the water and chloride permeability of alkali-activated slag mortars and concretes. *Constr. Build. Mater.* **2013**, *48*, 1187–1201. [CrossRef]
38. Grilo, J.; Faria, P.; Veiga, R.; Santos Silva, A.; Silva, V.; Velosa, A. New natural hydraulic lime mortars—Physical and microstructural properties in different curing conditions. *Constr. Build. Mater.* **2014**. [CrossRef]
39. Najafi Kani, E.; Allahverdi, A.; Provis, J.L. Efflorescence control in geopolymer binders based on natural pozzolan. *Cem. Concr. Compos.* **2012**, *34*, 25–33. [CrossRef]
40. *Earth Plasters—Terms and Definitions, Requirements, Test Methods*; DIN 18947; Deutsche Norm: Berlin, Germany, 2013.
41. Papayianni, I.; Stefanidou, M. Microstructural Analysis of Old Mortars of Byzantine Period. In Proceedings of the International Conference STREMAH, Bologna, Italy, 28–30 May 2001; pp. 45–52.
42. Eires, R.; Camões, A.; Jalali, S. Enhancing water resistance of earthen buildings with quicklime and oil. *J. Clean. Prod.* **2017**, *142*, 3281–3292. [CrossRef]
43. Yaghoubi, M.; Arulrajah, A.; Disfani, M.M.; Horpibulsuk, S.; Darmawan, S.; Wang, J. Impact of field conditions on the strength development of a geopolymer stabilized marine clay. *Appl. Clay Sci.* **2019**, *167*, 33–42. [CrossRef]
44. Slaty, F.; Khoury, H.; Wastiels, J.; Rahier, H. Characterization of alkali activated kaolinitic clay. *Appl. Clay Sci.* **2013**, *75–76*, 120–125. [CrossRef]

45. Edward, G. *Nawy Concrete Construction Engineering Handbook*; CRC Press: New York, NY, USA, 2008; ISBN 978-0-8493-7492-0.
46. Liptay, G. *Atlas of Thermoanalytical Curves: (TG-, DTG-, DTA-Curves Measured Simultaneously)*; Erdey, L., Paulik, F., Paulik, J., Eds.; Heyden and Son: London, UK; New York, NY, USA, 1971; ISBN 0855010525.
47. Davidovits, J. *Geopolymer: Chemistry & Applications*; Geopolymer Institute: Saint-Quentin, France, 2011; ISBN 9782951482098.
48. Pacheco-Torgal, F.; Castro-Gomes, J.; Jalali, S. Alkali-activated binders: A review. Part 2. About materials and binders manufacture. *Constr. Build. Mater.* **2008**, *22*, 1315–1322. [CrossRef]

materials

MDPI

Article

Effect of the Combination of Superabsorbent Polymers for Autogenous Shrinkage Control with Steel Fibers of High-Performance Concrete under Uniaxial Tension Using DIC

Karyne Ferreira dos Santos [1,*], António Carlos Bettencourt Simões Ribeiro [2], Eugênia Fonseca da Silva [1], Manuel Alejandro Rojas Manzano [3], Leila Aparecida de Castro Motta [4] and Romildo Dias Toledo Filho [5]

1 Postgraduate Program in Structural Engineering and Construction–PECC, Department of Civil and Environmental Engineering, SG-12 Building, University of Brasilia–UnB, 70910-900 Brasília-DF, Brazil; eugenia@unb.br

2 Materials Department, Nacional Laboratory for Civil Engineering–LNEC, Av. do Brasil 101, 1700-075 Lisbon, Portugal; bribeiro@lnec.pt

3 Department of Civil and Industrial Engineering, Faculty of Civil Engineering and Science, Pontifical Xavierian University of Cali, Calle 18 N° 118-250, Edificio Central, 2do. Piso, Cali 760031, Colombia; alejandro.rojas@javerianacali.edu.co

4 Faculty of Civil Engineering–FECIV, Federal University of Uberlândia–UFU, Av. João Naves de Ávila, 2121, Santa Mônica Campus 1Y, 38400-902 Uberlândia-MG, Brazil; lacastro@ufu.br

5 The Alberto Luiz Coimbra Institute for Graduate Studies and Research in Engineering–COPPE, Department of Civil Engineering, University of Rio de Janeiro–UFRJ, Av. Horácio Macedo 2030, Ilha do Fundão, 21941-972 Rio de Janeiro, RJ, Brazil; toledo@coc.ufrj.br

* Correspondence: karyne.ferreira@aluno.unb.br; Tel.: +351-919282525

Received: 1 September 2020; Accepted: 13 October 2020; Published: 17 October 2020

Abstract: This paper presents a study of the effect of a superabsorbent polymer (SAP) for autogenous shrinkage control on the uniaxial tensile behavior of steel fiber reinforced concrete (SFRC). The use of fibers and SAP potentially increases the durability of the concrete, preventing cracking by autogenous shrinkage and enhancing post-cracking behavior. Furthermore, SAP can provide further hydration for self-healing purposes and improve the ductility of the SFRC. In order to evaluate the effect of the addition of SAP in SFRC, dog-bone SFRC specimens with different dosages of superabsorbent polymers were cast and tested under uniaxial tension. The digital image correlation (DIC) technique was used to understand the effect of SAP on the steel fibers' crack-bridging mechanisms. Surface strains and crack openings were inferred using the DIC technique. The effect of SAP and fibers on fresh and hardened concrete was individually investigated by flow tests and compressive strength tests. Autogenous shrinkage was measured in plain concrete to investigate the minimum SAP content required to mitigate autogenous shrinkage of 0.3%. The use of 0.3% SAP was also sufficient to reach multiple cracking behavior. This content of SAP completely suppressed the autogenous shrinkage with minimal side effects on compressive strength. An analytical formulation for the tensile behavior of SFRC was developed using the variable engagement model, presenting a mean correlation of R^2 of 0.97 with the experimental results.

Keywords: superabsorbent polymer; steel fibers; high-performance fiber reinforced concrete; digital image correlation; autogenous shrinkage; tensile behavior

1. Introduction

The present investigation focused on the tensile behavior of steel fiber reinforced concrete (SFRC) with superabsorbent polymer (SAP). In general, SAP is used to control autogenous shrinkage without significantly affecting strength and workability [1–9]. Fibers are used in concrete for cracking control [10–15].

The synergic effect of SAP and fibers has been mainly studied with synthetic fibers [16–22]. The studies of concrete with synthetic fibers and SAP shows that the presence of the two constituents modifies the tensile behavior of the composite, and improvements on the durability of the material can be achieved [23–25]. The improvements are due to the reduction in the size of the crack by increasing the number of cracks. This multi-cracking behavior is obtained not only from the presence of fibers, but also by the multi-flaws provided by SAP [16].

Studies of concrete with fibers and SAP are scarce. The study presented by [26] showed promising results on the use of SAP to densify the interfacial transition zone and reduce micro-cracks around the fibers. Wang et al. [27] conducted a splitting tensile test study with SAP and steel fiber for cellular concrete applications, although the concrete was not high strength. None of these studies were performed with concrete direct tension.

The present document intends to contribute to the understanding of the overall behavior of the composite subjected to direct tension by analyzing the crack formation and crack pattern with the digital image analysis technique, which allows for simultaneously observing the behavior of a set of fibers and flaws produced by the SAP addition, and the interaction between them. This type of behavior cannot be observed when testing only one fiber in tension [26] nor with the splitting test [27].

The investigation also deals with other specific subjects: (1) Investigate the maximum SAP content to be incorporated in high strength concrete (HSC) and steel fiber reinforced concrete (SFRC) regarding the loss of compressive strength and workability; (2) Investigate a minimum SAP content that controls the autogenous shrinkage of HSC; (3) Characterize the tensile properties of SFRC with different SAP contents and complement the regular analysis with the crack pattern with full-field strain measurement using DIC; and (4) Develop an analytical model for predicting the tension behavior of HSC with varying SAP dosage.

2. Materials and Methods

2.1. Materials

Portland cement of high initial strength conforming to [28] Type CPV-ARI (CIPLAN, Brasília, Brazil) was used for all mixes in this study. A silica fume, of the non-densified type, meeting the requirements of the standard [29] was supplied by the national company Silmix (Breu Branco, Pará, Brazilcountry). The physical and chemical properties of the cement and silica fume are shown in Table 1. Locally available sand of the Corúmba River, with the maximum size of 4.75 mm and gradation conforming to [30] standard usable zones, was used. The sand fineness module was 2.73, and the specific mass was 2.65 kg/dm^3. A water reducer of high-efficiency, superplasticizer ADVA CAST 129 from Grace Company (Sorocaba, São Paulo, Brazil), based on polycarboxylates, was used to maintain the fluidity of the mortar within a fixed range for all mixes.

Table 1. Basic chemical composition of cement and silica fume.

Component	Cement (%)	Silica Fume (%)
SiO_2	24.41	93.95
Al_2O_3	7.09	0.16
CaO	53.74	0.74
Fe_2O_3	3.02	0.27
MgO	4.36	0.86
SO_3	3.28	-
Na_2O	0.28	0.37
K_2O	0.77	0.84
Density (g/cm^3)	3.03	2.21

2.1.1. Steel Fibers

Steel fibers of DRAMIX OL 13/.16 mm from BEKAERT (Zwevegem, Belgium), illustrated in Figure 1, with the content of 1.28% in volume, were used. These were made of smooth steel of 13 mm in length, 0.16 mm in nominal diameter, aspect ratio of 81.25, Young's modulus of 200 GPa, and tensile strength of 2750 MPa.

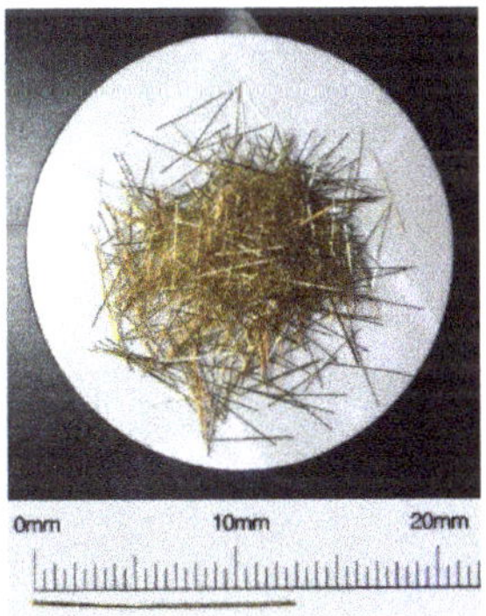

Figure 1. Dramix Ol 13/.16 steel fiber from Bekaert Company.

2.1.2. Superabsorbent Polymers (SAP)

The superabsorbent polymer used was an acrylic acid/acrylamide (Technical University of Denmark, Lyngby, Denmark) with covalent cross-links produced by the reverse suspension polymerization technique, with a mean particle size (D_{50}) of 66.3 μm. It has an absorption in the cementitious medium of 18 g of water for 1 g of SAP, and the density of 1.456 g/cm^3. The particle size curve of the SAP characterized by the portion of particles with diameters smaller than 27.3 μm is 10% (D_{10}), and the portion of particles with diameters below 101.7 μm is 90%. (D_{90}). It was developed for particular use in a high alkaline environment such as the cement suspension. It was supplied by Prof. Ole Mejlhede Jensen and developed at the Technical University of Denmark (DTU). In Figure 2, the polymer is presented in the dry and swollen state by using the scanning electron microscope and digital microscope H1000X / S1000X, respectively.

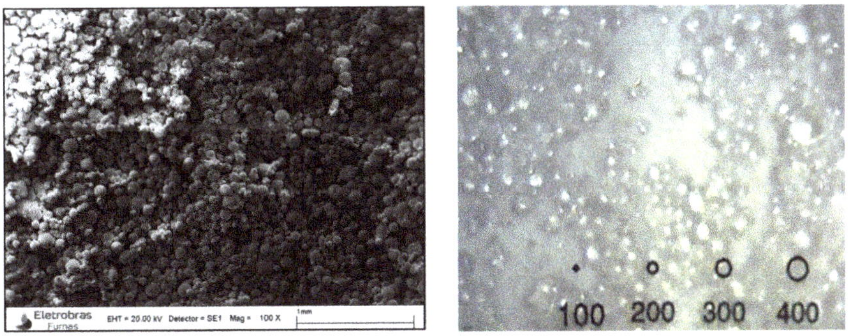

(**a**) (**b**)

Figure 2. Superabsorbent polymer: (**a**) dry; (**b**) swollen. The spheres have a diameter in micrometers.

The water absorption capacity of SAP was measured by the slump flow consistency method and by means of graduated cylinders proposed by [31]. Absorption of SAP was tested for deionized water and cement filtrate since the absorption of the mixing water by the SAP takes place with the addition of water to the remaining components of the concrete mix, in a way that the activation of SAP takes place in a highly active electrochemical environment. The absorption capacity in deionized water was 96 g of water per 1 g of SAP.

The smaller absorption of cement filtrate, when compared to absorption of water, caused by calcium and magnesium ions present in the cementitious fluid (increase the cross-linking in SAP) [32,33], is an advantage for self-healing purposes. This effect leads to a smaller void creation in the hardened mortar and, after crack formation, when the SAP gets into contact with clean water, it will have swollen larger and temporarily seal the crack. The saturated environment can then provoke self-healing of the crack by hydration of the anhydrous cement under certain conditions and quantities of SAP.

2.1.3. Mix Proportions

In order to investigate the combined effect of the SAP and steel fibers, eight concrete mixtures were prepared according to Table 2. These included three reference compositions, two without SAP or fibers (REF-035 and REF-040), and another with fibers and without SAP (REF-035F). The remaining five mixtures included SAP or SAP and fibers.

Table 2. Composition of the concrete mixtures, values in kg/m^3.

Mixture (kg/m^3)	Cement	Silica	Sand	SAP	Fiber	Superplasticizer	$w/c_{(basic)}$	$w/c_{(total)}$ [1]	Flow (mm)
REF-035	675.8	67.6	1306.5	0	0	11.15	0.35	0.35	183
REF-040	654.2	65.4	1264.5	0	0	8.50	0.40	0.40	200
SAP-0.2	675.8	67.6	1306.5	1.35	0	12.50	0.35	0.38	183
SAP-0.3	675.8	67.6	1306.5	2.03	0	12.50	0.35	0.40	180
REF-035F	675.8	67.6	1306.5	0	100	13.50	0.35	0.35	307
SAP-0.2F	675.8	67.6	1306.5	1.35	100	13.50	0.35	0.38	243
SAP-0.3F	675.8	67.6	1306.5	2.03	100	13.50	0.35	0.40	230
SAP-0.6F	675.8	67.6	1306.5	4.06	100	13.50	0.35	0.46	208

[1] Including water absorbed by SAP.

All mixtures presented low w/c ratio and high strength. The total water–cement (w/c) ratio was set at 0.35 or water–binder of 0.32. A reference mixture with w/c of 0.4 was also prepared for comparison with the mixture where additional water was placed due to SAP. In the mixtures without fibers, the content of the superplasticizer was adjusted in order to reach the required workability.

The mortar compositions all had the same proportions of binder and aggregate and the content of SAP varied from 0 to 0.6%. In the mixtures with fibers, the amount of superplasticizer was fixed at 2% and the fibers content was set constant at 1.28% in volume.

The mortars were prepared in a mechanical mixer (Solotest São Paulo, Brazil) under the following steps. The first step was dry mixing the cement, the SAP, and the sand for 5 min at a slow speed. The second step was the water and superplasticizer addition for 2 min, and then mixing at a high speed for 5 min. Then, the mixer was stopped, the edges of the bowl were scraped, and followed by mixing for another 3 min at high speed. In the correspondent mixtures, the fibers were then added constantly in motion for 5 min at slow speed, preventing the ball bearing effect, then the edges were scraped again for 3 min. Finally, the mortar was mixed for 5 min at high speed to ensure a uniform consistency. For proper compaction, the mixes were cast into molds placed on a vibrating table. The specimens were moved to a room with 100% relative humidity and 22 ± 1 °C. They were demolded after 24 h and continued in this room until the testing date.

2.2. Experimental Program

2.2.1. Measurement of Flowability and Compressive Strength

Slump flow tests were carried out right after mixing, following the procedure [34]. The fresh mortar was poured into a Hagerman cone (larger base diameter of 100 mm, height of 60 mm, and a smaller base of 70 mm) into two layers. Each layer was tamped 10 times evenly distributed, and then the cone was gently removed, and the spread was measured. The largest diameter was measured along with the diameter in the perpendicular direction.

Compressive tests were carried out according to the standard [35]. Cylinder specimens (Ø 50 mm × L 100 mm) were used, and an average of three tested samples was reported. The base of the specimens was ground to ensure a smooth and plane surface. The test adopted a Microcomputer-controlled electromechanical universal test systems (MTS), and the loading rate was 0.5 mm/min.

2.2.2. Autogenous Shrinkage

The autogenous shrinkage was assessed according to the methodology developed by [36], which is a modified method of [37] using the [38] apparatus, shown in Figure 3. The test consists of the measurement of the deformation of mortars, cast in a prismatic mold with nominal dimensions of 7.5 × 7.5 × 28.5 cm. Each mold was previously prepared, with polystyrene layers inside to decrease friction, allowing the concrete to move freely. Additionally, two threaded metal pins were placed in the extremities of the mold aligned with the specimen's longitudinal axis. Strain gauges are couples in these pins and connected to a computer to collect and store the data. The distance between these pins is called G, and the measurement of this distance determines the shrinkage.

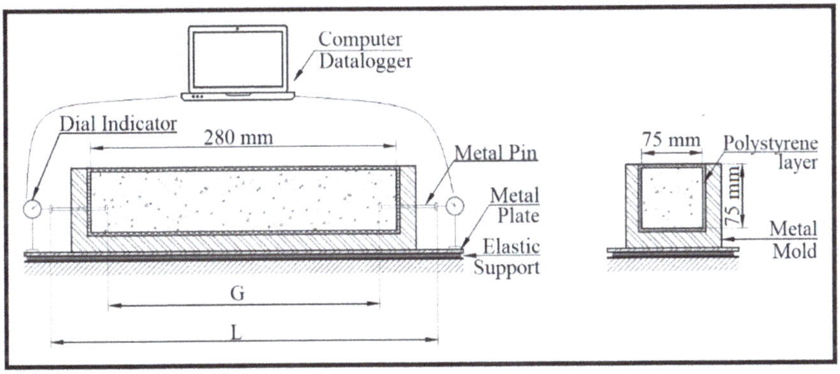

(**a**)

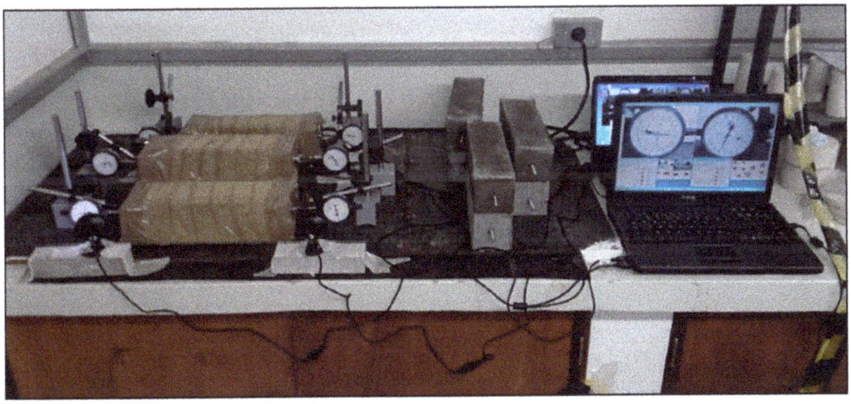

(**b**)

Figure 3. Autogenous shrinkage test: (**a**) test setup scheme and (**b**) on going test.

After casting, to prevent the loss of water to evaporation, the mold with the mortar was wrapped with multiple layers of polystyrene and adhesive tape. The weight of the mold with mortar was measured, and they were stored in a chamber with 50 ± 2% of humidity and 21 ± 2 °C for 28 days, where the straingauges continuously recorded the G distance.

2.2.3. Uni-Axial Testing Equipment with Digital Image Correlation (DIC) Arrangement

In order to verify the strain-hardening behavior of all HPC mixtures at 28 days, a series of direct tensile tests were performed following [39]. An INSTRON (Instron, São José dos Pinhais, Paraná, Brazil) electronic universal testing machine with displacement control and load capacity of 100 kN was used at a constant speed of 0.3 mm/min. The loading force was measured on a computerized data recording system as for the strain was measured by two linear variable displacement transducers (LVDT, HBM, São Paulo, Brazil), placed on both sides of the specimen. Additionally, the strain was also measured by digital image correlation (DIC, Correlated Solutions, Irmo, United States of America). The tensile setup and geometry of the specimen are shown in Figure 4.

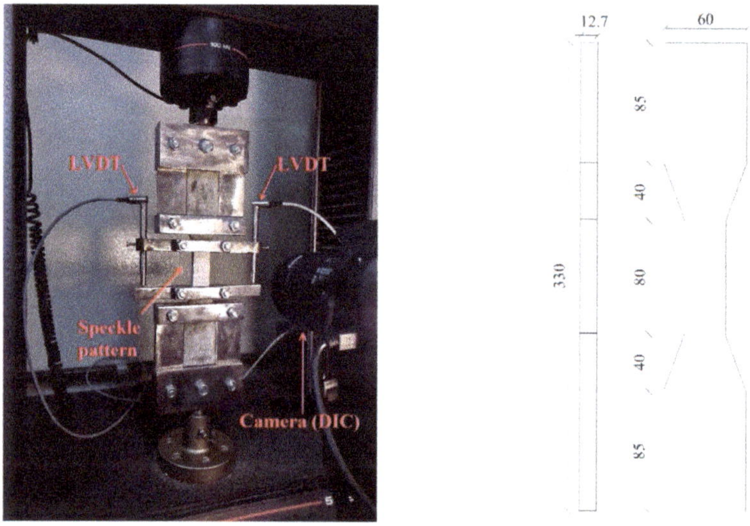

(a) (b)

Figure 4. The direct uniaxial tensile test: (**a**) test setup with DIC and (**b**) specimen dimension (mm).

The DIC was used to analyze the crack pattern and the continuous deformation of the specimen concerning the applied load to assist the investigation of the strain hardening behavior with the addition of the SAP and steel fibers. The DIC setup included a digital camera, sufficient light in the specimen, and the sample preparation. The sample preparation consisted of painting the specimen with white paint and aleatory and heterogeneously painting dark dots to form distinct patterns that can be recognized by the image correlation program. For the program to work well, these dots, the camera configuration, and position were adjusted for each dot to have four to six pixels in the picture. Before the test began, a calibration image was taken for each test to convert the pixel scale to a millimeter scale. The camera, testing machine, and LVDT were all started simultaneously so that the data could be correlated later on.

Image processing software VIC-2D was used in this study to correlate different images and the corresponding deformations. The software relates the deformed images by dividing the area of interest (AOI) into many small regions, called subsets, where each subset is unique and identified by the program via the dots pattern. The program detects the change in the first image subset, set as a reference, with the images taken during the test and calculates the distance, which is used to calculate the full-field displacement and strains by interactive techniques [40].

Due to the brittle behavior, it is not possible to obtain the strain–stress curve of plain concrete specimens with the direct tension test with displacement control. In this case, the splitting tensile strength test was carried out based on [41]. This method was used mainly to evaluate the effect of SAP incorporation on tensile strength. Three cylindrical samples (Ø 100m × 200 mm) were used to measure the splitting tensile strength of concrete on the seventh and 28th day. The machine for testing the splitting tensile strength was a MTS microcomputer-controlled electromechanical universal test systems, with the loading speed of 0.2 MPa/s.

3. Results and Discussions

3.1. Influence on Mortar Flow

The flow values of each mix and the corresponding superplasticizer content are presented in Table 2. The results of mixture REF-35 and SAP-02, without fibers, showed that to keep the same flow (183 mm) in both mixtures, 0.2% of SAP required an increase of 12% in the superplasticizer dosage. Mixture SAP-0.3, with the same superplasticizer dosage but another 0.1% of SAP, led to a reduction in the flow (180 mm). This effect on flow is expected, as the increase of SAP content also increased the particle concentration, but the effect was limited and manageable. Paiva [42] proposed that a water-reducing agent could be efficient at maintaining the flowability since SAP particles do not interfere with the plasticizer chains.

The adverse effects of steel fibers on the workability of concrete have been widely discussed by [43], and reinforced by recent publications with high strength concrete such as [44,45]. In order to have good workability after the incorporation of fibers, the superplasticizer content was increased to 2% of the cement weight, which led to a reference flow of 307 mm (REF-035F).

The effect of SAP on the flow can be more clearly seen in the last four mixtures of Table 2, since the water content and the superplasticizer dosage were kept constant. The flow progressively decreases as the addition of SAP increases. The reduction was 20.85%, 25.1%, and 32.2% for 0.2%, 0.3%, and 0.6% of SAP, respectively.

The decrease in the workability suggests that the additional water provided to fill the SAP is actually being absorbed by the SAP, and the flow decrease is due to lesser free water per unit of volume. Some extra water absorption by the SAP may also be occurring, these findings were supported by [46]. The opposite effect was reported by [47–49], which leads to a gap in the literature as the effect of the SAP in the workability is not entirely understood. The authors in [4,31,50] explained that the broad diversification of results would depend on the methodology used to accurately estimate the amount of water absorbed by SAP in the cementitious environment. The over or underestimated amount of additional water can affect the workability and the total w/c. Another hypothesis of the loss of workability was provided by [51], who believed that the swollen SAP particles behaved as soft aggregates and offered a restraining effect in the rheology of the mortar. Nevertheless, all the mixtures maintained good workability and no signs of segregation.

3.2. Compressive Strength

A summary of the compressive strength results for the fiber reinforced concrete and the plain concrete at 28 days is provided in Figure 5 and Table 3. Each compressive strength result is the average of six specimens.

Increases in SAP dosage for the same $w/c_{(basic)}$ tended to almost linearly decrease the compressive strength for both mixture series (with and without fiber reinforcement). However, comparing the compressive strength of mixtures with same $w/c_{(total)}$, REF-040, and SAP-0.3, the values were similar. This indicates the major role of the total volume of pores on strength, regardless of the presence or absence of SAP.

Strictly speaking, in order to individually evaluate the influence of the SAP, a specific reference mixture should be manufactured containing the same total w/c ratio, but for the purposes of the present work, this information was not considered necessary.

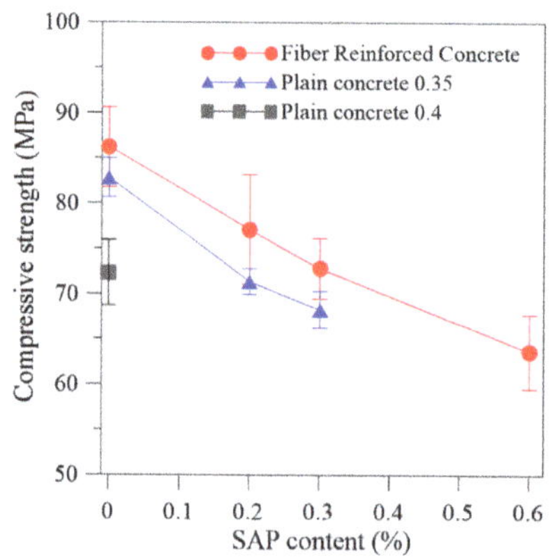

Figure 5. Compressive strength results at 28 days of the fiber reinforced concrete of the $w/c_{(basic)}$ of 0.35; plain concrete without fiber reinforcement with the $w/c_{(basic)}$ of 0.35, and the plain concrete without fiber reinforcement with the $w/c_{(basic)}$ of 0.40.

Table 3. Mechanical properties of the high strength concrete and steel fiber reinforced concrete.

Mixture	Mean Compressive Strength (MPa)	Strength Reduction Considering the $w/c_{(basic)}$ (%)	Mean Tensile Stress (MPa–First Cracking	Ratio of Tensile to Compressive Strength
REF-035	82.80 ± 2.1	-	5.39	0.065
REF-040	72.32 ± 3.6	-	4.92	0.068
SAP-0.2	71.36 ± 1.4	13.82	5.04	0.071
SAP-0.3	68.24 ± 2.0	17.58	5.07	0.074
REF-035F	86.20 ± 4.4	-	6.01 ± 0.45	0.070
SAP-0.2F	77.09 ± 4.8	10.57	4.59 ± 0.25	0.060
SAP-0.3F	72.78 ± 4.8	16.47	5.92 ± 0.04	0.081
SAP-0.6F	63.60 ± 4.1	26.22	3.23 ± 0.23	0.051

This subject leads to a discussion presented in the literature that has yet to be enlightened. Many authors have reported the loss of compressive strength in the literature. However, as published by [50], this loss of strength could be provoked by the excess of water addition due to a misleading measure of the SAP absorption in the cementitious environment. Simple methods have been used to estimate the SAP absorption capacity and could be overestimating the water of absorption, increasing the water in the mixture, and lowering the compressive strength. However, this hypothesis was not validated by the slump results obtained in this research. If there were an excess of water in the fresh state, the slump would not decrease since it would facilitate the workability.

Another explanation and more common for the loss of strength, supported by [4,51–53], and others, determined that the initial swelling of SAP creates a reasonable amount of macropores due to the SAP swelling. Snoeck et al. [51] further explained that in the fresh mix, macropores spontaneously form and become occupied by swollen SAP particles. Following this, the concrete pore solution is consumed by cement hydration, which decreases the ambient moisture where the SAP is located. Afterward, SAP slowly releases the inside water, causing the SAP to shrink. After SAP voids form, they result in increases in the total porosity of the concrete system. However, some studies have reported a straight gain due to effective internal curing, where the later hydration of the cement provided by the SAP entrapped water densified the pore structure, which was not the case in this study.

The addition of steel fibers in the mixture increased the compressive strength in 4.11%, 8.03%, and 5.51% when compared with the reference without fiber reinforcement and $w/c_{(basic)}$ of 0.35 for the 0.2%, 0.3%, and 0.6% of SAP incorporation, respectively. The 0.6% of SAP was determined to be the highest SAP incorporation for this mix design. Given that, according to [54,55], the lower limit of strength to be classified as HSC is 55 MPa, for this research, it was stipulated to reach a minimum value of 60 MPa so that the concrete can be classified as high strength.

3.3. Autogenous Shrinkage

The deformation measured in the test was considered as autogenous shrinkage since minimum moisture exchange occurred between the specimens and environment due to the coat of aluminum and plastic tape applied to the specimens before starting the test.

The scope of the research was to find the content of SAP that could mitigate or control autogenous shrinkage. Therefore, the worst-case scenario was to carry out the test without fiber reinforcement. The autogenous shrinkage for the mixtures without fiber reinforcement up to 28 days are shown in Figure 6. Each value represents an average of three specimens. As displaced in Figure 6, the reference, REF-035, presented autogenous shrinkage much higher than that of ordinary concrete, and increased significantly in the first seven days due to the absence of coarse aggregate and low water/binder ratio. REF-035 presented a maximum deformation of 424 μm/m and an initial expansion of 106 μm/m, which was overcome by autogenous shrinkage after 10 h.

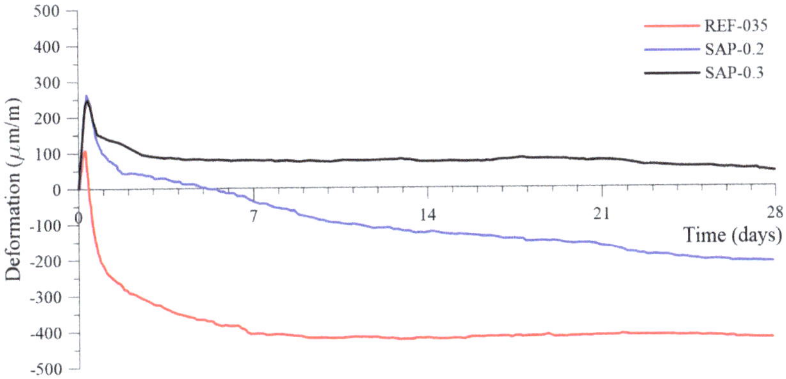

Figure 6. Autogenous strain (um/m) for cement mortar mixtures without fibers $w/c_{(total)}$ of 0.35, with SAP additions of 0.2% and 0.3%, determined from time zero up to 28 days.

The general trends of the curves in Figure 6 tended to be stable after 10 days. When studying the SAP-containing mixtures, the percentage of 0.3% completely mitigated the autogenous shrinkage and presented a maximum expansion of 248 μm/m at six hours after setting. The shrinkage did not counterbalance the expansion and, after 28 days, still presented 42 μm/m of expansion. The expansion phenomenon is not yet fully understood, but there are several attempts at explanation, for example, involving expansive pressure by forming hydration products (the high MgO content of the cement used may be a source of early expansion). This can be beneficial for some prestressed applications since the concrete compressive strength is higher than its tensile strength. The material is likely to withstand the maximum compression efforts induced by the expansion. Additionally, this expansion can be helpful and contribute to preventing cracking from drying shrinkage.

The content of 0.2% of SAP addition reduced 90% of the autogenous shrinkage compared to the reference at the age of seven days and reduced 50% at 28 days. It also presented an expansion of 262 μm/m at four hours after time 0. The authors in [1,2,56] described the water releasing mechanism of the SAP after setting of the cement-based material, which explained the reduction of autogenous shrinkage. The incorporation of SAP leads to the formation of controlled water-filled microscope inclusions, which prevent internal moisture evaporation from compensating water loss for curing, promote the hydration of unhydrated cement, and reduce the autogenous shrinkage.

The use of more than 0.3% of SAP addition is considered to ensure autogenous shrinkage control, with beneficial properties such as the expansion, which can avoid cracking and produce a more durable concrete, as seen in [57].

3.4. Tensile Properties

For the unreinforced specimens, the splitting tensile test was performed, and the tensile strength with the ratio of tensile to compressive strength for all mixtures is presented in Table 3. The specimens failed as expected, releasing almost all the energy soon after the peak load. The average tensile strength of the specimens without fiber reinforcement was 5.11 ± 0.2 MPa.

Typical stress-strain/load-displacement curves of the developed SFRC with SAP particles at 28 days are presented in Figure 7a. A diagrammatic sketch of the strain-softening behavior presented by [10] is shown in Figure 7b, who classified the composites based on their tensile response. The parameters regarding Naaman (2006), chosen to characterize the tensile behavior and to implement the analytical model described in the next section, were: first structural cracking stress (σ_{cc}) and force (F_{cc}); first structural cracking strain (ε_{cc}) and displacement (d_{cc}); maximum post-cracking stress (σ_{pc}) and force (F_{pc}); crack opening (w_{pc}); and tension toughness index (TTI), as presented in Table 4. Before the crack opening, the acquired displacement was calculated as a strain of the composite; after the first crack, it was evaluated as a crack opening.

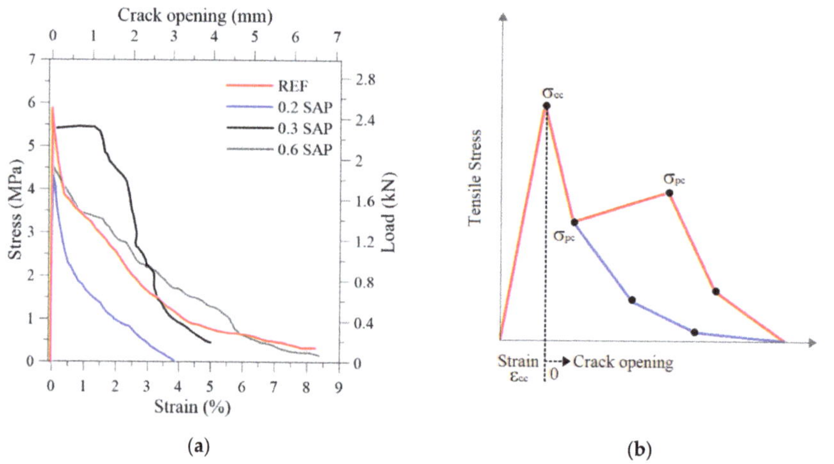

Figure 7. Stress x strain curve of: (**a**) SFRC with varying SAP content; (**b**) typical curves of strain softening behavior.

Table 4. Tensile properties of SFRC specimens at 28 days.

Mixture	F_{cc} (kN)	F_{pc} (kN)	σ_{cc} (MPa)	σ_{pc} (MPa)	d_{cc} (mm)	ε_{cc} (%)	w_{pc} (mm)	TTI (MPa)
REF-035F	2.45	1.63	5.87	3.86	0.023	0.029	0.325	0.1336
SAP-0.2F	1.85	0.99	4.29	2.31	0.050	0.064	0.404	0.0491
SAP-0.3F	2.38	2.26	5.62	5.35	0.035	0.044	1.169	0.1623
SAP-0.6F	1.92	1.41	4.50	3.31	0.038	0.048	1.235	0.1491

The main aspect to be witnessed by the SAP incorporation was the increase in the ductility of the composite. This behavior could be better observed by the DIC analysis shown in Figure 8 as the crack pattern of the SFRC. According to [10], the higher the strength, the lower the strain at the peak stress. This phenomenon could be observed in this experiment. Therefore, the general trade-off that exists in most materials between strength and ductility also applies to the developed composites. The tension toughness index (TTI) is a measurement of toughness calculated by area under the stress × strain curve until the specimen ultimately failed. In general, higher tension toughness (or energy absorption) was tightly related to higher SAP content, except for the SAP-0.2F. Compared to the reference, the TTI for SFRC increased by 21% for SAP-03F, 12% for SAP-06F, and decreased 37% for SAP-02F.

Mix	First Crack (A)	Crack Propagation (B)	Moment in the graphic
REF-035F	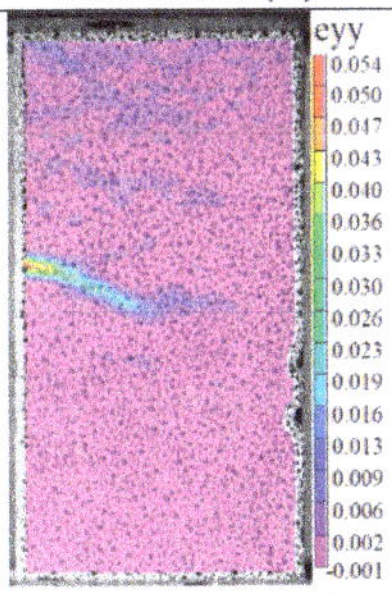 f_a = 2.45 kN, d_a = 0.023 mm σ_a = 5.87 MPa, ε_a = 0.00029	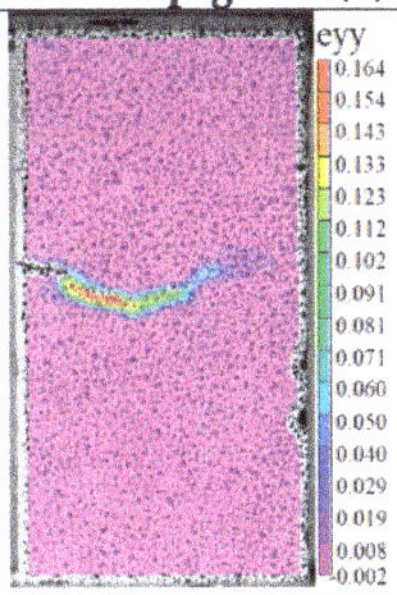 f_b = 2.25 kN, d_b = 0.075 mm σ_b = 5.33 MPa, ε_b = 0.00095	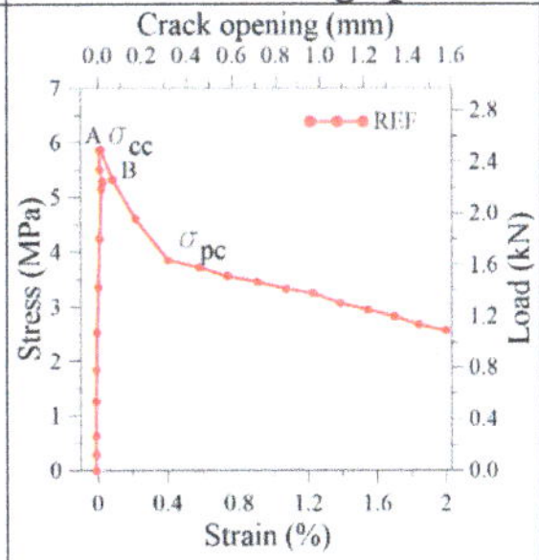 σ_{cc} = 5.87 MPa, ε_{cc} = 0.00029 σ_{pc} = 3.86 MPa, w_{pc} = 0.325 mm
SAP-0.2F	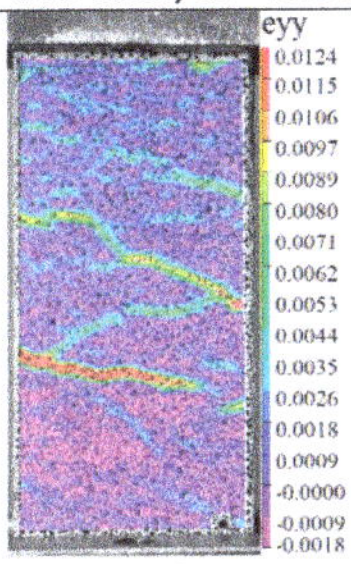 f_a = 1.84 kN, d_a = 0.050 mm σ_a = 4.29 MPa, ε_a = 0.00064	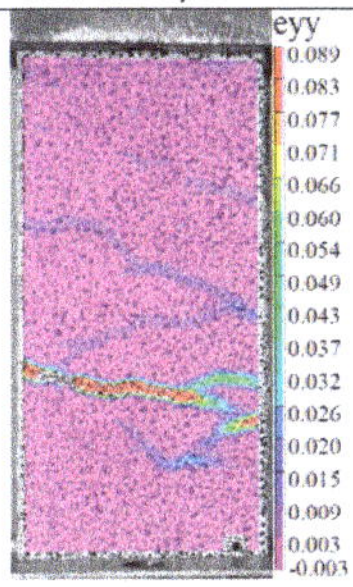 f_b = 1.54 kN, d_b = 0.139 mm σ_b = 3.60 MPa, ε_b = 0.00177	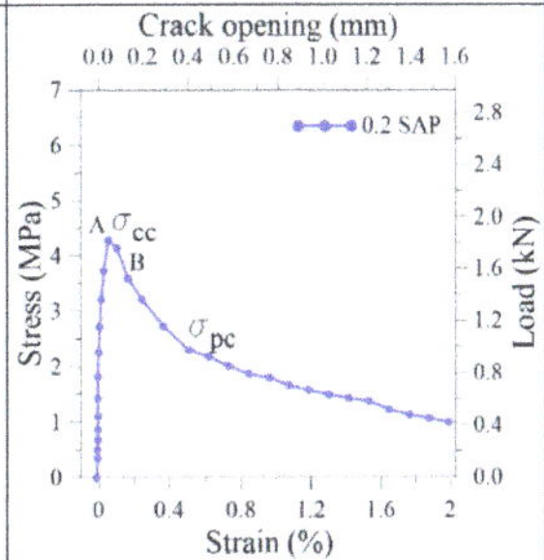 σ_{cc} = 4.29 MPa, ε_{cc} = 0.00064 σ_{pc} = 2.31 MPa, w_{pc} = 0.404 mm
SAP-0.3F	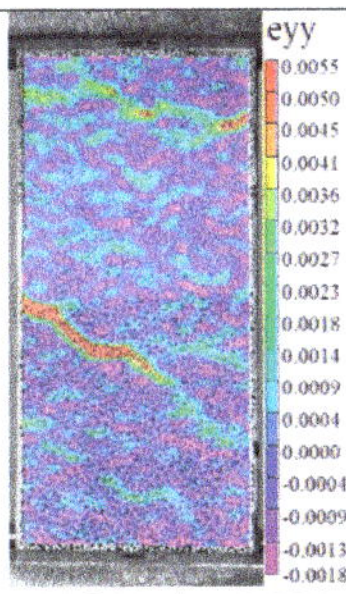 f_a = 2.38 kN, d_a = 0.035 mm σ_a = 5.62 MPa, ε_a = 0.00044	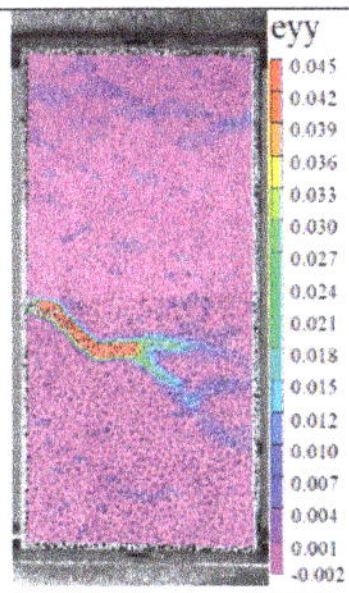 f_b = 2.29 kN, d_b = 0.056 mm σ_b = 5.41 MPa, ε_b = 0.00071	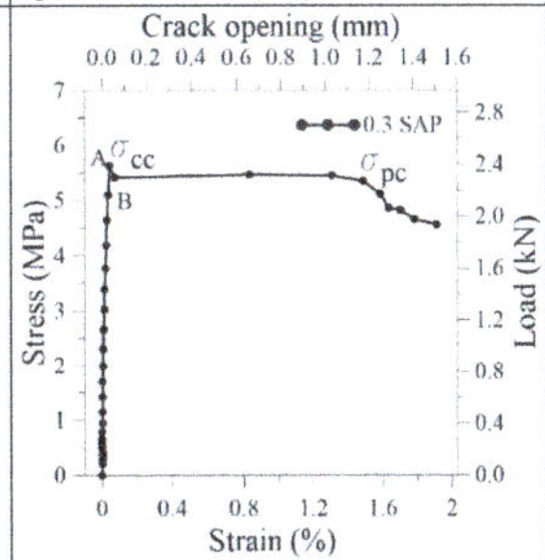 σ_{cc} = 5.62 MPa, ε_{cc} = 0.00044 σ_{pc} = 5.35 MPa, w_{pc} = 1.169 mm
SAP-0.6F	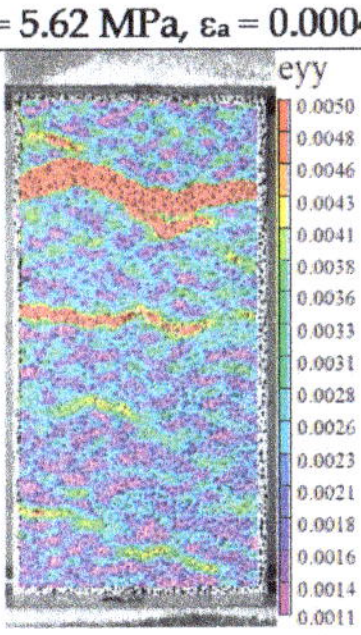 f_a = 1.92 kN, d_a = 0.038 mm σ_a = 4.50 MPa, ε_a = 0.00049	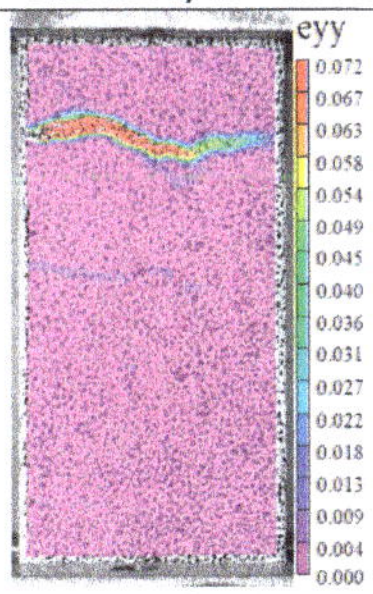 f_b = 1.48 kN, d_b = 0.674 mm σ_b = 3.48 MPa, ε_b = 0.00859	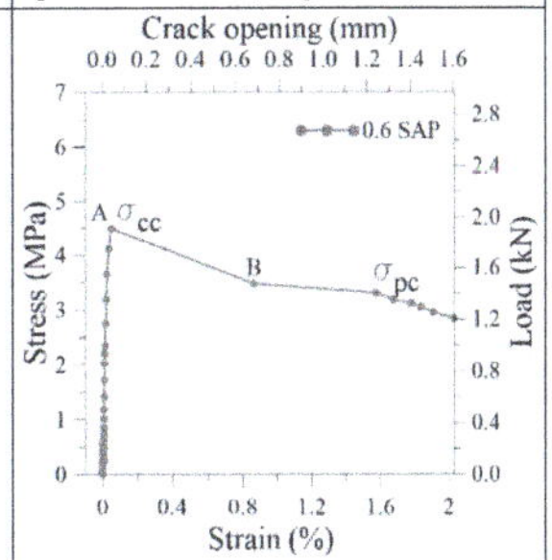 σ_{cc} = 4.50 MPa, ε_{cc} = 0.00048 σ_{pc} = 3.31 MPa, w_{pc} = 1.235 mm

Figure 8. Morphology of cracking apparent on the face of a specimen of SFRC, where A is the first cracking point, and B is the crack propagation, both presented in the respective graphic for each mix.

From Figure 8, the images of the DIC in the first cracking point were named A. The point soon after the cracking occurrence was named B. These points were correlated between the camera and the data acquisition. It was possible to observe that at the first cracking stress (σ_{cc}), point A, the composite without SAP produced one transverse localized crack and propagated later on at point B. As for the composites with SAP, they produced a different damage mode, leading to the creation of several cracks in the first peak stress (σ_{cc}). Moreover, when the stress concentration opens a single crack, the cracking evolution, shown in point B, produces a different morphology with more branches, releasing more energy and in accordance with the TTI results.

The authors in [16] incorporated SAP to control and improve the performance of fiber-reinforced concrete with polyvinyl alcohol (PVA) fibers. SAP improved the ductility of the material by the insertion of a mechanical flaw. According to the micromechanical theory developed by [24,25], one of the criteria to increase the toughness and multiple cracking in the cementitious composite is when $\sigma_{cc} \leq \sigma_{pc}$. This criterion can be achieved by decreasing the strength of the matrix by inserting a flaw in the matrix. The matrix tensile strength equals the stress of the bond between the fiber and matrix, and the composite develops a more ductile behavior.

This behavior was intensely studied with PVA fibers. However, for steel fibers, it was observed that as soon as the specimen cracked on reaching the σ_{cc}, there was a sudden drop in stress resistance, leading to extensive cracking, widening before the stresses were transferred from the matrix to the fibers (σ_{pc}). Nevertheless, it was also noted from Figure 8 that the SAP incorporation enhanced the toughness of the composite as it increased the multiple-crack behavior when the difference of σ_{cc} and σ_{pc} was lower. The 0.3% SAP addition achieved the best behavior because the regain of strength after the first cracking was significant. The difference between the two stresses was 0.3 MPa. However, for the developed SFRC, the σ_{pc} was not higher than the σ_{cc}, and the multiple-cracking behavior was not achieved. Still, the fibers fulfilled their purpose to increase the toughness and produce a progressive, yet gradual decrease in the load-carrying capacity.

The overall trend of the tensile strength presented a decrease as the SAP addition increased. The mixture with a content of 0.2% SAP was the exception. Despite following the general tendency to decrease the tensile strength by the insertion of SAP, it presented, in both cases of the concrete with and without fiber reinforcement, a higher reduction than the ones with more SAP insertion. This behavior was not expected and no reasonable explanation was found since, for the results of compressive strength and slump, it was inside the pattern of decrease. As for the autogenous shrinkage, it also presented results inside the expected trend.

Liu, Farzadnia and Shi [27] along with Wang et al. [26], are the few articles investigating the tensile behavior of steel fiber reinforced concrete with superabsorbent polymers. Wang et al. [27] focused on investigating different steel fibers types and contents with an established SAP content in the mixture, as opposed to our research that fixed the fiber type and content to investigate the influence of different SAP additions.

As reported by [26], the addition of SAP increased the flexural to compressive strength ratio of their ultra-high performance concrete (UHPC). For the UHPC with small size SAP, the increase in the ratio regarding the reference was 8% with 0.3% of SAP and 22% with 0.6% of SAP. Despite the difference in the tensile measurement, flexural strength is an indirect measure of the tensile strength. This phenomenon could also be observed for the concrete without fiber reinforcement (Table 3). The ratio increased by 9% for the addition of 0.2% SAP and 14% for the increase with 0.3%. The increment in the tensile to compressive strength ratio means that the SAP's internal cure is more beneficial for the development of tensile than compressive strength. Liu, Farzadnia and Shi [26] raised the hypothesis that the addition of SAP increased the interstitial bonding strength of steel fibers. Additionally, this could be one of the critical factors in increasing the tensile to compressive results. However, for SFRC, only the 0.3% of SAP addition showed an improvement in the ratio of 16%; as for the other content, no improvement was observed.

Analytical Tensile Evaluation of SAP Incorporation

There are many analytical models to describe the behavior of fiber reinforced concrete, especially regarding steel fiber reinforcement, for example, [58–60]. It was not within the scope of this work to evaluate the best analytical model to describe the experimental obtained curves. Instead, a more contemporary and consolidated model was chosen to describe and predict the SFRC. The variable engagement model (VEM) has been extensively used to investigate fiber reinforcement, even in DIC analyses of direct tensile test such as in [61], more details of the model can be found in [62,63]. This model gives an approach for modeling strain softening behavior on uniaxial tension, where the fibers are randomly orientated in three dimensions.

For modeling the developed SFRC, the fibers were straight, so the mechanical anchorage was dismissed for the model. The total tensile stress developed by the model was composed by the sum of the stress of the matrix and the stress provided by the frictional bond between fiber and matrix, as exemplified in Figure 9a and Equation (1).

$$f_t = f_{ct} + f_{st} \tag{1}$$

where f_t is the total tension stress carried by the fiber reinforced concrete; f_{ct} is the stress carried by the matrix; and f_{st} is the stress carried by the fiber.

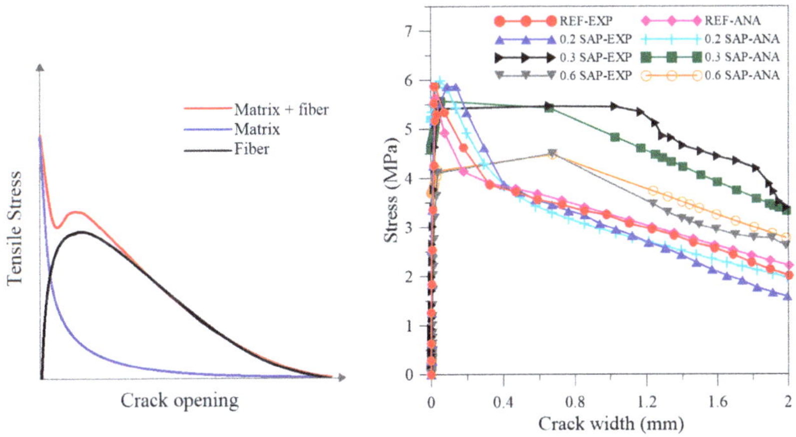

(a) (b)

Figure 9. (**a**) Typical stress versus crack opening after cracking for the fiber–matrix composite. (**b**) Analytical model with experimental results for the SFRC with SAP additions.

VEM considers that the embedded fiber is pulled out from the crack's side with the shorter embedded fiber length, and ignores the axial elastic deformation of the fibers. An exponential tension softening relationship is provided as:

$$f_{ct} = f_t' \cdot \exp(-c \cdot w) \tag{2}$$

where f'_t is the tensile strength of the specimens without fibers; w is the crack width of the specimen at a given load; and c is an attenuation factor for the concrete matrix undergoing tension decay after cracking. The c parameter was a variable to the model fitting into the experimental curves.

The stress carried by the fiber is given by:

$$f_{st} = K_f \alpha_f \rho_f \tau_b \tag{3}$$

where K_f is the global orientation factor; α_f is the aspect ratio; ρ_f is the volumetric ratio; and τ_b is the mean shear stress between the fiber and the matrix.

The aspect ratio is given by:

$$\alpha_f = \frac{l_f}{d_f} \quad (4)$$

where l_f is the length of the fiber and d_f the diameter, and the global orientation factor is given by:

$$K_f = \frac{(\tan^{-1}(w/\alpha))}{\pi}\left(1 - \frac{2w}{l_f}\right)^2 \quad (5)$$

where α is the engagement parameter, which for fiber composites with straight or end hooked steel fibers is of $\alpha = d_f/3.5$.

For the model to work without the fiber fracture, the following equation must be satisfied. For the present case, it was fulfilled, and the model without fiber fracture was chosen. Additionally, by observing the cracked section after the tensile test was performed, no fiber fracture was observed.

$$l_f < l_c = \frac{d_f}{2}\frac{\sigma_{fu}}{\tau_b} \quad (6)$$

where l_c is the critical fiber length and σ_{fu} is the ultimate tensile strength of the fiber.

The analytical model presented using Equations (1)–(6) was plotted with the experimental results in Figure 9b, where EXP designates the experimental and ANA represents the analytical curves.

Since the composites presented the same fiber content and type, in order for the model to fit the experimental results, only three parameters could vary: the parameter c that is dependent of the type of concrete; the stress of the fiber bond; and the tensile stress of each composite at 28 days. The tensile stress was provided by the results obtained. The other parameters were initially based on the literature and then modified to adjust the model to the experimental curves better.

The fitting of these parameters revealed that the bond between the fiber and matrix was enhanced with the SAP addition, except for the 0.2% as already discussed, to eventually be an out layer. The initial value set for the bond strength was 10 MPa, tested by [64]. The interpolation to better fit the experimental curve to the model presented the following results: for the reference, the stress bond was 9 MPa, and SAP-0.2F, SAP-0.3F, and SAP-0.6F were 8, 13.5, and 11.2 MPa, respectively. This confirmed that SAP-0.3F presented a superior bond between the fiber and matrix, consequently enhancing the toughness and presenting the best behavior. Furthermore, SAP-0.6F validated that the SAP addition enhanced the bond between the matrix and steel fiber.

The parameter c can be physically interpreted as the behavior of the concrete or mortar undergoing tension decay after cracking; the typical value of concrete is 15 and 30 for mortar. Furthermore, the interpolation revealed that the SAP addition modified the matrix to be more likely to be a mortar than concrete. The REF-035 was 15 as a result of the fitting, which is the typical value of concrete. The addition of 0.2, 0.3, and 0.6 presented the values of 6, 18, and 22, respectively.

Overall, the proposed models presented a good correlation, indicated by a mean correlation R^2 of 0.97. Future work to investigate the pull-out behavior and confirm the indicated values should be performed. Furthermore, the analytical model can describe the behavior of the SFRC with SAP and can be applied in the future analysis of elements under uniaxial tension.

4. Conclusions

This study established the framework to investigate the tension behavior of a durable steel fiber reinforced concrete with increasing dosage of SAP and to develop a HSC with SAP to mitigate autogenous shrinkage and develop an analytical model for predicting the tension behavior of HSC with varying SAP dosage. Based on the results presented, the following conclusions can be drawn:

- HSC can incorporate a dosage of 0.6% of SAP, keeping high strength and workability.
- Autogenous shrinkage can be adequately mitigated with the content of 0.3% of SAP. However, a content of 0.2% may give sufficient reduction to avoid cracking.
- The content of 0.3% of SAP was shown to be beneficial in different aspects. It successfully mitigated autogenous shrinkage, can be used without major influence on workability, is compatible with high compressive strength, and significantly enhanced the SFRC's ductile performance.
- The variable engagement model used was capable of describing the behavior of the SFRC with SAP. Moreover, it can be used in future finite element applications.

Author Contributions: Conceptualization, K.F.d.S., A.C.B.S.R., and E.F.d.S.; Methodology and Data Acquisition, K.F.d.S., M.A.R.M., and L.A.d.C.M.; Supervision, A.C.B.S.R. and E.F.d.S.; Writing-Original Draft Preparation, K.F.d.S., A.C.B.S.R. and E.F.d.S.; Writing-Review & Editing, K.F.d.S., A.C.B.S.R., E.F.d.S., and R.D.T.F. All authors have read and agreed to the published version of the manuscript.

Funding: This study was financed in part by the Coordenação de Aperfeiçoamento de Pessoal de Nível Superior–Brasil (CAPES)–Finance Code 001.

Acknowledgments: The authors gratefully acknowledge the financial support given by CAPES to the execution of the present research project.

Conflicts of Interest: The authors declare no conflict of interest.

References

1. Jensen, O.M.; Hansen, P.F. Water-entrained cement-based materials:I. Principles and theoretical background. *Cem. Concr. Res.* **2001**, *31*, 647–654. [CrossRef]
2. Jensen, O.M.; Hansen, P.F. Water-entrained cement-based materials: II. Experimental observations. *Cem. Concr. Res.* **2002**, *32*, 973–978. [CrossRef]
3. Schröfl, C.; Mechtcherine, V.; Gorges, M. Relation between the molecular structure and the efficiency of superabsorbent polymers (SAP) as concrete admixture to mitigate autogenous shrinkage. *Cem. Concr. Res.* **2012**, *42*, 865–873. [CrossRef]
4. Hasholt, M.T.; Jensen, O.M.; Kovler, K.; Zhutovsky, S. Can superabsorent polymers mitigate autogenous shrinkage of internally cured concrete without compromising the strength? *Constr. Build. Mater.* **2012**, *31*, 226–230. [CrossRef]
5. Mechtcherine, V.; Reinhardt, H. *Application of Super Absorbent Polymers (SAP) in Concrete Construction: State-of-the-Art Report Prepared by Technical Committee 225-SAP*, 1st ed.; Mechtcherine, V., Reinhardt, H., Eds.; Springer: Dordrecht, The Netherlands, 2012; ISBN 978-94-007-2733-5.
6. Mechtcherine, V.; Gorges, M.; Schroefl, C.; Assmann, A.; Brameshuber, W.; Ribeiro, A.B.; Cusson, D.; Custódio, J.; Da Silva, E.F.; Ichimiya, K.; et al. Effect of internal curing by using superabsorbent polymers (SAP) on autogenous shrinkage and other properties of a high-performance fine-grained concrete: Results of a RILEM round-robin test. *Mater. Struct. Constr.* **2014**, *47*, 541–562. [CrossRef]
7. Shen, D.; Wang, X.; Cheng, D.; Zhang, J.; Jiang, G. Effect of internal curing with super absorbent polymers on autogenous shrinkage of concrete at early age. *Constr. Build. Mater.* **2016**, *106*, 512–522. [CrossRef]
8. Wyrzykowski, M.; Igarashi, S.I.; Lura, P.; Mechtcherine, V. Recommendation of RILEM TC 260-RSC: Using superabsorbent polymers (SAP) to mitigate autogenous shrinkage. *Mater. Struct. Constr.* **2018**, *51*, 1–7. [CrossRef]
9. Reis, P.F.O.; Evangelista, F.; Silva, E.F. Profile of internal relative humidity and depth of drying in cementitious materials containing superabsorbent polymer and nano-silica particles. *Constr. Build. Mater.* **2020**, *237*, 117412. [CrossRef]
10. Naaman, A.E.; Reinhardt, H.W. Proposed classification of HPFRC composites based on their tensile response. *Mater. Struct. Constr.* **2006**, *39*, 547–555. [CrossRef]
11. Kwon, S.; Nishiwaki, T.; Mihashi, H. Self-healing Capability of Ultra-High-Performance Fiber-Reinforced Concrete (UHP-FRC). In Proceedings of the Fifth International Conference on Self-Healing Materials, Durham, NC, USA, 22–24 June 2015.
12. Van Zijl, G.P.A.G.; Wittmann, F.H.; Oh, B.H.; Kabele, P.; Toledo Filho, R.D.; Fairbairn, E.M.R.; Slowik, V.; Ogawa, A.; Hoshiro, H.; Mechtcherine, V.; et al. Durability of strain-hardening cement-based composites (SHCC). *Mater. Struct. Constr.* **2012**, *45*, 1447–1463. [CrossRef]

13. Afroughsabet, V.; Biolzi, L.; Ozbakkaloglu, T. High-Performance Fiber-Reinforced Concrete: A Review. *J. Mater. Sci.* **2016**, *51*, 6517–6551. [CrossRef]
14. Naaman, A.E. Fiber reinforced concrete: Five decades of progress. In Proceedings of the 4th Brazilian Conference on Composite Materials, io de Janeiro, Brazil, 22–25 July 2018; pp. 35–56.
15. Di Prisco, M.; Plizzari, G.; Vandewalle, L. Fibre reinforced concrete: New design perspectives. *Mater. Struct. Constr.* **2009**, *42*, 1261–1281. [CrossRef]
16. Yao, Y.; Zhu, Y.; Yang, Y. Incorporation superabsorbent polymer (SAP) particles as controlling pre-existing flaws to improve the performance of engineered cementitious composites (ECC). *Constr. Build. Mater.* **2012**, *28*, 139–145. [CrossRef]
17. Snoeck, D.; Van Tittelboom, K.; Steuperaert, S.; Dubruel, P.; De Belie, N. Self-healing cementitious materials by the combination of microfibres and superabsorbent polymers. *J. Intell. Mater. Syst. Struct.* **2014**, *25*, 13–24. [CrossRef]
18. Tosun-Felekoğlu, K.; Felekoğlu, B.; Ranade, R.; Lee, B.Y.; Li, V.C. The role of flaw size and fiber distribution on tensile ductility of PVA-ECC. *Compos. Part B Eng.* **2014**, *56*, 536–545. [CrossRef]
19. Snoeck, D.; De Belie, N. Effect of fibre type and superabsorbent polymers on the self-healing properties of strain-hardening cementitious materials. In Proceedings of the 11th International Symposium on Brittle Matrix Composites (BMC), Warsaw, Poland, 28–30 September 2015; pp. 213–222.
20. Sherir, M.A.A.; Hossain, K.M.A.; Lachemi, M. Interaction of Superabsorbent Polymers and Admixtures on the Properties of Engineered Cementitious Composites. *Can. Soc. Civ. Eng.* **2015**, *1*, 191–200.
21. Snoeck, D. Self-Healing and Microstructure of Cementitious Materials with Microfibres and Superabsorbent Polymers. Ph.D. Thesis, Ghent University, Ghent, Belgium, 15 December 2015.
22. Mignon, A.; Snoeck, D.; Dubruel, P.; Van Vlierberghe, S.; De Belie, N. Crack Mitigation in Concrete: Superabsorbent Polymers as Key to Success? *Materials* **2017**, *10*, 237. [CrossRef]
23. Li, V.C. On Engineered Cementitious Composites (ECC) A Review of the Material and Its Applications. *J. Adv. Concr. Technol.* **2003**, *1*, 215–230. [CrossRef]
24. Wang, S.; Li, V.C. Tailoring of pre-existing flaws in ECC matrix for saturated strain hardening. *Fract. Mech. Concr. Struct.* **2004**, *8*, 1005–1012.
25. Li, V.C.; Wang, S.; Wu, C. Tensile strain-hardening behavior of polyvinyl alcohol engineered cementitious composite (PVA-ECC). *ACI Mater. J. Am. Concr. Inst.* **2001**, *98*, 483–492.
26. Liu, J.; Farzadnia, N.; Shi, C. Effects of superabsorbent polymer on interfacial transition zone and mechanical properties of ultra-high performance concrete. *Constr. Build. Mater.* **2020**, *231*, 117142. [CrossRef]
27. Wang, X.; Zhang, S.; Wang, C.; Cao, K.; Wei, P.; Wang, J. Effect of steel fibers on the compressive and splitting-tensile behaviors of cellular concrete with millimeter-size pores. *Constr. Build. Mater.* **2019**, *221*, 60–73. [CrossRef]
28. ABNT NBR 16697. *Cimento Portland—Requisitos*; Brazilian Association of Technical Standards: São Paulo, Brazil, 2018; p. 9.
29. ABNT NBR 13956. *Sílica Ativa Para uso com Cimento Portland em Concreto, Argamassa e Pasta Parte 1: Requisitos*; Brazilian Association of Technical Standards: São Paulo, Brazil, 2012; p. 6.
30. ABNT NBR 7211. *Agregados Para Concreto—Especificação*; Brazilian Association of Technical Standards: São Paulo, Brazil, 2009; p. 9.
31. Jensen, O.M. Water absorption of superabsorbent polymers in a cementitious environment. In *International RILEM Conference On Advances in Construction Materials Through Science and Engineering*; RILEM Publications SARL: Hong Kong, China, 2011; pp. 22–35.
32. Snoeck, D.; Velasco, L.F.; Mignon, A.; Van Vlierberghe, S.; Dubruel, P.; Lodewyckx, P.; De Belie, N. The effects of superabsorbent polymers on the microstructure of cementitious materials studied by means of sorption experiments. *Cem. Concr. Res.* **2015**, *77*, 26–35. [CrossRef]
33. De Rooij, M.; Van Tittelboom, K.; De Belie, N.; Schlangen, E. *RILEM TC 221-SHC: Self-Healing Phenomena in Cement-Based Materials*; Springer Science & Business Media: Berlim, Germany, 2013; Volume 1.
34. *DIN 18555-2 Testing of Mortars Containing Mineral Binders—Freshly Mixed Mortars Containing Aggregates of Dense Structure (Heavy Aggregates)—Determination of Consistence, Bulk Density and Air Content*; DIN/EN: Berlin, Germany, 1982.
35. ABNT NBR 5739. *Concrete—Compression Test of Cylindrical Specimens*; Brazilian Association of Technical Standards: São Paulo, Brazil, 2018; p. 71.

36. Silva, E.F. Variações Dimensionais em Concretos de alto Desempenho Contendo Aditivo Redutor de Retração. Ph.D. Thesis, Universidade do Rio de Janeiro, Rio de Janeiro, Brazil, 2007.
37. Tawaza, E. *Autogenous Shrinkage of Concrete*, 1st ed.; CRC Press: London, UK, 1999.
38. *ASTM C490 Standard Practice for Use of Apparatus for the Determination of Length Change of Hardened Cement Paste, Mortar, and Concrete*; ASTM Standard: West Conshohocken, PA, USA, 2000.
39. Yokota, H.; Rokugo, K.; Sakata, N. JSCE Recommendations for Design and Construction of High Performance Fiber Reinforced Cement Composite with Multiple Fine Cracks. In *High Performance Fiber Reinforced Cement Composites*; Springer: Tokyo, Japan, 2008.
40. Schreier, H.; Orteu, J.J.; Sutton, M.A. *Image Correlation for Shape, Motion and Deformation Measurements: Basic Concepts, Theory and Applications*; Springer: Berlin, Germany, 2009.
41. BS EN 12390-6. *Testing Hardened Concrete. Tensile Splitting Strength of Test Specimens*; British Standard Institution: London, UK, 2009; p. 14.
42. Paiva, H.; Esteves, L.; Cachim, P.; Ferreira, V.F. Rheology and hardened properties of single-coat render mortars with different types of water retaining agents. *Constr. Build. Mater.* **2009**, 1141–1146. [CrossRef]
43. Bayasi, M.Z.; Soroushian, P. Effect of steel fiber reinforcement on fresh mix properties of concrete. *ACI Mater. J.* **1992**, *89*, 369–374.
44. Ferrara, L.; Meda, A. Relationships between fibre distribution, workability and the mechanical properties of SFRC applied to precast roof elements. *Mater. Struct. Constr.* **2006**, *39*, 411–420. [CrossRef]
45. Holschemacher, K.; Mueller, T.; Ribakov, Y. Effect of steel fibres on mechanical properties of high-strength concrete. *Mater. Des.* **2010**, *31*, 2604–2615. [CrossRef]
46. Filho, R.D.T.; Silva, E.F.; Lopes, A.N.M.; Mechtcherine, V.; Dudziak, L. Effect of superabsorbent polymers on the workability of concrete and mortar. In *Application of Super Absorbent Polymers (SAP) in Concrete Construction: State-of-the-Art Report Prepared by Technical Committee 225-SAP*; Springer: Berlin, Germany, 2012; pp. 39–50.
47. Liu, J.; Khayat, K.H.; Shi, C. Effect of superabsorbent polymer characteristics on rheology of ultra-high performance concrete. *Cem. Concr. Compos.* **2020**, *112*, 103636. [CrossRef]
48. Faping, L.; Jiesheng, L. Study on the Properties and Mechanism of Mortars Modified by Super Absorbent Polymers. *J. Test. Eval.* **2019**, *47*, 20170374. [CrossRef]
49. Dudziak, L.; Mechtcherine, V. Mitigation of volume changes of ultra-high performance concrete (UHPC) by using super absorbent polymers. In Proceedings of the 2nd International Symposium on Ultra High Performance Concrete, Kassel, Germany, 5–7 March 2008; pp. 425–432.
50. Zhao, S.; Jensen, O.M.; Hasholt, M.T. Measuring absorption of superabsorbent polymers in cementitious environments. *Mater. Struct. Constr.* **2020**, *53*, 1–16. [CrossRef]
51. Snoeck, D.; Schaubroeck, D.; Dubruel, P.; De Belie, N. Effect of high amounts of superabsorbent polymers and additional water on the workability, microstructure and strength of mortars with a water-to-cement ratio of 0.50. *Constr. Build. Mater.* **2014**, *72*, 148–157. [CrossRef]
52. Chindasiriphan, P.; Yokota, H.; Pimpakan, P. Effect of fly ash and superabsorbent polymer on concrete self-healing ability. *Constr. Build. Mater.* **2020**, *233*, 116975. [CrossRef]
53. Lee, H.X.D.; Wong, H.S.; Buenfeld, N.R. Self-sealing of cracks in concrete using superabsorbent polymers. *Cem. Concr. Res.* **2016**, *79*, 194–208. [CrossRef]
54. ACI Committee 363. *363R-10 Report on High-Strength Concrete*; ACI Mater. J.: Farmington Hills, MI, USA, 2010.
55. *Concreto Para Fins Estruturais-Classificação pela Massa Específica, por Grupos de Resistência e Consistência*; Assoc. Bras. Normas Técnicas: São Paulo, Brazil, 2015; ABNT NBR 8953:2015.
56. Lura, P.; Jensen, O.M.; Van Breugel, K. Autogenous shrinkage in high-performance cement paste: An evaluation of basic mechanisms. *Cem. Concr. Res.* **2003**, *33*, 223–232. [CrossRef]
57. Wyrzykowski, M.; Terrasi, G.; Lura, P. Expansive high-performance concrete for chemical-prestress applications. *Cem. Concr. Res.* **2018**, *107*, 275–283. [CrossRef]
58. Aveston, J.; Kelly, A. Theory of multiple fracture of fibrous composites. *J. Mater. Sci.* **1973**, *8*, 352–362. [CrossRef]
59. Gray, R.J. Analysis of the effect of embedded fibre length on fibre debonding and pull-out from an elastic matrix—Part 1 Review of theories. *J. Mater. Sci.* **1984**, *19*, 861–870. [CrossRef]
60. Rossi, P.; Richer, S. Numerical modelling of concrete cracking based on a stochastic approach. *Mater. Struct.* **1987**, *20*, 334–337. [CrossRef]

61. Rasheed, M.A.; Prakash, S.S. Behavior of hybrid-synthetic fiber reinforced cellular lightweight concrete under uniaxial tension—Experimental and analytical studies. *Constr. Build. Mater.* **2018**, *162*, 857–870. [CrossRef]
62. Voo, J.Y.L.; Foster, S. Tensile-fracture of fibre-reinforced concrete: Variable engagement mode. In *6th International RILEM Symposium on Fibre Reinforced Concretes*; RILEM Publications SARL: Varenna, Italy, 2004; pp. 875–884.
63. Lei, Y.; Dura, V.; Sdn, T.; Foster, S.; Sydney, U.; Voo, Y.L. Variable Engagement Model for the Design of Fibre Reinforced Concrete Structures. In Proceedings of the Advanced Materials for Construction of Bridges, Buildings, and Other Structures III, Davos, Switzerland, 7–12 September 2005.
64. Orange, G.; Acker, P.; Vernet, C. A new generation of UHP Concrete: Ductal damage resistance and micromechanical analysis. In Proceedings of the Fifth RILEM Symp. Fiber-Reinforced Concr., Lyon, France, 13–15 September 2000.

Publisher's Note: MDPI stays neutral with regard to jurisdictional claims in published maps and institutional affiliations.

materials

MDPI

Article

The Influence of Curing Regimes in Self-Healing of Nano-Modified Cement Pastes

Maria Stefanidou *, Eirini-Chrysanthi Tsardaka and Aspasia Karozou

Laboratory of Building Materials, Aristotle University of Thessaloniki, 54635 Thessaloniki, Greece; extsardaka@gmail.com (E.-C.T.); akarozou@gmail.com (A.K.)

* Correspondence: stefan@civil.auth.gr; Tel.: +30-2310996535

Received: 30 October 2020; Accepted: 20 November 2020; Published: 23 November 2020

Abstract: The present study proposes nano-calcium oxide (NC) and nano-silica (NS) particles as healing agents in cement pastes, taking into account the curing conditions. Two series of specimens were treated in water and under wetting-drying cycles. The addition of NC (1.5%wt of binder) triggered early healing since cracks were healed within 14 days in underwater immersion and before 28 days at wetting-drying cycles. Attenuated Total Reflectance (ATR) spectroscopy and SEM analysis revealed that the healing products were mainly aragonite and calcite in water conditions and more amorphous carbonates under wetting-drying cycles. The combination of NS and NC (3.0%wt in total) offered healing under both curing conditions before 28 days. The presence of NS assisted toward porosity refinement and NC increased the carbonates' content. The newly formed material was dense, and its elemental analysis by SEM revealed the C-S-H compounds that were also verified by ATR.

Keywords: nano-calcium oxide; nano-silica; self-healing; cement pastes; curing regime

1. Introduction

The formation of cracks in infrastructures creates potential pathways for aggressive agents to insert into the bulk cementitious materials. This fact can cause significant problems that require high-cost repair solutions. Self-healing can reduce the deterioration rate, extend the structure's service life, and reduce the repair frequency and cost over the life cycle [1]. As a result, self-healing methodologies for cement-based materials were extensively studied in recent decades. Among other methods, incorporating mineral admixtures, such as fly ash or slag, is a well-studied method that can trigger self-healing [2,3]. The primary condition is the homogeneous distribution of the admixtures to react under the presence of humidity [4]. The mineral admixtures absorb and interact with water to penetrate the cracks and lead to product deposition that fills the cracks [5].

Huang et al. [5] suggested no particular method for self-healing, but one self-healing method could be the most suitable for a particular situation. Therefore, mineral admixtures proposed for self-healing should be accompanied by specific parameters and application conditions. The impact of environmental conditions on the self-healing of cement-based materials has been studied mostly by curing in water immersion and wetting-drying cycles, as the humidity and water availability plays a crucial role in the progress of autogenous self-healing and the mineral admixtures mechanism [4]. Water immersion constitutes a very effective curing method, as it brings in contact the calcium cations and the $HCOOH^-$ anions that lead to calcite formation and then calcite precipitation [6–8]. This mechanism functions in cement-based materials [9–11] and air lime and lime-pozzolan ones [12]. Encouraging results have also been reported for wetting-drying cycles conditioning. Jiang et al. [13] combined different mineral admixtures and promoted self-healing of cementitious composites under wet-dry cycles and high pH environments. Kan et al. [14] identified autogenous healing before five cycles, and the self-healing products included C-S-H compounds and calcite. On the other hand, there have been reports that mention no autogenous healing under wet-dry cycles, but only in water submersion, and still, the healing was limited to the surface of the cracks [15]. Also, Roig-Flores et al. [9] reported that the same specimens exhibited different self-healing behavior in different environments, while no self-healing was recorded for constant humidity and constant temperature.

The nanoparticles that have been widely studied in cement-based materials [16,17] have the potential to work as mineral admixtures. Nano-silica (NS) and nano-calcium oxide (NC) are metal oxides with chemical affinity to cement and increased reactivity due to their high specific surface [18,19]. Moreover, small additions could ensure the homogeneous distribution of numerous particles throughout the material's volume, under specific protocols. Huseien et al. [1] referred to nanoparticles as materials produced through chemical synthesis, and therefore, they can generate uniform structured products and nano-sized crystals with "perfect atomic and molecular ordering." So far, the NS and NC nanoparticles have been used to self-healing cement pastes, providing promising results. An effort was made by Stefanidou et al. [20] when 5%wt addition of NC offered self-healing of cracked cement pastes after seven days in water immersion. Accordingly, Stefanidou et al. [21] proved that 1.5%wt NC favored the calcite precipitation and reduced the open porosity of cement pastes in water immersion remarkably.

The present work is an effort to accent the self-healing ability of NS and NC in cement pastes under different curing conditions. A novel approach, by using reactive nano-particles in an efficient amount, in order to promote the healing mechanisms is presented. This procedure is tested under different curing regimes in order to conclude on the optimum conditions affecting self-healing. The optimum curing regime for these nano-particles' best performance on healing has not been investigated before. Cement pastes with 1.5%wt NC and cement pastes with combined 1.5%wt NC and 1.5%wt NS were subjected to water immersion conditioning and wetting-drying cycles. The samples were cured for 90 days to record mechanical and physical properties, as well as the healing process of the pastes. At 28 days, controlled cracks were formed at prismatic samples using the 3-point bending method. That period was chosen in order to avoid early cement hydration. The crack pattern was recorded and the samples were placed again in the different curing regimes to record the healing process in the formed cracks.

2. Materials and Methods

Nanoparticles used in this experimental work were supplied by Sigma-Aldrich (Saint Louis, MO, USA). The mineralogical composition of nano-calcium oxide (NC) is calcium oxide (<180 nm), and that of nano-silica (NS) (0.007 μm) is amorphous silicon dioxide diffraction patterns that have been given in previous works [22].

For the X-ray Fluorescence elemental analysis of the cement I42.5N, an S8 Tiger, Bruker Instruments (Karlsruhe, Germany) was used, and the content of oxides is given in Table 1. Table 2 shows the cement granulometry recorded using a particle size analyzer, Mastersizer 2000, Malvern Instruments (Malvern, UK). It is observed that 50% of the cement powder is below 4.667 μm. The density of the cement was measured according to ASTM-C188-95.

Table 1. Elemental analysis of cement, expressed in oxides.

	Na_2O	K_2O	CaO	MgO	Fe_2O_3	Al_2O_3	SiO_2	SO_3
CEM I42.5N	0.02	1.10	61.70	1.20	3.07	3.14	15.01	3.34

Table 2. Granulometry and density (g/cm^3) of cement powder.

	d (0.1) (μm)	d (0.5) (μm)	d (0.9) (μm)	Density (g/cm^3)
cement	1.275	4.667	222.03	3.105

Compressive strength (loading rate: 0.5 KN/s) was tested at 28 and 90 days, using a Technik ToniNorm device (Toni Technik GmbH, Berlin, Germany), and the mean value of six measurements was calculated. Open porosity was tested at the same ages, according to RILEM CPC 11.3 method, in water under vacuum and the mean value of three specimens was calculated. For Scanning Electron Microscopy, a Jeol JSM-6390LV, Oxford Instruments (Abingdon-on-Thames, UK), was used to observe the microstructure at 28 days. The thermogravimetric (TG) method was used to determine $Ca(OH)_2$%wt and carbonates %wt content of the samples at 28 days. A dynamic method was applied to determine the mass loss of the samples at a function of temperature. A Neutsch F5 Jupiter instrument (Juliet, TN, USA) was used in an N_2 atmosphere (50 mL per minute) from 50 °C to 1000 °C and 20 °C per minute step. Differencial thermogravimetric analysis (DTG) curves were extracted from TG thermograms.

The composition of cement pastes and nano-modified cement pastes is given in Table 3. A capital letter "R" is used for reference cement pastes. The letters "NL" are used for cement pastes with NC addition, and letters "NSL" are used for cement pastes with a combination of NC and NS additions. The small letter "w" is used for the specimens cured in water immersion, and the small letter "c" is used for the specimens cured in wetting-drying cycles. The NS particles were added in a pre-weight quantity of water and were subjected to ultrasonication for 30 min. The suspensions were directly added to the cement powder and stirred up to homogenization. The NC particles were added in water and were directly used in the mixture. The ultrasonication step was not needed as NC is instantly hydrated in a pre-weighted amount of water. The pastes' consistency was determined using a Vicat apparatus (according to EN 196-3:2016). Superplasticizer (SP) was used in nano-modified mixtures to maintain the systems' consistency (Vicat 6 ± 1 mm). Pastes were cast in (25 × 25 × 50) mm^3 molds to test the mechanical properties and the microstructure. Moreover, specimens of (40 × 40 × 160) mm^3 were produced for pre-cracking and to observe the self-healing effect. All the samples, regardless of their size, were cured under two different conditions. One condition was the water immersion until the performing of the tests. The second condition was the conduction of wetting-drying cycles, where the samples daily were subjected to water for six hours and set on specific conditions for 18 h (temperature ~20 °C and RH 65%).

Table 3. Composition of cement pastes.

Acronym	CEM I42.5N	NC %wt	NS %wt	w/b	SP %wt	Curing
Rw	1	-	-	0.30	-	Water immersion
NLw	1	1.5	-	0.23	1.5	
NSLw	1	1.5	1.5	0.30	1.5	
Rc	1	-	-	0.30	-	Wetting-drying
NLc	1	1.5	-	0.23	1.5	
NSLc	1	1.5	1.5	0.30	1.5	

The "3-point bending method" [12,21,23] was used for the formation of the pre-cracks. The specimens were marked and notched in the middle of their length, and they were carefully cracked after 28 days of curing (Figure 1). A low displacement rate (0.3 mm/min) was used to form cracks of less than 0.5 mm in width. The observation and measurement of cracks were performed with the Dino-Lite2 microscope and the DinoCapture2.0 software. An optical microscope was used to evaluate the healing of the cracks under the two different curing conditions. The width of the cracks was measured at three different ages. Firstly, immediately after the cracking, and at 14 and 28 days after the cracking. The cracked specimens were subjected to their previous curing conditions (water immersion and cycles) until testing.

After the optical observation of the healed cracks, the newly formed material was cautiously removed and collected in small containers. These samples were measured using ATR (Attenuated Total Reflectance) Spectroscopy to identify the compounds. A Cary 630 FTIR instrument with an ATR probe of Agilent Technologies was used to determine reflectance patterns from 375 cm^{-1} to 4000 cm^{-1} wavenumber.

Figure 1. Specimen notched in the middle of its length and pre-cracked, at 28 days from their production.

3. Results

3.1. Compressive Strength and Open Porosity

The compressive strength values and the variation of these values compared to the reference are given in Figure 2. Each nano-modified system was compared to the reference cement paste. The equation used for the calculation of the variation (V) is V (%) = (N − R) × 100/R, where N is the value of the nano-modified paste, and R is the value of the reference cement paste.

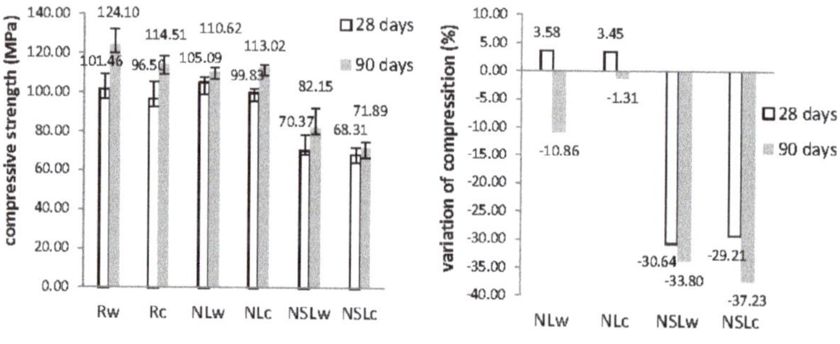

Figure 2. (**a**) Compressive strength values of the pastes at 28 and 90 days; (**b**) Compressive strength variation of nano-modified cement pastes, in respect to the reference.

The curing conditions slightly affected compressive strength values. At three months, the strength was higher in relation to the values recorded at 28 days in all cases. Also, neat cement pastes have lower strength when subjected to cycles in relation to water immersion. In other reports, the curing regimes and w/c influenced the results concerning the physicomechanical properties of the cement pastes. Abusnina et al. suggested that both compressive strength and porosity were modified when cement pastes with varying proportions of oil-contamination were evaluated in the age of curing [24].

Nevertheless, the cement pastes cured in water immersion had slightly greater compressive strength than the pastes cured under wetting-drying cycles, mainly for the reference and the cement pastes with NS presence. On the other hand, the presence of the nanoparticles also affected this property. The addition of NC slightly improved compressive strength at 28 days recording, by 3.58% and 3.45%, at 28 and 90 days, respectively. On the contrary, the combination of NS and NC decreased the compressive strength significantly in both curing conditions. This reduction was approximately 30% at 28 days and more than 30% at 90 days. The increased water demand in these systems could have influenced the strength. Another possible cause could be the high amount of the incorporated nanoparticles (1.5%wt NC + 1.5%wt NS). This condition did not favor the distribution of them, causing agglomerations [25,26].

The open porosity evolution and the variation of open porosity at 28 and 90 days are given in Figure 3. The open porosity values are typical of cement pastes and in agreement with previous works [17,21]. Porosity decreases with time. The curing conditions seem to influence this property. NC-cement paste presented higher open porosity when cured in water than wetting-drying cycles, as the nano-calcium oxide hydrates and forms calcium hydroxide. When the samples remain in the water, the calcium hydroxide content remains hydrated.

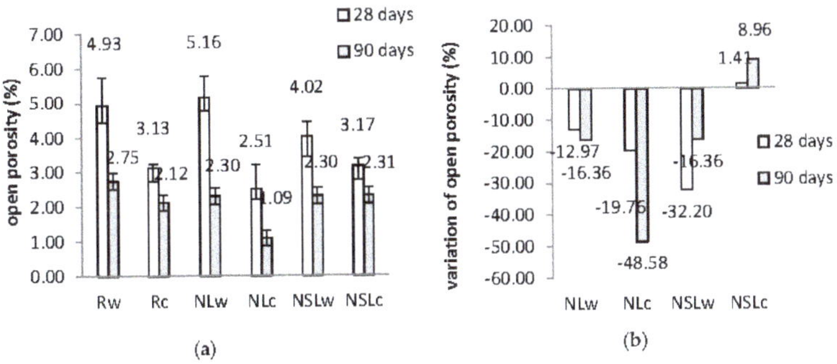

Figure 3. (**a**) Open porosity evolution of the pastes at 28 and 90 days; (**b**) Open porosity variation of nano-modified cement pastes, in respect to the reference.

On the other hand, in wetting-drying cycles, NC hydrates and then dries in the air and forms calcium carbonate. As a result, the porosity of NLw was found to increase at 28 days. Additionally, all the systems cured in cycles have smaller porosity values and smaller variations than the reference (Figure 3b). The addition of NC favored open porosity closure at 90 days, as the variation recorded was −16.36% when cured in water and −48.58% when cured in cycles. In contrast, the combination of nanoparticles did not benefit the reduction of porosity, both at 28 and 90 days, but still, the open porosity values of NSLw and NSLc are low.

The open porosity results are in agreement with compressive strength results. The physicomechanical properties of NC-cement pastes were benefited, compared to neat cement paste. In previous work, when NC was added in cement pastes with a w/b ratio of 0.32, the open porosity results had demonstrated an extreme reduction of porosity by −50% [21]. The combined nanoparticles addition leads to a compressive strength decrease and a slight open porosity increase, especially when cured in water. With time, the porosity is similar, regardless of the curing regime. In neat samples, water assists cement hydration and the porosity is quickly reduced.

3.2. *Differential Thermogravimetric Analysis*

The DTG curves of the samples are displayed in Figures 4 and 5. The calcium hydroxide and carbonated species content %wt of the samples, measured at 28 days, are given in Table 4. The results were calculated from the TG curves.

At early age cement pastes analysis, the portlandite quantity represents the hydration of the calcium silicate compounds [27]. The curing conditions did not affect the content of neat cement pastes. By contrast, the nano-modified pastes presented lower portlandite content, compared to the references in both curing conditions. After 28 days in water immersion, the NC addition presented lower portlandite content and higher carbonates content. Like NL, the combination of nanoparticles (NSL) led to higher carbonated species content (Figure 4). The shifted peak temperature from 720 °C to 752 °C could indicate crystallinity modification of carbonates or the presence of additional carbonated species [28]. After 28 days of wetting-drying cycles, the portlandite content was found reduced compared to the reference, and the carbonates of NC-cement pastes were found increased (Figure 4).

Consequently, the curing conditions seem to strongly affect the quantities of portlandite and carbonates, depending on the nanoparticle that has been incorporated. This fact must be connected to the microstructure to verify the nanoparticles' behavior in different curing conditions in cement pastes.

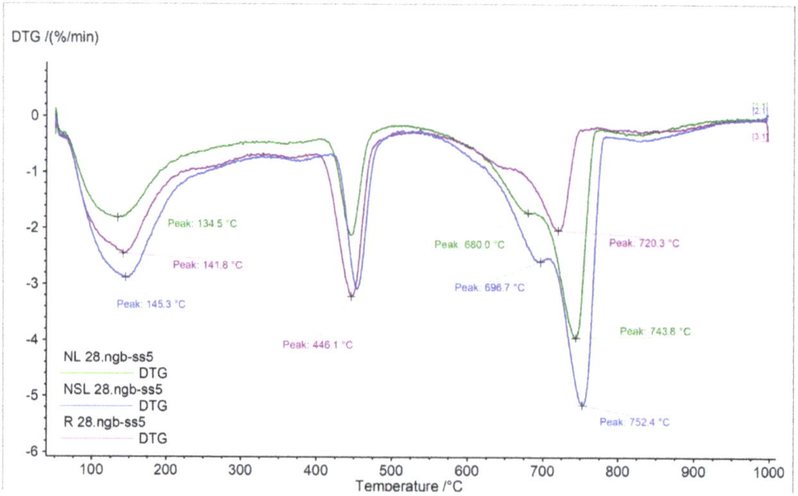

Figure 4. DTG curves of neat cement paste and nano-modified cement pastes cured in water immersion, 28 days.

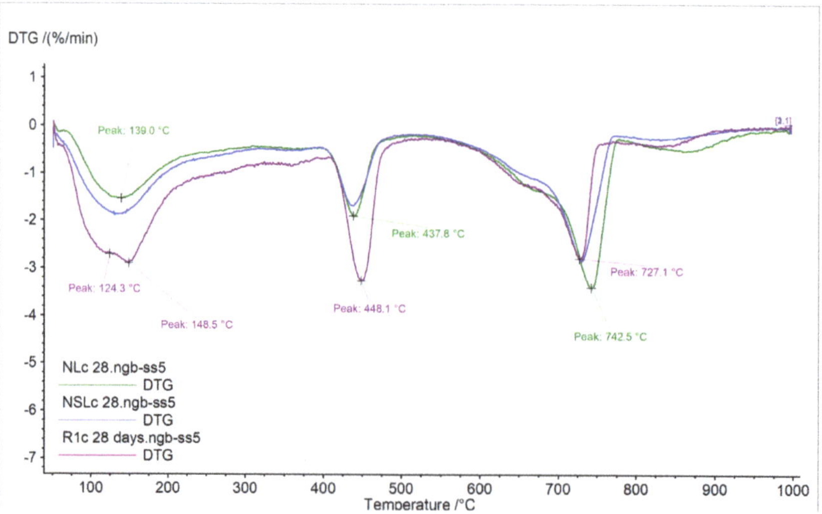

Figure 5. DTG curves of neat cement paste and nano-modified cement pastes cured under wetting-drying cycles at 28 days.

Table 4. Calcium hydroxide and carbonate species content, as a percentage by mass, at 28 days.

Sample	$Ca(OH)_2$ %wt	Carbonates Calculated as $CaCO_3$ %wt
Rw	31.45	22.81
NLw	27.14	37.09
NSLw	26.15	52.02
Rc	31.57	28.25
NLc	26.60	34.30
NSLc	18.91	28.41

3.3. Microstructure Observation under SEM

The microstructure observation revealed the different behavior of each nanoparticle in different curing conditions, concerning their potential contribution to self-healing. Representative SEM images of the systems are displayed in Figure 6, and the elemental analysis (EDS) of the samples is given in Table 5.

The cement pastes presented a typical dense structure, where few large pores were observed. In the case of water immersion, ettringite formation was observed inside the pores, as elemental analysis confirmed with the presence of Al/S = 1/2 (Rw, Sp.2, Table 5). In NLw samples, more $CaCO_3$ is formed in relation to NLc. Nevertheless, this difference does not seem to affect the compressive strength or porosity. In samples with combined nanoparticles, the $CaCO_3$ is much higher in the water immersion regime in relation to the cycle's conditions. Also, in this case, the $CaCO_3$ formation at 28 days does not affect either compressive strength or porosity. On the edge of the pore, low silicon content was recorded, while the Ca/Si ratio as 11/1 indicates an area where calcite content prevails. Similarly, in the case of wetting-drying cycles, the edges of the pores presented high calcium content. A small number of deposits inside the pores had a Ca/Si ratio equal to 2/1.

The action of NC offered different results depending on the curing conditions. In water immersion, the NC-cement pastes presented healed pores with newly formed material that had its micro-porosity and a "sponge"-like structure. The additional micro-porosity created assisted in recording high open porosity, as shown in Figure 3a. This material has a high quantity of silicon and aluminum, according to NLw, Sp.2, Table 5. The previous open pores can hardly be recognized, as they have been filled with this rich in Si/Al material. This fact also explains the strength recorded for NLw samples at 28 days. In the absence of NC, the pores of cement pastes did not present such behavior, so it is believed that NC has contributed to these formations.

When NC-cement pastes were cured under wetting-drying cycles, the diameter of the pores was found smaller and this was recorded as low open porosity in Figure 3a. In this case, the pores were not filled, though some formations were started to develop. The elemental analysis showed high calcium content of these deposits (NLc, Sp.1 and Sp.3, Table 5).

The results are reversed when NS and NC are combined in cement pastes. The microstructure of the pastes treated in water immersion was dense. Small and few pores were present in the structure. This fact explains and the low porosity values recorded for these compositions. Indications of newly crystallized material were observed in the pores. In the case of NSLw, precipitation of newly formed crystals was observed within the pores but did not manage to fill the pores. The edges of the pores were hardly observed due to the material formed that had a ratio of Ca/Si/Mg approximately equal to 4/0.6/1. The indications of NS-NC-cement pastes' microstructure were very promising for self-healing capacity under wetting-drying cycles, probably, due to the initial size of the pores formed into the structure. In this case, the open pores were filled with newly formed material. The micro-porosity of this material was also observed, and the morphology was rougher than the neat cement. This material had a Ca/Si ratio of 3/1, which might have played an essential role in the density and the morphology of the formed material. This proportion could be an indication of calcium silicate hydrated compounds that compose the new material.

(a) (b)

Figure 6. SEM-micrographs of samples cured; (**a**) in water immersion and; (**b**) under wetting- drying cycles, at 28 days.

3.4. Optical Observation of Cracks

The prisms produced for this research were pre-cracked after 28 days of curing under the two regimes. Representative images of optical observation under the curing regimes are given in Figure 7. The images were captured when the cracks were formed and correspond to 0 days. The images captured at 14 and 28 days after cracking correspond to 28 + 14 and 28 + 28 days total age. Table 6 shows representative measurements of the crack width evolution during the curing.

The images of neat cement depict the autogenous self-healing that the system could perform under these specific conditions. The width decrease is not significant (Table 6) in both curing conditions.

At 14 days, there are open parts in the cracks that have not healed, while at 28 days, the situation is improved. However, unhealed areas still remain.

The water immersion promoted the formation of crystals in NC-cement pastes and allowed the healing of up to 14 days. The mean crack width was 0.156 mm at 0 days and is healed, as depicted in Figure 7. The newly formed material also overflowed on the surface of very small cracks. The 28 days captions showed a progression of this crystallization. A greater quantity of the material was observed on the surface of the cracks and in a broader area around them. The system presented healing to a smaller degree when it was subjected to wetting-drying cycles. The crack width decreased significantly up to 14 days, from 0.106 mm to 0.031 mm, and it was eliminated at 28 days.

The combination of nanoparticles showed effloresing phenomena after 14 days of cracking, but the cracks were not healed at that age. Healing was recorded up to 28 days for samples cured in the water, and a satisfactory crack closure was recorded for samples kept in the wetting-drying regime.

Table 5. EDS analysis of the micrographs shown in Figure 6.

Sample	Spectrum	Mg (%)	Al (%)	Si (%)	S (%)	K (%)	Ca (%)	O (%)
Rw	Sp.1	0.05	1.24	4.86	0.1	7.61	56.29	29.85
	Sp.2	0.89	1.92	3.25	3.94	3.13	58.77	28.09
	Sp.3	2.77	3.2	12.45	1.97	5.82	36.91	36.87
NLw	Sp.1	1.19	0.02	13.21	0.31	0.08	49.28	35.91
	Sp.2	1.65	15.61	30.35	0.16	1.29	2.53	48.41
NSLw	Sp.1	4.86	1.17	10.44	1.89	3.00	44.73	33.90
	Sp.2	10.25	1.80	6.35	2.35	3.55	39.93	35.76
Rc	Sp.1	1.43	1.29	5.62	0.36	2.29	40.67	48.19
	Sp.2	0.20	0.40	5.89	0.05	1.42	30.90	61.12
	Sp.3	0.22	0.76	12.19	0.22	1.24	21.99	62.54
NLc	Sp.1	0.31	0.51	4.29	5.12	0.94	38.05	50.75
	Sp.2	0.50	2.02	17.10	10.78	1.74	29.36	38.51
	Sp.3	0.54	1.10	10.75	12.71	1.65	30.51	42.75
	Sp.4	0.70	0.71	8.13	1.14	0.00	37.37	51.95
	Sp.5	0.1	0.9	52.78	0.50	0.57	10.18	34.97
NSLc	Sp.1	0.13	1.1	39.02	0.01	0.61	17.23	41.9
	Sp.2	0.1	0.02	12.83	0.01	0.05	35.09	51.94
	Sp.3	0.02	1.84	34.09	0.02	0.04	20.57	43.44
	Sp.4	0.01	0.86	10.72	0.01	0.1	39.58	48.78

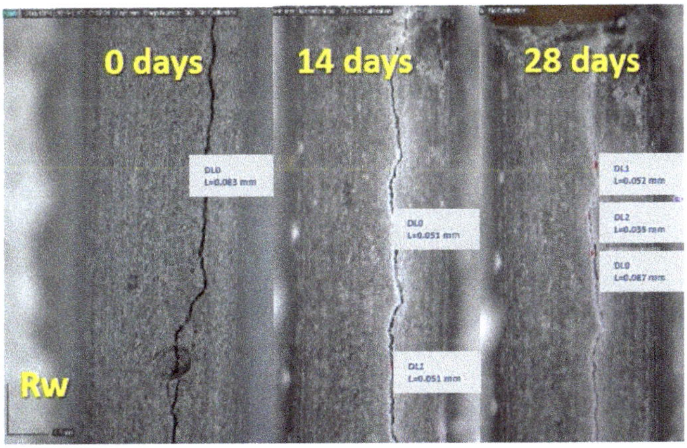

Figure 7. *Cont.*

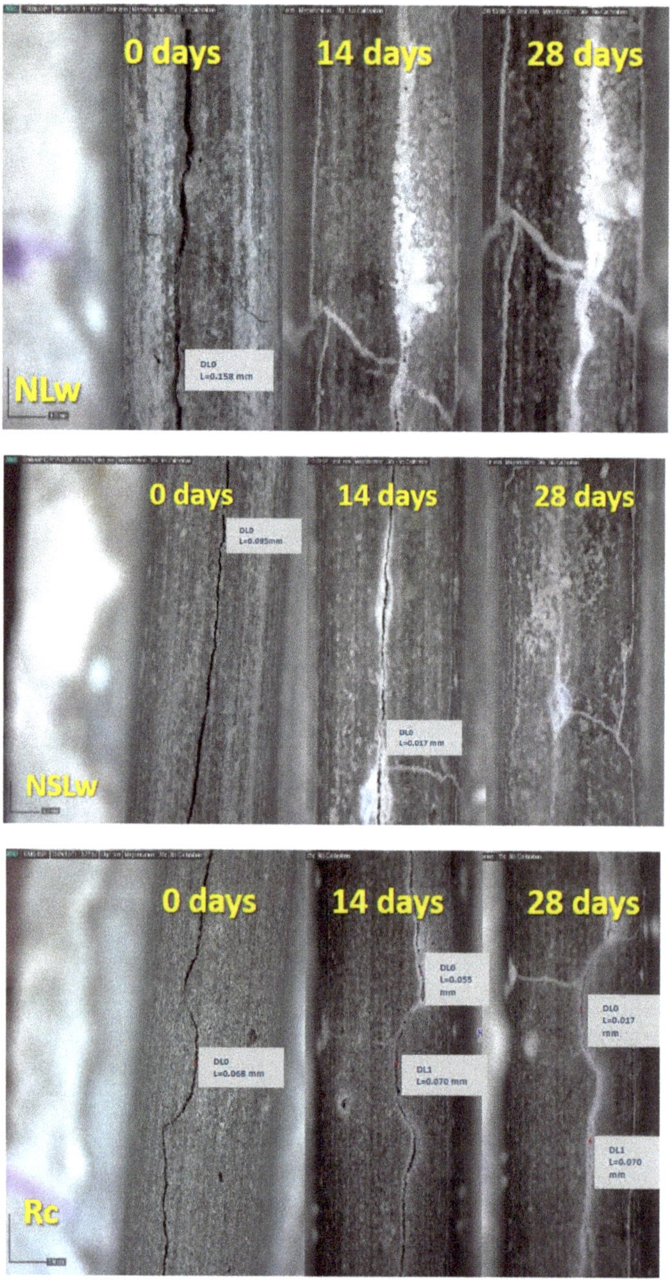

Figure 7. *Cont.*

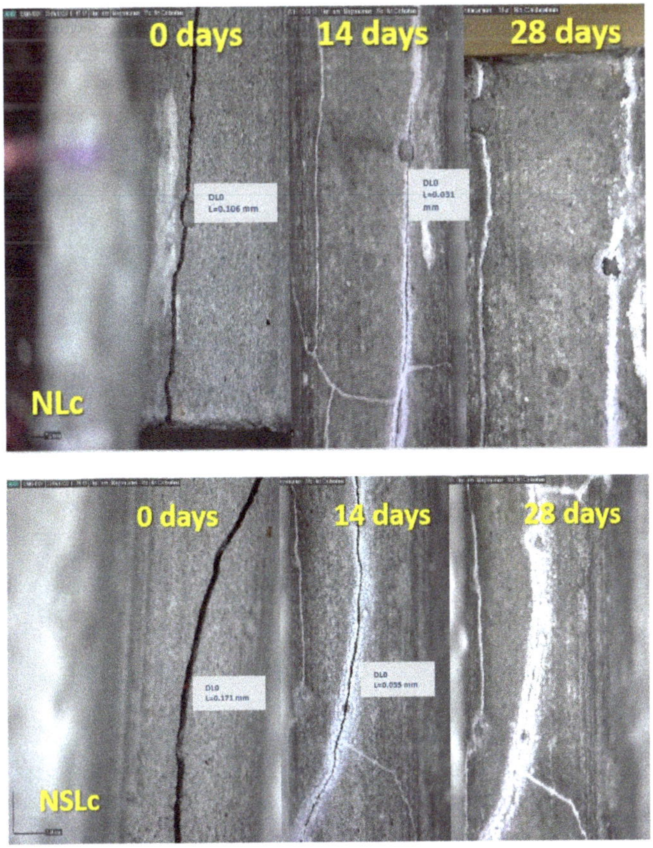

Figure 7. Optical microscopy observation of the cement pastes at 28 + 0 days of the cracking, at 28 + 14 days and 28 + 28 days, cured in water immersion (w) and wetting-drying cycles (c) (Scale equals to 1 mm).

Table 6. Crack width evolution (mean values of 14 measurements).

Sample	0 Days (mm)	14 Days (mm)	28 Days (mm)
Rw	0.083	0.051	0.057
Rc	0.068	0.070	0.070
NLw	0.156	healed	healed
NLc	0.106	0.031	healed
NSLw	0.095	0.017	healed
NSLc	0.071	0.036	healed

3.5. ATR Spectroscopy

ATR spectra of the newly formed materials in the cracks are displayed in Figures 8 and 9. All the systems that were cured in water immersion presented aragonite and calcite peaks. The bands at 690 cm^{-1}, 855 cm^{-1}, 1082 cm^{-1} and 1456 cm^{-1} are bond vibrations of aragonite [29]. The calcite bond vibrations correspond to ~1400 cm^{-1}, 875 cm^{-1} and 712 cm^{-1} [29]. The band of carbonates (Figure 8) has its maxima at 1456 cm^{-1} (aragonite) but is broad and includes the $CO_3{}^{-2}$ response of calcite at 1400 cm^{-1}. The asymmetric stretching vibration of Si-O-M responses between 900–1100 cm^{-1} [30], where M is a metal cation. At 1003 cm^{-1}, the response of calcium silicate is possible in the healing material of the nano-modified systems. In the reference healing material (Rw), this band was not observed.

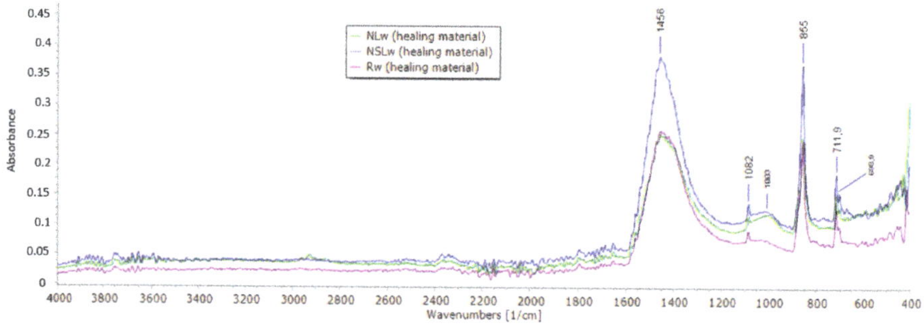

Figure 8. ATR spectra of the healing material obtained from the crack of the samples cured in water immersion.

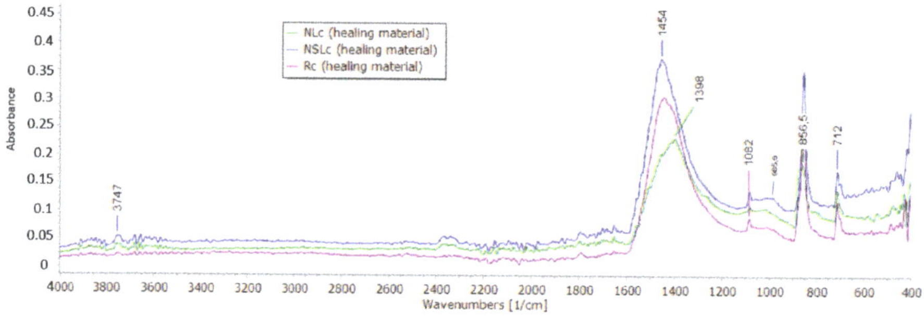

Figure 9. ATR spectra of the healing material obtained from the crack of the samples cured under wetting-drying cycles.

Under wetting-drying cycles (Figure 9), the reference healing material and the healing material from combined nanoparticles showed similar behavior with the water immersion healing materials. Aragonite and calcite presence prevailed in the newly formed material, though the presence of calcium silicate products is possible for NSLc. The addition of NC led to the formation of calcite rather than aragonite. The cycles have played a crucial role in the precipitation of the calcium carbonates in this case. The band at 1398 cm^{-1} corresponds to carbonates. As for the silicate compounds, their band is shifted to 1017 cm^{-1}.

4. Discussion

The contribution of nanoparticles NC and NS to the self-healing of cement pastes was tested under two different curing regimes. The results were slightly differentiated depending on both the kind of nanoparticle and the curing conditions.

The characterization of NC-cement pastes cured in water immersion showed that the system could form new material in empty spaces, such as pores and voids in humid conditions, at an early age. SEM-micrographs revealed the deposition of the porous newly formed material. This material assists in satisfactory strength with time. EDS analysis revealed that this material has calcium-silicon-aluminum content. DTG curves revealed a lower portlandite quantity than the reference, indicating its consumption and its participation in other formations. Accordingly, the calcite proportion is greater than the reference and the material shows a more significant strain to carbonation. Nano-calcium oxide corresponded under both curing conditions. At 28 + 14 days, the crack width was reduced by 70.75% compared to the 28 + 0 days. The small particle size of NC achieved early healing due to the accelerated carbonation of the calcium hydroxide with time.

Under wetting-drying cycles, the optical observation displayed the healing crystals formed at 28 + 28 days, even though the characterization of the microstructure at 28 days showed limited filling of the pores. ATR spectroscopy showed aragonite and calcite presence (Figure 9). The needle-like shaped aragonite has large micro-porosity, according to Shen et al. [31].

Combined nanoparticles decreased the compressive strength of cement pastes by almost 30%, possibly having to do with the increased water demand during the preparation. Additionally, the system depicted high carbonation after water immersion accompanied by dense microstructure with pores slightly closed by the newly formed material. The elemental analysis showed high calcium and magnesium content. The healing of the cracks of this system occurred progressively. At 28 + 14 days, the crack width decreased by 82.10%, and at 28 + 28 days, the crack was closed with solid and compact material. ATR spectroscopy revealed that the healed crack material was calcite and aragonite, while calcium silicate indications were also revealed (Figures 8 and 9).

The mineral admixtures are divided into two categories, depending on their behavior [5]. On the one hand, the expansive additives absorb water and form large-volume products [32]. On the other hand, crystalline additives react with calcium hydroxide and form crystalline products [5,33]. According to this categorization, NC could be considered an expansive agent because it forms calcium hydroxide when absorbs water and leads to calcite precipitation when it is dried. Accordingly, NS could be considered that works as a semi-crystalline agent, as its mechanism of action is to participate in new C-S-H compounds by interacting with available calcium hydroxide [34].

Qureshi et al. [35] indicated that the different expansive minerals influenced the microstructure of healing products added, among quick lime, bentonite, and magnesium oxide. Accordingly, in comparing two different systems with nanoparticles, it was depicted that the healing products were mainly calcite and aragonite. SEM images showed the structure's potential to close its pores with newly formed material of different roughness and micro-porosity during different curing of the same system. Additionally, EDS results and DTG quantifications showed that the systems performed different hydration degrees (according to the available portlandite) and different carbonated species formation (according to available carbonates) during different curing regimes. Consequently, it could be assumed that the different curing conditions of these systems influence the microstructure of the healing products.

The use of nanoparticles seems beneficial for the healing process as they contribute to microstructure changes and the production of calcitic products, promoting the healing of empty spaces. The nanoparticles were used as additives in small percentages and it was proven that even in 1.5%wt of binder, they could be effective. This is important as the cost of nanoparticles is a parameter that should be taken into account when designing such composites. The fact that the cost is reduced in time due to new, low-cost techniques engaged for their production and the low cost of the raw materials used is also important. The fact that nanoparticles were supplied in powder (and not as suspension) is also due to the easy and accurate use of the needed quantity. This fine powder involves the risk of inhalation, as all fine binders, during processing and all the necessary measures for protection (such as mask and gloves) should be taken into account during the preparation of the suspension. The exact impact of the nano-particles' use is not clear due to the lack of available toxicity information [36,37].

The results of this study can be used by materials scientists to improve the development of self-healing cement-based materials and engineers involved in works based on criteria of durable materials with minimum intervention requirements in the future. The system is easy to be prepared, both in a laboratory or in a plant. As demonstrated in the present study, nano-particles can offer early age self-healing, making these systems suitable for applications in humid or wetting-drying environments.

5. Conclusions

Two different nanoparticles; nano-calcium oxide (NC) and nano-silica (NS) were used in cement pastes, in small amounts, as additives. The specimens produced were cured under two different curing regimes; water immersion and wet-dry cycles. Tests were performed at 28 days of curing as well at 90 days. The aim was to record the self-healing procedure and its influence on the microstructure and the mechanical and physical properties of the pastes. Autogenous healing was recorded for neat cement pastes, mainly in samples cured in the water. Nevertheless, the effect of this healing was negligible and did not affect any of the tested properties.

The NC presence in the pastes resulted in early healing and open spaces such as pores and cracks were filled with a newly formed porous material of Ca-Si-Al composition. The healing material had a "sponge-like" form and this fine porosity. Sufficient healing was recorded within 14 days. The role of NC in cycling conditions also had a positive effect but needs more time to be effective. Low porosity and high strength were recorded at 90 days, as drying periods assist the $CaCO_3$ precipitation.

The combination of nanoparticles was misleading, as the total amount of these fine materials was high (3.0%wt total), leading to increased water demand (w/c = 0.30). The strength was reduced due to w/c, but NS assisted on the porosity refinement, and the open spaces into the microstructure were smaller in relation to the reference samples. The curing regime had a small impact on the healing procedure, as in both conditions, the healing was completed within 28 days.

When adding nanoparticles in cement pastes, healing is accelerated, and the presence of water is catalytic. The precipitation of calcite and the presence of aragonite and C-S-H compounds are filling the open spaces and assisting toward low porosity materials and low permeability.

Author Contributions: Conceptualization, methodology; validation, editing M.S.; analysis, investigation; data curation; writing, E.-C.T. and A.K. All authors have read and agreed to the published version of the manuscript.

Funding: This research is co-funding by Greece and the European Union (European Social Fund—ESF) through the Operational Programme «Human Resources Development, Education and Lifelong Learning 2014–2020» in the context of the project "Use of nanoparticles and inorganic admixtures in producing self-healing cement products" MIS 5047560.

Acknowledgments: The authors would like to grant special thanks to the Greece and the European Union (European Social Fund—ESF) that have funded this research through the Operational Programme «Human Resources Development, Education and Lifelong Learning 2014–2020» in the context of the project "Use of nanoparticles and inorganic admixtures in producing self-healing cement products".

Conflicts of Interest: The authors declare no conflict of interest.

Acronyms:

NC	nano-calcium oxide
NS	nano-silicon oxide
ATR	Attenuated Total Reflection
SEM	Scanning Electron Microscopy
EDS	Energy Dispersive Spectroscopy
DTG	Differential Thermogravimetric Analysis
C-S-H	calcium silicate hydrated compounds
Rw	cement pastes cured in water immersion
Rc	cement pastes cured in wetting-drying cycles
NLw	cement pastes with NC 1.5%wt, cured in water immersion
NLc	cement pastes with NC 1.5%wt, cured in wetting-drying cycles
NSLw	cement pastes with NC 1.5%wt and NS 1.5%wt, cured in water immersion
NSLc	cement pastes with NC 1.5%wt and NS 1.5%wt, cured in wetting-drying cycles

References

1. Abousnina, R.; Manalo, A.; Ferdous, W.; Lokuge, W.; Benabed, B.; Saif Al-Jabri, K. Characteristics, strength development and microstructure of cement mortar containing oil-contaminated sand. *Constr. Build. Mater.* **2020**, *252*, 119155. [CrossRef]
2. Ahn, T.H.; Kishi, T. Crack self-healing behavior of cementitious composites incorporating various mineral admixtures. *J. Adv. Concr. Technol.* **2010**, *8*, 171–186. [CrossRef]

3. Bakand, S.; Hayes, A. Toxicological considerations, toxicity assessment, and risk management of inhaled nanoparticles. *Int. J. Mol. Sci.* **2016**, *17*, 929. [CrossRef] [PubMed]
4. Bakand, S.; Hayes, A.; Dechsakulthorn, F. Nanoparticles: A review of particle toxicology following inhalation exposure. *Inhal. Toxicol.* **2012**, *24*, 125–135. [CrossRef]
5. Edvardsen, C. Water permeability and autogenous healing of cracks in concrete. *ACI Mater. J.* **1999**, *96*, 448–454.
6. Hartley, P.A.; Parfitt, G.D.; Pollack, L.B. The role of the van der Waals force in the agglomeration of powders containing submicron particles. *Powder Technol.* **1985**, *42*, 35–46. [CrossRef]
7. Huang, H.; Ye, G.; Qian, C.; Schlangen, E. Self-healing in cementitious materials: Materials, methods and service conditions. *Mater. Des.* **2016**, *92*, 499–511. [CrossRef]
8. Huang, H.; Ye, G.; Shui, Z. Feasibility of self-healing in cementitious materials - By using capsules or a vascular system? *Constr. Build. Mater.* **2014**, *63*, 108–118. [CrossRef]
9. Huseien, G.F.; Shah, K.W.; Sam, A.R.M. Sustainability of nanomaterials based self-healing concrete: An all-inclusive insight. *J. Build. Eng.* **2019**, *23*, 155–171. [CrossRef]
10. Jiang, Z.; Li, W.; Yuan, Z. Influence of mineral additives and environmental conditions on the self-healing capabilities of cementitious materials. *Cem. Concr. Compos.* **2015**, *57*, 116–127. [CrossRef]
11. Kan, L.L.; Shi, H.S. Investigation of self-healing behavior of Engineered Cementitious Composites (ECC) materials. *Constr. Build. Mater.* **2012**, *29*, 348–356. [CrossRef]
12. Kusters, K.A.; Pratsinis, S.E.; Thoma, S.G.; Smith, D.M. Ultrasonic fragmentation of agglomerate powders. *Chem. Eng. Sci.* **1993**, *48*, 4119–4127. [CrossRef]
13. Li, G.; Liu, S.; Niu, M.; Liu, Q.; Yang, X.; Deng, M. Effect of granulated blast furnace slag on the self-healing capability of mortar incorporating crystalline admixture. *Constr. Build. Mater.* **2020**, *239*, 117818. [CrossRef]
14. Liu, F.; Gao, Y.; Zhao, S.; Shen, Q.; Su, Y.; Wang, D. Biomimetic fabrication of pseudohexagonal aragonite tablets through a temperature-varying approach. *Chem. Commun.* **2010**, *46*, 4607–4609. [CrossRef] [PubMed]
15. Liu, H.; Huang, H.; Wu, X.; Peng, H.; Li, Z.; Hu, J.; Yu, Q. Effects of external multi-ions and wet-dry cycles in a marine environment on autogenous self-healing of cracks in cement paste. *Cem. Concr. Res.* **2019**, *120*, 198–206. [CrossRef]
16. Nasim, M.; Dewangan, U.K.; Deo, S.V. Effect of crystalline admixture, fly ash, and PVA fiber on self-healing capacity of concrete. *Mater. Today Proc.* **2020**. [CrossRef]
17. Perić, J.; Vučak, M.; Krstulović, R.; Brečević, L.; Kralj, D. Phase transformation of calcium carbonate polymorphs. *Thermochim. Acta* **1996**, *277*, 175–186. [CrossRef]
18. Puligilla, S.; Mondal, P. Co-existence of aluminosilicate and calcium silicate gel characterized through selective dissolution and FTIR spectral subtraction. *Cem. Concr. Res.* **2015**, *70*, 39–49. [CrossRef]
19. Qureshi, T.; Kanellopoulos, A.; Al-Tabbaa, A. Autogenous self-healing of cement with expansive minerals-I: Impact in early age crack healing. *Constr. Build. Mater.* **2018**, *192*, 768–784. [CrossRef]
20. Reches, Y. Nanoparticles as concrete additives: Review and perspectives. *Constr. Build. Mater.* **2018**. [CrossRef]
21. Roig-Flores, M.; Moscato, S.; Serna, P.; Ferrara, L. Self-healing capability of concrete containing crystalline admixtures in different exposure conditions. *Concr. Innov. Des. Fib Symp. Proc.* **2015**, *86*, 230–231.
22. Roig-Flores, M.; Pirritano, F.; Serna, P.; Ferrara, L. Effect of crystalline admixtures on the self-healing capability of early-age concrete studied by means of permeability and crack closing tests. *Constr. Build. Mater.* **2016**, *114*, 447–457. [CrossRef]
23. Sanchez, F.; Sobolev, K. Nanotechnology in concrete—A review. *Constr. Build. Mater.* **2010**, *24*, 2060–2071. [CrossRef]
24. Singh, L.P.; Karade, S.R.; Bhattacharyya, S.K.; Yousuf, M.M.; Ahalawat, S. Beneficial role of nanosilica in cement based materials—A review. *Constr. Build. Mater.* **2013**, *47*, 1069–1077. [CrossRef]
25. Stefanidou, M.; Tsampali, E.; Karagiannis, G.; Amanatiadis, S.; Ioakim, A.; Kassavetis, S. Techniques for recording self-healing efficiency and characterizing the healing products in cementitious materials. *Mater. Des. Process. Commun.* **2020**. [CrossRef]
26. Stefanidou, M.; Tsardaka, E. Influence of nano-CaO in self-healing properties of cement-based mortars. In Proceedings of the 6th International Conference on Self-Healing Materials, Friedrichshafen, Germany, 25–28 June 2017.

27. Stefanidou, M.; Tsardaka, E.; Tsampali, E. The role of nano-particles in Self-healing process of cementitious materials. In Proceedings of the SynerCrete18: Interdisciplinary Approaches for Cement-based Materials and Structural Concrete: Synergizing Expertise and Bridging Scales of Space and Time, Funchal, Portugal, 24–26 October 2018.
28. Steiner, S.; Lothenbach, B.; Proske, T.; Borgschulte, A.; Winnefeld, F. Effect of relative humidity on the carbonation rate of portlandite, calcium silicate hydrates and ettringite. *Cem. Concr. Res.* **2020**, *135*, 106116. [CrossRef]
29. Suleiman, A.R.; Nehdi, M.L. Effect of environmental exposure on autogenous self-healing of cracked cement-based materials. *Cem. Concr. Res.* **2018**, *111*, 197–208. [CrossRef]
30. Taylor, H.F.W. *Cement Chemistry*; Thomas Telford: London, UK, 1997. [CrossRef]
31. Tsampali, E.; Tsardaka, E.C.; Pavlidou, E.; Paraskevopoulos, K.M.; Stefanidou, M. Comparative study of the properties of cement pastes modified with nano-silica and Nano-Alumina. *Solid State Phenom.* **2019**. [CrossRef]
32. Tsardaka, E.-C.; Stefanidou, M. Application of an alternative way for silica fume dispersion in cement pastes without ultrasonication. *Cem. Concr. Res.* **2019**, *115*. [CrossRef]
33. Tsardaka, E.C.; Pavlidou, E.; Stefanidou, M. The role of nanoparticles on the durability of lime-pozzolan binding system. *Solid State Phenom.* **2019**, *286*, 119–132. [CrossRef]
34. Wang, X.; Fang, C.; Li, D.; Han, N.; Xing, F. A self-healing cementitious composite with mineral admixtures and built-in carbonate. *Cem. Concr. Compos.* **2018**, *92*, 216–229. [CrossRef]
35. Yang, Y.; Lepech, M.D.; Yang, E.H.; Li, V.C. Autogenous healing of engineered cementitious composites under wet-dry cycles. *Cem. Concr. Res.* **2009**, *39*, 382–390. [CrossRef]
36. Yuan, L.; Chen, S.; Wang, S.; Huang, Y.; Yang, Q.; Liu, S.; Wang, J.; Du, P.; Cheng, X.; Zhou, Z. Research on the improvement of concrete autogenous self-healing based on the regulation of cement particle size distribution (PSD). *Materials* **2019**, *12*, 2818. [CrossRef] [PubMed]
37. Zhang, W.; Zheng, Q.; Ashour, A.; Han, B. Self-healing cement concrete composites for resilient infrastructures: A review. *Compos. Part B Eng.* **2020**, *189*, 107892. [CrossRef]

Publisher's Note: MDPI stays neutral with regard to jurisdictional claims in published maps and institutional affiliations.

materials

MDPI

Article

Lightweight Gypsum Materials with Potential Use for Thermal Insulations

Cristina Dima, Alina Badanoiu *, Silviu Cirstea, Adrian Ionut Nicoara and Stefania Stoleriu

Faculty of Applied Chemistry and Materials Science, University Politehnica of Bucharest, Str. Ghe. Polizu 1-7, 011061 Bucharest, Romania; dima_cryss@yahoo.com (C.D.); cirstea.silviu94@gmail.com (S.C.); adrian.nicoara@upb.ro (A.I.N.); stefania.stoleriu@upb.ro (S.S.)

* Correspondence: alina.badanoiu@upb.ro

Received: 25 October 2020; Accepted: 26 November 2020; Published: 30 November 2020

Abstract: This article presents the influence of three additions i.e., hydroxyethyl methyl cellulose (HEMC), sodium bicarbonate and flue gas desulfurization (FGD) gypsum on the porosity of gypsum-based materials. The specific microstructure for a material with good thermal insulation properties i.e., numerous closed pores distributed in the binding matrix, was achieved using HEMC (0.3 wt.%) and sodium bicarbonate (0.5–2 wt.%). The addition of HEMC to the gypsum binder determines, as expected, an increase of the porosity due to its ability to stabilize entrained air. In the case of a sodium bicarbonate addition, the pores are formed in the binding matrix due to the entrapment of the gas (CO_2) generated by its reaction. Sodium bicarbonate addition delays the setting of gypsum binder therefore in this study FGD gypsum (waste produced in the desulfurization process of combustion gases generated in power plants) was also added to the mixture to mitigate this negative effect. The decrease of geometrical density (up to 13%, in correlation with the additive nature and dosage) correlated with the increase of the porosity, determines, as expected, the decrease of flexural and compressive strengths (33–75%), but improves the thermal properties i.e., decreases the thermal conductivity (9–18%).

Keywords: gypsum binder; additives; light-materials; porosity; thermal insulation

---

1. Introduction

Porous materials have numerous applications which depend on the composition of material, as well as number of pores, distribution, shape, size, connectivity, etc. [1]. Among the numerous applications of porous materials are thermal and acoustic insulations, filtration membranes, heavy metal absorption, catalysis, electro-magnetic interference shielding, energy and health (scaffolds, substrates for controlled drug release, wound healing, etc.) [1–3].

Porous materials used for thermal insulations are numerous and possess a wide range of properties. Nowadays, the most common thermal insulation materials are those based on organic compounds such as polyurethane and polystyrene; these organic materials have low densities and low values of thermal conductivity, but their main disadvantage is reduced fire resistance. Therefore, part of the research currently performed in this area focusses on the development of inorganic, fire-resistant thermal insulation materials [4–6]. Another topic related to the development of highly efficient and resistant thermal insulation materials is the incorporation of nanomaterials such as graphene oxide, nanocellulose or aerogels [2,7–9]. The high manufacturing cost of these nanomaterials is the main drawback which limits their current application on a large scale in the construction industry [9].

Another material which can be used for the development of cost-effective light weight porous materials with good thermal and acoustical insulation properties is gypsum [4,10].

Gypsum binders are obtained by the thermal treatment of gypsum rock ($CaSO_4{\cdot}2H_2O$–CsD) at temperatures above 105 °C, when it can be transformed in hemi-hydrate ($CaSO_4{\cdot}0.5H_2O$–CsH)

or in anhydrite ($CaSO_4$)—at higher temperatures [11]. Due to the relatively low thermal treatment temperatures and the possibility to recycle it in close loops, the gypsum binder is considered environmentally friendly as compared with other types of inorganic cements [12]. Nowadays, one of the main utilizations of gypsum binder is the production of gypsum plasterboards (dry-walls) [11,12]; another utilization of gypsum binder is to produce thermal and sound insulation materials [4,11,13–18]. According to Dolezelova et al. [10], gypsum-based light materials can be used in construction to replace autoclaved aerated concrete (AAC) and the energy consumption for the manufacture of gypsum materials is lower as compared with the one used for the manufacture of AAC.

The thermal insulation materials based on gypsum can be obtained either by its mixing with lightweight aggregates, including various waste such as expanded polystyrene [18], rubber, polyurethane foam and chopped electric cables waste [15,16], or direct foaming using various types of gas generators (aluminum sulphate with citric acid, calcium carbonate or sodium bicarbonate [4,13,14]). The thermal conductivity of this type of materials can vary from 0.085 up to 0.416 W/(m.K) depending on the nature and dosage of the additions and the assessment method [4,10,14,15].

In this paper gypsum-based materials were prepared using two foaming additions: hydroxyethyl methyl cellulose (HEMC) and sodium bicarbonate. In contact with water, sodium bicarbonate generates CO_2, which can be trapped in the binding matrix, thus generating supplementary porosity. The presence of sodium bicarbonate in gypsum paste delays the binders setting, and so therefore in the materials presented in this paper, flue gas desulfurization gypsum (FGD gypsum) was used as a setting accelerator.

FGD gypsum is a waste produced in the desulfurization process of combustion gases generated in power plants. The desulphurization process of exhaust gasses from combustion plants is imposed by the EU legislative framework regarding emissions [19]. This process consists of SO_2 (from flue gas) reaction with an alkaline substance (such as dolomite, limestone, lime, or hydrated lime) to produce sulphite or sulphate [20]. The presence of $CaSO_4 \cdot 2H_2O$ (CsD) as a main component in FGD gypsum justifies its use as a setting accelerator in gypsum binders. The CsD particles from FGD gypsum can act as nucleation sites for newly formed CsD crystals (by $CaSO_4 \cdot 0.5H_2O$ hydration), therefore accelerating the setting process [14,16].

Hydroxyethyl methyl cellulose addition was used to produce supplementary porosity, due to its ability to stabilize the air entrained during the mixing of the components (high affinity for the water-air interface) [21,22].

This paper presents the influence of these three additions (sodium bicarbonate, FGD gypsum and hydroxyethyl methyl cellulose) on the main properties of gypsum binders i.e., setting time, geometrical density, open porosity, compressive strength, and thermal conductivity.

2. Materials and Methods

The materials used in this research were:

- Gypsum binder (Saint-Gobain, Bucharest, Romania) with the following characteristics: calcium sulphate content over 50%, setting time 15-20 min and a fineness corresponding to 99% passing through 315 μm sieve;
- Flue gas desulfurization (FGD) gypsum (CET, Romania) resulted in the desulfurization process of gases emitted during coal combustion in a power plant; the as-received FGD gypsum was moist (approx. 40% water) therefore it was dried at 40 °C and grinded up to a fineness corresponding to 98% passing through 200 μm sieve, before mixing with the other components; the main compound assessed by X ray diffraction (XRD) in FGD gypsum is $CaSO_4 \cdot 2H_2O$ (CsD)—Figure 1.
- Sodium bicarbonate ($NaHCO_3$) (Sigma-Aldrich, Darmstadt, Germany) chemical reagent;
- Hydroxyethyl methyl cellulose (HEMC) (Dow, Midland, MI, USA)—chemical product.

The composition of studied materials is presented in Table 1.

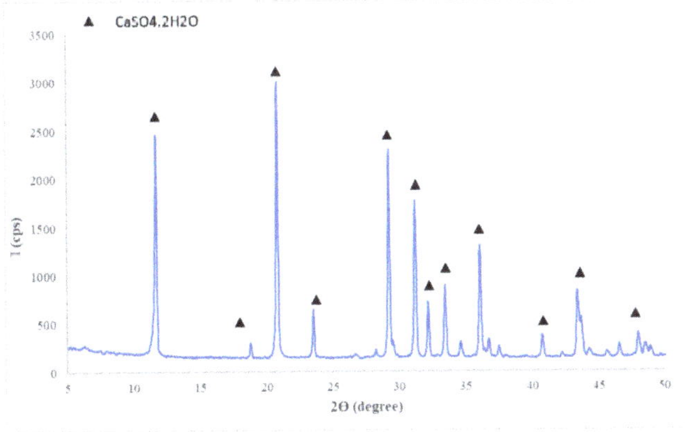

Figure 1. XRD patterns of flue gas desulfurization (FGD) gypsum.

Table 1. Composition of studied materials.

Specimens	Dosage (wt.%)			
	Gypsum Binder (I)	FGD Gypsum (G)	$NaHCO_3$ * (B)	HEMC * (C)
I	100	-	-	-
IC	100	-	-	0.3
IB0.5	100	-	0.5	-
IB1	100	-	1	-
IB2	100	-	2	-
IG5B0.5	95	5	0.5	-
IG5B1	95	5	1	-
IG5B2	95	5	2	-
IG10B0.5	90	10	0.5	-
IG10B1	90	10	1	-
IG10B2	90	10	2	-
IG3B0.5	97	3	0.5	-
IG1B0.5	99	1	0.5	-
IG1B1	99	1	1	-

* calculated with reference to I+G mixture; B and HEMC were added to I+G mixture.

The compositions were labelled as follow: I—gypsum binder; G—FGD gypsum (1 wt.%, 3 wt.%, 5 wt.% or 10 wt.%); B—sodium bicarbonate (0.5 wt.%, 1 wt.% or 2 wt.%); C—HEMC (0.3 wt.%).

For all compositions, the water to binder ratio was 0.6.

The following analysis were performed on binder pastes:

- Setting times were assessed according to the methods presented in ASTM C 472—99 and GB/T17669.4-1999 [23,24];
- Compressive or/and flexural strengths were assessed on paste specimens (cubes—20 × 20 × 20 mm and cuboids—40 × 40 × 160 mm) hardened for 7 days in air; the cuboid specimens were dried at 40 ± 2 °C up to constant mass before the test; minimum 3 flexural strength values and minimum 6 compressive strength values were considered for the calculation of average values of flexural and compressive strengths; the mixing procedure for paste preparation, the curing conditions (air, RH = 50 ± 5%, 23 ± 2 °C) and the procedure for mechanical testing are described in EN 13279-2:2014 [25];
- Geometrical density was assessed on cubes (20 × 20 × 20 mm) and was calculated as mass to volume ratio based on minimum 5 values (assessed on specimens cured in similar conditions);

- Open porosity was measured on fractured specimens, using the liquid saturation method under vacuum [26]; the working liquid was xylene (ρ = 0.866 g/mL).
- Thermal conductivity was assessed according to European Standard EN 12667 [27] on boards (300 × 300 × 20 mm) hardened for 7 days in air (RH = 50 ± 5%, 23 ± 2 °C). Before testing the specimens were dried at 40 °C up to constant mass. The HESTO-Lambda-CONTROL A90 (HESTO Elektronik GmbH, Steinbach, Germany) equipment measures the heat flow through a specimen placed between two plates with different temperatures [27]. All tests were performed at 23 ± 2°C and the values at 10 °C were calculated with the equipment's software.
- X ray diffraction analyses were performed on a Shimadzu XRD 6000 (Shimadzu, Kyoto, Japan) with monochromatic radiation CuKα (λ = 1.5406 Å);
- Scanning Electron Microscopy (SEM) analyses were performed on pastes hardened in air, using a high-resolution electronic scanning microscope equipped with a Schottky emission electron beam FEI Inspect F50 (Thermo Fisher—former FEI, Eindhoven, Nederland) with a resolution of 1.2 nm at 30 kV and 3 nm at 1 kV (BSE). All the SEM images were acquired on freshly fractured samples, fixed with carbon tape on an aluminium holder, covered with gold by metallization for 45 s, and then visualized in a vacuum using a 30 kV acceleration voltage and spot 3.5.

3. Results and Discussion

The setting times of the compositions based on gypsum binder are presented in Figure 2. As can be seen, the addition of both HEMC and sodium bicarbonate strongly delay the setting times. The delay caused by the HEMC addition could be due to the formation of a polymer film on the surface of gypsum plaster grains which prevents their hydration.

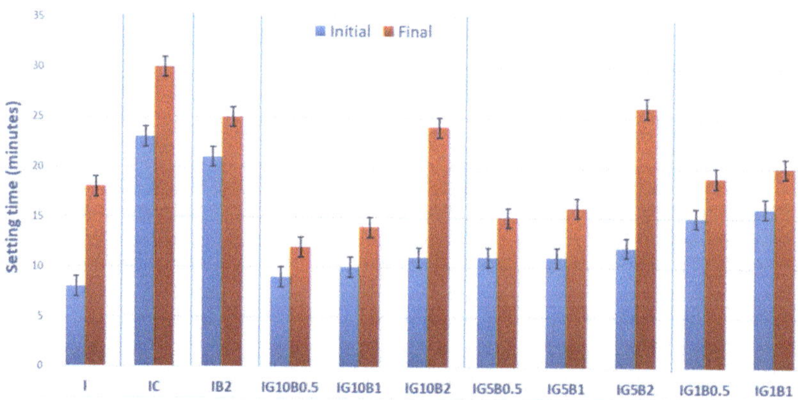

Figure 2. Setting times of gypsum binders with 0.3% HEMC (IC) and various amounts of sodium bicarbonate (0.5%, 1% and 2%) and FGD gypsum (1%, 5% and 10%).

The presence of sodium bicarbonate addition also inhibits the hydration of gypsum binder and delays the setting time; these data are in good correlation with those reported by Umponpanarat and Wansom [14].

To mitigate this effect FGD gypsum was added to the binding system. Previous results reported by our research group showed an important decrease of setting time when FGD gypsum is used as addition to gypsum binder [16]. The calcium sulphate dihydrate (CsD) crystals present in FGD gypsum can act as nucleation sites for the new CsD crystals, formed by the hydration of calcium sulphate hemihydrate (CsH) from gypsum plaster, therefore shortening the setting time (Figure 2).

As can be seen in Figure 2, the decrease of sodium bicarbonate (B) dosage from 2% up to 0.5% in the compositions with FGD gypsum (1%, 5% and 10%), decreases the setting times. The increase of FGD gypsum dosage from 5% up to 10% decreased the setting times.

The relative geometrical density and relative compressive strength (calculated with reference to the values obtained for I specimen) for the gypsum binder with 0.3% HMEC and various amounts of sodium bicarbonate and FGD gypsum are presented in Figure 3.

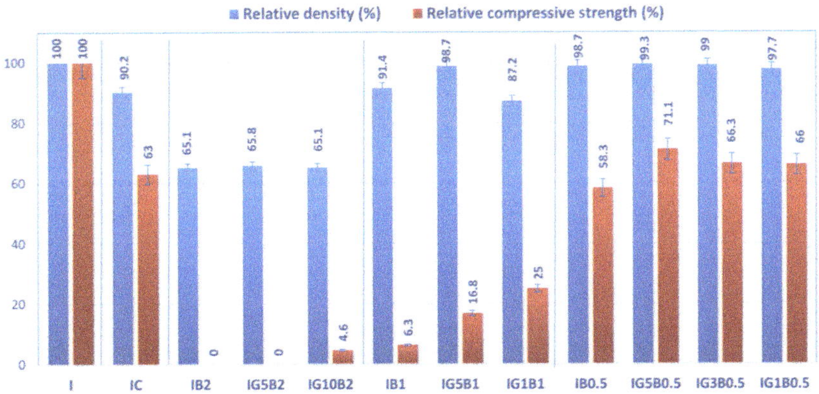

Figure 3. Influence of FGD gypsum (G), sodium bicarbonate (B) and HEMC (C) additions on the normalized geometrical density and compressive strength.

The addition of HEMC to the gypsum binder determines an increase of the porosity (as will be further presented), and so therefore the values of geometrical density and compressive strengths are smaller for IC as compared with I.

The addition of sodium bicarbonate also decreases the geometrical density and compressive strengths in correlation with its dosage (Figure 3); for the highest dosage of sodium bicarbonate (2 wt.%) the compositions IB2 and IG5B2 have lower values of geometrical density (as compared with I), but no recordable compressive strength. The increase of FGD gypsum dosage up to 10 wt.% (IG10B2) determines a small increase of compressive strength values for a similar value of geometrical density.

As expected, the decrease of sodium bicarbonate dosage determines the increase of compressive strengths correlated with higher values of geometrical density. This evolution is in good correlation with the porosity of these materials (as will be presented further).

The microstructure of gypsum pastes hardened for 7 days in air was assessed by scanning electron microscopy (Figures 4–9).

In the SEM images of I specimen one can notice the presence of large round pores with sizes comprised between 0.1–0.25 microns (see arrow in Figure 4a) formed by the air entraining during the mixing operation. In Figure 4b one can notice the presence of smaller pores with various shapes and sizes (see arrows) specifically for the binding matrices formed by the hydration of CsH; the long CsD crystals are interlocked and forms the binding matrix [13,15,28].

For the IC specimen (Figure 5) the presence of HEMC addition determines an important increase of pores numbers and sizes, up to 0.75–1 mm. A close-up on the binding matrix reveals the presence of interlocked needle-like CsD crystals specific for hardened gypsum binder (see arrow in Figure 5c,d). The increase of the number of round closed pores assessed on SEM images (see arrows in Figure 5a) when HEMC is added in the binding system, can be explained by its specific ability to stabilize the entrained air (high affinity for the water-air interface [21,22]).

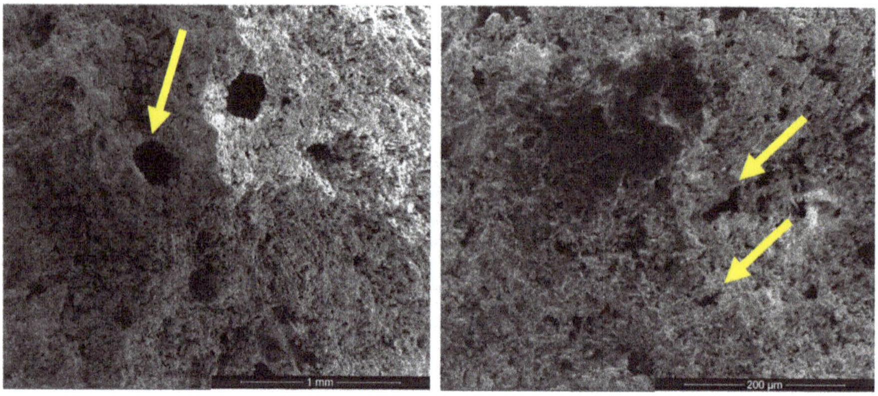

(**a**) (**b**)

Figure 4. SEM images of gypsum binder (I) after 7 days of hardening in air, with various magnifications: (**a**) ×100; (**b**) ×500.

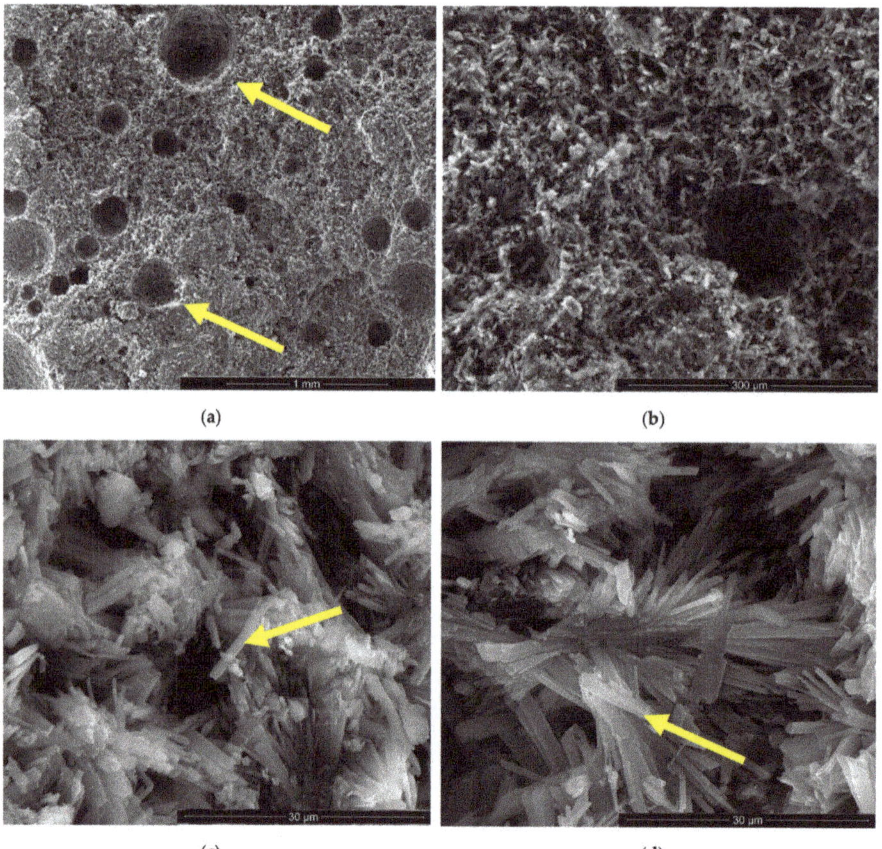

(**a**) (**b**)

(**c**) (**d**)

Figure 5. SEM images of gypsum binder with HEMC (IC) after 7 days of hardening in air with various magnifications: (**a**) ×100; (**b**) ×500; (**c**) ×5000; (**d**) ×5000.

For the specimens with sodium bicarbonate an important increase of volume was noticed shortly after the pouring of the paste in the mold. This phenomenon is due to the CO_2 generation in the reaction of sodium bicarbonate with water and calcium sulphate hemihydrate (CsH) [13]. This process is more intense for a higher dosage of sodium bicarbonate and the resulted porosity is higher, as can be noticed for the SEM images presented in Figures 6–9.

For the gypsum plaster with 2% sodium bicarbonate (IB2), one may assess in the SEM images the presence of big pores (over 1 mm) formed in the binding matrix by the released CO_2 gas (Figure 6a); the irregular shapes of these pores suggest the coalescence of smaller pores.

It is interesting to note the shape of crystals in IB2 i.e., short thick crystals and plate like crystals (Figure 6c,d). The modification of CsD crystal shape and size can be due to the modification of reaction conditions in the presence of sodium bicarbonate addition [28] or/and to an oriented growth of crystals due to selective adsorption of retarding addition [24].

In the SEM images of plaster with 5% FGD gypsum and 2% sodium bicarbonate (IG5B2) one may notice the presence of numerous round pores with sizes comprised between 0.25–1 mm (Figure 7a) as well as smaller pores (10–20 microns) formed between the CsD crystals (Figure 7b,d). Short and plate-like CsH crystals also present in this composition (Figure 7c,d).

As expected, the reduction of sodium bicarbonate content reduces the average size of pores assessed on SEM images (Figures 8a and 9a). The CsD crystals continues to be shorter with an average size of 10–20 microns. The interlocking of gypsum crystals increases (Figure 9b,c) and this contributes to the increase of mechanical strengths.

Rubio Avalos et al. [13] reported the presence of Na_2SO_4 as secondary phase in the $CaSO_4{\cdot}0.5H_2O$–$NaHCO_3$–H_2O system; according to these authors, the sodium sulphate (small round crystals) precipitates inside the gypsum crystals bulk. This phase was not detected in this study by XRD in the specimens with 0.5% and 1% sodium bicarbonate (Figure 10); this could be due to the low dosage of sodium bicarbonate in these compositions.

Nevertheless, the EDX analysis presented in Figure 11 shows the presence of sodium in high quantity in some specific areas, which could be associated to a phase with sodium content.

The European norm EN 13279-1 sets the requirements for gypsum binders and plasters; these requirements refer to the flexural strength (higher than 1 N/mm^2) and compressive strength (higher than 2 N/mm^2) [29].

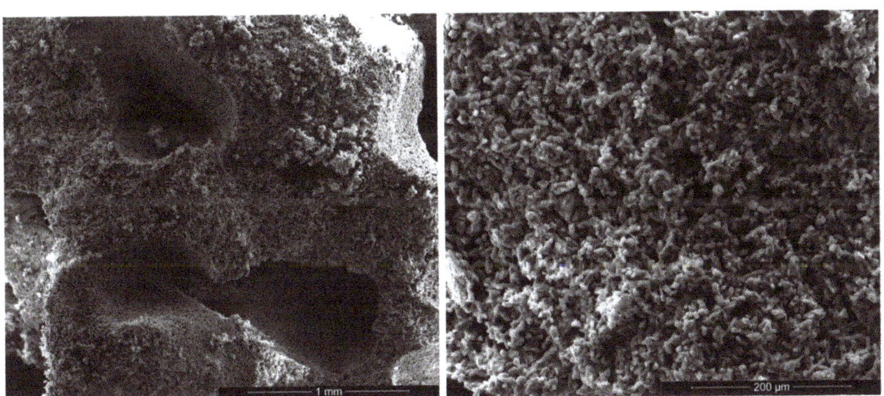

(**a**) (**b**)

Figure 6. *Cont.*

(c) (d)

Figure 6. SEM images of gypsum plaster with 2% sodium bicarbonate addition (IB2) with various magnifications: (**a**) ×100; (**b**) ×500; (**c**) ×5000; (**d**) ×5000.

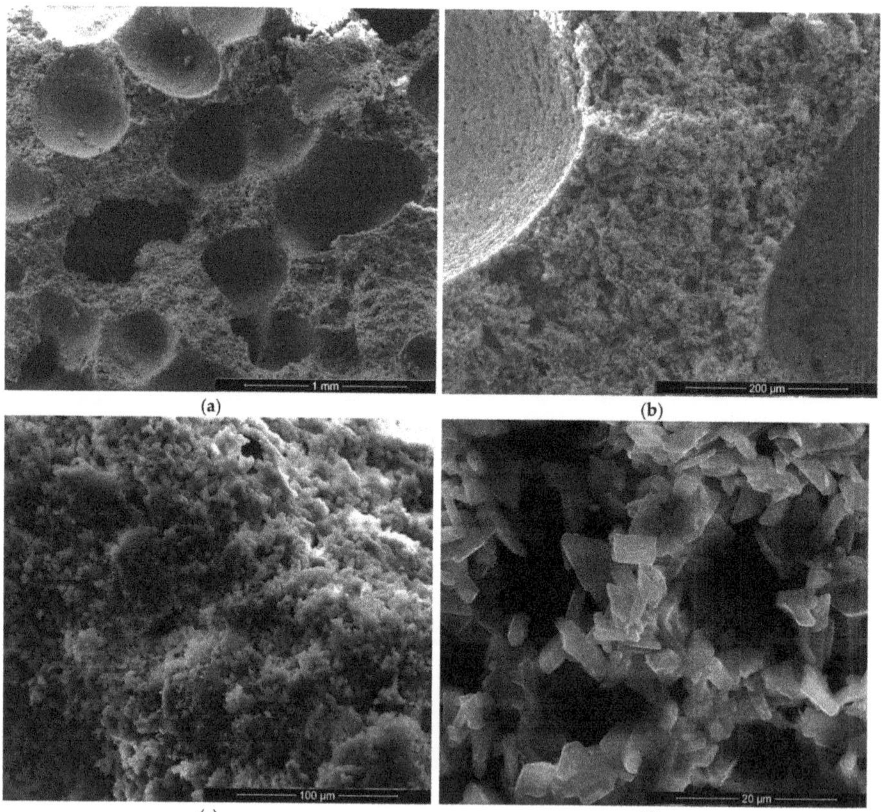

(a) (b) (c) (d)

Figure 7. SEM images of gypsum plaster with 5% FGD and 2% sodium bicarbonate addition (IG5B2) with various magnifications: (**a**) ×100; (**b**) ×500; (**c**) ×1000; (**d**) ×5000.

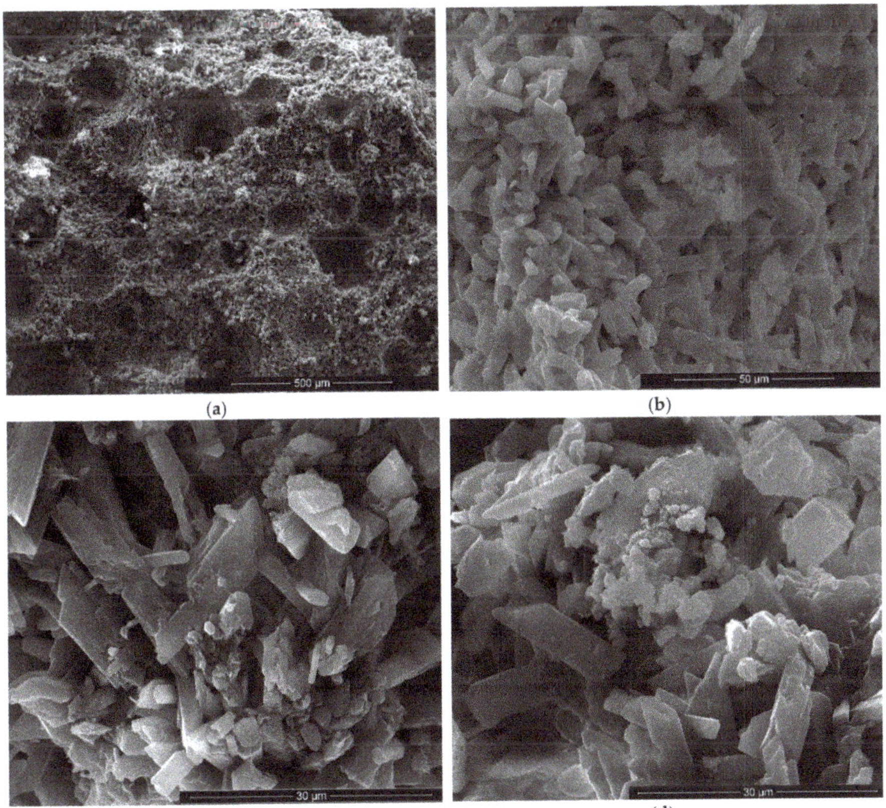

Figure 8. SEM images of gypsum binder with 1% FGD and 1% sodium bicarbonate addition (IG1B1) after 7 days of hardening in air with various magnifications: (**a**) ×200; (**b**) ×2000; (**c**) ×5000; (**d**) ×5000.

The compositions IC and IG1B1 fulfill the above-mentioned requirements (Table 2), and therefore thermal conductivity was assessed on these specimens. The thermal conductivity at 10 °C (set I), assessed in accordance with the norm EN ISO 10456 [30], is the thermal conductivity value usually declared by the European producers of this type of construction material based on the fact that 10 °C is considered the average yearly temperature at which the thermal insulation of buildings must operate.

As expected, the decrease of the geometrical density and increase of open porosity of IC and IG1B1, due to the presence of hydroxyethyl methyl cellulose (HEMC) and sodium bicarbonate, improves the thermal properties of these materials (i.e., reduces the values of thermal conductivity at 10 °C).

Table 2. Geometrical density, open porosity, flexural and compressive strengths and thermal conductivity of gypsum-based materials.

Specimens	Geometrical Density (kg/m^3)	Open Porosity (%)	Flexural Strength (N/mm^2)	Compressive Strength (N/mm^2)	Thermal Conductivity at 10 °C (W·m^{-1}·K^{-1})
I	1253 ± 25	29.76 ± 0.30	5.35 ± 0.26	21.06 ± 1.05	0.2112 ± 0.0002
IC	1130 ± 22	32.87 ± 0.33	3.39 ± 0.15	13.25 ± 0.65	0.1917 ± 0.0002
IG1B1	1093 ± 21	37.38 ± 0.38	2.07 ± 0.05	5.26 ± 0.22	0.1736 ± 0.0002

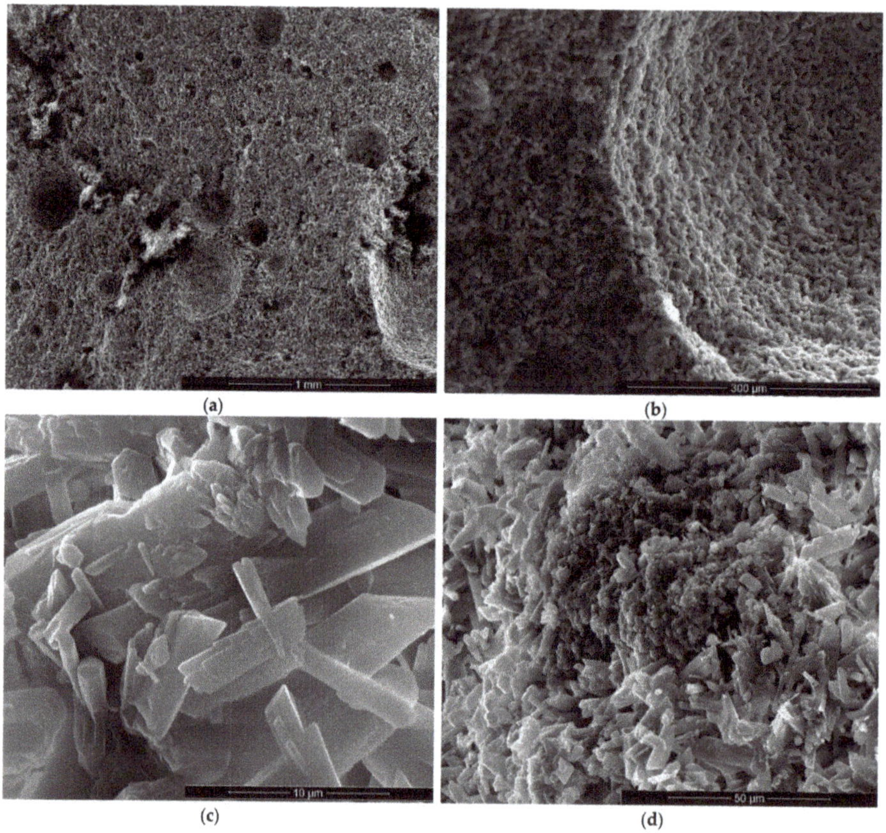

Figure 9. SEM images of gypsum binder with 1% FGD and 0.5% sodium bicarbonate addition (IG1B0.5) after 7 days of hardening in air, with various magnifications: (**a**) ×100; (**b**) ×500; (**c**) ×10000; (**d**) ×2000.

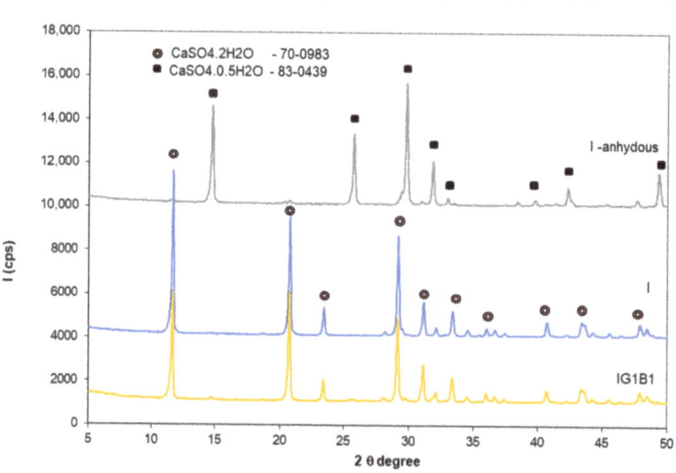

Figure 10. XRD patterns of gypsum binder I—anhydrous and I and IG1B1 hardened pastes.

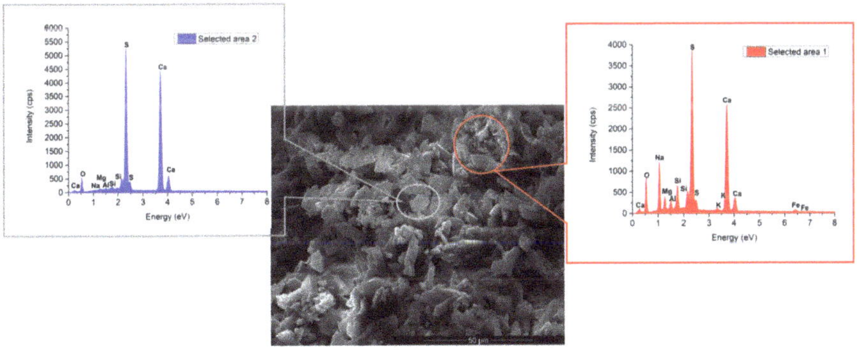

Figure 11. SEM and EDX analyses of IG1B1 paste.

In conformity with European norm EN 13279-1 [29], gypsum plasters or gypsum binders are classified as reaction to fire Class A1 when they contain less than 1% by weight or volume of organic materials, without supplementary testing. The only material presented in this paper which contains organic material is the one with hydroxyethyl methyl cellulose but its amount (0.3%) is below the previously mentioned limit; therefore, the gypsum based materials obtained in this study can be classified as Class A1 (reaction to fire).

Based on the properties assessed for the studied gypsum-based materials a potential practical application could be for the manufacture of light gypsum blocks/boards for non-load-bearing walls with improved thermal insulation properties and good fire behavior.

4. Conclusions

In this study light gypsum-based materials were obtained using hydroxyethyl methyl cellulose (HEMC) and sodium bicarbonate. To reduce the setting time of gypsum binders with sodium bicarbonate addition, FGD gypsum was also added to the mixture.

The following conclusions can therefore be drawn:

- The porosity of the gypsum pastes with HEMC increases due to its specific ability to stabilize the entrained air during the mixing of solid and liquid components (high affinity for the water-air interface); the average size of the pores assessed by scanning electron microscopy (SEM) in these materials is comprised between 0.25–0.75 mm.
- The density of the materials with sodium bicarbonate decreases with the increase of $NaHCO_3$ dosage. Numerous pores (assessed by SEM) are formed in the binding matrix due to the entrapment of the gas (CO_2) generated by the $NaHCO_3$ reaction. For the specimens with 0.5% and 1% sodium bicarbonate, pores with sizes comprised between 10 microns and 1 mm are assessed by SEM. For the compositions with 2% $NaHCO_3$, big pores (sizes over 1 mm) with irregular shapes are present in the material; the irregular shapes of these pores suggest the coalescence of smaller pores.
- As expected, the decrease of density correlated with the increase of porosity determines the decrease of flexural and compressive strengths, with 37% for IC and 12–60% (flexural strength) and 33–75% (compressive strength) for the materials with sodium bicarbonate and FGD gypsum additions. Nevertheless, for an adequate dosage of the additives, flexural strengths higher than 2 N/mm^2 and compressive strength higher than 5 N/mm^2 were achieved.
- The values of thermal conductivity of the gypsum binders with the studied additives, decreases (9–18% with reference to plain gypsum). Moreover, because these materials contain less than 1% by weight of organic materials, in conformity with European norm EN 13279-1 they can be classified as Class A1—reaction to fire.

Author Contributions: Conceptualization, C.D. and A.B.; Investigation, C.D., S.C., A.I.N.; Methodology: C.D. and A.B.; Supervision, A.B. and S.S.; Writing—original draft, C.D. and A.B.; Writing—review & editing, A.B. and S.S. All authors have read and agreed to the published version of the manuscript.

Funding: The SEM analyzes were possible due to EU-funding project POSCCE-A2-O2.2.1-2013-1/Prioritary Axe 2, Project No. 638/12.03.2014, ID 1970, SMIS-CSNR code 48652.

Conflicts of Interest: The author declares no conflict of interest.

References

1. Nishihora, K.; Rachadel, P.L.; Quadri, M.G.N.; Hotza, D. Manufacturing porous ceramic materials by tape casting—A. review. *J. Eur. Ceram. Soc.* **2018**, *38*, 988–1001. [CrossRef]
2. Chenga, H.; Li, L.; Wang, B.; Feng, X.; Mao, Z.; Vancso, G.J.; Sui, X. Multifaceted applications of cellulosic porous materials in environment, energy, and health. *Prog. Polym. Sci.* **2020**, *106*, 101253. [CrossRef]
3. Singh, A.K.; Shishkin, A.; Koppel, T.; Gupta, N. A review of porous lightweight composite materials for electromagnetic interference shielding. *Compos. B. Eng.* **2018**, *149*, 188–197. [CrossRef]
4. Vimmrova, A.; Keppert, M.; Svoboda, L.; Cerny, R. Lightweight gypsum composites: Design strategies for multi-functionality. *Cem. Concr. Compos.* **2011**, *33*, 84–89. [CrossRef]
5. Bai, C.; Colombo, P. Processing, properties and applications of highly porous geopolymers: A review. *Ceram. Int.* **2018**, *44*, 16103–16118. [CrossRef]
6. Stoleriu, S.; Vlasceanu, I.N.; Dima, C.; Badanoiu, A.I.; Voicu, G. Alkali activated materials based on glass waste and slag for thermal and acoustic insulation. *Mater. Constr.* **2019**, *69*, e194, ISSN-L 0465-2746. [CrossRef]
7. Shang, L.; Lyu, Y.; Han, W. Microstructure and Thermal Insulation Property of Silica Composite Aerogel. *Materials* **2019**, *12*, 993. [CrossRef]
8. Stefanidou, M.; Pachta, V. Influence of perlite and aerogel addition on the performance of cement-based mortars at elevated temperatures. In Proceedings of the IOP Conf. Series: Earth and Environmental Science, Thessaloniki, Greece, 23–25 October 2019; IOP Publishing Ltd.: Bristol, UK, 2020; Volume 410, p. 012111.
9. Bostanci, L. Synergistic effect of a small amount of silica aerogel powder and scrap rubber addition on properties of alkali-activated slag mortars. *Constr. Build. Mater.* **2020**, *250*, 118885. [CrossRef]
10. Doleželová, M.; Krejsová, J.; Vimmrova, A. Lightweight gypsum based materials: Methods of preparation and utilization. *Int. J. Sus. Dev. Plann.* **2017**, *12*, 326–335. [CrossRef]
11. Pedreno-Rojas, M.A.; Flores-Colen, I.; Brito, J.; Rodríguez-Linan, C. Influence of the heating process on the use of gypsum wastes in plasters: Mechanical, thermal and environmental analysis. *J. Clean. Prod.* **2019**, *215*, 444–457. [CrossRef]
12. GotoGo Project. Available online: https://gypsumtogypsum.org/gtog/gtog-project/ (accessed on 8 June 2020).
13. Rubio-Avalos, J.C.; Manzano-Ramırez, A.; Yanez-Limo, J.M.; Contreras-Garcıa, M.E.; Alonso-Guzman, E.M.; Gonzalez-Hernandez, J. Development and characterization of an inorganic foam obtained by using sodium bicarbonate as a gas generator. *Constr. Build. Mater.* **2005**, *19*, 543–549. [CrossRef]
14. Umponpanarat, P.; Wansom, S. Thermal conductivity and strength of foamed gypsum formulated using aluminum sulfate and sodium bicarbonate as gas-producing additives. *Mater. Struct.* **2016**, *49*, 1115–1126. [CrossRef]
15. Dima, C.; Badanoiu, A.I.; Stoleriu, S.; Danes, C. Composite Materials Based on Gypsum Plaster and Industrial Wastes. *Rev. Rom. Mater.* **2018**, *48*, 147–153.
16. Dima, C.; Badanoiu, A.I.; Stoleriu, S.; Luca, D. Influence of various types of waste on the main properties of gypsum based composites for thermal insulations. *Rev. Chim.* **2019**, *70*, 1750–1753. [CrossRef]
17. Du, Z.; She, W.; Zuo, W.; Hong, J.; Zhang, Y.; Miao, C. Foamed gypsum composite with heat-resistant admixture under high temperature: Mechanical, thermal and deformation performances. *Cem. Concr. Compos.* **2020**, *108*, 103549. [CrossRef]
18. Bicer, A.; Kar, F. Thermal and mechanical properties of gypsum plaster mixed with expanded polystyrene and tragacanth. *Therm. Sci. Eng. Prog.* **2017**, *1*, 59–65. [CrossRef]
19. Directive (EU) 2015/2193 of the European Parliament and of the Council of 25 November 2015 on the Limitation of Emissions of Certain Pollutants into the Air from Medium Combustion Plants. Available online: https://eur-lex.europa.eu/legal-content/EN/TXT/PDF/?uri=CELEX:32015L2193 (accessed on 8 June 2020).

20. Lisnic, R.; Jinga, S.I. Study on current state and future trends of flue gas desulphurization technologies: A review. *Rev. Rom. Mater.* **2018**, *48*, 83–90.
21. Ciobanu, C.; Iluc, S.; Lazau, I.; Pacurariu, C. Some physico—Mechanical properties of dry mortars, with cellulose ethers additives. *Rev. Rom. Mater.* **2011**, *41*, 30–41.
22. Jenni, A.; Holzer, L.; Zurbriggen, R.; Herwegh, M. Influence of polymers on microstructure and adhesive strength of cementitious tile adhesive mortars. *Cem. Concr. Res.* **2005**, *35*, 35–50. [CrossRef]
23. ASTM C 472—99. *Standard Test Methods for Physical Testing of Gypsum, Gypsum Plasters and Gypsum Concrete*; ASTM International: West Conshohocken, PA, USA, 2020.
24. Zhang, Y.; Yang, J.; Cao, X. Effects of several retarders on setting time and strength of building gypsum. *Constr. Build. Mat.* **2020**, *240*, 117927. [CrossRef]
25. European Standard EN 13279-2. *Gypsum Binders and Gypsum Plasters. Part 2: Test Methods*; BSI: London, UK, 2014.
26. ISO 10545-3. *Ceramic Tiles—Part 3: Determination of Water Absorption, Apparent Porosity, Apparent Relative Density and Bulk Density*; ISO: Geneva, Switzerland, 1995.
27. European Standard EN 12667. *Thermal Performance of Building Materials and Products—Determination of Thermal Resistance by Means of Guarded Hot Plate and Heat Flow Meter Methods—Products of High and Medium Thermal Resistance*; BSI: London, UK, 2001.
28. Inoue, M.; Hirasawa, I. The relationship between crystal morphology and XRD peak intensity, on $CaSO_4 \cdot 2H_2O$. *J. Cryst. Growth.* **2013**, *380*, 169–175. [CrossRef]
29. European Standard EN 13279-1. *Gypsum Binders and Gypsum Plasters. Part 1: Definitions and Requirements*; BSI: London, UK, 2009.
30. European Standard EN ISO 10456. *Building Materials and Products. Hygrothermal Properties. Tabulated Design Values and Procedures for Determining Declared and Design Thermal Values*; CEN: Brussels, Belgium, 2007.

materials

MDPI

Article

The Influence of HEMC on Cement and Cement-Lime Composites Setting Processes

Edyta Spychał * and Przemysław Czapik

Faculty of Civil Engineering and Architecture, Kielce University of Technology, 25-314 Kielce, Poland; p.czapik@tu.kielce.pl
* Correspondence: espychal@tu.kielce.pl

Received: 20 November 2020; Accepted: 19 December 2020; Published: 20 December 2020

Abstract: In this article, the effect of hydroxyethyl methyl cellulose (HEMC), which is a polymeric viscosity modifying admixture on the mineral based composites setting processes, was studied. Previous studies available in the literature included the evaluation of the influence of this admixture on the hydration processes of cement or lime pastes. In this paper, the analysis of this issue was extended to include cement-lime composites. The composition of the pastes and mortars differed in the type of binder (the tests were carried out on cement-based and cement-lime-based materials, in which the cement was replaced in 50% with hydrated lime), as well as the amount and viscosity of the admixture. The study of mortars setting processes and hardening processes using the ultrasonic method was supplemented in the work with calorimetric measurements and phases analysis by the X-ray diffraction method. Finally, it was found that the HEMC reduces the rate of a hydration reaction in cement and cement-lime pastes. The amount of admixture used has a greater influence on the changes taking place during the setting process than the admixture viscosity or the type of binder.

Keywords: cement; hydrated lime; HEMC; mortar; paste; hydration; setting process; ultrasonic technique

1. Introduction

The use of chemical admixtures in modern mortar technology is a prerequisite for obtaining materials of high quality and durability that meet the standard requirement. The aim is to ensure that the mortars are characterized by good workability, ease of application, adequate adhesion to the masonry, and lowest possible shrinkage deformation.

Cellulose ethers (CE) are a group of polymer admixtures used in dry pre-mix mineral-based binder materials [1–3]. Hydroxyethyl methyl cellulose (HEMC) in addition to hydroxypropyl methyl cellulose (HPMC) is one of the most popular cellulose ethers [3–5]. They are a group of admixtures, with the use of which it is possible to modify the viscosity of pastes and mortars, and, thus, change their rheological or application properties [1,2,5–14], which is important to know how this admixture influences the properties of fresh mortars. The research [2,3,7,8,14,15] showed that the cellulose ethers have an impact on the increase of water retention. Bülichen et al. [3] found that, at low dosages, HEMC achieves water retention by intramolecular sorption of water and concomitant swelling while, at a higher amount of admixture, the molecules agglomerate into large, hydrocolloidal microgel particles. This admixture changes the material rheological properties including plastic viscosity and yield stress [1,3,5,6,12,14]. The high-water retention capacity of the mortar, which can be obtained by using these polymers, prevents rapid water loss from the material and the drainage of some water from the mortar to the substrate. Thanks to this, it is possible to obtain greater adhesion and reduce shrinkage deformations [16,17]. On the other hand, cellulose ethers can cause a delay of hydration processes, as a result of which the setting and hardening time of the binders and mortars become longer [1,18–27]. According to Hua et al. [25], the retardation effect of cellulose ethers favors the increase of operating time and the decrease of the consistency or slump loss of freshly mortars, but possibly leads to unexpected delay during construction. Study of the influence of this admixture on cement hydration kinetics can facilitate when controlling cement hydration, which is important, especially for application. Izaguirre et al. [27] tested two different additives (cellulose ether and guar gum), which were added to lim-based mortars. Among other things, the setting time was tested. The obtained results showed an increase in setting time from 3 h 15 min for the reference mortar to 10 h 30 min for mortar with an admixture. An adsorption mechanism of this admixture on the $Ca(OH)_2$ crystals was reported to reduce its entanglement between chains and, hence, the viscosity of the pastes as well as its water retention ability. Similar results of the influence on the delay of setting time were observed in ceramic tile adhesive mortars by Petit and Wirquin [1]. Initial setting time increased from 4 h for the reference mortar to 12 h 25 min for mortar modified with an admixture, but final setting time changed from 5 h 25 min to 18 h. Pourchez et al. [19] studied the influence of cellulose ethers on the hydration of C_3A and C_3A-sulphate systems. It was observed that this admixture leads to a gradual slowing down of the C_3A hydration depending on cellulose ether chemistry. Two types of cellulose ethers were used in their research: hydroxypropyl methyl cellulose (HPMC) and hydroxyethyl cellulose (HEC). They observed that HECs include a higher delay than HPMCs. In their opinion, the substitutions group seems to be more important for controlling factors on C_3A hydration rather than molecular mass. Betioli et al. [24] observed that cellulose ether HEMC did not change the typical profile of isothermal calorimetry curves. However, it decreased the rate of heat evolution during the acceleration period and decreased the maximum peak. It was expected that the induction period was importantly extended with a hydroxyethyl methyl cellulose (HEMC) admixture. The total heat output after 4 h of hydration was reduced (even 54% for paste with 0.5% HEMC). Oscillatory rheometry and isothermal calorimetry were used for showing the influence of cellulose ether on a hydration process. A steric dispersant barrier effect during the first two hours of hydration associated with a cement retarding nature was displayed, which resulted in reducing the setting speed. Authors of this article observed that the rheology of cement paste changes during the induction period and it was caused by the agglomeration of the particles. Pichniarczyk [22] concluded that methylcellulose has a significant influence on the early hydration of cement. However, that effect practically disappears after a longer period. The XRD (X-Ray Diffraction) pattern of cement paste with cellulose ether admixture peaks of portlandite and ettringite were of significantly lower intensity after 24 h of hydration, but gypsum peaks were of high intensity. These test results indicate a slower hydration process in pastes modified with methylcellulose. After 168 h, hydration differences in the intensity of portlandite and ettringite peaks were lower. X-ray diffraction was complemented with microstructure observations under the SEM. Knapen and Gemert [23] observed that the presence of water-soluble polymers in cement mortars influences the rate and degree of cement hydration as well as the nature and amount of hydration products. The heat evolution was measured by isothermal calorimetry measurements. It was found that the start of the acceleration stage of hydration was postponed by about 30 min in the MC (methyl cellulose) modified pastes (1% MC, HEC cement pastes) and by more than 5 h in the HEC modified paste. The induction period and the dormant period were extended. The rate of the following hydration reactions was slowed down. What was confirmed by lower values of the maximum heat release and broader exothermal peaks in the isothermal calorimetric curves of the cellulose ethers modified pastes.

The correct selection of the admixture, especially its quantity and viscosity, affects the parameters of the finished product. Knowledge of the processes taking place during the setting and hardening of mortars modified with this admixture allows for a correct and comprehensive assessment of its effectiveness and applicability.

Based on the publications available in the literature, it can be concluded that the research results presented so far focus mainly on determining the influence of cellulose ethers on the rheological, physical-mechanical properties of cement [3,5,6,24] or lime [7,8,13,28] pastes and mortars. The cement hydration processes or its impact on the individual phases are also assessed. The subject of setting processes of mortars modified with a cellulose ether admixture focuses on cement-based materials.

There is no explanation of the mechanism of action of this admixture and its influence on the hydration of pastes with a cement-lime binder. Therefore, this issue is the subject of these studies. Moreover, in publications for the evaluation of hydration processes, calorimetric measurements, XRD method, or conductometric measurements are usually used [21,22,24,28,29]. In this publication, the authors performed an analysis of the influence of the admixture on the setting processes of mortars, using the ultrasonic technique. Applying the proposed method of examinations, it is possible to describe the rate of setting and hardening of materials based on cement and cement-lime mortars with high accuracy due to the fact that measurements are made on an ongoing basis, which can make it possible to monitor and control the properties of mortars in the plastic state. Using the knowledge gained on the basis of an ultrasonic technique, it is possible to evaluate the influence of individual components on the parameters of the mortar or paste. The results of the ultrasonic technique complement the standard tests. In addition, they allow us to predict the mechanical properties of mortars [30].

2. Materials and Methods

2.1. Materials

In the studies, Portland cement CEM I 42.5 R (Cemex, Chełm, Poland), hydrated lime deeply separated (Alpol, ZSChiM "PIOTROWICE II", Sitkówka, Poland), two fractions of quartz sand 0.1–0.5 mm and 0.2–0.8 mm (Grudzeń Las, Grudzeń Las, Poland), and polymer admixture in the form of HEMC (WALOCEL, The Dow Chemical Company, Midland, MI, USA) were used. Cellulose ether used in tests is a hydroxyethyl methyl cellulose with viscosity of 25,000 mPa·s and 45,000 mPa·s (for measurements using a Brookfield rheometer at 20 °C) with low chemical modification. This admixture has the form of white powder with a grain size below 0.063 mm. The pH of its 1% solution in water (at 20 °C) is 7.0. The characteristics of cement and hydrated lime are given in Table 1 and in Table 2, respectively.

Table 1. Chemical composition and physical properties of cement.

Chemical Composition (wt. %)		Physical Properties	
SiO_2	20.22	Water requirement of normal consistency/%	28.8
Al_2O_3	4.43	Initial setting/min	173
Fe_2O_3	3.65	Final setting/min	237
CaO	64.06	Specific surface area/(m^2/kg)	387.9
Na_2O	0.29	2 d compressive strength/MPa	28.9
MgO	1.24	28 d compressive strength/MPa	59.5
SO_3	3.31	2 d flexural strength/MPa	5.1
Cl	0.093	28 d flexural strength/MPa	8.4
K_2O	0.54	Loss on ignition/%	3.81
free CaO	1.83	-	-

Table 2. Chemical composition of hydrated lime.

Chemical Composition (wt. %)	
CaO + MgO	95.17
MgO	0.80
CO_2	1.86
SO_3	0.41

2.2. Preparation of Mortars and Pastes

Mortars and pastes differing in their composition with the type of binder, amount, and viscosity of HEMC admixture were prepared for the tests. As a reference material, a cement mortar without admixture, designated as C-0, was used. In addition, cement composites modified with admixture in various amounts and viscosity (with symbols C-0.05-MV, C-0.3-MV, and C-0.175-HV) and composites modified with cellulose ether, in which cement was replaced with 50% hydrated lime (marked with symbols CL-0.05-MV, CL-0.3-MV, and CL-0.175-HV). The MV symbol stands for Medium Viscosity (cellulose ether with a viscosity of 25,000 mPa·s), whereas the HV symbol stands for High Viscosity (cellulose ether with viscosity of 45,000 mPa·s). The admixture was added in the amount of 0.05%, 0.175%, and 0.3% in the relation to the dry ingredients of the mortar (binder and fine aggregate). The detailed composition of mortars is given in Table 3. The article is a part of a wider study concerning the analyzed issue in which the influence of cellulose ethers and hydrated lime on selected properties of plastering mortars was studied [31]. Some research in this direction has been published in articles [14,15,30]. The composition of pastes was similar to the composition of mortars, excluding fine aggregate.

Table 3. Composition of mortars.

Mortar/Paste	Binder (Cement + Lime) (g)	Fine Aggregate (g)		Cellulose Ether (wt.%)	Viscosity of Cellulose Ether	Water (g)
		0.1–0.5 mm	0.2–0.8 mm		(mPa·s)	
C-0	96	437	467	0.00	-	195
C-0.05-MV	96	437	467	0.05	25,000	148
C-0.3-MV	96	437	467	0.30	25,000	220
C-0.175-HV	96	437	467	0.175	45,000	185
CL-0.05-MV	48 + 48	437	467	0.05	25,000	176
CL-0.3-MV	48 + 48	437	467	0.30	25,000	225
CL-0.175-HV	48 + 48	437	467	0.175	45,000	195

Cement, hydrated lime, and fine aggregate were weighed with an accuracy of 0.1 g, and the chemical admixture was weighed with an accuracy of 0.0001 g. The ingredient mixing procedure was as follows: 90 seconds of mechanical mixing at 45 rpm, 30 s of rest, and 90 s of mechanical mixing at 57 rpm. The time and method of mixing were constant (the same in each case).

All mortars were performed, maintaining a constant weight ratio of the binder to the fine aggregate of 1:10. The amount of water was chosen so as to obtain a constant consistency of 165 mm, measured using a flow table, according to the PN-EN 1015-3:2000 standard [32]. All samples were prepared and tested in an air-conditioned laboratory at a temperature of 20 ± 2 °C and a relative air humidity of 65 ± 5%.

2.3. Methods

Determination of the effect of hydroxyethyl methyl cellulose on the setting processes of cement and cement-lime composites was carried out using three methods, which included:

- Testing of setting processes by the ultrasonic method,
- Calorimetric measurements,
- Phases analysis by the XRD method.

Using the first method, mortar samples were tested, whereas, using the second method, paste samples were tested. In each case, the experiments were carried out for the first 48 h of material maturation. Additionally, using X-ray diffraction analysis, measurements after 40 days of maturation were made to verify whether there are differences in the leaven hydration products compared to the results obtained after two days.

The setting process was determined using the ultrasonic wave velocity method by an IP-8 Ultrasonic Measuring System (UltraTest GmbH, Achim, Germany). The essence of the study consisted in passing the ultrasonic wave through the material from one end to the other and recording its velocity over time. Ultrasonic heads and a temperature sensor were mounted on both sides for all forms. Each time, a sample of fresh mortar was taken (a cylindrical sample with a height of 5 cm and a diameter of 5 cm), which was placed for the entire duration of the test in the sampler (Figure 1). The upper surface of the tested material was protected with a glass plate to limit its water loss through drying.

The change in the velocity of the ultrasonic wave and the temperature inside the samples were recorded every 60 s for a period of 48 h of measurements and saved automatically in the computer's memory.

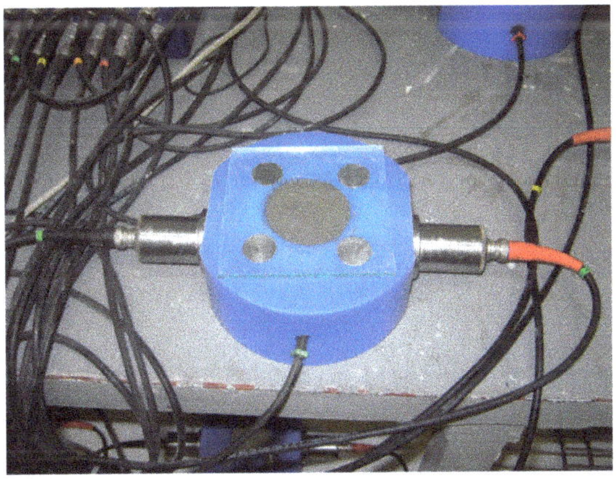

Figure 1. Mortar sample during ultrasonic testing.

The BT2.15CS low-temperature differential scanning microcalorimeter by SETARAM (SETARAM, Plan-les-Ouates, Geneva, Switzerland) was used for micro-calorimetric measurements. Binder samples (with admixture or without it) were mixed with distilled water and put into small PE zip-bags and, immediately after mixing, placed in the calorimeter. Calorimetric curves were recorded with an electronic measuring unit, connected to a personal computer.

XRD images were obtained in an X-ray diffractometer Empyrean (PANalytical, Almelo, Netherlands), equipped with a copper lamp. FT scanning was carried out at a 2θ range of 5–65° and the step length was 0.0167°. The PANalytical XRD analysis software HighScore 4.6 (PANalytical, Almelo, Netherlands) with the International Center for Difration Data (ICDD) database PDF-2 was used for phase identification. X-ray diffractometric studies were carried out on powdered samples of pastes after the time of 2 and 40 days of setting.

3. Results and Discussion

3.1. Effect of Setting Processes by the Ultrasonic Method of Mortars

The ultrasonic technique used in the research is based on the registration of the speed of the ultrasonic wave, which is related to the change in the viscosity and elasticity of the medium. The publications [14,30,33–36] show the possibilities of using this method. Since the author rightly confirmed [33], the ultrasonic technique is used to evaluate the quality of ready products. A relatively new area of research where this method can be used is the examination of the setting and hardening processes of cement-based materials. According to Zych [33], due to changes in visco elastic properties caused by hardening, the velocity of wave propagation starts growing. Applying this method, it is possible to describe the rate of setting and hardening of mortars based on cement and cement-lime binders and it is possible to assess the influence of the admixture on the setting processes, which might be useful for assessing the processing time of plastering mortars when choosing a method or establishing the conditions of building, repair, and assembly works [14,33]. The results of measurements obtained with the use of the ultrasonic method, presented in Figure 2 and in Table 4, prove the different properties of mortars. There is a clear influence of the admixture on the reduction of the ultrasonic wave velocity during the entire tested measurement. Thus, it can be concluded that cellulose ether affects the mortar setting processes. In the case of cement mortars, Kulesza et al. [36] obtained similar results. However, the authors focused their research on evaluating the influence of re-dispersible powders on setting and hardening processes of thin-bed mortars. The influence of cellulose ether was assessed by them as a secondary effect. As for the monitoring of the setting processes of cement-lime mortars, no reference articles were found in the literature.

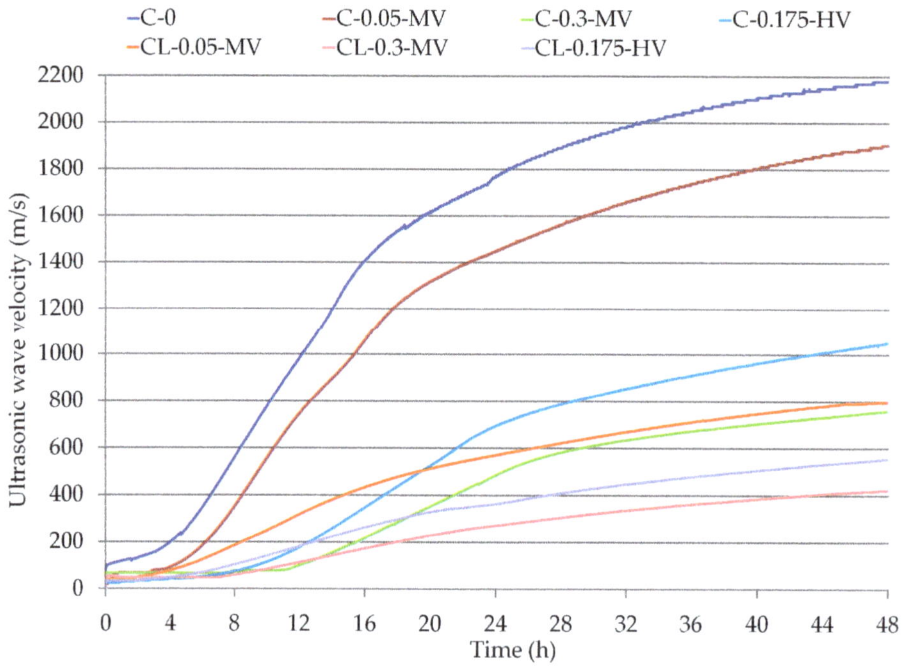

Figure 2. Ultrasonic wave velocity changes in all mortars (according to Table 3).

Table 4. Comparison of results read from the graphs (Figure 2).

Mortar	Induction Time (h)	Ultrasonic Wave Velocity			
		after 12 h (m/s)	after 24 h (m/s)	after 36 h (m/s)	after 48 h (m/s)
C-0	0 h 00	980.6	1765.7	2047.5	2180.1
C-0.05-MV	0 h 00	822.1	1478.7	1740.4	1905.3
C-0.3-MV	10 h 54	98.9	484.0	673.2	760.0
C-0.175-HV	3 h 40	174.9	693.3	911.9	1050.4
CL-0.05-MV	0 h 00	315.6	570.0	712.3	797.3
CL-0.3-MV	7 h 18	112.0	270.4	363.2	423.3
CL-0.175-HV	2 h 04	181.0	363.2	479.3	556.4

The setting process varies in intensity depending on the binder used and the amount and viscosity of the cellulose ether. It is clearly visible that the use of an admixture in the amount of 0.3% delays the setting process of both cement and cement-lime mortars (C-0.3-MV and CL-0.3-MV mortars). This is especially true when comparing the induction period of individual materials. This parameter is defined as the time counted from the setting of the measurement to the moment of a sudden change in the velocity of the ultrasonic wave. It should also be noted that, in the case of all modified mortars, the maximum velocity of the ultrasonic wave has been significantly reduced, from 2180.1 m/s, even to 423.3 m/s (more than 5 times), which may be related to the different microstructure (among other porosities) of these materials, resulting from the replacement of cement binder on lime and the use of admixtures. Mortar CL-0.3-MV modified with cellulose ether, in which 50% cement was replaced with hydrated lime, was characterized not only by the lowest ultrasonic wave velocity after 48 h of maturation, but also the ultrasonic wave velocity recorded during all setting processes was the lowest. The smallest differences in the setting process, in relation to the reference sample (C-0), are visible for the cement mortar with the admixture of 0.05% (sample C-0.05-MV) and viscosity of 25,000 mPa·s. In other cases, it is clearly visible that both the lime and the admixture significantly reduce the velocity of the ultrasonic wave during every period of time for measurements. The induction period of 10 h 54 min and 7 h 18 min indicates that the setting processes of C-0.3-MV and CL-0.3-MV mortars were delayed due to the use of an admixture in the amount of 0.3% and an admixture viscosity of 25,000 mPa·s. The induction time is visible only in the case of mortars modified with a cellulose ether of 0.175% and 0.3%. Despite the fact that the induction time is clearly visible only in the case of four mortars (Table 4), a slow increase in the ultrasonic velocity was recorded in the initial periods for the remaining materials. It is clearly visible that the most important factor in delaying the setting process is the use of admixtures, which is not so noticeable, even in the case of replacing a part of the cement binder with hydrated lime. When testing with the ultrasonic method, the amount of the admixture had a greater influence on the setting process than its viscosity. When analyzing this process for C-0.3-MV and CL-0.05-MV mortar, it is worth noting that, although the ultrasonic wave velocity recorded after 48 h is at a similar level, the induction period and the setting course in the first 12 h are different, which only confirms the conclusions regarding the influence of the amount of the admixture on the processes taking place during the maturation of these materials.

3.2. *Effect of Hydration by the Calorimetry Method of Pastes*

The changes in the rate of heat release and the amount of heat released during the hydration of pastes as a function of time are shown in Figures 3 and 4. In the case of C-0 paste, there is a clear peak in the heat flow diagram (Figure 3). In the first 16 h, a rapid hydration reaction was noted, which was confirmed by using the ultrasonic method. In Figure 2, the setting course of the C-0 sample was characterized by a rapid increase in the ultrasonic wave velocity. Based on the micro-calorimetric measurements, it can be concluded that the admixture delays the hydration processes and extends the induction period, which was also observed in literature [10,22,24,37]. The influence of the admixture on the delay of the hydration process is related to the specificity of this admixture–its water retention capacity [2,3,6,10]. Adding a cellulose ether changes the viscosity of the pastes [5,6,9,16] as well as the processes of cement hydration [19,23,24]. The amount of cellulose ether is essential when changing heat flow. Replacing a part of the cement binder with hydrated lime also reduces the heat release energy throughout the measurement and the induction period is also extended. When analyzing Figure 3, it can be noticed that pastes C-0 and C-0.05-MV have three distinct peaks in the curve, pastes C-0.3-MV and C-0.175-HV–two peaks, but pastes CL-0.05-MV, CL-0.3-MV, and CL-0.175-HV have only one clear peak. One peak on curves for this pastes may be caused by replacing part of the cement binder with hydrated lime.

The curve of cumulative heat of hydration (Figure 4) shows that cellulose ether reduces the maximum rate of heat release in 48 h, which was observed in the research [10,24,25]. Cumulative heat after 48 h was reduced by up to 76% in the case of cement pastes and up to 79% in the case of cement-lime pastes (Table 5). The polymer admixture extended the induction time from 2h 50 min to 13 h (in the case of cement pastes) and from 2 h 50 min to 5 h 15 min (in the case of cement-lime pastes). Adsorption between cellulose ethers and hydration products is thought of as the major cause for the retardation, which acts between OH groups of cellulose ethers and metal hydroxides on the surface of the hydration products [25].

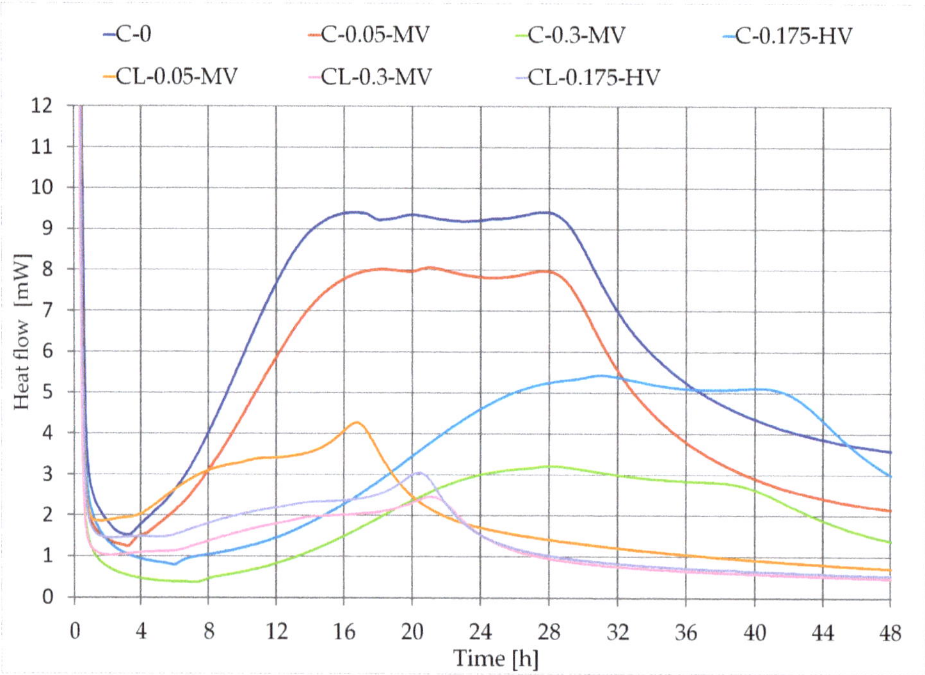

Figure 3. Isothermal calorimetry curves of all pastes (according to Table 3).

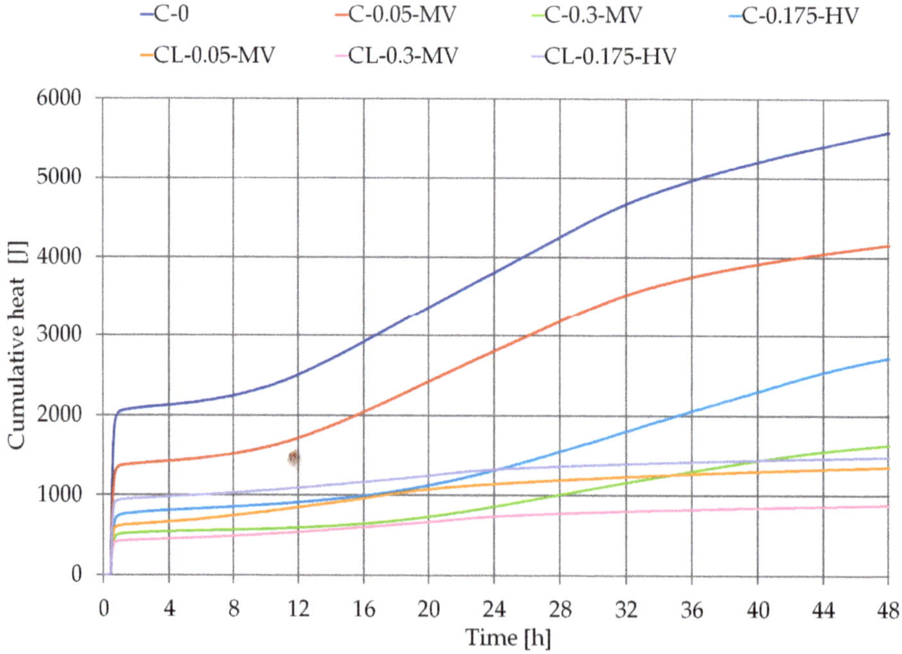

Figure 4. Cumulative curves of all pastes (according to Table 3).

Table 5. Characteristic parameters of heat evolution curves for all pastes (Figures 3 and 4).

Pastes	Induction Time (h)	Time of Second Maximum (h)	Cumulative Heat			
			after 12 h (J)	after 24 h (J)	after 36 h (J)	after 48 h (J)
C-0	2 h 50	16 h 35	2555.49	3870.52	5004.47	5583.49
C-0.05-MV	3 h 10	18 h 15	1753.67	2855.17	3781.15	4160.72
C-0.3-MV	13 h 00	28 h 10	598.18	878.04	1316.79	1625.28
C-0.175-HV	4 h 45	31 h 00	919.00	1341.37	2088.41	2718.38
CL-0.05-MV	4 h 30	16 h 40	861.95	1146.33	1271.30	1351.04
CL-0.3-MV	5 h 15	21 h 05	545.89	740.88	826.38	878.06
CL-0.175-HV	4 h 50	20 h 20	1100.67	1327.26	1418.96	1476.46

In Table 5, some parameters relating to the heat evolution curves were noted. Induction time was read based on Figure 3, but the time of maximum and cumulative heat were read based on Figure 4.

Comparing the results obtained on the basis of measurements using the ultrasonic technique with the results of calorimetric tests, there can be a similarity of the course of the curves showing the change in the velocity of the ultrasonic wave in time (Figure 2, Table 4) and the course of the curves of cumulative heat of hydration (Figure 4, Table 5). Mortar C-0.3-MV and paste C-0.3-MV have the longest induction time, which was measured by an ultrasonic technique and calorimetric measurements. The differences visible in the diagrams of changes in the velocity of the ultrasonic wave of individual mortars (Figure 2) are similar to the diagram of accumulated heat during the hydration processes of pastes (Figure 4).

3.3. *Effect of Hydration by the X-ray Diffraction of Pastes*

The hydration of the reference paste (cement paste without admixture of cellulose ether) is model-like, which means that, in this case, significant amounts of Portlandite are formed (Figure 5). The largest amount of Portlandite in the C-0 paste and the smallest intensity of the peaks of the phases of alite and belite are related to their faster reaction as a result of the hydration process, which was confirmed in calorimetric studies [38,39]. Diffraction patterns of pastes modified with cellulose ether after 48 h of setting indicate that the admixture inhibits the hydration process of cement and cement-lime pastes. In the case of the diffractograms of the samples from C-0.3-MV to CL-0.175-HV, a lower intensity of Portlandite can be seen compared to the diffractogram of the C-0 and C-0.05-MV pastes. It can be seen that the hydration of the cement paste with an admixture of 0.05% with a viscosity of 25,000 MPa·s (C-0.05-MV) does not differ from the hydration of the C-0 paste. Hence, the conclusion that HEMC added in the amount of 0.05% does not have such a large impact on the setting process as in other cases. The lowest amount of Portlandite was observed for the cement paste modified with an admixture of 0.3% with a viscosity of 25,000 MPa·s. These experiments confirmed that the admixture viscosity does not have such a large influence on the changes in the hydration process as its amount, regardless of the type of binder.

For cement-lime pastes, the increase of calcite peaks is also characteristic, which may indicate of the partial lime binder carbonation. Due to a binder composition, alite and belite have less intense peaks in these pastes.

The diffractograms of the pastes after 40 days of maturation were shown in Figure 6. It can be seen that the process of setting individual pastes is similar. There are few significant differences (between the individual pastes) in the intensity of portlandite, alite, and belite peaks, as could be observed for the measurements carried out after two days of hydration of the samples. A similar conclusion for cement pastes was observed in literature reports [22]. Only for the C-0.3-MV and C-0.175-HV pastes, differences for the characteristic peaks of Portlandite are clearly visible. In both cases, the peak for 18.04° 2θ is significantly higher and the peak for 34.05° 2θ is lower than in the other pastes. The least intense Portlandite peak was obtained for CL-0.3-MV.

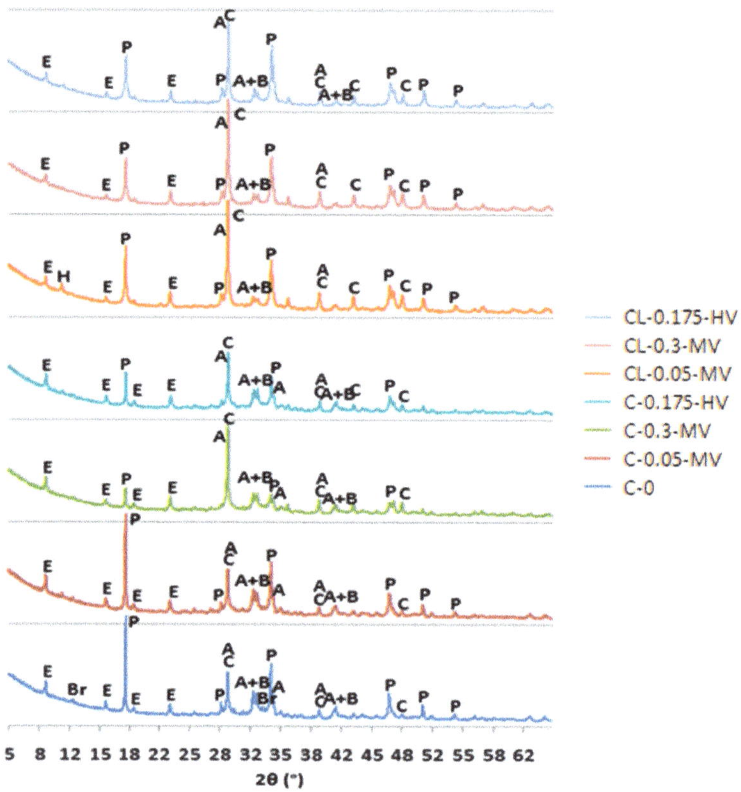

Figure 5. X-ray patterns of all pastes for 48 h. Denotation: A–alite, B–belite, Br–brownmillerit, C–calcite, E–ettringite, P–portlandite, H–C-S-H.

However, comparing the diffraction patterns after two days and after 40 days, it is possible to notice the different intensity of Portlandite and calcite. A greater amount of Portlandite after 40 days may be caused by the hydration process, and an increased amount of calcite is associated with the formation of calcium carbonate $CaCO_3$ and the progressive carbonation process. In Table 6, the percentage changes of the intensity of Portlandite peaks for 18.04° 2θ after 40 days of pastes hydration were shown. In the case of calcite, the difference in peak intensity between the diffractograms made after two days and after 40 days is also visible, although not as large as in the case of Portlandite. However, the results are consistent with the studies carried out by Izaguirre et al. [27]. In the case of the sample with cellulose ether, a significant increase in the amount of free water, $Ca(OH)_2$ and $CaCO_3$ were seen in TG analysis. Percent content H_20, $Ca(OH)_2$, and $CaCO_3$ after 28 and 91 days was bigger than after 7 days maturation (compared to a reference sample), which confirm the ongoing carbonation processes. On the other hand, percent content of $Ca(OH)_2$ and $CaCO_3$ after 7 days was smaller in the case of the sample with cellulose ether than the reference sample, which may indicate a delay of reaction of carbonation processes caused by the admixture.

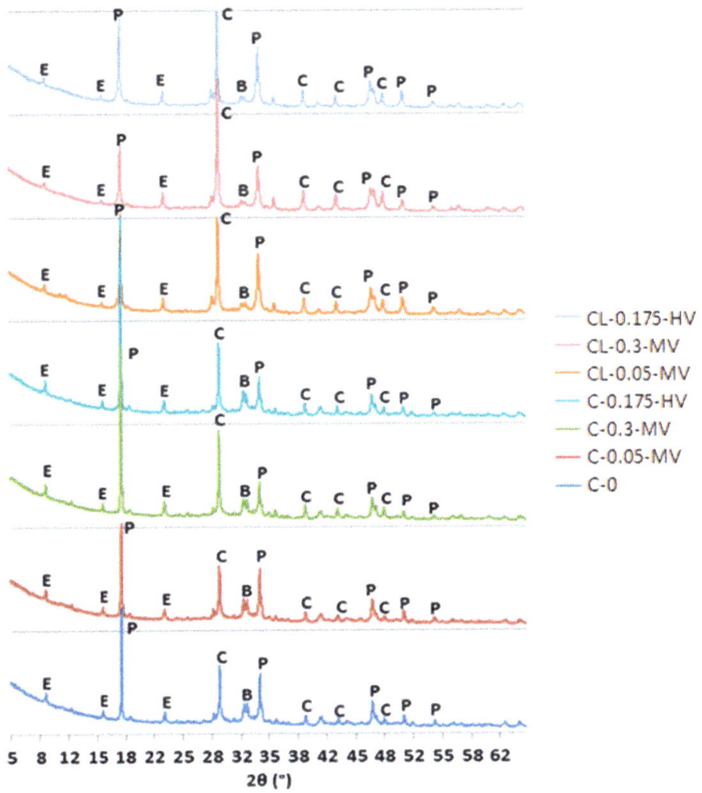

Figure 6. X-ray patterns of all pastes for 40 days. Denotation: A–alite, B–belite, Br–brownmillerit, C–calcite, E–ettringite, P–portlandite.

Table 6. Percentage changes in Portlandite intensity after 2 and 40 days of maturation for all pastes (according to Table 3).

Pastes	Peak Intensity after 2 Days (%)	Peak Intensity after 40 Days (%)	Changes in Peak Intensity (%)
C-0	100.00	100.00	+15.60
C-0.05-MV	98.58	83.43	−2.17
C-0.3-MV	19.44	149.50	+788.85
C-0.175-HV	33.77	172.03	+488.81
CL-0.05-MV	58.96	83.26	+63.25
CL-0.3-MV	44.87	50.65	+30.48
CL-0.175-HV	44.58	73.40	+90.33

4. Conclusions

The following conclusions could be drawn from the research.

- The addition of hydroxyethyl methyl cellulose influences the setting processes of cement and cement-lime mortars. The addition of HEMC with a viscosity of 25,000 mPa·s in the amount of 0.3% significantly delays the setting process regardless of the type of binder (cement, hydrated lime), which is not so visible in the case of the admixture in the amount of 0.05%.

- HEMC reduces the rate of the hydration reaction in cement and cement-lime pastes, which has been confirmed in calorimetric tests. The induction time on the heat evolution curve becomes significantly longer (by a maximum of 10 h 10 min). The changes are the most visible with an increasing dosage of the admixture.
- In the XRD analysis, significant effects of the presence of HEMC were found on the resulting Portlandite peaks. Limiting their intensity during the period of the first two days shows that the hydration process is lower.
- The differences in the hydration of pastes after 40 days of maturation are much smaller than the hydration of the same samples after two days of maturation. This shows that significant phase changes take place at a later stage of hydration (after 48 h). The addition of HEMC has a greater effect on the hydration of cement and cement-lime pastes in the initial hours of maturation. After a longer time, this effect practically disappears. Therefore, control of hydration kinetics of cement and cement-lime pastes on early stages is important.
- The effect of HEMC admixture on the inhibition of setting processes is more visible in pastes and mortars containing only cement binder, especially by analyzing the differences in the induction time from the ultrasonic technique and calorimetric measurements.
- The amount of the admixture used is the most important in the process of setting pastes and mortars. A much-reduced impact is visible when taking into account the viscosity of the admixture or the type of binder used.
- The ultrasonic technique in combination with calorimetric measurements and XRD method gives the possibility of a more complete and broader assessment of the setting mortars. It also enables the assessment of the influence of a chemical admixture on the hydration and setting processes of pastes and mortars. It can be useful in monitoring and controlling for application plastering mortars.

Author Contributions: Conceptualization, E.S. Methodology, E.S. and P.C. Validation, E.S. and P.C. Formal analysis, E.S. and P.C. Investigation, E.S. and P.C. Resources, E.S. Data curation, E.S. and P.C. Writing—original draft preparation, E.S. and P.C. Writing—review and editing, E.S. and P.C. Visualization, E.S and P.C. Supervision, E.S. Project administration, E.S. Funding acquisition, E.S. All authors have read and agreed to the published version of the manuscript.

Funding: This research received no external funding.

Acknowledgments: This research was funded by the Kielce University of Technology research program number 02.0.05.00/2.01.01.00.0000 SUBB.BKTO.20.002.

Conflicts of Interest: The authors declare no conflict of interest.

References

1. Petit, J.-Y.; Wirquin, E. Evaluation of various cellulose ethers performance in ceramic tile adhesive mortars. *Int. J. Adhes. Adhes.* **2013**, *40*, 202–209. [CrossRef]
2. Brachaczek, W. Influence of cellulose ethers on the consistency, water retention and adhesion of renovating plasters. In *IOP Conference Series: Materials Science and Engineering*; IOP Publishing: Bristol, UK, 2019; Volume 471, p. 032020.
3. Bülichen, D.; Kainz, J.; Plank, J. Working mechanism of methyl hydroxyethyl cellulose (MHEC) as water retention agent. *Cem. Concr. Res.* **2012**, *42*, 953–959. [CrossRef]
4. Szczygielska, A.; Rudnik, E.; Polaczek, J. Cellulose ethers—Preparation, application and market situation. *Przem. Chem.* **2002**, *81*, 704–707.
5. Paiva, H.; Esteves, L.; Cachim, P.; Ferreira, V. Rheology and hardened properties of single-coat render mortars with different types of water retaining agents. *Constr. Build. Mater.* **2009**, *23*, 1141–1146. [CrossRef]
6. Brumaud, C.; Baumann, R.; Schmitz, M.; Radler, M.; Roussel, N. Cellulose ethers and yield stress of cement pastes. *Cem. Concr. Res.* **2014**, *55*, 14–21. [CrossRef]
7. Vyšvařil, M.; Hegrová, M.; Žižlavský, T. Influence of cellulose ethers on fresh state properties of lime mortars. *Solid State Phenom.* **2018**, *276*, 69–74. [CrossRef]

8. Vyšvařil, M.; Bayer, P. Cellulose ethers as water-retaining agents in natural hydraulic lime mortars. In Proceedings of the 13th International Conference "Modern Building Materials, Structures and Techniques" (MBMST 2019), Vilnius, Lithuania, 16–17 May 2019; pp. 194–200.
9. Lachemi, M.; Hossain, K.; Lambros, V.; Nkinamubanzi, P.-C.; Bouzoubaâ, N. Performance of new viscosity modifying admixtures in enhancing the rheological properties of cement paste. *Cem. Concr. Res.* **2004**, *34*, 185–193. [CrossRef]
10. Chen, N.; Wang, P.; Zhao, L.; Zhang, G. Water retention mechanism of HPMC in cement mortar. *Materials* **2020**, *13*, 2918. [CrossRef]
11. Wang, Z.; Zhao, Y.; Zhou, L.; Xu, L.; Diao, G.; Liu, G. Effects of hydroxyethyl methyl cellulose ether on the hydration and compressive strength of calcium aluminate cement. *J. Therm. Anal. Calorim.* **2019**, *140*, 545–553. [CrossRef]
12. Sobala, M.; Nosal, K.; Pichniarczyk, P. Effect of methyl cellulose on the properties of adhesive cement mortars. *Cem. Lime Concr.* **2010**, *6*, 359–365.
13. Seabra, M.; Labrincha, J.A.; Ferreira, V.M. Rheological behaviour of hydraulic lime-based mortars. *J. Eur. Ceram. Soc.* **2007**, *27*, 1735–1741. [CrossRef]
14. Spychał, E. The effect of lime and cellulose ether on selected properties of plastering mortar. *Procedia Eng.* **2015**, *108*, 324–331. [CrossRef]
15. Kotwa, A.; Spychał, E. The influence of cellulose ethers on the chosen properties of cement mortar in the plastic state. *Struct. Environ.* **2016**, *8*, 153–159.
16. Gołaszewski, J.; Cygan, G. Wpływ domieszek zwiększających lepkość na skurcz wczesny zapraw. *Civ. Envir. Eng.* **2011**, *2*, 263–266.
17. Messan, A.; Ienny, P.; Nectoux, D. Free and restrained early-age shrinkage of mortar: Influence of glass fiber, cellulose ether and EVA (ethylene-vinyl acetate). *Cem. Concr. Compos.* **2011**, *33*, 402–410. [CrossRef]
18. Pourchez, J.; Govin, A.; Grosseau, P.; Guyonnet, R.; Guilhot, B.; Ruot, B. Alkaline stability of cellulose ethers and impact of their degradation products on cement hydration. *Cem. Concr. Res.* **2006**, *36*, 1252–1256. [CrossRef]
19. Pourchez, J.; Grosseau, P.; Ruot, B. Current understanding of cellulose ethers impact on the hydration of C3A and C3A-sulphate systems. *Cem. Concr. Res.* **2009**, *39*, 664–669. [CrossRef]
20. Pourchez, J.; Peschard, A.; Grosseau, P.; Guyonnet, R.; Guilhot, B.; Vallée, F. HPMC and HEMC influence on cement hydration. *Cem. Concr. Res.* **2006**, *36*, 288–294. [CrossRef]
21. Pourchez, J.; Grosseau, P.; Guyonnet, R.; Ruot, B. HEC influence on cement hydration measured by conductometry. *Cem. Concr. Res.* **2006**, *36*, 1777–1780. [CrossRef]
22. Pichniarczyk, P. Influence of methylcellulose on the hydration of cement. *Cem. Lime Concr.* **2015**, *3*, 186–192.
23. Knapen, E.; Van Gemert, D. Cement hydration and microstructure formation in the presence of water-soluble polymers. *Cem. Concr. Res.* **2009**, *39*, 6–13. [CrossRef]
24. Betioli, A.M.; Gleize, P.J.P.; Silva, D.; John, V.M.; Pileggi, R.G. Effect of HMEC on the consolidation of cement pastes: Isothermal calorimetry versus oscillatory rheometry. *Cem. Concr. Res.* **2009**, *39*, 440–445. [CrossRef]
25. Ou, Z.H.; Ma, B.G.; Jian, S.W. Influence of cellulose ethers molecular parameters on hydration kinetics of Portland cement at early ages. *Constr. Build. Mater.* **2012**, *33*, 78–83. [CrossRef]
26. Qu, X.; Zhao, X. Influence of SBR latex and HPMC on the cement hydration at early age. *Case Stud. Constr. Mater.* **2017**, *6*, 213–218. [CrossRef]
27. Izaguirre, A.; Lanas, J.; Alvarez, J.I. Characterization of aerial lime based mortars modified by the addition of two different water-retaining agents. *Cem. Concr. Compos.* **2011**, *33*, 309–318. [CrossRef]
28. Arizzi, A.; Cultrone, G. Aerial lime-based mortars blended with a pozzolanic additive and different admixtures: A mineralogical, textural and physical-mechanical study. *Constr. Build. Mater.* **2012**, *31*, 135–143. [CrossRef]
29. Dima, C.; Badanoiu, A.; Cirstea, S.; Nicoara, A.I.; Stoleriu, S. Lightweight gypsum materials with potential use for thermal insulations. *Materials* **2020**, *13*, 5454. [CrossRef]
30. Spychał, E. Ultrasonic method in research of plastering mortars' setting and hardening process. In *Współczesny stan wiedzy w inżynierii lądowej. Prace naukowe doktorantów*; Bzówka, J., Ed.; Wydawnictwo Politechniki Śląskiej: Gliwice, Poland, 2015; pp. 491–500.
31. Spychał, E. Evaluation of Selected Properties of Plastering Mortar Modified with Hydrated Lime and Cellulose Ether. Ph.D. Thesis, Kielce University of Technology, Kielce, Poland, 2016.

32. *PN-EN 1015-3:2000+A2:2007 Test Methods of Mortars for Walls. Determining the Consistency of Fresh Mortar (Using a Flow Table)*; Comite Europeen de Normalisation: Warsaw, Poland, 2007.
33. Zych, J. Application of ultrasonic technique in examination of the setting and hardening process of cement-bonded materials. *Cem. Lime. Concr.* **2006**, *1*, 5–16.
34. Trtnik, G.; Valič, M.I.; Kavčič, F.; Turk, G. Comparison between two ultrasonic methods in their ability to monitor the setting process of cement pastes. *Cem. Concr. Res.* **2009**, *39*, 876–882. [CrossRef]
35. Reinhardt, H.W.; Grosse, C.U. Continuous monitoring of setting and hardening of mortar and concrete. *Constr. Build. Mat.* **2004**, *18*, 145–154. [CrossRef]
36. Kulesza, M.; Dębski, D.; Fangrat, J. Effect of redispersible polymer powder on setting time of thin-bed mortars. In Proceedings of the MATEC Web of Conferences, MATBUD'2018 8th Scientific-Technical Conference on Material Problems in Civil Enginnering, Cracow, Poland, 25–27 June 2018; EDP Sciences: Les Ulis, France, 2018; Volume 163, p. 04005.
37. Zhang, G.; Zhao, J.; Wang, P.; Xu, L. Effect of HEMC on the early hydration of Portland cement highlighted by isothermal calorimetry. *J. Therm. Anal. Calorim.* **2015**, *119*, 1833–1843. [CrossRef]
38. Kurdowski, W. *Cement and Concrete Chemistry*; Springer Science and Business Media: Berlin/Heidelberg, Germany, 2014.
39. Owsiak, Z. The hydration of Portland cement with fly ash. *Cem. Lime Concr.* **2000**, *5*, 29–31.

materials

MDPI

Article

Effect of Natural Graphite Fineness on the Performance and Electrical Conductivity of Cement Paste Mixes for Self-Sensing Structures

Ioanna Papanikolaou *, Chrysoula Litina, Amir Zomorodian and Abir Al-Tabbaa

Department of Engineering, University of Cambridge, Trumpington St., Cambridge CB2 1PZ, UK; cl519@cam.ac.uk (C.L.); az412@cam.ac.uk (A.Z.); aa22@cam.ac.uk (A.A.-T.)
* Correspondence: ip324@cam.ac.uk

Received: 24 November 2020; Accepted: 14 December 2020; Published: 21 December 2020

Abstract: Cementitious composites are the most widely used construction materials; however, their poor durability necessitates frequent monitoring and repairs. The emergence of self-sensing composites could reduce the need for costly and time-consuming structural inspections. Natural graphite, due to its low cost and wide availability, is a promising additive to generate an electrically conductive network which could ultimately lead to a self-sensing mechanism. Despite several studies using natural graphite as a conductive additive, the effect of its fineness on the cementitious composite's performance has not been explored. This study experimentally investigated the effect of three graphite products of varying fineness on the early age, mechanical, and electrical conductivity performance of cement pastes. The fluidity of the graphite-cement paste reduced significantly with increasing graphite fineness, and graphite did not affect the cement hydration. The finer the graphite, the lower the effect on the mechanical performance, as confirmed by compressive strength testing and micro-indentation. Electrical conductivity testing showed that the percolation threshold depended on the graphite fineness and was found at ~20 wt % for the fine and medium graphite, while it increased to 30–40 wt % for the coarse graphite. This is the first study that has investigated holistically the effect of graphite fineness on the performance of cement pastes and will pave the way for using this material as an additive for self-sensing structures.

Keywords: cement; graphite; additives; electrical conductivity; self-sensing

1. Introduction

Cement composites, including pastes, mortars, and concrete are the most widely used construction materials; however, cementitious structures often suffer from poor durability that necessitates repair and maintenance (R&M) to be carried out. From 2011 to 2015, approximately a fifth of the civil engineering works in the UK were due to R&M [1], while in the US, Americans undertake over 200 million trips a day across deficient bridges [2]. This inadequate durability has resulted in a need for structural inspections which are usually carried out visually [3,4]; however, due to the inherent uncertainties and risks with visual surveys, the use of sensors has started to flourish. Despite the increasing use of structural health monitoring systems, the use of external sensors has often resulted in high costs, low sensitivity, need for frequent calibration, and incompatibility with structural materials [5–7].

Recent advances in material science have generated increasing interest in the use of smart, biomimetic materials that mimic natural systems and could monitor their own condition [4]. One such example, is self-sensing concrete, which refers to a material that can sense its condition and identify any damage, whilst maintaining or improving the structural performance [8]. A self-sensing mechanism can be achieved by passing electric current through the structure and monitoring changes in the electrical conductivity. The non-reversible change in conductivity can be used to monitor damage, while the reversible change can be used to monitor strain (loading) [8–10]. Nevertheless, cementitious composites behave as insulators to electricity and do not allow the passage of electric current [11]. Thus, the use of a conductive additive is necessary to ensure that an electrically conductive network can be formed within the cementitious structure.

In terms of conductive additives, at least 10 different types, along with hybrid combinations, have been investigated, such as steel fibers, carbon fibers, graphite powder, carbon nanotubes, and graphene nanoplatelets [8,12,13]. The minimum additive dosage that is needed to form continuous electrical paths inside the composite is known as the percolation threshold, which depends on many parameters, including additive composition (size and shape), concentration, and degree of aggregation [12]. For example, fibrous additives with a high aspect ratio, reach a percolation threshold at a lower dosage (~1.5 wt %), compared to particle ones (>5 wt %) [8].

One such functional additive is natural graphite, which due to its wide availability and low cost, could have promising applications in cementitious composites for self-sensing structures. Graphite powder has a layered planar structure, rendering it relatively soft due to its anisotropy and weak inter-planar forces, ability to conduct electricity and heat well, resistance to chemical attack, and stability under standard conditions [8,14]. Graphite has been used as a conductive filler in some studies, and it was found that it could improve the electrical conductivity performance [15–18]. However, graphite size can vary, and very often, this property is not reported, while at the same time, the percolation threshold could depend on the size of the conductive additive. In one study, it was found that when graphite was added in cement, the composite became a conductor; however, the mechanical and electrical properties depended on the water content and the setting process [15]. When dry-mixing graphite and cement powder, a minimum threshold of 2 wt % graphite was needed, below which the insulating cement prohibited the formation of conductive graphite pathways and the conductivity leveled off at around 10 wt % graphite [16]. Similarly, the DC electrical resistivity decreased rapidly between the percolation threshold at 2 and 10 wt %, where it plateaued with a graphite that had a particle size of 10–20 μm [17]; however, another study found a conductivity threshold at ~30 to 40 wt % of graphite with a d_{50} size of 4.5 μm [18].

Despite some promising early findings in the literature around the improvement in electrical conductivity with graphite, the effect of its fineness has not been investigated in detail. Most studies do not report the inherent material properties, such as the size; thus, it is not possible to understand the effect of the additive on the properties of cement composites and use the graphite material at its full potential. Furthermore, there is a lack of a holistic assessment of the effect of natural graphite, not only on the electrical conductivity, but also on other properties that are essential for cementitious composites, such as rheology, hydration, and mechanical performance. Therefore, this study experimentally investigated three natural graphite products of varying sizes to understand the effect of graphite fineness on the electrical conductivity, microstructure, hydration, and mechanical performance of cementitious pastes that could then be used for self-sensing applications.

2. Materials and Methods

2.1. Materials

Portland Cement CEMI 52.5N, supplied by Hanson and conforming to BS EN 197-1:2011 [19] was used and its chemical composition is tabulated in Table 1. The cement had a size distribution of 5–30 μm, a surface area of 0.3–0.4 m^2/g, a density of 2.7–3.2 g/cm^3 and a loss on ignition (LOI%) of 2.2, as provided by the supplier in the material safety data sheet. Three commercial products of natural graphite were used to create the graphite-cement paste. A coarse graphite powder (−10 mesh, 2 mm) was supplied by AlfaAesar (Haverhill, MA, USA), and two finer graphite powders (−100 mesh, 0.150 mm and −325 mesh, 44 μm) were supplied by Sigma-Aldrich (St. Louis, MO, USA). The suppliers only provided the mesh size of the graphite rather than the particle size distribution and hence only the mean size is used here. These graphite products are referred to as coarse (2 mm), medium (0.15 mm) and fine (44 μm). The graphite powders were used at varying concentrations needed to reach the percolation threshold (from 0 to 40 wt %). The cement paste had a constant *w/c* = 0.45. It should be noted that the graphite powders were added as additions rather than cement replacement; therefore, the cement content and *w/c* ratio are the same in all mixes.

Table 1. Chemical composition of the CEMI.

Oxides (%)	CEMI 52.5N
CaO	63.4
SiO_2	20.4
Al_2O_3	4.7
Fe_2O_3	2.7
MgO	1.0
K_2O	0.6
SO_3	3.1
Cl	0.02

2.2. Sample Preparation

To prepare the cement paste mixes, the graphite powder and cement were dry-mixed first, in a laboratory bench-scale mixer, for 2 min, and then water was added. Immediately, when the water was added, the mixer was started at a slow speed for 60 s and high speed for 30 s. Then, the mixer was stopped for 90 s, and during the first 30 s, the paste adhering to the walls and bottom of the bowl was removed with a scraper and placed in the middle of the bowl. Mixing was then continued at high speed, for a further 60 s. It should be noted that no chemical admixtures were used to aid the dispersion of the graphite. When mixing was completed, the fresh paste was cast in three layers of equal thicknesses, in oiled stainless-steel moulds. For compressive strength, the specimens had a size of 40 × 40 × 40 mm^3; for micro-indentation, the cylindrical specimens were 50 mm diameter and 100 mm height; and for electrical conductivity, the prismatic specimens had a size of 40 × 40 × 160 mm^3. The specimen size was reduced by half (20 × 20 × 80 mm^3) for electrical impedance spectroscopy testing. The specimens were demolded after 24 h and cured in a water tank, at temperatures of 20 ± 2 °C and a relative humidity 60%, until testing.

2.3. Experimental Programme

2.3.1. Material Characterization, Dispersion, and Microstructure

To assess the dispersion of the natural graphite in the cement paste and the microstructure of the composite, Scanning Electron Microscopy (SEM) was carried out, using a ZEISS EVO LS 15 SEM–EDX (Oberkochen, Germany). Chipped pieces were extracted from the cracked specimen surfaces, following the compressive strength testing, and all samples were gold-coated before testing.

Thermogravimetric analysis (TGA, PerkinElmer, Waltham, MA, USA) was used to characterize the graphite powder materials and understand their decomposition. The TGA experiment was performed in air. The temperature ranged from 40 to 1000 °C, at a steady rate increase of 10 °C/min, and the gas flow rate was kept constant at 30 mL/min.

X-ray computed tomography in the form of micro-CT scanning (μCT) was carried out, using a XT H 225 ST CT scan device by Nikon (Tokyo, Japan). A ~5 mm sample was extracted from the cracked surface of the cement paste specimens containing natural graphite and analyzed.

2.3.2. Rheology Measurement

To assess the rheology of the graphite-cement pastes, a smooth-walled Brookfield DV3T Rheometer (SC4-27 spindle, Middleboro, MA, USA) was used. Measurements were carried out every 15 s at room temperature and a shearing profile that is representative of the typical concreting process, such as hauling and casting, was followed [20]. Each sample was pre-sheared for 1 min, and it was then left for 30 s, to stabilize. The shear speed increased progressively from 0 to 150 rpm, in 25 rpm intervals. It was then kept constant at 175 rpm and then decreased progressively from 150 to 0 rpm, again, in 25 rpm intervals (descending rates). The Bingham model was used to calculate the viscosity of the samples, using the descending shearing rates.

2.3.3. Hydration, Compressive Strength and Micro-Indentation

An isothermal calorimeter I-CAL 2000 HPC was used and ASTM C1679-08 [21] was followed to assess the effect of natural graphite on the hydration of cement paste. A CONTROLS Advantest9 machine (Milan, Italy) was used for the unconfined compressive strength at a loading rate of 2400 N/s according to BS EN 196-1 [22], and three cubic specimens of 40 × 40 × 40 mm^3 were tested at 2, 7, and 28 days. Micro-indentation testing was also performed to establish the effect of graphite on the hardness and elastic modulus of the specimens according to ASTM E384–16 [23]. Cylindrical specimens of 50 mm diameter and 100 mm height were prepared and cut into discs of ~25 mm thickness. The testing was carried out after 56 days of curing and only one concentration of each graphite was tested (20 wt %). To ensure a smooth surface, a polishing protocol was followed. A P240 silicon carbide (SiC) paper was first used for 30 s for both sides of the disc. One side was then polished progressively, using P400-P800-P2500 SiC paper, for 30 s in each stage. The polished discs were then tested with an Anton Paar MHT micro-indentation tester (Graz, Austria) with a Vickers tip. An indentation force of 8 N was used, and 15 measurements were taken for each sample following a protocol suitable for cement paste [24]. The loading and unloading times were 15 s, and the indentation time at maximum force (8 N) was 20 s, while a Poison's ratio of 0.25 was used.

2.3.4. Electrical Conductivity

The electrical conductivity was tested by using a 4-probe DC setup, using perforated steel sheet electrodes. A DC of 10 V was supplied to the outer two electrodes, and the inner two electrodes were used to measure the voltage (Figure 1), which was recorded with a datalogger every second and illustrated in the LABView software.

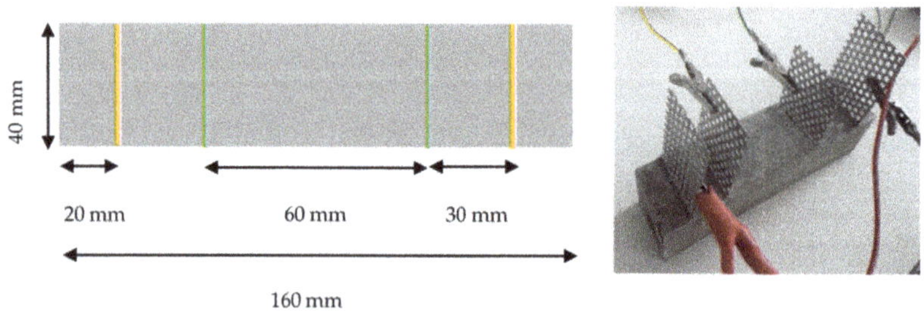

Figure 1. Four-probe setup for electrical resistivity measurements. Outer electrodes supplied a DC = 10 V, and inner electrodes measured the change in voltage.

Electrical impedance spectroscopy (EIS) was used to characterize the electrical parameters of the materials and to investigate the frequency-dependent response and material–electrode interface. By applying an AC current with a set amplitude over a range of frequencies, the response of the specimen could be measured in terms of magnitude and phase angle. A 2-probe setup with embedded electrodes was used with specimens of 20 × 20 × 80 mm^3 size. Two dosages were selected for each graphite type, one at 10 wt % and one at the start of the percolation threshold (either 20 or 40 wt % for the coarse graphite), and triplicate tests were run at 7 and 28 days. The results of the EIS testing are shown in the form of Nyquist plots, with the real part of the impedance shown in the x-axis ($Z'(\Omega)$), and the imaginary part in the y-axis ($-Z''(\Omega)$); the average of three samples is represented with each

Nyquist plot. The Nyquist plot contains semicircles or arcs, and their diameters correspond to the resistances of the different components in the composite microstructure [25].

3. Results and Discussion

3.1. Material Characterization, Dispersion, and Microstructure

The TGA results in Figure 2, show that the fine and medium graphite materials decomposed completely (lost 100% of their weight) between 40 and 1000 °C, while the coarse graphite did not. This could be due to the particle size of the coarse graphite (2 mm) as the flakes were too large for all their minerals to decompose. The smaller the particle size, the faster the minerals decompose with increasing temperature, and therefore the curves shift to a lower temperature as the material size reduces. None of the materials showed any mass loss before 100 °C; therefore, they are all expected to be stable during cement mixing and hydration. Furthermore, the DTG curves do not show the presence of any other minerals so the graphite products are of high purity.

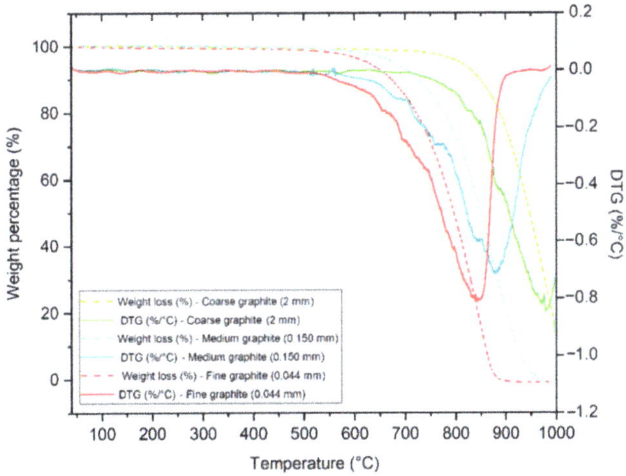

Figure 2. Characterization of the three graphite products by thermogravimetric analysis (TGA), showing the weight loss from 100 to 1000 °C.

SEM was used to understand the morphology of the graphite materials. In Figure 3a,b the coarse graphite and some large flakes with a size of at least 1 mm can be observed, which is expected due to the inaccuracies when sieving. For the medium graphite (Figure 3c,d), flakes are of varying sizes but are all between 150 and 200 µm. In Figure 3e,f, the fine graphite particles can be seen with a wrinkled and folded structure. As the graphite became finer, some agglomerates could be seen, and it was difficult to isolate individual flakes. SEM–EDX was also used to assess the dispersion effectiveness of the dry mixing protocol. Only a coarse graphite-cement paste specimen at 28 days of hydration was used and the coarse graphite particles are shown as green in Figure 3g, whilst the cement paste matrix is shown as pink. The EDS analysis (Figure 3h) confirms that the main elements are carbon and calcium, which was expected due to the presence of graphite and because calcium silicate (C–S–H) is the main reaction product of cement hydration. Other typical elements from cement hydration, such as silicon, oxygen, sulfur, and magnesium, were also present.

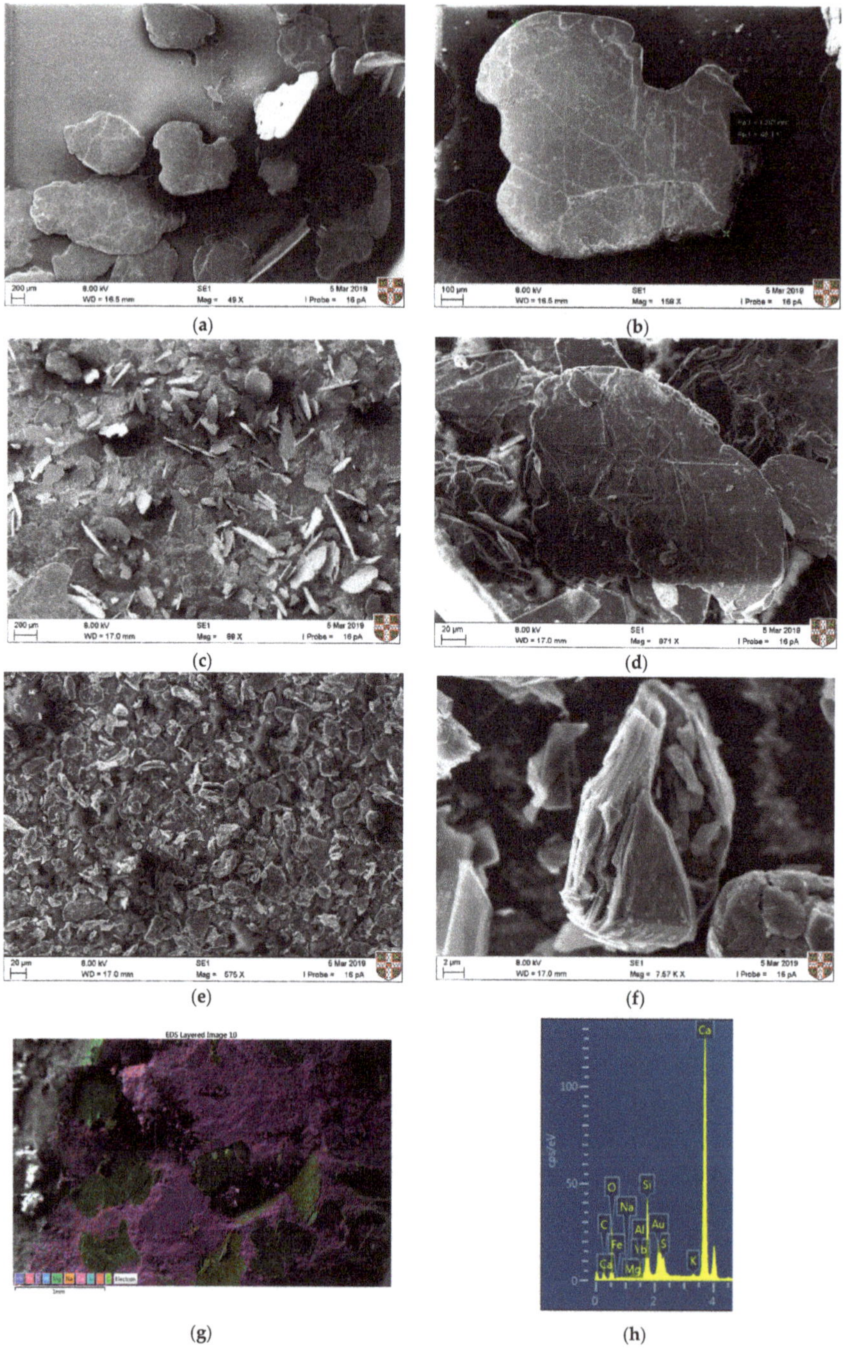

Figure 3. SEM images of the three graphite products, with two different magnification images for each: coarse graphite (**a**,**b**), medium graphite (**c**,**d**), fine graphite (**e**,**f**), and SEM–EDX confirming good dispersion of the coarse graphite flakes at 28 days (**g**,**h**).

SEM testing was also undertaken after five months of curing, to assess the interaction of the graphite particles with the hydrated cement paste and to understand the effect on microstructure. In Figure 4, the individual graphite flakes could be identified in all cases, irrespective of the graphite size. Flakes were observed in proximity, which was expected, as graphite concentrations were high (the percolation threshold samples are illustrated below), but no obvious agglomeration was present. Therefore, dispersion of graphite particles was assumed to be adequate with dry mixing, which is a technique that has also been followed in the literature [16]. Observing the interaction between the graphite particles and the cement hydration products, we can see an interfacial transition zone similar to that of aggregates for the coarse graphite (Figure 4a), which could lead to planes of weakness and a reduction in mechanical performance. These results are not directly comparable with other studies, as the microstructural interaction could also be affected by the cement type, dispersion technique and graphite morphology. Overall, none of the three natural graphite products that were used here was found to alter the microstructure of the CEMI paste.

A μCT-scan was also used to assess the dispersion of a 30 wt % coarse graphite dosage in the cement matrix with the dry mix protocol. Figure 5a,b shows the 3D reconstructed image of the 5 mm graphite-cement paste sample and Figure 5c shows a slice through the 3D image. A grayscale analysis was used to detect the different substances based on their density. Three distinct peaks were observed in the μCT-scan, with each representing the different elements including air, bulk cement paste, and graphite particles. In Figure 5, the graphite particles are represented with a pink color, whilst the bulk cement paste is shown as black. The graphite flakes were found to be well dispersed within the matrix and near each other which was due to the high graphite concentration (30 wt %). The orientation of the flakes appeared to vary, even though some clusters were oriented in the same direction. The key finding from the μCT-scan was that graphite flakes were well dispersed with the dry mix method, and even at a high graphite concentration, no agglomerates were observed.

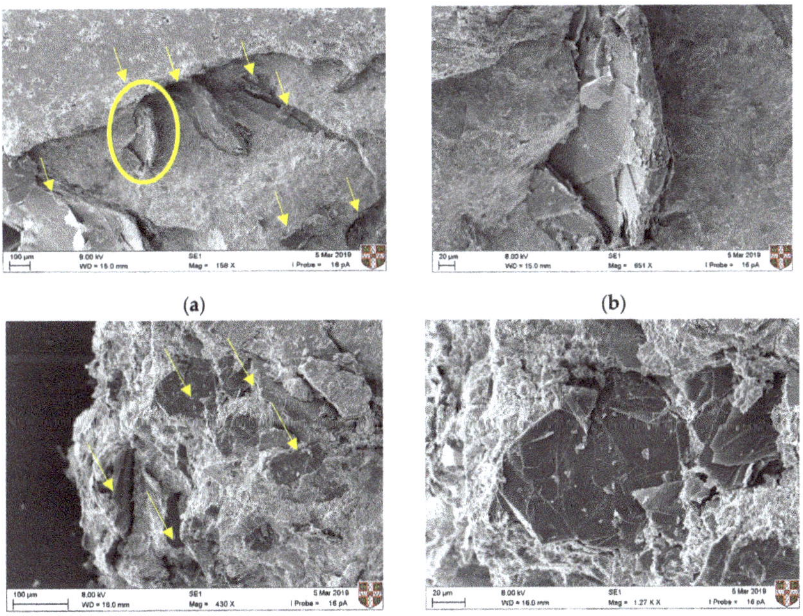

(a) (b) (c) (d)

Figure 4. *Cont.*

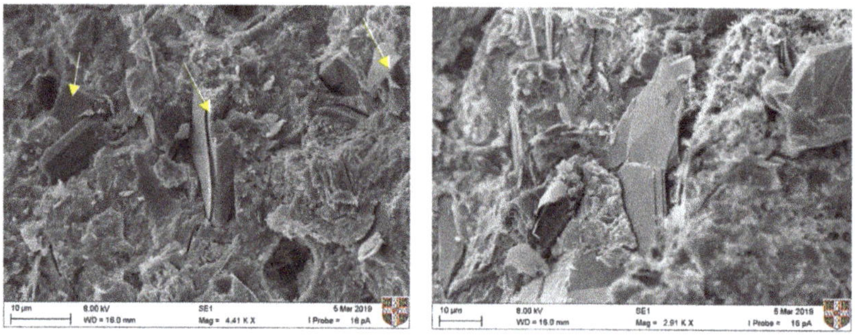

(**e**) (**f**)

Figure 4. Microstructural characterization of the three graphites-cement paste after ~5 months by SEM (**a**,**b**) coarse, (**c**,**d**) medium, and (**e**,**f**) fine graphite.

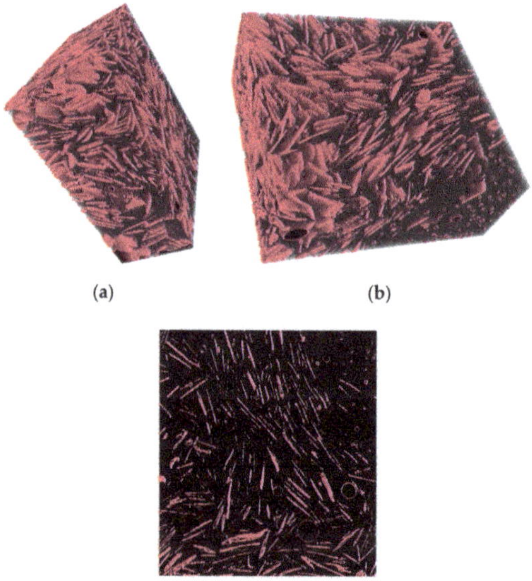

(**a**) (**b**) (**c**)

Figure 5. X-ray computed tomography (CT-scan) of an approximately 5 mm cement paste specimen with a 30 wt % coarse graphite addition. The graphite flakes (pink) are dispersed in the matrix (gray): (**a**,**b**) 3D reconstructed image of the specimen and (**c**) a slice through the 3D image.

The microstructural characterization showed that the dry-mixing protocol that was followed here was effective, while graphite had no pronounced effect on the microstructure of cement paste, irrespective of its fineness. This means that graphite is acting as an inert filler in the mix and does not alter the chemical structure of the hydrated cement paste.

3.2. Rheology of Graphite-Cement Pastes

Rheology testing was carried out to assess the effect of graphite fineness on the fluidity of the pastes. A sufficient fluidity is necessary from a practical perspective for the material to be mixed, pumped, and placed without any bleeding or segregation. From Figure 6, the graphite addition resulted in an increase in the viscosity of cement paste in all cases. For the coarse graphite, the viscosity increased progressively by 76% to 171% compared to the control when graphite was added at concentrations of 10–40% by weight. For the medium graphite, viscosity also increased dramatically, and it was not possible to mix after the 20 wt % concentration. Specifically, the viscosity increased by 126% and 148% for the 10 and 20 wt % graphite, which was much higher than the coarse graphite at the same dosages. For the fine graphite at 10 wt %, the viscosity increase was comparable to that of the medium graphite but at 20 wt % dosage, the viscosity was almost three times higher compared to the control. Furthermore, the error bar was large at that concentration, due to difficulties in mixing this high fine graphite concentration. Due to the very high viscosity at the 20 wt % dosage, it was not possible to mix the higher 30 and 40 wt % concentrations with the fine graphite.

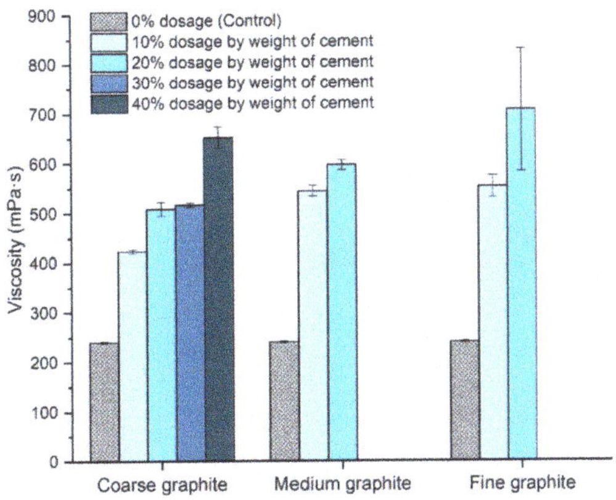

Figure 6. Effect of natural graphite size and concentration on the viscosity (mPa·s) of cement paste.

From rheology testing, it was found that increasing graphite fineness led to a dramatic reduction in fluidity, due to the inter-particle friction with cement particles, as well as because of the low hydrophilicity of graphite [26]. The finer the graphite, the more dramatic the reduction in fluidity (increase in viscosity) for a given % concentration; the fine graphite for the same weight dosage would have more particles that caused inter-particle friction with cement, and therefore the viscosity increased dramatically. Moreover, the smaller size graphite is expected to have a higher surface area, which would require more water to cover its surface. This effect on viscosity has also been confirmed in other studies that used additives in cementitious composites and can be explained by the *crowding* phenomenon, where an increase in the additive population obstructs the movement of water around them, thus increasing the viscosity [27,28]. Overall, the observed reduction in fluidity could introduce practical limitations when using graphite as a conductive additive, and the mix design and the water content would need to be adjusted, to ensure sufficient flowability.

3.3. *Effect of Graphite on the Hydration of Cement Paste*

Isothermal calorimetry testing was carried out to assess the effect of graphite on the hydration of cement paste and for comparative reasons; all graphites were tested at 10 and 20 wt %, while the control (CEMI paste, w/c = 0.45) remained constant. The mass to what the heat/power related to in the y-axis (mW/g and J/g) was that of cement only. The reason that cement was chosen as the mass unit was to isolate and identify the specific contribution of the graphite content and particle size on the hydration kinetics. As illustrated in Figure 7, the same hydration peaks were observed in all cases; therefore, the cement hydration was not affected by graphite addition, in agreement to another study that showed that graphite did not directly participate in cement hydration [29]. However, the three graphite products had a somewhat different effect on the hydration. The coarse graphite depressed and widened the main hydration peak (Figure 7a), while the cumulative heat release was lower than the control. It was also observed that the hydration lines for the 10 and 20 wt % coarse graphite overlapped; hence, the increasing coarse graphite dosage did not affect the cement hydration further. These changes in the cumulative heat and progress of hydration may impact and influence the latter development of the hardened properties of the graphite-cement paste. However, the effect on the mechanical performance cannot be solely deducted from isothermal calorimetry testing; therefore, explicit mechanical testing was carried out to confirm this hypothesis and the results are presented in Section 3.4. As the graphite became finer, the effect on hydration was less pronounced. For the 10 wt % medium graphite, the impact on the hydration was insignificant (Figure 7b), and when the dosage increased to 20 wt %, the peak power increased slightly from 3.82 mW/g for the control to 4.35 mW/g. For the fine graphite (Figure 7c), the effect on hydration was even less pronounced, with the hydration curves and the total cumulative heat of hydration remaining almost unaltered, as compared to the control.

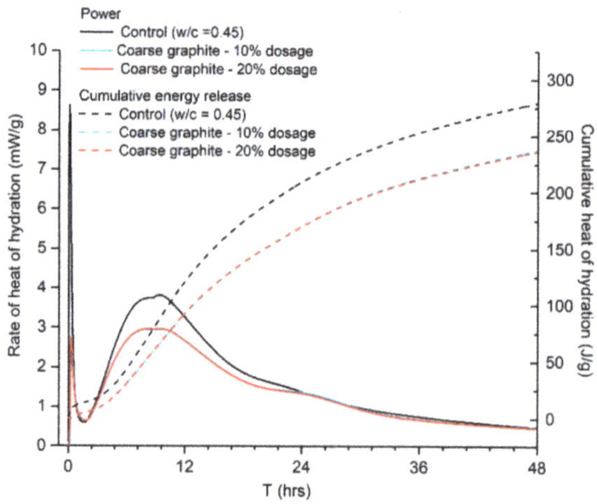

(a)

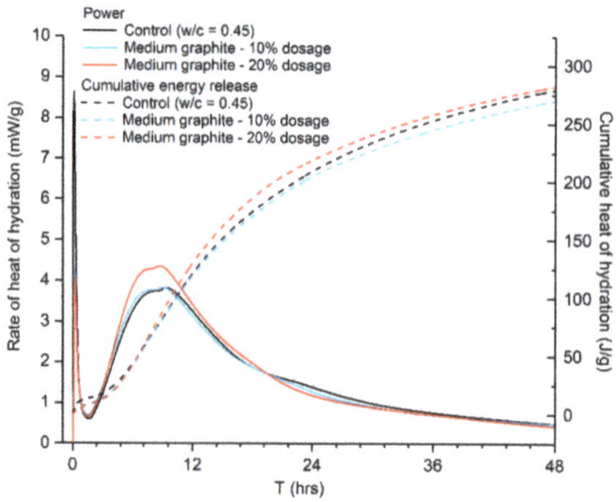

(b)

Figure 7. *Cont.*

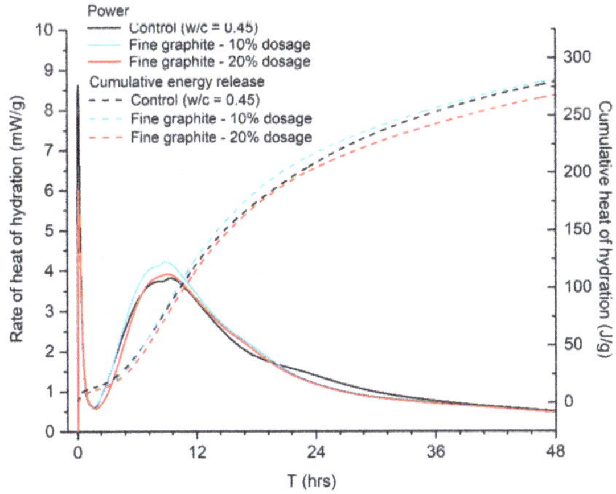

(c)

Figure 7. Effect of graphite size and concentration on the hydration of CEMI paste, measured by isothermal calorimetry: (**a**) coarse, (**b**) medium, and (**c**) fine graphite.

Hydration testing confirmed the hypothesis that graphite is acting as an inert filler and does not participate directly in cement hydration. The differences found between the different fineness products could be explained by a physical mechanism. The w/c was fixed, and therefore the addition of graphite resulted in an increase of the water/solids ratio. The effect of water on the hydration of Portland cement has been widely reported with higher water contents resulting in accelerated cement hydration and increased cumulative heat [11,30]. In the case of coarse graphite, the large graphite particles could have acted as physical blockers for the water to reach the cement grains and therefore resulted in a depression of the main hydration peak at ~10 h. With increasing graphite fineness, a filler effect started being present [18,31] and graphite particles helped in improving the packing density without physically blocking the water from reaching the individual cement grains. Graphite particles are also slightly hydrophobic and would push the water towards the cement grains, therefore promoting hydration. Overall, the graphite materials were not found to participate in the hydration process directly; however, the coarse graphite could block the water from reaching the cement grains, which may result in a strength reduction.

3.4. Mechanical Performance of Graphite-Cement Pastes

Figure 8 shows the effect of graphite fineness on the compressive strength of cement paste. The three graphite materials were used at 10 and 20 wt % concentrations. At all test ages, the compressive strength reduced with graphite, irrespective of its fineness. Furthermore, the higher the graphite dosage, the lower the compressive strength, with the 20 wt % concentration samples always resulting in a lower compressive strength than the 10 wt % samples. The summary of the % reduction in strength comparing to the control is presented in Table 2. The compressive strength had an inverse relationship with graphite size; the coarse graphite produced the lowest strength, whilst the fine graphite had the least reduction, as compared to the control.

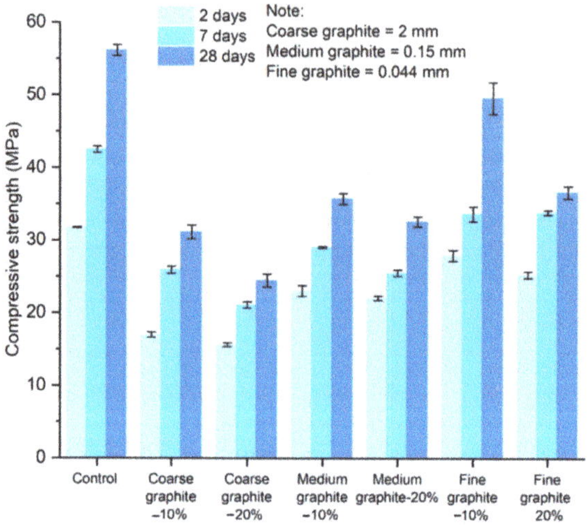

Figure 8. Effect of graphite fineness on the compressive strength of cement paste (*w/c* = 0.45) at 2, 7, and 28 days.

Table 2. Reduction in compressive strength (%) with the three different graphite materials.

Graphite and Dosage %	2 Days	7 Days	28 Days
Coarse graphite–10%	−47%	−39%	−45%
Coarse graphite–20%	−51%	−50%	−56%
Medium graphite–10%	−28%	−32%	−36%
Medium graphite–20%	−31%	−40%	−42%
Fine graphite–10%	−12%	−21%	−12%
Fine graphite–20%	−21%	−20%	−35%

Micro-indentation testing was also carried out to assess the effect of graphite fineness on the mechanical performance of cement paste, by assessing the hardness of the specimens. Graphite–cement paste samples were tested at 56 days, with a 20 wt % graphite concentration, and the error bars show the average of 15 measurements. The compressive strength results would suggest that the hardness would reduce with graphite addition and the effects would be more pronounced for the coarse graphite. Indeed, as seen in Figure 9a, the coarser the graphite, the lower the hardness. The fine graphite almost maintained the hardness of the control specimen, while, on the contrary, the coarse graphite reduced the hardness by 28%. The results can be explained by the graphite softness, which was expected to reduce the hardness of the paste, and also because of packing density, where the finer graphite resulted in a more compact mix, which improved the overall hardness of the sample. The hardness results also provide a further indication of sufficient dispersion of graphite. The error bars were small, and the variance was similar to the control, further supporting the SEM and μCT scan findings. Young's modulus in Figure 9b reduced for the coarse graphite by 15% but increased for the two finer graphite materials, as compared to the control (14% improvement for fine graphite). The modulus of elasticity is expected to increase with increasing compressive strength [11]; therefore, the finer the graphite, the higher the Young's modulus (as strength was also higher). The increase in stiffness with the fine graphite could also be explained by changes in porosity, where due to better packing density, the fine graphite-cement paste had a lower porosity, and therefore it was stiffer compared to the control.

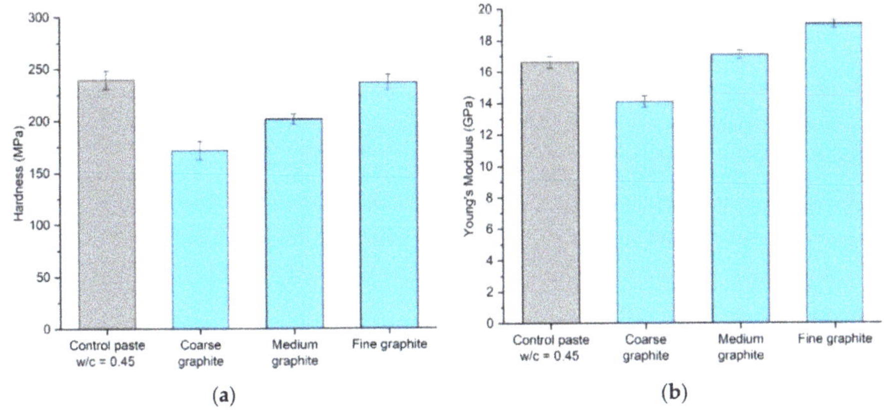

Figure 9. Micro-indentation results of the effect of 20 wt % graphite addition on the cement paste, at 56 days, in terms of (**a**) hardness (**b**) and Young's modulus.

Mechanical performance testing indicated that increasing graphite fineness was beneficial for mechanical performance. The fine graphite was more effective in maintaining the compressive strength and hardness of the specimens, while the use of the coarse graphite led to significant reductions in compressive strength and hardness. Therefore, when using coarse graphite materials, the mechanical performance could be significantly compromised, and this is a key limitation in their use as conductive fillers for self-sensing application. From a practical viewpoint, it would not be feasible to use higher than 10 wt % graphite additions as the compressive strength is significantly compromised. This is particularly the case for the coarse graphite powders, that more than halved the compressive strength when added as a 20 wt % addition. To mitigate the impacts on mechanical performance, a more realistic perspective would be the targeted use of the material in locations more prone to damage. Potentially, the graphite-cement paste could be locally used as a coating of the structure rather than in the bulk of the matrix, ensuring, in this way, that the structural performance is retained, while the graphite-cement paste coating could yield the sensing capabilities, assisting with maintenance and ensuring the resilience of the structure.

3.5. *Electrical Conductivity of Graphite-Cement Pastes*

The effect of graphite fineness on the electrical conductivity of cement paste was investigated to establish whether graphite could be used for self-sensing applications. Tests were carried out at 2, 7, and 28 days, and the results in terms of electrical conductivity vs. graphite content are illustrated in Figure 10. Irrespective of graphite fineness and test age, before the sudden increase in electrical conductivity, the conductivities of all samples were less than 2 S/m. Moreover, the electrical conductivity clearly reduced as curing progressed. The two-day samples (solid lines) had higher conductivities than those tested at 7 or 28 days (dashed lines). The reason that the conductivity of the samples reduced over time was due to the free water content available in the mix. Electric current can travel both through the conductive additive (termed as electronic conduction) and through the free water available, which is termed as electrolytic conduction. As the hydration progressed from 2 to 28 days, the free water in the mix reduced; thus, it was more difficult for the electric current to pass the matrix. At the same time, at low graphite dosages, the conductive filler content was not sufficient for electronic conduction to take place and form uninterrupted travel paths. This means that, at low graphite dosages, the specimens were acting as insulators, and, over time, the electrical conductivity would diminish. Furthermore, the percolation threshold was not affected by curing age.

By observing the effect of graphite fineness, the coarse graphite (blue lines) had a percolation threshold between 30 and 40 wt % dosages, and, at 28 days, the conductivity at 40 wt % dosage was over nine times higher than the control. For the coarse graphite, it was also found that the conductivity was compromised at low graphite concentrations, and this can be explained by changes in porosity and water content in the mix. Even though the *w/c* remained constant, as graphite was added in the mix, the effective water/solids ratio was reduced, which resulted in less water available for electrolytic conduction and at the same time, the graphite concentration was not sufficient to create electric conduction paths through the specimen. Hence, the conductivity of the sample was compromised due to a reduction in the effective water in the mix. By observing the medium graphite (green lines), the percolation threshold was found at a lower concentration, compared to the coarse graphite, and at 30 wt %, the electrical conductivity was ~17 times higher, compared to the control. Therefore, the medium graphite resulted in much higher conductivity at a lower concentration, compared to the coarse graphite. For the fine graphite (red lines), the percolation threshold could be found between 20% and 30%. At 30 wt % concentration, the fine graphite had a conductivity twenty-eight times higher than the control and a 37% higher conductivity compared to the medium graphite at the same dosage.

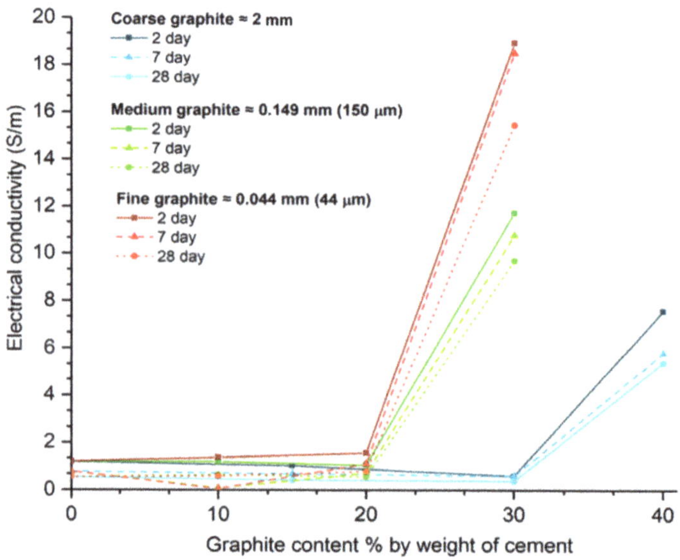

Figure 10. Effect of graphite fineness and dosage on the electrical conductivity of CEMI pastes (*w/c* = 0.45).

To better understand the effect of graphite fineness on the electrical behavior of the graphite-cement pastes, electrical impedance spectroscopy (EIS) testing was undertaken. The control mix refers to a cement paste with *w/c* = 0.45, and it is the same in all cases. Coarse graphite was tested at 10 and 40 wt %, with the former being below and the latter above the percolation threshold. The Nyquist plots for 7 and 28 days are shown in Figure 11a,b, whilst the resistance vs. frequency plot is illustrated in Figure 11c. As the hydration progressed, the Nyquist plots shifted to the right, at higher true resistance values on the *x*-axis, which was expected as electrical resistance increases with curing age. The incomplete electrode arc on the right side of the plot, was very clear for the 10 wt % graphite, meaning that the resistance measurement comprised of both the inherent electrical conductivity of the material and that of the electrode. However, at 40 wt % coarse graphite (above the percolation threshold), the Nyquist plot was obviously different, and the electrode arc was not present. The measured electrical response corresponded only to the bulk response of the material, meaning that a fully conductive path was formed through the cement composites. The arc of the 40 wt % graphite remained almost unchanged as curing progressed from 7 to 28 days and no increase in resistance over time was observed. Therefore, when graphite was added at a concentration below the percolation threshold, the electrical conductivity depended on the water content, and therefore the resistivity increased over time as the hydration progressed. When the percolation threshold was exceeded, the electrolytic conduction mechanism became irrelevant and electric current traveled primarily due to the conductive network that was formed with the graphite particles. In this case, the continuous cement hydration, which reduces the free water, had an insignificant effect on the electrical resistivity.

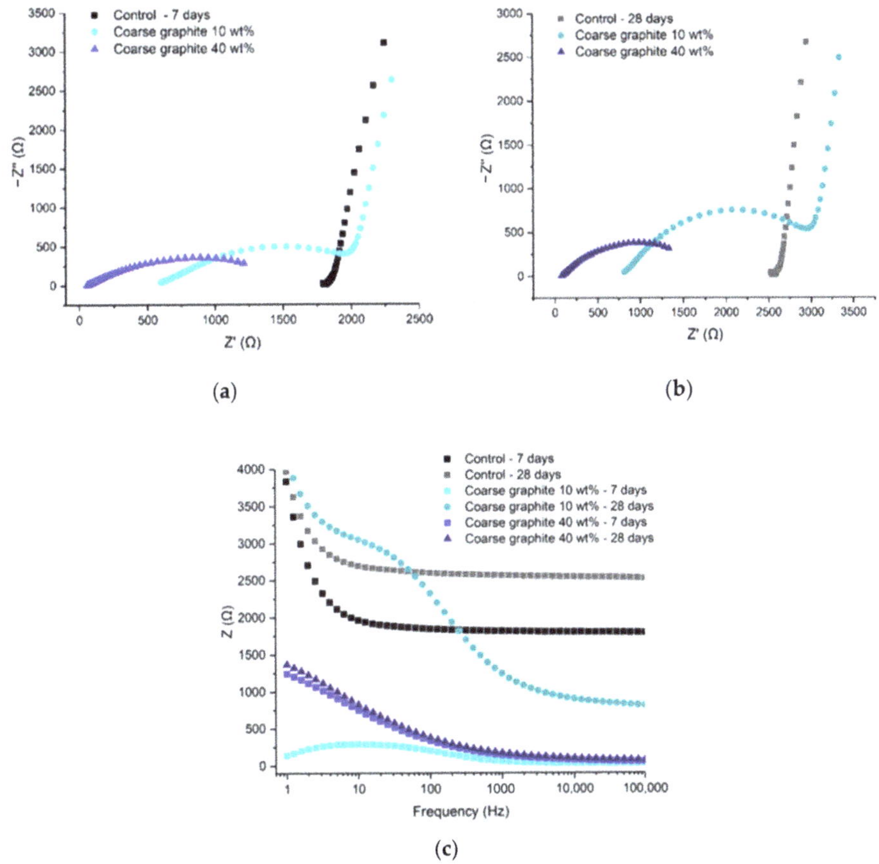

Figure 11. Cement paste with coarse graphite: (**a**) 7-day Nyquist plot, (**b**) 28-day Nyquist plot, and (**c**) frequency-dependent resistance.

The cement pastes with the medium graphite (0.150 mm), at 10 and 20 wt %, were examined. As shown in Figure 12, the Nyquist plots at 7 and 28 days showed similar arcs, with the electrode effect present in all cases and illustrated by the incomplete rightmost arc. The resistance increased with hydration age, due to the consumption of free water, which meant that less water was available for electrolytic conduction. The fact that the arcs showed both the electrode effect and the bulk material response meant that the conductive network was not fully formed at 20 wt % and that the percolation threshold was slightly higher for this medium graphite. From Figure 12c, it can be observed that the total resistance reduced with increasing graphite content, irrespective of the frequency.

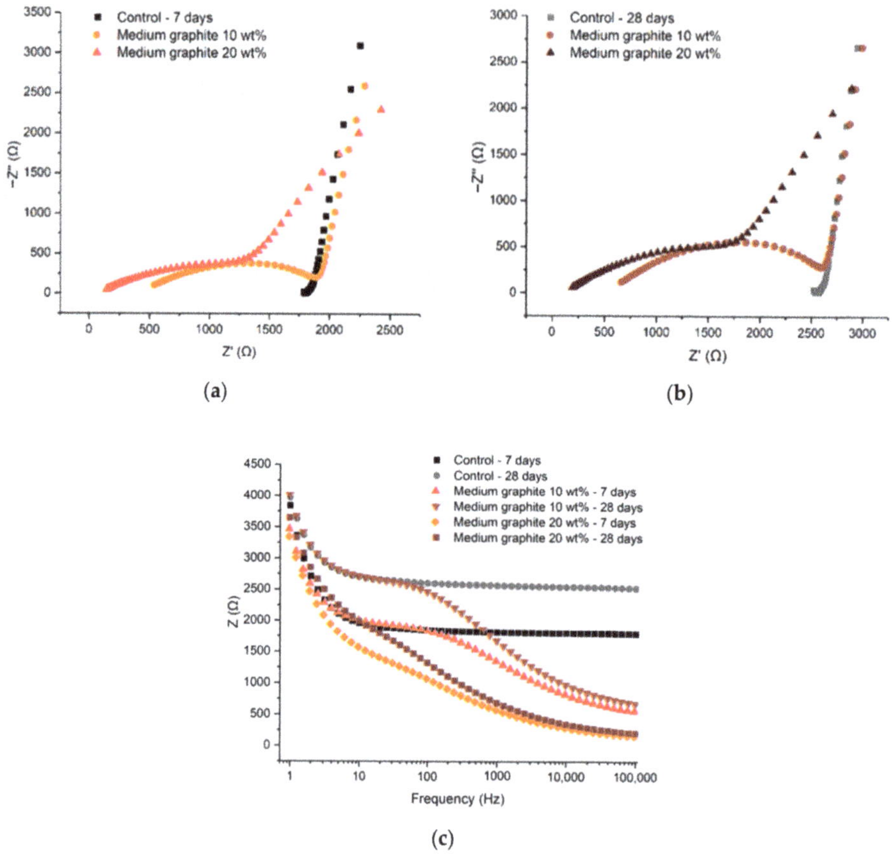

Figure 12. Cement paste with medium graphite: (**a**) 7-day Nyquist plot, (**b**) 28-day Nyquist plot, and (**c**) frequency-dependent resistance.

Figure 13 illustrates the findings for the fine graphite (44 μm) at 10 and 20 wt %, where the Nyquist plots between the two graphite dosages were different. The 10 wt % graphite was characterized by two arcs, a full semicircle representing the bulk material response, and an incomplete arc on the right size, which showed the electrode response to the electric current. For the 10 wt % fine graphite, the resistance increased with age due to cement hydration; therefore, the electrical conduction was due to both the presence of water (electrolytic) and conductive filler (electronic). Instead, at 20 wt %, only a semicircle arc was found at the Nyquist plots, which remained almost unchanged with age, meaning that the percolation threshold was reached. Comparing to the medium graphite, the percolation threshold was reached at 20 wt % concentration, which was not the case for the medium graphite where a higher graphite dosage was needed. Similar to the coarse and medium graphite materials, the electrical response was frequency dependent, especially below the percolation threshold. The response stabilized at ~1000 Hz for 20 wt % fine graphite but only after ~10,000 Hz for the lower graphite concentration.

The effect of graphite fineness on electrical conductivity was investigated, with an outlook that the graphite-cement pastes could be used for self-sensing applications. It was found that the coarse graphite had a percolation threshold at ~30 to 40 wt %, which reduced to between 20 and 30 wt % when a medium and a fine graphite were used. Therefore, increasing graphite fineness leads to a lower percolation threshold, which is beneficial in terms of material usage and in maintaining the mechanical performance. It should be noted that these filler dosages are higher, compared to the dosages traditionally used for inert fillers in cementitious composites. The reason that high filler dosages were selected was to ensure that an uninterrupted electrically conductive path was successfully formed in the cementitious matrix. Contrary to other inert fillers that are primarily used to improve the packing density of the mix, the purpose here was to reach the percolation threshold, which necessitated the higher additive concentrations. EIS testing showed that a combination of an electrolytic and electronic conduction mechanism was present when the conductive additive was used below the percolation threshold. Instead, when the percolation threshold was reached and an uninterrupted conductive path was formed, the electronic conduction became the dominant mechanism, and the conductivity did not depend on the presence of water. The main finding from the electrical conductivity testing was that, the finer the graphite, the lower the dosage that is needed to establish a percolation threshold, and the higher the conductivity at that dosage, as compared to a coarser graphite. These results can be explained by packing density principles. The formation of conduction paths relates to the particle size and aspect ratio of the conducting graphite filler [32]. The finer graphite particles tend to stabilize in dense configurations, resulting in more inter-particle contacts, as compared to coarser materials, creating inevitably more paths for current to pass through [33].

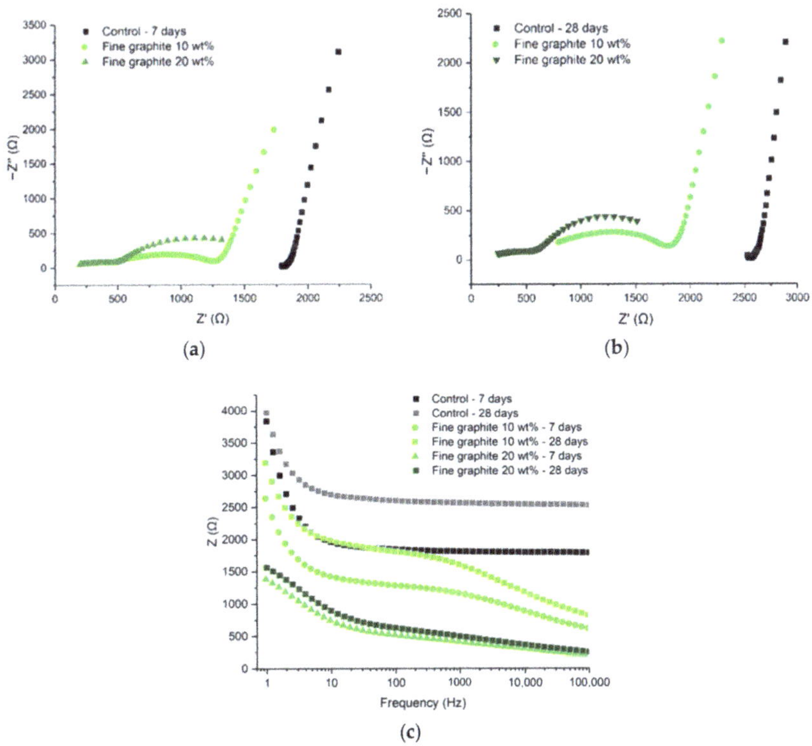

Figure 13. Cement paste with fine graphite: (**a**) 7-day Nyquist plot, (**b**) 28-day Nyquist plot, and (**c**) frequency-dependent resistance.

4. Conclusions

This study examined the effect of graphite fineness on the performance of cement paste to be used in self-sensing applications. Three graphite products of varying fineness were tested, and it was found that graphite fineness can greatly affect the performance of cementitious composites. In terms of early age performance, increasing graphite fineness led to a dramatic reduction in the fluidity of the paste, which could introduce practical limitations. On the other hand, graphite was not found to affect the hydration of the cement paste, meaning that it acts as an inert conductive filler. Compressive strength testing and micro-indentation showed that, the finer the graphite, the lesser the effect on mechanical performance. Electrical conductivity testing then showed that increasing graphite fineness reduces the percolation threshold for electrical conductivity. The finer graphite had a percolation threshold at 20 wt % concentration, which increased to 30–40 wt % for the coarse graphite. However, the use of graphite as a conductive additive in cement paste introduces some practical limitations. The reduction in fluidity could be a significant barrier for using this composite material in situ, while the reductions in compressive strength could make its use prohibiting for structural applications. Overall, it is recommended to use a graphite of greater fineness, as this leads to better maintenance of mechanical performance, and it reduces the dosage required to reach an electrical percolation threshold. Practical limitations around the fluidity of the paste could be overcome by adjusting the mix design and dispersion protocol.

Author Contributions: Experimental investigation, data processing and analysis, and writing original draft, I.P.; data analysis and writing—review and editing, C.L. and A.Z.; conceptualization and supervision, A.A.-T. All authors have read and agreed to the published version of the manuscript.

Funding: This work was supported by the Engineering and Physical Sciences Research Council (Grant No. EP/L016095/1-EPSRC Centre for Doctoral Training in Future Infrastructure and Built Environment and Grant No. EP/P02081X/1-Resilient Materials for Life (RM4L)) and Costain Group.

Conflicts of Interest: The authors declare no conflict of interest. The funders had no role in the design of the study; in the collection, analyses, or interpretation of data; in the writing of the manuscript; or in the decision to publish the results.

References

1. Treasury, H.M. Infrastructure Cost Review: Main Report. 2010. Available online: http://www.hm-treasury.gov.uk/d/cost_review_main211210.pdf (accessed on 15 October 2018).
2. Whiteley, D.; Goethert, K.; Goodwin, F.; Golter, H.P.; Kennedy, J.; Komas, T.W.; Meyer, J.; Petree, M.; Smith, B.; Trepanier, S.; et al. Sustainability for Repairing and Maintaining Concrete and Masonry Buildings. 2015. Available online: http://c.ymcdn.com/sites/www.icri.org/resource/collection/1023A08D-21D0-4AE9-8F9A-5C0A111D4AC9/ICRICommittee160-Sustainability_whitepaper.pdf (accessed on 5 November 2019).
3. Gardner, D.; Lark, R.; Jefferson, T.; Davies, R. A survey on problems encountered in current concrete construction and the potential benefits of self-healing cementitious materials. *Case Stud. Constr. Mater.* **2018**, *8*, 238–247.
4. Paine, K.; Al-Tabbaa, A.; Gardner, D.; Jefferson, A. Resilient materials for life: Biomimetic self-healing and self-diagnosing concretes. In *UKIERI Concrete Congress Concrete*; The Global Builder: Jalandhar, India, 2019.
5. Ou, J.; Han, B. Piezoresistive cement-based strain sensors and self-sensing concrete components. *J. Intell. Mater. Syst. Struct.* **2009**, *20*, 329–336.
6. Spencer, B.F. *Structural Health Monitoring of Civil Infrastructure: From RESEARCH to Engineering Practice*; University of Cambridge: Cambridge, UK, 2009.
7. Webb, G.; Vardanega, P.; Middleton, C. Categories of SHM Deployments: Technologies and Capabilities. *J. Bridg. Eng.* **2015**, *20*, 04014118.
8. Han, B.; Yu, X.; Ou, J. *Self-Sensing Concrete in Smart Structures*; Elsevier Inc.: Oxford, UK, 2014.
9. Chen, P.W.; Chung, D.D.L. Concrete as a new strain stress sensor. *Compos. Part B Eng.* **1996**, *27*, 11–23.
10. Chung, D. Electrical conduction behavior of cement-matrix composites. *J. Mater. Eng. Perform.* **2002**, *11*, 194–204.
11. Neville, A. *Properties of Concrete*, 5th ed.; Pearson Education Limited: London, UK, 2011.
12. Han, B.; Ding, S.; Yu, X. Intrinsic self-sensing concrete and structures: A review. *Meas. J. Int. Meas. Confed.* **2015**, *59*, 110–128.
13. Horszczaruk, E.; Sikora, P.; Lukowski, P. Application of Nanomaterials in Production of Self-Sensing Concretes: Contemporary Developments and Prospects. *Arch. Civ. Eng.* **2016**, *62*, 61–74.

14. Graphenea. Graphene & Graphite—How Do They Compare? 2017. Available online: https://www.graphenea.com/pages/graphene-graphite#.Wc5PnmhSzD5 (accessed on 5 November 2019).
15. Peinado, F.; Roig, A.; Vicente, F.; Vilaplana, J.; López, J. Electrochemical Characterization of Cement Graphite and Cement Aluminum Materials. *J. Mater. Sci. Lett.* **1994**, *13*, 609–612.
16. Sachdev, V.K.; Sharma, S.K.; Bhattacharya, S.; Patel, K.; Mehra, N.C.; Gupta, V.; Tandon, R.P. Electromagnetic shielding performance of graphite in cement matrix for applied application. *Adv. Mater. Lett.* **2015**, *6*, 965–972.
17. Bhattacharya, S.; Sachdev, V.K.; Chatterjee, R.; Tandon, R.P. Decisive properties of graphite-filled cement composites for device application. *Appl. Phys. A Mater. Sci. Process.* **2008**, *92*, 417–420.
18. Frattini, D.; Accardo, G.; Ferone, C.; Cioffi, R. Fabrication and characterization of graphite-cement composites for microbial fuel cells applications. *Mater. Res. Bull.* **2017**, *88*, 188–199.
19. BS EN 197-1:2011. Cement. Composition, Specifications and Conformity Criteria for Common Cements. Available online: https://shop.bsigroup.com/ (accessed on 5 November 2019).
20. Nanthagopalan, P.; Santhanam, M. A new empirical test method for the optimisation of viscosity modifying agent dosage in self-compacting concrete. *Mater. Struct. Constr.* **2010**, *43*, 203–212.
21. ASTM C1679-08. *Standard Practice for Measuring Hydration Kinetics of Hydraulic Cementitious Mixtures Using Isothermal Calorimetry*; ASTM: West Conshohocken, PA, USA, 2008; pp. 1–4.
22. BS EN 196-1. Methods of Testing Cement. Determination of Strength. 2016. Available online: https://shop.bsigroup.com/ (accessed on 15 March 2020).
23. ASTM E384-16. *Standard Test Method for Microindentation Hardness of Materials*; ASTM: West Conshohocken, PA, USA, 2016; pp. 1–28.
24. Zhang, Q.; Le Roy, R.; Vandamme, M.; Zuber, B. Long-term creep properties of cementitious materials: Comparing microindentation testing with macroscopic uniaxial compressive testing. *Cem. Concr. Res.* **2014**, *58*, 89–98.
25. Wansom, S.; Kidner, N.J.; Woo, L.Y.; Mason, T.O. AC-impedance response of multi-walled carbon nanotube/cement composites. *Cem. Concr. Compos.* **2006**, *28*, 509–519.
26. Kozbial, A.; Li, Z.; Sun, J.; Gong, X.; Zhou, F.; Wang, Y.; Xu, H.; Liu, H.; Li, L. Understanding the Intrinsic Water Wettability of Graphite. *Carbon* **2015**, *31*, 218–225.
27. Chong, J.S.; Christiansen, E.; Baer, A. Rheology of concentrated suspensions. *J. Appl. Polym. Sci.* **1971**, *15*, 2007–2021.
28. Kanellopoulos, A.; Giannaros, P.; Al-Tabbaa, A. The effect of varying volume fraction of microcapsules on fresh, mechanical and self-healing properties of mortars. *Constr. Build. Mater.* **2016**, *122*, 577–593.
29. Yuan, H.W.; Lu, C.H.; Xu, Z.Z.; Ni, Y.R.; Lan, X.H. Mechanical and thermal properties of cement composite graphite for solar thermal storage materials. *Sol. Energy* **2012**, *86*, 3227–3233.
30. Scrivener, K.L.; Juilland, P.; Monteiro, P.J.M. Advances in understanding hydration of Portland cement. *Cem. Concr. Res.* **2015**, *78*, 38–56.
31. Hamad, H. Modification and Performance of Activated Carbon for CO_2 Sequestration in Pervious Concrete. Ph.D. Thesis, University of Cambridge, Cambridge, UK, October 2019.
32. Nagata, K.; Iwabuki, H.; Nigo, H. Effect of particle size of graphites on electrical conductivity of graphite/polymer composite. *Compos. Interfaces* **1999**, *6*, 483–495.
33. Sbia, L.A.; Peyvandi, A.; Soroushian, P.; Balachandra, A.M.; Sobolev, K. Evaluation of modified-graphite nanomaterials in concrete nanocomposite based on packing density principles. *Constr. Build. Mater.* **2015**, *76*, 413–422. [CrossRef]

materials

MDPI

Article

Effect of Multi-Walled Carbon Nanotubes on Strength and Electrical Properties of Cement Mortar

Elena Cerro-Prada *,†, Rosalía Pacheco-Torres † and Fernando Varela †

Construction, Infrastructure and Transportation Department, Universidad Politécnica de Madrid (UPM), Alfonso XII, 3, 28014 Madrid, Spain; rosalia.pacheco@upm.es (R.P.-T.); fernando.varela@upm.es (F.V.)
* Correspondence: elena.cerro@upm.es; Tel.: +34-91-067-4612
† These authors contributed equally to this work.

Abstract: This work aims to investigate the effects of multi-walled carbon nanotubes (MWCNTs) on the strength and electrical properties of cement mortar. MWCNTs were added to cement mortar in four different concentrations: 0.00 wt.%, 0.01 wt.%, 0.015 wt.%, and 0.02 wt.% by the mass of cement. The consistency, density, setting time and compressive and flexural strength of mixes were tested and analyzed at 28 and 90 days curing time. Mechanical performance tests confirm an increase of 25% and 20% in the ultimate compressive and flexural strength respectively, which results from MWCNT 0.02 wt.% loading at 90 days curing time. The resistivity measurements in mortars with 0.01 and 0.015 wt.% MWCNT loading result up to 10% decrement at both 28 and 90 days curing. Activation energy calculations show fully accordance with these statements, resuming that 0.01 wt.% MWCNT appears to be the most effective loading scheme to produce certain conductivity enhancement in cement mortar.

Keywords: multi-walled carbon nanotubes; cement-based materials; flexural strength; compressive strength; electrical resistivity; activation energy

Citation: Cerro-Prada, E.; Pacheco-Torres, R.; Varela, F. Effect of Multi-Walled Carbon Nanotubes on Strength and Electrical Properties of Cement Mortar. *Materials* **2021**, *14*, 79. https://dx.doi.org/10.3390/ma14010079

Received: 29 November 2020
Accepted: 22 December 2020
Published: 26 December 2020

1. Introduction

Concrete is a key material in the construction industry. It is in fact the most used element both in building and infrastructures construction such as bridges, hydraulic works, pavements, etc. Its attributes include easy and affordable manufacturing [1,2], good durability if properly executed [3,4], and remarkable structural capabilities [5–7], which make this material the most widely consumed globally. Concrete materials are designed solely to carry compressive loads, although their continuos presence in structures may induce to seek for additional properties which could make concrete a more exploitable material. However, the very nature of cement based materials imposes limitations to develop highly customizable features with added functionalities.

One of the properties that cement microstructure prevents from developing is the electrical conductivity. Wet concrete behaves as a semiconductor, with resistivity in the range of 10^5 Ω/mm. However, dry concrete has resistivity in the range of 10^{12} Ω/mm, which makes the material be considered as an insulator. The variation in the measured electrical resistivity in wet and dry concrete can be interpreted to find that concrete electrical conductivity is an effect of the evaporable water present in the material [8].

The volume of evaporable water found in saturated concrete varies from 60% at the time of mixing to 40% when cement is fully hydrated. This water contains ions whose concentration varies over time, directly affecting concrete conductivity. When the concentration of these ions is very high, ionic association begins, giving rise to C-S-H gel formation and ettringite, which in turns generates an electrical insulation layer in the cement grains. This leads to a decrease in ions mobility, therefore increasing the resistivity. The increase in resistivity with time is also due to less porosity and more tortuosity [9]. After this abrupt rise, resistivity continues to increase at a much slower rate, due to the decrease in hydration reactions. In a porous medium such as concrete, resistivity reflects the ability to transport electric charge throughout the ions dissolved in the aqueous phase of a certain volume, assuming that aggregates are electrically inert since their resistivity turns out to be several orders of magnitude higher than that of the porous solution.

This physico-chemical approach to analyze the capacity to carry electrical charge will not only make it possible to develop a concrete with conductive properties-which will surely lead to very interesting applications-but also, it will provide information on the mechanical performance of the material. Upon mixing cement with water, a suspension is obtained whose resistivity is very low, however, as the cement hydrates and the concrete sets and hardens, the resistivity increases. The evolution of resistivity is therefore parallel to the evolution of strength. In this way, resistivity allows to predicting performance and can act as an indicator of the "age factor", which is essential for certain durability models.

At this point, we incorporate into our study carbon nanotubes (CNTs) within the cementitious material microstructure. CNTs are emerging nanomaterials that have captured the interest of researchers in recent years due to their attractive physical and chemical properties. CNTs mechanical properties, such as strength, stiffness and toughness, have been broadly studied and reported. In particular, Young's modulus of CNTs is in the range of 1–1.2 TPa, which indicates elastic behavior, with a tensile strength in the order of 36 GPa, denoting yield strains of up to 10% [10–14]. These outstanding mechanical characteristics provide a promising horizon in the achievement of CNT-composites with very high performance strength.

As nanoadditions incorporated into the cement microstructure, CNTs have been widely studied expecting that, due to CNTs superb mechanical performance, an impressive reinforcement in cement-based composites will be achieved. There are indeed a significant number of researchers that have demonstrated great enhancements in flexural and compressive strengths in cement composites by adding low concentration of CNTs. In particular, considering 28 days curing, Li et al. [15] achieved up to 19% and 25% increases in compressive and flexural strength respectively, by adding CNTs 0.5 wt.% to Portland cement paste. Hawreen and Bogas [16] incorporated CNTs additions varying between 0.05% and 0.5% by cement weight to concrete, obtaining improvement in the compressive strength of concrete up to 21%. CNTs were added in ultra high strength concrete by Lu et al. [17], achieving 4.63% increase in compressive strength with 0.05 wt.% CNTs loading. More recently, Hu et al. [18] achieved 2.4% and 9.6% improvements in compressive and flexural strength respectively with the addition of 0.05% CNTs in cement mortar at 28 days curing. However, these researchers reported detrimental effects in compressive and flexural strength with the addition of 0.5% CNTs.

General discussions from these and other reported studies conclude that low loadings of carbon nanotubes are very effective in improving the performance of cement-based materials. However, the incorporation of higher amounts of 0.5% CNTs seems to lead to lower compressive strengths. Furthermore, even the addition of 0.5 wt.% appears to produce contradictory strength results in different cement-based materials, as indicated in the previous paragraph. It seems necessary, therefore, to study in depth the incorporation of CNTs in cementitious materials in order to refine the amounts of CNTs to be added without producing detrimental effects.

However, researchers are clearly in agreement on the existence of a major downside to incorporating CNTs into cementitious composites. CNTs are atomic crystals and therefore do not dissolve in water. Furthermore, the large surface area of nanotubes induces strong attractive forces between the CNTs themselves. These facts result in the formation of clusters and agglomerations of CNTs when mixed with water. These clusters remain even after their insertion into the clinker, due to the van der Waals forces between carbon nanotubes. Furthermore, as CNTs are chemically inert, they do not participate in the hydration process. Consequently, agglomeration of CNTs around the cement grains hinders clinker hydration promoting defects formation in cement composites during the development of their microstructure [19]. All this results in deficiencies in the mechanical properties of the cement-base material, since optimal resistance is developed during the correct hydration of the cement grains. In short, good CNTs dispersion is crucial and helps to obtain maximum mechanical performance by filling the pores and increasing adhesion with cementitious and hydration products [20–23].

Consequently, some authors have promoted surface modification processes to ensure an adequate dispersion of CNT and a homogeneous inclusion in the cement-water mixture [23–25]. Another interesting approach consists of using sonication to achieve a homogeneous dispersion of CNTs in the cementitious matrix. Many authors combine sonication with other shear mixing methods, such as mechanical, magnetic, or/and hand-stirring mixing methods [26–28]. On the other hand, cement type influences CNTs dispersion within the matrix, mainly due to the clinker particle size. In fact, most of the tests are carried out with cement of the CEM I 52.5 R type corresponding to Portland cement with a finer particle size and therefore more similar to the size scale of the nanomaterials, as they can be worked together in a more appropriate way. In any case, the hydrophobility of these nanostructures is maintained in the absence of electric field, since the water molecules cannot enter into the graphene layer spontaneously [29]. Consequently, due to their hydrophobility, we can expect that the presence of CNTs in cement microstructure will give rise to movements of evaporable water in the porous structure, and therefore variations in the intrinsic electrical resistivity of the cement composite.

Apart from ionic current due to water presence, electrical conductivity can also be developed in cement microstructure by means of CNTs connectivity, as this nanomaterial exhibits electrical conductivity similar to those of metallic materials [30]. However, it is clear that uniform CNTs distribution is essential to produce a continuous electrically conductive network. In fact, Kim et al. [31] reported that CNTs agglomerations induce damage in the electrically conductive pathways that CNTs should otherwise create. Remarkable improvement in electrical conductivity was achieved by Kim at al. [32], obtaining more than 1000 times reduction in electrical resistivity of well dispersed CNTs-cement paste compared to that of plain cement paste at 28 days curing, whereas poorly-dispersed CNTs-cement paste only achieves electrical resistivity approximately 2 times lower than that of control.

Very large number of CNT modified cement-based materials studies in the current literature aims at assessing the optimal CNT dosage to produce remarkable improvements in strength performance. Despite their importance, these studies use small-scale specimens that are not likely to reflect the actual mechanical behavior of large-scale structures. Taking into account that the electrical resistivity of cement-based materials can be used in quality control or for service life prediction of full size elements, this paper aims at providing more insights at important features that may need to be captured in the ongoing development of standard test methods to be used for CNT modified cement-based materials.

In this work, we present an experimental study on the effect of multi-walled carbon nanotubes (MWCNTs) on the microstructure of cement mortars, in terms of mechanical and electrical properties. For this, standard mortar prismatic samples will be modified with different proportions of MWCNTs, i.e., 0.00 wt.%, 0.01 wt.%, 0.015 wt.%, and 0.02 wt.% by the mass of cement. The nanomaterials will be introduced into the mortar material by priorly prepared water-MWCNT nanofluids following a combined mechanical and sonication dispersion procedure. The consistency, density, setting time and compressive and flexural strength of mixes will be tested and analyzed at 28 and 90 days curing time, in order to account for possible effects due to MWCNT presence in fully hydrated mortar. Finally, this paper will also investigate the electrical resistivity of the different MWCNT-mortar composites and the influence of temperature on the samples resistivity, to confirm the correlation of mechanical and electrical properties of CNT modified cement-based materials.

2. Experimental

2.1. Materials

2.1.1. Carbon Nanotubes

Multi-walled carbon nanotubes, provided by Sigma-Aldrich in Spain, were used. Their properties are given in Table 1. Figure 1 shows its transmission electron microscope (TEM) image [33].

Table 1. Properties of multi-walled carbon nanotubes (MWCNTs).

External diameter	6–13 nm
Length	2, 5–20 μm
Size	10 μm
Form	powder
Purity	>98% carbon basis
Special Surface Area	220 m^2/g
Density	2.1 g/mL at 25 °C
Average Wall Thickness	7–13 graphene layers

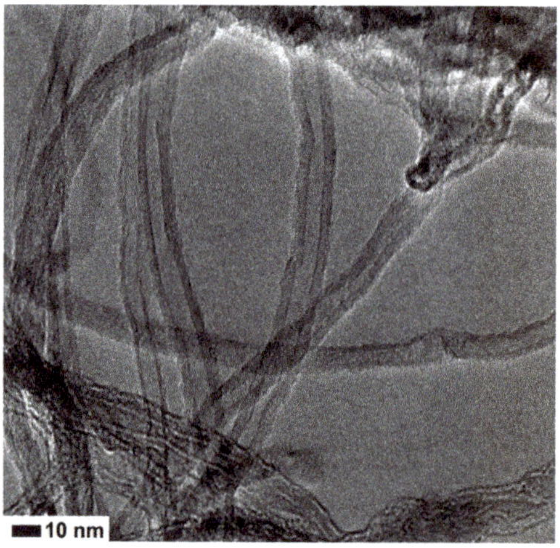

Figure 1. Transmission electron microscope image of MWCNT, adapted with permission from Sikora et al. (2019) [33].

2.1.2. Cement, Sand and Mortar

CEM I 42.5 R type cement was used to manufacture the mortar. Table 2 summarises composition and some physical-chemical properties of the cement used in this work, suppled by Portland Valderrivas, in Spain. Siliceous sand from a nearby quarry, with continuous granulometry and 2 mm maximum size was used. The sand equivalent was 87.35% and the density was 2.63 g/cm^3.

Mortar mixes with 0.00 wt.%, 0.01 wt.%, 0.015 wt.% and 0.02 wt.% of MWCNTs were prepared. The mix proportions are shown in Table 3. All samples were mixed in a laboratory drum mixer. Then, the fresh mortar samples were placed in 40 × 40 × 160 mm^3 prismatic moulds and homogenized using a vibrating table. Three specimens of each mix were manufactured. The specimens were cured in a curing chamber at 100% humidity and 20 °C temperature. Two curing periods were considered: 28 and 90 days. Additionally, to study the thermal variation of electrical resistivity of MWCNT-mortars, one sample of each MWCNT proportion, for every curing period, was manufactured with a thermocouple inserted at the center of the beam. These specimens were cast and cured in the same conditions as stated before.

Table 2. Elemental composition (wt.%) of source cement and physical-chemical properties (according to the supplier).

Elemental Content (%)	Si	7.78
	Ca	48.2
	K	0.28
	Mg	0.49
	Fe	3.53
	Al	1.32
	P	0.07
	S	1.55
	Ti	0.08
Physical	Blaine number (cm^2/g)	3.75
	Setting start time (min)	170
	Initial setting time (min)	170
	Final setting time (min)	220
Loss on Ignition (976 °C)		3.2

Table 3. Mortar mix proportions.

Reference	MWCNTs Content (wt.%)	Sand (g)	Cement (g)	Water (g)
Control	0.00	1350	450	225
CNT-0.01	0.01	1350	450	225
CNT-0.015	0.015	1350	450	225
CNT-0.02	0.02	1350	450	225

Inclusion and dispersion of MWCNTs in the mortar were achieved by prior preparation of MWCNT nanofluids. The corresponding amount of MWCNTs was dispersed in 100 g mixing water by mechanical stirring for 3 min, followed by magnetic stirring for 30 min and sonication for 1 h. Model 505 Sonic Dismembrator (Fisher Scientific, Waltham, MA, USA) was used for sonication. This probe is designed as 20 kHz operating frequency and a maximum power of 500 W. During the sonication, 50% amplitud, and 2 s ON and 2 s OFF pulses were applied. This so-prepared nanofluid was then incorporated to the cement-sand-remain water mix, that kept mixing in the drum mixer for the necessary time (approximately 4 min) until homogeneity.

2.2. Methods

2.2.1. Testing Procedures

Technological tests were performed to evaluate the mechanical behavior and durability properties of the cement mortars, namely: consistency, density and setting time of fresh mortars, and flexural strength and compressive strength of hardened mortars at the ages of 7, 28 and 90 days, following UNE-EN 1015-3:2000/A2:2007, UNE-EN 12390 3:2020 and UNE-EN 12390-5:2020 [34–36], respectively.

2.2.2. Electrical Resistivity Measurements

The electrical resistivity was measured by means of the four point probe Wenner technique [37] which is an electrical resistivity determination test performed on the surface of the structure or test body. According to Wenner method, by using four equally-distanced electrodes placed on the mortar surface, an alternating current I (A) is sent through the two outer electrodes and the voltage V (V) is measured through the two inner electrode to finally calculate the apparent resistivity ρ_a ($k\Omega\cdot$ cm) through Ohm's law, as expressed in Equation (1)

$$\rho_a = \frac{2\pi a V}{I} \tag{1}$$

where a (cm) is the distance between electrodes.

This experimental methodology to calculate resistivity assumes that the mortar is semi-infinite and homogeneous. As mortar confined in laboratory specimens cannot be considered semi-infinite, a correction should be made according to the geometry of the specimen [38], by means of a cell constant K, so that the real resistivity ρ_{real} is obtained from the following expression given in Equation (2):

$$\rho_{real} = \frac{\rho_a}{K} \tag{2}$$

In this work, the electrode spacing used was 2.5 cm, and all measured resistivity values were corrected for the geometry effect with a cell constant $K = 3.1$, according to [39]. An alternating current was applied by a waveform generator (National Instrument, 9263) which imposed a peak amplitude voltage of ± 10 V at 500 kHz frequency. Values of the voltage and current intensity were recorded by National Instrument 9227 and 9222 respectively.

Figure 2A shows a graphic scheme of the described measurement method. In addition to the thermocouple, four pure copper pins were fixed in the mortar samples surface using a conductive epoxy. Figure 2B shows a schematic of the beam dimensions, spacing and position of the electrodes and temperature probe.

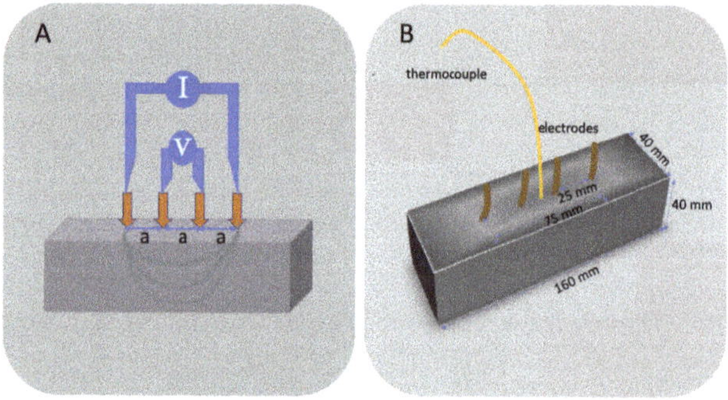

Figure 2. Diagram of Wenner method for resistivity measurements in mortar specimens (**A**), and (**B**) electrode spacing and dimensions.

To investigate the electrical transport within the MWCNT-mortar microstructure in terms of activation energy, we introduce the effect of temperature on the resistivity of MWCNT-mortar composites. The resistivity/temperature response is presented in an Arrhenius format thereby allowing evaluation of the activation energy for electrical conduction processes. Specimens were placed in a temperature controlled heater and their resistivity tested over the range 20–60 °C, following the so-called dynamic temperature test (DTT) as described by Liu and Presuel–Moreno [40]. This procedure consists of the monitoring the thermal-electrical behavior of the mortar samples when a thermal gradient is applied. The test was carried out in a controlled environment of constant temperature and relative humidity.

3. Results and Discussion

3.1. Consistency, Density and Setting Time

Consistency of the fresh MWCNT-mortar samples was determined by flow table. The loading of MWCNTs in mortars resulted in consistency values lower than that of the control mortar. As the MWCNT addition ratio increased, the consistency values decreased (Table 4). However, density values increased as the MWCNT loading increased (Table 4).

Both density and consistency are related to MWCNTs dispersion, since these fresh mortar properties are indicative of homogeneity and porosity. Well dispersed MWCNTs are expected to contribute to better homogeneity and promote filling effects on the porous structure of cement composites. Our results show an increase in density with MWCNTs content, which can be attributed to the reduction of air content in the cement matrix as CNTs are added and occupy empty sites within the pore structure. This itself is an indicative of good MWCNTs dispersion. Coherently, increase in consistency with MWCNTs content should be expected, leading to more fluid mixes as porous structure is refined and mix fine particles get better dispersed. However, taking in account that no superplastifier nor surfactant agent was added along with MWCNTs, the decrease in consistency displayed in our results should be understood as an effect of large water retention in the mortar, mainly due to the higher water-demand of the pastes because of the MWCNTs' high specific surface.

Table 4. Physical properties of the fresh mortars.

Reference	MWCNTs Content (wt.%)	Consistency (mm) UNE-EN 1015-3	Density (g/cm^3) UNE-EN 1015-6	Setting Time (min) UNE-EN 196-3	
				Initial	Final
Control	0.00	188.00	1939	140	242
CNT-0.01	0.01	162.50	2001	130	237
CNT-0.015	0.015	158.50	2035	123	228
CNT-0.02	0.02	144.00	2056	119	224

Regarding the setting time, it was observed that for MWCNT-mortar both the initial and final setting times are shorter than that of the control mortar. Some researchers [41] state the same results. However, initial setting time differences are more significant than final time, although changes in the MWCNT loading did not significantly modify the overall trend. According to [42], this behavior can be explained by a rapid nucleation during the initial steps which later leads to a barrier effect, slowing the formation of additional hydration products in the MWCNT-mortar [14].

3.2. Compressive and Flexural Strength

The compressive and flexural strength results obtained for each mixture are shown in Figure 3. The maximum failure load and its failure pattern is noted and results are presented in Table 5 below.

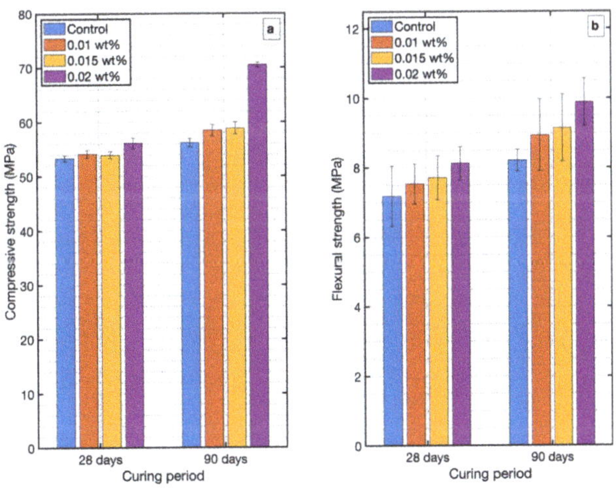

Figure 3. Compressive (**a**) and flexural strength (**b**) of MWCNT-mortar specimens.

Table 5. Compressive and flexural loads and strength.

	MWCNTs Content (wt.%)	Compressive			Flexural		
		Failure Load (N)	Strength (MPa)	Increase %	Failure Load (N)	Strength (MPa)	Increase %
28 days	0.00	8.74	53.31	–	3.12	7.18	–
	0.01	8.88	54.15	1.6	3.27	7.53	4.9
	0.015	9.0	54.93	3.0	3.35	7.71	7.4
	0.02	9.20	56.11	5.3	3.53	8.12	13.3
90 days	0.00	9.21	56.20	–	3.57	8.21	–
	0.01	9.59	58.48	4.1	3.88	8.93	8.8
	0.015	9.65	58.85	4.7	3.97	9.14	11.3
	0.02	11.55	70.48	25.4	4.30	9.88	20.3

The results confirm an increase in compressive strength in all the specimens with MWCNT content in the cement matrix for the two curing periods, in agreement with related works which also studied low CNT loading [17,43,44]. In particular, all mortars manufactured with 0.01 wt.% MWCNT showed 1.6% and 4.1% higher compressive strength values than those manufactured with ordinary cement (control) for 28 and 90 days curing periods, respectively. Furthermore, the improvements were significant in mortars prepared with 0.015 wt.% MWCNT, which showed 3.0% and 4.7% increases in compressive strength at 28 and 90 days curing period respectively. This increase in strength can be explained by the CNTs-induced formation of highly dense zones in the cement microstructure, as seen in the strength-density relationship shown in Figure 4. The densification of cement, forming virtual aggregate clusters (cement-MWCNTs) with presumably rounded shapes, minimizes crack initiation, propagation and distribution [45]. More remarkable effect was shown in cement mortars with 0.02 wt.% MWCNT loading at the age of 90 days, that exhibited an improvement of 25.4% in compressive strength.

Flexural strength results show the same trends in increasing flexural strength when low MWCNT loading in present in mortar. MWCNT 0.02 wt.% demonstrates to be the optimum dosage in accommodating in the cement matrix, achieving improvements up to 20.3% in flexural strength at 90 days curing period. Additions 0.01 and 0.015 wt.% also showed satisfactory results, with an increase of the ultimate strength of about 8.8% and 11.3%, respectively, at the same curing period of 90 days.

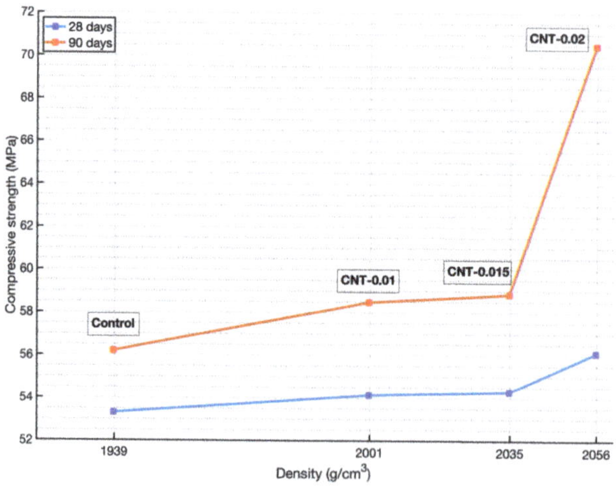

Figure 4. Relationship between compressive strength and density of MWCNT-mortar specimens.

These findings show that the modification of cement microstructure after inclusion of low dosage of MWCNTs results in an increase in the ultimate strength, both compressive and flexural, specially at long curing time. These results also lead to the importance of caring cement paste hydration parameters at long ages, when microstructure development is almost completed, and CNTs can contribute to a filled porous structure with enhanced strength properties. If mortar curing has developed properly under cared curing conditions, hardened material can take advantage of CNTs for an improved loads distribution.

3.3. Electrical Resistivity Analysis

Figure 5 shows the average results of the electrical surface resistivity of MWCNT-mortars versus the amount of MWCNTs for each curing time. In agreement with Konsta-Gdoutos and Aza [46], a decrease in resistivity is observed with the presence of MWCNTs in the cement matrix even for very low nanomaterial additions. In general, similar decrement range is observed regarding the curing times, regardless of the MWCNT loading. For example, in the case of mortar specimens with 0.01 wt.%, the registered resistivity values were 3.32 kΩ · cm and 3.82 Ω · cm, at curing times of 28 and 90 days respectively, which corresponds to a decrease of 11% and 6%, respectively. A similar behavior happened in the case of mortar specimens with 0.015 wt.% MWCNT, where electrical resistivity values reached decrements of 8% and 5% at curing times of 28 and 90 days respectively. Thus, in the case of both MWCNT loadings, average values of electrical resistivity are in a similar range, although our findings demonstrate that the curing time does have an influence on the electrical resistivity of the specimens. This statement is in agreement with relevant works where findings on nearly linear increment of resistivity as degree of hydration increases are reported [47,48]. Surprisingly, the sample with the addition of MWCNTs 0.02%, manages to increase the resistivity by 27% at 90 days curing time. These very MWCNT-mortar specimens did show a remarkable increase in compressive strength test at 90 days curing time, which in indeed should explain this different trend at electrical resistivity variation, that is presumed to be a physical property related to mechanical performance of cement-based materials.

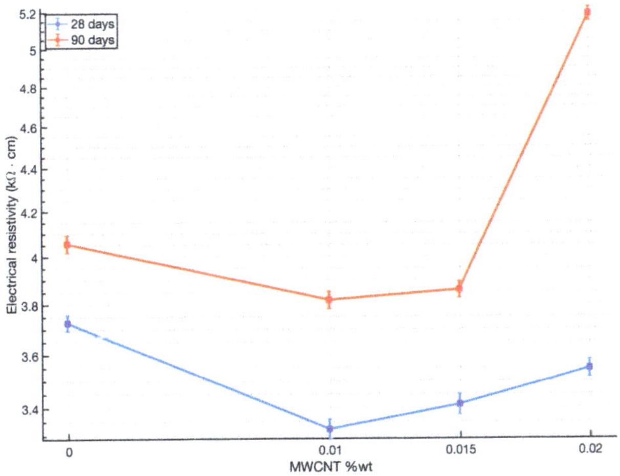

Figure 5. Average electrical surface resistivity of MWCNTs-mortar samples.

Regarding the variation of resistivity with respect to the MWCNTs content in the samples, it is clear from the results that there is a slight detriment in conductivity as the addition of MWCNTs increases. No significant improvement in conduction properties is obtained from an increasing loading of MWCNTs. Once electrical conduction is established in the material, the presence of a greater amount of nanomaterial does not imply a gain in electrical flux. Although in this work we cannot warranty good MWCNTs dispersion within the matrix of our samples, we can derive, however, that good dispersion was achieved from the fresh properties results. Therefore, we believe that conduction paths do not break due to bad MWCNT dispersion, but that resistivity increases due to polarization effects caused by the amount of water and the dissolved ions in the mixes [46], as consistency results proved large water retention in our samples.

Measurements of the dependence of resistivity with temperature are presented in Figure 6a. All samples were tested at the age of 90 days. Temperature significantly influences the measured resistivity, with higher temperatures leading to lower resistivity values, as demonstrated in Figure 6a. From these experimental findings, it is clear that all the profiles of resistivity versus temperature follow a trend of exponential decay, as previously reported [49].

To reach a better insight in this topic, we analyze the activation energy as provided by the Arrhenius equation, given by Equation (3)

$$\rho = A \cdot \exp \frac{E_a}{R_g T} \quad (3)$$

where ρ is the electrical resistivity measured at temperature $T(K)$, A is the pre-exponential factor and represents the nominal resistivity at infinite temperature, R_g is Universal Gas constant (8.314 $\text{J} \cdot \text{mol}^{-1} \cdot \text{K}^{-1}$), and E_a is the activation energy for the conduction process ($\text{J} \cdot \text{mol}^{-1}$). In Figure 5b below, natural logarithm of the inverse of resistivity (natural logarithm of conductivity σ), is plotted against 1000/T, therefore multiplying the slope of the plot by the Universal Gas constant R_g, E_a is obtained in $\text{kJ} \cdot \text{mol}^{-1}$. Results of the calculated activation energy of MWCNT-mortar specimens are presented in Table 6.

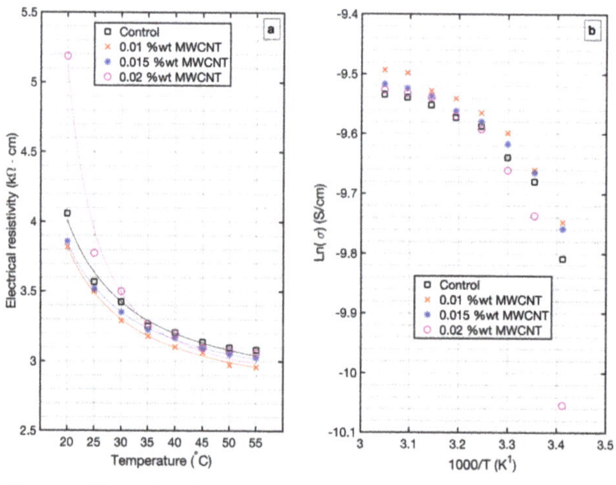

Figure 6. Thermal evolution of electrical resistivity of MWCNT-mortar at 90 days curing time (**a**), and Arrhenius plots for MWCNT-mortar at 90 days curing time presented over the temperature range 20–55 °C (**b**).

Table 6. Summary of activation energy values, calculated with Arrhenius equation of MWCNTs-mortar at 90 days curing time. Conductivity of the MWCNT-mortar at 20 °C, σ_{20} is also presented.

MWCNT wt.%	E_a **(KJ · mol^{-1})**	$\sigma_{20} \times 10^{-4}$ **(S · cm^{-1})**
Control	8.68	2.46
0.01 wt.%	7.83	2.62
0.015 wt.%	7.38	2.59
0.02 wt.%	17.14	1.93

The above figures are in agreement with resistivity results as plotted in Figure 6a; high resistivity correlates with high activation energy and vice versa. The activation energy slightly decreases for mortars containing MWCNTs 0.01 and 0.015 wt.%, which implies that in these mortars the energy barrier that must be overcome for an ion to conduct is lesser than the control specimen.

Further understanding of the above facts can be obtained from the analysis of electrical resistivity behavior in porous materials, as mortar, which can be described by using the following Equation (4):

$$\rho_T = \rho_o \frac{1}{\phi\beta} \tag{4}$$

where ρ_T is the total resistivity, ρ_o is the resistivity of the pore solution, which in turns is a function of the ions composition and concentration in solution; ϕ is the porosity of the system that is accessible to fluids, and β is the connectivity of the pores in the system [48]. The term $1/\phi\beta$ is ofter called Formation Factor F, which comprises the material microstructural characteristics, such as the water-to-cement ratio, volume of paste, and degree of hydration.

Since electrical conduction through saturated cement mortar occurs via ions in the continuous (percolated) aqueous phase between the electrodes [50]; the fact that samples with MWCNT 0.01 wt.% loading show better conductivity than samples with 0.015 wt.% could, in part, be attributed to microstructural changes resulting from further hydration in sample 0.01 wt.% (being hydration reactions thermally activated processes). Another plausible explanation could be the effect of a smaller formation factor in this specimens, which reflects either a higher porosity or a higher connectivity. Higher porosity directly comes out of a more hydrated mortar, as introduced before. Higher connectivity, however, can only be explained by the presence of MWCNTs in the microstructure. From the results, it seems that 0.01 wt.% be more efficient in developing conductive paths within the microstructure. Nevertheless, these findings indicate that the MWCNT loadings studied in this work are insufficient to produce conductive cement mortar.

Similarly, the increase in activation energy, and subsequently increased resistivity, with respect to the control sample for mortars with MWCNT 0.02 wt.% wt can be explained considering that these specimens have better developed their hydration products, as confirmed by the results of mechanical resistance, therefore the formation factor has been enlarged as result of decrease porosity.

4. Conclusions

The effect of 0.01, 0.015 and 0.02 wt.% MWCNT loading on mechanical strength and electrical resistivity of cement mortar is studied in the present work. MWCNTs are dispersed in mortar microstructure and tested through mechanical performance analysis at 28 and 90 curing period. The electrical resistivities of MWCNT-mortar are measured at different temperatures up to 60 °C, using the four point probe Wenner setup and dynamic temperature test (DTT). Activation energies corresponding to different MWCNT loadings are calculated using Arrhenius equation.

It is found that an increase in both compressive and flexural strength is achieved in mortars with MWCNT contents. Remarkable improvements of 25.4 and 20.3% at 90 days are found in compressive and flexural strength respectively, for an addition of 0.02 wt.% MWCNTs. Shorter initial and final setting times are obtained for 0.02 wt.% MWCNT loading, which implies a denser microstructure and finer porous structure.

The resistivity measurements in mortars with 0.01 and 0.015 wt.% MWCNT loading result up to 10% decrement to respect to control mortar at both 28 and 90 days curing. Mortar with 0.02 wt.% MWCNTs, however, depicts a sudden 27% increase in resistivity at 90 days curing time, due to well hydrated microstructure as confirmed by the mechanical tests. Activation energy calculations show fully accordance with these statements, resuming that 0.01 wt.% MWCNT seems to be the most effective loading scheme to produce certain conductivity enhancement in cement mortar.

Author Contributions: Conceptualization, E.C.-P.; Data curation, F.V.; Formal analysis, E.C.-P.; Funding acquisition, F.V.; Investigation, E.C.-P. and R.P.-T.; Methodology, E.C.-P., F.V. and R.P.-T.; Resources, E.C.-P. and R.P.-T.; Validation, E.C.-P. and R.P.-T.; Writing original draft preparation, E.C.-P. and R.P.-T., writing review and editing, E.C.-P. All authors have read and agreed to the published version of the manuscript.

Funding: This research received no external funding.

Institutional Review Board Statement: Not applicable.

Informed Consent Statement: Not applicable.

Data Availability Statement: The data presented in this study are available on request from the corresponding author.

Conflicts of Interest: The authors declare no conflict of interest.

References

1. Gambhir, M.L. *Concrete Technology: Theory and Practice*; Tata McGraw-Hill Education: New York, NY, USA, 2013.
2. Naik, T.R. Sustainability of concrete construction. *Pract. Period. Struct. Des. Constr.* **2008**, *13*, 98–103. [CrossRef]
3. Armaghani, J.M.; Larsen, T.J.; Romano, D.C. Aspects of Concrete Strength and Durability. *Transp. Res. Rec.* **1992**, *1335*, 63–69.
4. Bentz, D.P.; Clifton, J.R.; Ferraris, C.F.; Garboczi, E.J.; Torrents, J.M. *Transport Properties and Durability of Concrete: Literature Review and Research Plan*; Technical Report; U.S. Department of Commerce: Washington, WA, USA, 1999.
5. Vu, C.C.; Plé, O.; Weiss, J.; Amitrano, D. Revisiting the concept of characteristic compressive strength of concrete. *Constr. Build. Mater.* **2020**, *263*, 120126. [CrossRef]
6. Chen, X.; Wu, S.; Zhou, J. Variability of compressive strength of concrete cores. *J. Perform. Constr. Facil.* **2014**, *28*, 06014001. [CrossRef]
7. Zech, B.; Wittmann, F. Variablility and Mean Value of Strength of Concrete as Function of Load. *J. Proc.* **1980**, *77*, 358–362.
8. Madhavi, T.C.; Annamalai, S. Electrical conductivity of concrete. *ARPN J. Eng. Appl. Sci.* **2016**, *11*, 5979–5982.
9. Levita, G.; Marchetti, A.; Gallone, G.; Princigallo, A.; Guerrini, G. Electrical properties of fluidified Portland cement mixes in the early stage of hydration. *Cem. Concr. Res.* **2000**, *30*, 923–930. [CrossRef]
10. Yu, M.F.; Lourie, O.; Dyer, M.J.; Moloni, K.; Kelly, T.F.; Ruoff, R.S. Strength and breaking mechanism of multiwalled carbon nanotubes under tensile load. *Science* **2000**, *287*, 637–640. [CrossRef]
11. Ajayan, P.M.; Schadler, L.S.; Giannaris, C.; Rubio, A. Single-walled carbon nanotube–polymer composites: strength and weakness. *Adv. Mater.* **2000**, *12*, 750–753. [CrossRef]
12. Demczyk, B.G.; Wang, Y.M.; Cumings, J.; Hetman, M.; Han, W.; Zettl, A.; Ritchie, R. Direct mechanical measurement of the tensile strength and elastic modulus of multiwalled carbon nanotubes. *Mater. Sci. Eng. A* **2002**, *334*, 173–178. [CrossRef]
13. Yu, M.F.; Files, B.S.; Arepalli, S.; Ruoff, R.S. Tensile loading of ropes of single wall carbon nanotubes and their mechanical properties. *Phys. Rev. Lett.* **2000**, *84*, 5552. [CrossRef] [PubMed]
14. Makar, J.; Margeson, J.; Luh, J. Carbon nanotube/cement composites-early results and potential applications. In Proceedings of the 3rd International Conference on Construction Materials: Performance, Innovations and Structural Implications, Vancouver, BC, Canada, 22–24, August 2005; pp. 1–10.
15. Li, G.Y.; Wang, P.M.; Zhao, X. Mechanical behavior and microstructure of cement composites incorporating surface-treated multi-walled carbon nanotubes. *Carbon* **2005**, *43*, 1239–1245. [CrossRef]
16. Hawreen, A.; Bogas, J. Creep, shrinkage and mechanical properties of concrete reinforced with different types of carbon nanotubes. *Constr. Build. Mater.* **2019**, *198*, 70–81. [CrossRef]
17. Lu, L.; Ouyang, D.; Xu, W. Mechanical properties and durability of ultra high strength concrete incorporating multi-walled carbon nanotubes. *Materials* **2016**, *9*, 419. [CrossRef]
18. Hu, S.; Xu, Y.; Wang, J.; Zhang, P.; Guo, J. Modification Effects of Carbon Nanotube Dispersion on the Mechanical Properties, Pore Structure, and Microstructure of Cement Mortar. *Materials* **2020**, *13*, 1101. [CrossRef]
19. Isfahani, F.T.; Li, W.; Redaelli, E. Dispersion of multi-walled carbon nanotubes and its effects on the properties of cement composites. *Cem. Concr. Compos.* **2016**, *74*, 154–163. [CrossRef]
20. Elkashef, M.; Abou-Zeid, M.N. Performance of carbon nanotubes in mortar using different surfactants. *Can. J. Civ. Eng.* **2017**, *44*, 619–625. [CrossRef]
21. Alrekabi, S.; Cundy, A.; Lampropoulos, A.; Savina, I. Experimental investigation on the effect of ultrasonication on dispersion and mechanical performance of multi-wall carbon nanotube-cement mortar composites. *Studies* **2016**, *9*, 11.
22. Zou, B.; Chen, S.J.; Korayem, A.H.; Collins, F.; Wang, C.M.; Duan, W.H. Effect of ultrasonication energy on engineering properties of carbon nanotube reinforced cement pastes. *Carbon* **2015**, *85*, 212–220. [CrossRef]

23. Collins, F.; Lambert, J.; Duan, W.H. The influences of admixtures on the dispersion, workability, and strength of carbon nanotube–OPC paste mixtures. *Cem. Concr. Compos.* **2012**, *34*, 201–207. [CrossRef]
24. Wang, B.; Han, Y.; Liu, S. Effect of highly dispersed carbon nanotubes on the flexural toughness of cement-based composites. *Constr. Build. Mater.* **2013**, *46*, 8–12. [CrossRef]
25. Kang, J.; Al-Sabah, S.; Théo, R. Effect of Single-Walled Carbon Nanotubes on Strength Properties of Cement Composites. *Materials* **2020**, *13*, 1305. [CrossRef] [PubMed]
26. Xu, S.; Liu, J.; Li, Q. Mechanical properties and microstructure of multi-walled carbon nanotube-reinforced cement paste. *Constr. Build. Mater.* **2015**, *76*, 16–23. [CrossRef]
27. Douba, A.; Emiroglu, M.; Kandil, U.F.; Taha, M.M.R. Very ductile polymer concrete using carbon nanotubes. *Constr. Build. Mater.* **2019**, *196*, 468–477. [CrossRef]
28. Parveen, S.; Rana, S.; Fangueiro, R.; Paiva, M.C. Microstructure and mechanical properties of carbon nanotube reinforced cementitious composites developed using a novel dispersion technique. *Cem. Concr. Res.* **2015**, *73*, 215–227. [CrossRef]
29. Xu, B.; Qiao, Y.; Zhou, Q.; Chen, X. Effect of electric field on liquid infiltration into hydrophobic nanopores. *Langmuir* **2011**, *27*, 6349–6357. [CrossRef]
30. Yamabe, T.; Fukui, K.; Tanaka, K. *The Science and Technology of Carbon Nanotubes*; Elsevier: Amsterdam, The Nederland, 1999.
31. Kim, G.M.; Yoon, H.N.; Lee, H.K. Autogenous shrinkage and electrical characteristics of cement pastes and mortars with carbon nanotube and carbon fiber. *Constr. Build. Mater.* **2018**, *177*, 428–435. [CrossRef]
32. Kim, G.; Yang, B.; Cho, K.; Kim, E.; Lee, H. Influences of CNT dispersion and pore characteristics on the electrical performance of cementitious composites. *Compos. Struct.* **2017**, *164*, 32 – 42. [CrossRef]
33. Sikora, P.; Abd Elrahman, M.; Chung, S.Y.; Cendrowski, K.; Mijowska, E.; Stephan, D. Mechanical and microstructural properties of cement pastes containing carbon nanotubes and carbon nanotube-silica core-shell structures, exposed to elevated temperature. *Cem. Concr. Compos.* **2019**, *95*, 193–204. [CrossRef]
34. Spanish Association for Standardization. *UNE-EN 1015-3:2000/A2:2007: Methods of Test for Mortar for Masonry-Part 3: Determination of Consistence of Fresh Mortar (by Flow Table)*; Spanish Association for Standardization: Madrid, Spain, 2007.
35. European Committee for Standardization. *EN 12390-3: 2020: Testing Hardened Concrete, Part 3: Compressive Strength of Test Specimens*; European Committee for Standardization: Brussels, Belgium, 2020.
36. European Committee for Standardization. *EN 12390-5: 2020: Testing Hardened Concrete, Part 5: Flexural Strength of Test Specimens*; European Committee for Standardization: Brussels, Belgium, 2020.
37. Gowers, K.; Millard, S. Measurement of concrete resistivity for assessment of corrosion. *ACI Mater. J.* **1999**, *96*, 536–541.
38. Broomfield, J.; Millard, S. Measuring concrete resistivity to assess corrosion rates. *Concrete* **2002**, *36*, 37–39.
39. Liu, Y. *Experiments and Modeling on Resistivity of Multi-layer Concrete with and Without Embedded Rebar*; Florida Atlantic University: Boca Raton, FL, USA, 2008.
40. Liu, Y.; Presuel-Moreno, F.J. Normalization of temperature effect on concrete resistivity by method using Arrhenius law. *ACI Mater. J.* **2014**, *111*, 433–442. [CrossRef]
41. Manzur, T.; Yazdani, N.; Emon, M.; Bashar, A. Potential of carbon nanotube reinforced cement composites as concrete repair material. *J. Nanomater.* **2016**, *2016*, 1421959. [CrossRef]
42. MacLeod, A.J.; Collins, F.G.; Duan, W.; Gates, W.P. Quantitative microstructural characterisation of Portland cement-carbon nanotube composites using electron and x-ray microscopy. *Cem. Concr. Res.* **2019**, *123*, 105767. [CrossRef]
43. Madhavi, T.C.; Pavithra, P.; Singh, S.B.; Vamsi Raj, S.; Paul, S. Effect of multiwalled carbon nanotubes on mechanical properties of concrete. *Int. J. Sci. Res.* **2013**, *2*, 166–168.
44. Tamimi, A.; Hassan, N.M.; Fattah, K.; Talachi, A. Performance of cementitious materials produced by incorporating surface treated multiwall carbon nanotubes and silica fume. *Constr. Build. Mater.* **2016**, *114*, 934–945. [CrossRef]
45. Suchorzewski, J.; Tejchman, J.; Nitka, M. Discrete element method simulations of fracture in concrete under uniaxial compression based on its real internal structure. *Int. J. Damage Mech.* **2018**, *27*, 578–607. [CrossRef]
46. Konsta-Gdoutos, M.S.; Aza, C.A. Self sensing carbon nanotube (CNT) and nanofiber (CNF) cementitious composites for real time damage assessment in smart structures. *Cem. Concr. Compos.* **2014**, *53*, 162–169. [CrossRef]
47. Castro, J.; Spragg, R.; Kompare, P.; Weiss, W.J. *Portland Cement Concrete Pavement Permeability Performance*; Indiana Department of Transportation and Purdue University: West Lafayette, Indiana, 2010.
48. Spragg, R.; Bu, Y.; Snyder, K.; Bentz, D.; Weiss, J. *Electrical Testing of Cement-Based Materials: Role of Testing Techniques, Sample Conditioning, and Accelerated Curing*; Indiana Department of Transportation and Purdue University: West Lafayette, Indiana, 2013.
49. Kane, C.; Mele, E.; Lee, R.; Fischer, J.; Petit, P.; Dai, H.; Thess, A.; Smalley, R.; Verschueren, A.; Tans, S.; et al. Temperature-dependent resistivity of single-wall carbon nanotubes. *EPL (Europhys. Lett.)* **1998**, *41*, 683. [CrossRef]
50. Whittington, H.; McCarter, J.; Forde, M. The conduction of electricity through concrete. *Mag. Concr. Res.* **1981**, *33*, 48–60. [CrossRef]

materials MDPI

Article

Effect of High Calcium Fly Ash, Ladle Furnace Slag, and Limestone Filler on Packing Density, Consistency, and Strength of Cement Pastes

Eleftherios K. Anastasiou

Department of Civil Engineering, Aristotle University of Thessaloniki, 541 24 Thessaloniki, Greece; elan@civil.auth.gr

Abstract: Environmental considerations and technical benefits have directed research towards reducing cement clinker content in concrete, and one of the best ways to do this is to replace cement with supplementary cementitious materials. High calcium fly ash, ladle furnace slag, and limestone filler were investigated as supplementary cementitious materials in cement pastes, and binary mixtures were produced at 10%, 20%, and 30% cement replacement rates for each material. The water requirement for maximum packing and for normal consistency were obtained for each paste, and strength development was determined at 3, 7, 28, and 90 days for the 20% replacement rate. Furthermore, two ternary mixtures at 30% cement replacement were also prepared for maximum packing density and tested for compressive strength development. The results showed that high calcium fly ash decreased cement paste packing and increased water demand but contributed to strength development through reactivity. Ladle furnace slag and limestone filler, on the other hand, were less reactive and seemed to contribute to strength development through the filler effect. The ternary paste with 70% cement, 20% high calcium fly ash, and 10% limestone filler showed equivalent strength development to that of the reference cement paste.

Keywords: supplementary cementitious materials; packing density; filler effect; high calcium fly ash; ladle furnace slag; limestone filler

Citation: Anastasiou, E.K. Effect of High Calcium Fly Ash, Ladle Furnace Slag, and Limestone Filler on Packing Density, Consistency, and Strength of Cement Pastes. *Materials* **2021**, *14*, 301. https://doi.org/10.3390/ma14020301

Received: 4 December 2020
Accepted: 4 January 2021
Published: 8 January 2021

1. Introduction

Supplementary cementitious materials (SCMs) are used increasingly in cement-based products, either for improving their properties or for reducing the carbon footprint of cement. Given that the hydration of ordinary Portland cement (OPC) is not yet understood in full, these materials bring even more complex reactions into the hydration process [1,2]. The benefit from utilization of SCMs lies either in their reactivity [3,4] or in the enhancement of cement hydration, as explained by the filler effect [5,6]. The environmental benefit from cement replacement with SCMs increases with the rate of replacement [7,8], but cement substitution must be limited to the extent that the performance of the final product is not undermined. The effect of SCM use could be beneficial to the performance of the produced concrete, depending on its fineness and on its cement substitution rate [9,10]. Cement substitution by a SCM is expected to affect the rate of strength development and the final strength, but also the water requirement and consistency of the cement paste [11,12].

SCM particles have a different size and specific surface area compared to Portland cement and, therefore, alter the microstructure and packing density of the cement paste. The particle size distribution of the cementitious materials has been found to influence both workability and hydration of cement pastes by improving their packing density [13,14]. Mixture design optimization by considering particle packing has been utilized in the design of ultra-high performance concrete [15,16]. There have been several mathematical packing models proposed to predict the packing density of multi-component mixes [17,18]. In order to predict the effect of cement substitution with SCMs, Yu et al. [19] proposed a linear packing model, considering the surface characteristics (sphericity) of particles. De Larrard [20] proposed the compressive packing model, considering the degree of compaction rather than particle surface characteristics, while Fennis et al. [21] proposed the compaction-interaction packing model, considering particle interaction and aggregate inclusion. According to Zhang et al. [22], the fresh cement paste can be seen as a suspension, with water either filling the voids between particles (filling water) or coating particles and providing fluidity (excess water). SCMs can act as fillers, improving the packing density of the suspension, which means that they can reduce the required amount of filling water. They can also provide space for the hydration of cement, accelerating strength development (filler effect). On the other hand, large SCM particles could block smaller cement particles from hydrating (wall effect) and a large proportion of small SCM particles could increase the distance between hydrated cement particles (loosening effect), affecting the consistency of cement pastes [23]. Mehdipour and Khayat [24] suggested that the presence of more fine particles than the amount required to fill the voids in the cement matrix contributes to the flowability of concrete. SCM particles themselves, on the other hand, may be contributing directly to strength development, if they exhibit hydraulic or pozzolanic properties.

There are several well-known SCMs, such as siliceous fly ash, silica fume, ground granulated blastfurnace slag, and limestone filler (LF). Their availability, however, varies locally, and several other materials are being researched, based on local availability, such as high calcium fly ash (HCFA), metakaolin, ladle furnace slag (LFS), and rice husk ash [25]. The present research investigates the use of HCFA, a by-product of lignite-fired power plants; LFS, a by-product of the steelmaking process; and LF, ground natural limestone, in cement pastes. HCFA is known to exhibit both pozzolanic and self-cementing properties and has been used for the past decades in blended cement manufacturing [26,27]. LFS is a weak pozzolan with some latent hydraulic properties and is mostly considered as filler [28,29]. LF is receiving increasing attention in the literature as a SCM since it seems to promote cement hydration [30–32]. A combination of SCMs in ternary systems is often proposed since there seems to be some synergy between alternative materials of different chemical composition and of different fineness [33,34].

The aim of the present research was to investigate the effect of cement replacement with the above SCMs, considering packing density, in binary and ternary mixtures. Since the fineness and reactivity of the SCMs have an impact on fresh paste consistency and strength development, it is important to understand how increasing cement replacement and altering packing density affects these properties. Furthermore, the ternary binders consisting of OPC, HCFA and LFS or OPC, HCFA, and LF were studied to identify possible synergistic effects. The goal was to identify ways of designing cement pastes with SCMs in the most beneficial way possible, since successful implementation can result in maximizing the positive effects of cement substitution and increasing SCM use.

2. Materials and Methods

CEM type I 42.5 N according to EN 197-1 [35] was used as OPC for all the tests. HCFA was used unprocessed, as received from the power plant. LFS was water-quenched and air-cooled and then sieved through the 100 μm sieve. LF was used as received from the cleaning of the limestone aggregate silos in a ready-mixed concrete plant. Table 1 shows the chemical composition of all the materials used, measured by atomic absorption spectroscopy (AAnalyst 400, Perkin Elmer, Waltham, MA, USA). The loss on ignition (LOI) and the chloride, sodium, and sulfate ion contents were determined by ionic chromatography (Thermo Scientific, Waltham, MA, USA, Dionex ICS-1100) for all the materials used.

Figure 1 shows the particle size distribution of the materials used, and Table 2 shows their median particle size diameter d50, specific surface area, and apparent specific density values. The particle size distribution, d50, and specific surface area of the fine materials were measured using a laser particle size analyzer (Malvern Mastersizer 2000, Worcestershire, United Kingdom). The apparent specific density of the fine materials was determined using a Le Chatelier flask, according to ASTM C188-14 [36].

Table 1. Chemical properties of the fine materials used in the present research (%wt.).

Constituents	OPC	HCFA	LFS	LF
CaO	66.80	47.20	50.70	51.30
SiO_2	19.60	33.10	32.40	3.80
Al_2O_3	3.74	7.29	1.36	1.00
Fe_2O_3	2.40	4.00	2.66	0.40
MgO	3.91	3.20	2.77	1.20
Na_2O	0.57	1.00	0.78	-
K_2O	1.08	0.53	0.06	-
LOI	1.91	3.75	6.72	41.00
Cl^-	0.03	0.04	0.02	-
NO_3^-	0.02	-	-	-
SO_4^{2-}	1.49	4.81	0.43	-

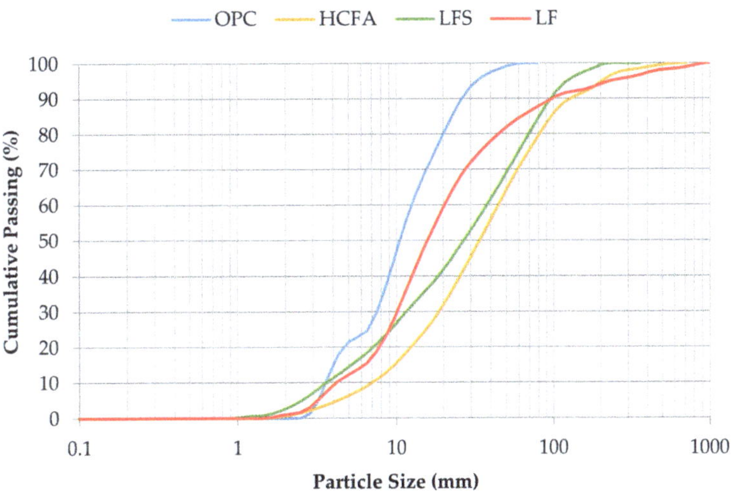

Figure 1. Particle size distribution of the fine materials used.

Table 2. Physical properties of the materials used.

Material	Median Particle Size Diameter d50 (μm)	Specific Surface Area (m^2/kg)	Apparent Specific Density (kg/m^3)
OPC	12.22	642	3140
HCFA	39.92	307	2420
LFS	31.47	496	2590
LF	18.45	480	2620

The first step in the experimental program was to identify water demand when substituting OPC with each of the three alternative binders. The required water to cementitious material (w/cm) ratios were determined both for maximum packing density and for equal consistency. The wet packing density approach, as proposed by Wong and Kwan [36], was followed in order to determine packing density for pastes with 100% OPC and for pastes with 10%, 20%, and 30% cement replacement with HCFA, LFS, and LF. The w/cm for maximum packing was recorded, referred to as optimum water demand. The reduction of the w/cm ratio increases packing density up to the point that the water fills the voids amongst solid particles, but further water reduction decreases packing. Thus, the optimum water demand is determined at the point where packing density is maximized. However, at maximum packing, the workability of the fresh paste is typically very low and serves as a measure of the effect of SCM use on packing and maximum strength development. In order to determine water demand for workable pastes, the w/cm ratio for normal consistency, according to the Vicat method as described in European Standard EN 196-3 [37], was also determined for the same replacement rates.

The cement pastes were prepared in a laboratory mixer by adding water first and then adding the dry-mixed binders and mixing for 120 s. Additional mixing time of 30 s was allowed if required. The fresh pastes were placed in the 40 mm deep truncated conical Vicat mold and compacted on a vibration table for elimination of entrapped air and weighed for the determination of wet packing. Wong and Kwan [38] have identified packing density as the solid concentration φ, which is calculated from Equations (1) and (2) as follows:

$$\varphi = \frac{V_c}{V} \quad (1)$$

where V_c is the solid volume of the cementitious materials and V is the volume of the mold. The solid volume V_c can be calculated from the following formula:

$$V_c = \frac{M}{\rho_w u_w + \rho_\alpha R_\alpha + \rho_\beta R_\beta} \quad (2)$$

where M is the mass of the paste in the mold; ρ_w, ρ_α, ρ_β are the densities of water and cementitious materials α and β, respectively; u_w is the water to cementitious material ratio by volume (w/cm$_V$); and R_α and R_β are the volumetric ratios of cementitious materials α and β, respectively.

After weighing, the truncated conical specimens, still in the mold, were subjected to measurement of Vicat plunger penetration, according to EN 196-3, and the depth of penetration was recorded. The depth of plunger penetration, with a minimum of 0 mm and a maximum of 40 mm, characterizes the consistency of the paste and was used as a measure of workability. Normal consistency is described in the standard as the consistency that allows the plunger to penetrate the specimen 34 mm. Lecomte et al. [39] followed a similar approach to characterize the packing ability of various cement pastes. A series of pastes, 8 to 12 for each paste formulation, was prepared for various w/cm ratios in order to identify the optimum water demand for maximum packing and to determine the w/cm ratio for a paste of normal consistency, which was selected as a suitable level of workability. The above procedure was carried out for 100% OPC as reference and for 10%, 20%, and 30% wt. OPC replacement with HCFA, LFS, and LF, resulting in ten different formulations.

Based on the w/cm ratios for optimum water demand and for normal consistency, binary pastes with 20% OPC replacement with HCFA, LFS, or LF were prepared and tested for compressive strength development at 3, 7, 28, and 90 days. At least six 40 mm cubic specimens were tested at each age and paste, after curing at 20 °C and 95% relative humidity. These were compared to reference 100% OPC cement pastes and were used to assess the contribution of each SCM to strength development, either at maximum packing (optimum water demand), or at equal fresh state workability (normal consistency). Since the tested SCMs had varying effects on workability and strength development, it was decided to test ternary binders to identify possible benefits from the interaction of binders. Based on the strength development test results, two ternary cement pastes, one with 70% OPC, 20% HCFA, and 10% LFS and one with 70% OPC, 20% HCFA, and 10% LF were prepared and tested for strength development at 3, 7, 28, and 90 days at optimum water demand. Scanning Electron Microscopy (SEM-JEOL 840A JSM, Tokyo, Japan) was also used to assess the crystals formation and the microstructure of the reference and ternary cement pastes.

3. Results

3.1. Material Properties

The chemical composition of the materials, as shown in Table 1, shows that all the binders used were rich in calcium. This is expected to influence the kinetics of hydration in a different way compared to traditional silica-rich SCMs (siliceous fly ash, silica fume, ground granulated blastfurnace slag), especially due to the presence of free-CaO in HCFA and LFS; HCFA has both hydraulic and pozzolanic properties, while LFS can be described as a weak pozzolan with latent hydraulic properties and is often regarded as filler [40,41]. HCFA shows a relatively high sulphate ion content, which is not expected to affect cement hydration negatively [41].

Although SCM use for increasing particle density often relies on materials finer than cement, the particle size distribution of the SCMs used shows that all of them were coarser than cement. According to the literature, however, coarser SCMs may still contribute to cement hydration through the filler effect [13,22,42]. HCFA is the coarsest material, considering both median particle size diameter d50 and specific surface area. LFS and LF have similar specific surface areas but different particle size diameters. This occurrence can be explained by their surface characteristics, as LF is ground natural stone, resulting in more spherical shaped particles compared to LFS, which is molten and then water-quenched, resulting in more irregularly shaped particles. According to Sakai et al. [43], spherical-shaped particles are expected to increase packing density and fluidity of cement-bound mixtures.

3.2. Effect on Wet Packing Density

Figures 2–4 show the relationship between packing density and w/cm ratio for various cement replacement rates with SCM. As can be seen in all cases, reducing w/cm ratio increases packing up to the point where the water is not sufficient to fill the voids between the particles. From the curves of the figures, it is easy to estimate the w/cm ratio for the maximum packing, but it is also possible to estimate the effect of each SCM on water demand and its impact on packing density of the cement paste. The use of HCFA, as shown in Figure 2, increases water demand considerably, even at 10% OPC replacement. It also reduces the solid concentration in the paste in all cases.

Cement replacement with LFS, on the other hand, seems to have a different effect. Although water demand was increased (to a lesser extent compared to HCFA), particle packing was equal or even slightly increased compared to that of the reference paste. As with HCFA, the rate of replacement, ranging from 10% to 30%, had little effect. Cement replacement with LF had a similar effect on packing to that of LFS, but more pronounced. At 10% and 20% OPC replacement rates with LF, packing density increased, while it decreased at the 30% replacement rate. Water demand increased, but only slightly.

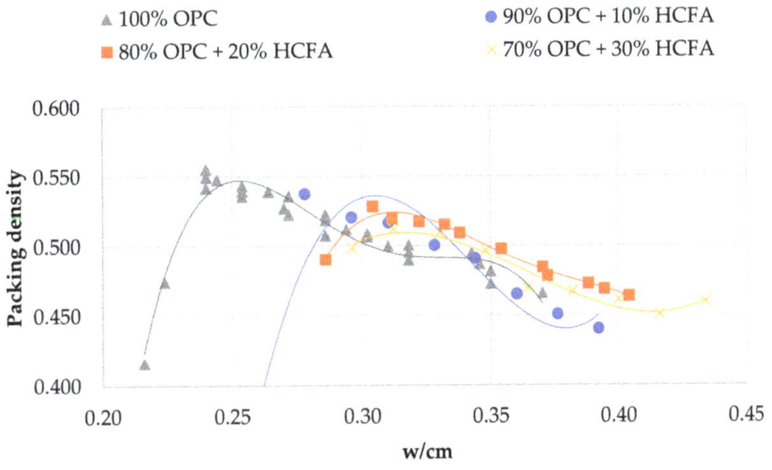

Figure 2. Packing density vs. w/cm in pastes with OPC and HCFA.

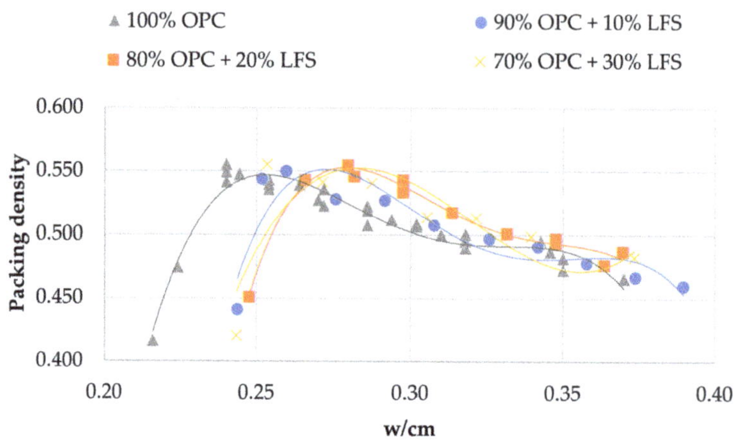

Figure 3. Packing density vs. w/cm in pastes with OPC and LFS.

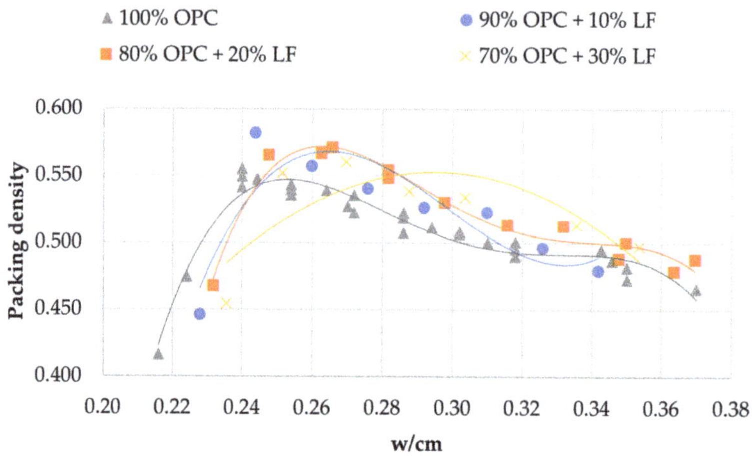

Figure 4. Packing density vs. w/cm in pastes with OPC and LF.

The effect of cement replacement with the various SCMs on packing density seems in all cases to be linked with their fineness. Indeed, an analysis of variance shows that the influence of fineness on packing density is statistically significant ($p < 0.05$) for all the materials. The finer materials (LFS and LF) contribute to the increase of solid concentration, while the coarser material (HCFA) decreases packing. OPC substitution with 10% and 20% LF results in the highest packing densities, while the lowest are observed with 30% HCFA substitution. HCFA, and LFS to a lesser extent, require increased w/cm ratios in order to reach maximum packing. An analysis of variance shows that the influence of w/cm ratio on packing density is statistically significant ($p < 0.05$), which means that increased water demand, mostly for HCFA, but also for LFS, results in lower packing density. This increase in water demand, however, does not seem to be linked with fineness, but can be associated with chemical composition, and more specifically, with free lime content [44]. LF shows a slight increase in water demand, which can be attributed to the water absorption of limestone particles.

3.3. Effect on Consistency

Figures 5–7 show the relationship between w/cm ratio and consistency, as expressed by the Vicat plunger penetration, for each SCM. Figure 5 shows that OPC replacement with HCFA increases the w/cm ratio for normal consistency, and the w/cm ratio needs to be increased further for higher replacement rates. This can be explained due to the higher water demand of HCFA but also due to its negative effect on packing density. OPC substitution with LFS (Figure 6), on the other hand, seems to have a minimal effect on consistency, and the w/cm ratio for normal consistency remains unchanged regardless of the rate of cement replacement. The same seems to be the case for OPC substitution with LF (Figure 7), where there seems to be a small increase in consistency at the 30% replacement rate. Again, the effect of LFS and LF on the consistency of fresh cement pastes can be linked with their effect on packing density, which agrees with the literature [45,46].

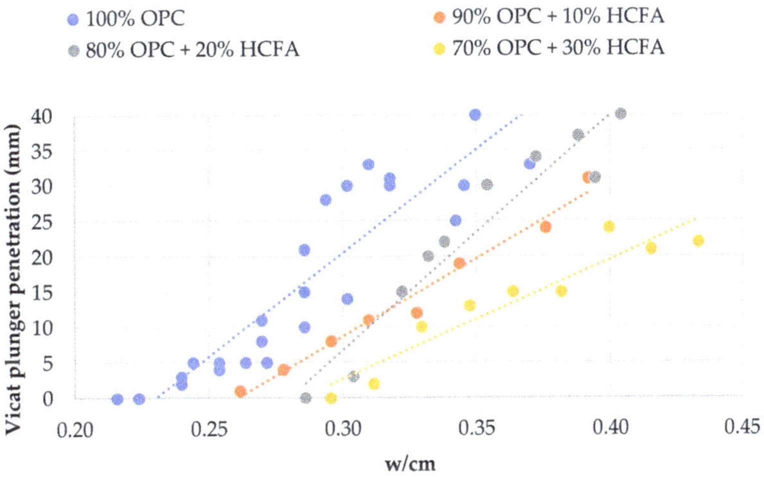

Figure 5. Effect of OPC replacement with HCFA on fresh paste consistency.

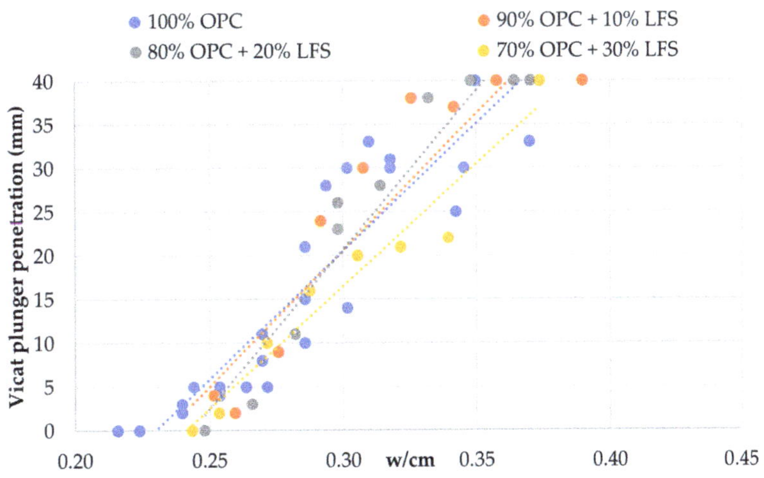

Figure 6. Effect of OPC replacement with LFS on fresh paste consistency.

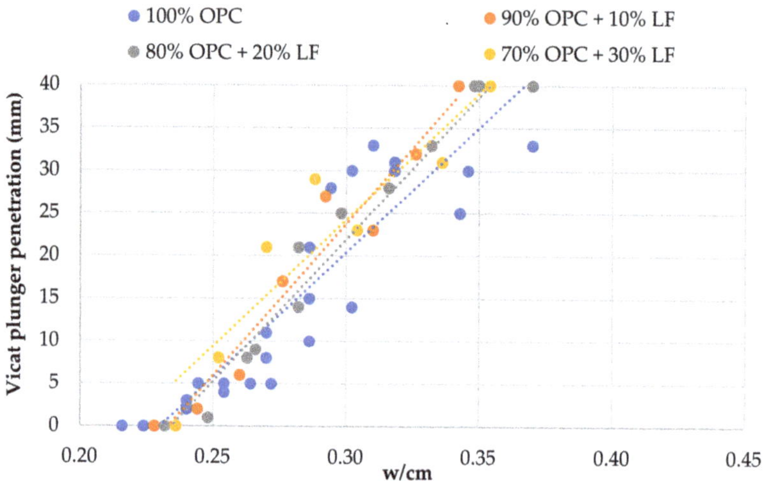

Figure 7. Effect of OPC replacement with LF on fresh paste consistency.

3.4. Effect on Strength Development in Binary and Ternary Pastes

Table 3 shows the cement pastes tested for compressive strength development. The reference and binary mixtures were prepared either with w/cm ratios for maximum packing (optimum water demand) obtained from Figures 2–4 or with w/cm ratios for standard consistence, obtained from Figures 5–7. Two ternary mixtures were prepared with w/cm for maximum packing, determined following the same procedure as for the binary mixtures.

Table 3 shows that 20% replacement with either HCFA, LFS, or LF at optimum water demand gave 92–94% of the 28-day compressive strength of the reference paste, despite lower packing density and increased w/cm. Figure 8 shows that the same pattern continued at 90 days, while HCFA had some accelerating effect at 7 days, which may be attributed to the presence of free lime. Strength development was similar for all of the SCMs in the binary systems, despite the fact that pastes with HCFA had higher w/cm ratios and reduced packing densities. For normal consistence, as shown in Figure 9, 20% OPC replacement with HCFA showed slightly better strength development compared to LFS and LF, again despite higher w/cm ratios and lower packing densities. The results show that the reactivity of HCFA had a greater effect on strength development than fineness, water demand, or packing. LFS and LF, on the other hand, seemed to be contribute to strength development mostly through the filler effect.

Table 3. Fresh and hardened paste properties.

Solid Constituents (% wt.)	w/cm Ratio	Packing Density	Consistency	28-Day Compressive Strength (MPa)
100% OPC	0.24	0.542	-	87.7
100% OPC	0.35	0.487	normal	76.4
80% OPC + 20% HCFA	0.31	0.520	-	82.4
80% OPC + 20% HCFA	0.39	0.469	normal	72.0
80% OPC + 20% LFS	0.28	0.549	-	81.8
80% OPC + 20% LFS	0.35	0.497	normal	68.8
80% OPC + 20% LF	0.26	0.567	-	80.6
80% OPC + 20% LF	0.35	0.496	normal	68.1
70% OPC + 20% HCFA + 10% LFS	0.30	0.539	-	77.0
70% OPC + 20% HCFA + 10% LF	0.29	0.534	-	84.4

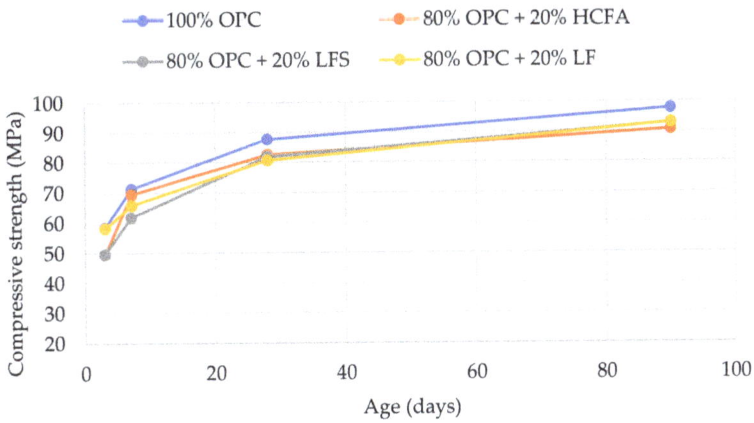

Figure 8. Compressive strength development at optimum water demand for binary pastes.

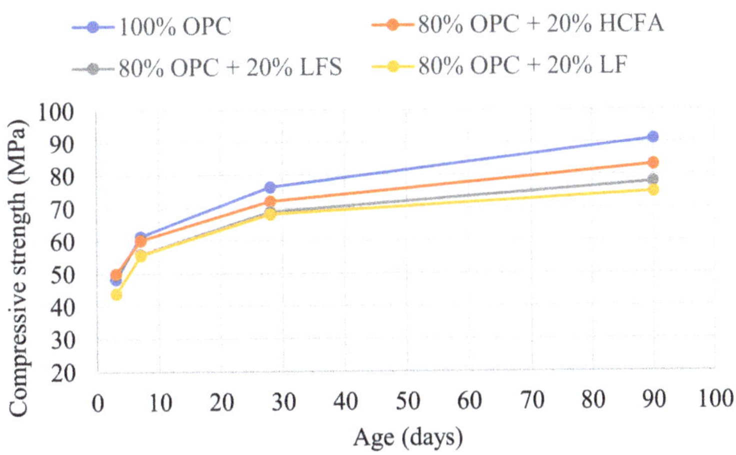

Figure 9. Compressive strength development at normal consistence for binary pastes.

Based on these results, it was decided to test ternary binders with 20% HCFA and 10% LFS or LF cement substitution. The results shown in Table 3 showed that the ternary binders reached a packing density close to that of the reference paste at a higher w/cm ratio; 28-day compressive strength, however, was higher when LF rather than LFS was used as the third constituent. The rate of strength development, as shown in Figure 10, shows that the accelerating effect at 7 days observed with 20% HCFA use was enhanced with an extra 10% LF replacement, while the 90-day compressive strength was equal to that of the reference.

The results show a good synergy between HCFA and LF, also identified by other researchers. De Weerdt et al. [47] suggest that limestone interacts with the hydration products of OPC–fly ash systems and increases compressive strength. Thongsanitgarn et al. [48] have shown that the increase in strength development of OPC–fly ash systems is higher when finer limestone is added. Other researchers point out that different SCMs may co-operate in the cement paste matrix, despite showing different chemical activity and physical characteristics [49]. The same synergy, however, did not take place when HCFA was combined with LFS, resulting in reduced strength development (88% of the reference at 28 and 90 days). A possible explanation for this is that the calcium in LF was in the form of $CaCO_3$, which is known to promote cement hydration [50], while in LFS, it was mostly in the form of CaO or $Ca(OH)_2$, as shown from the values of loss on ignition in Table 1.

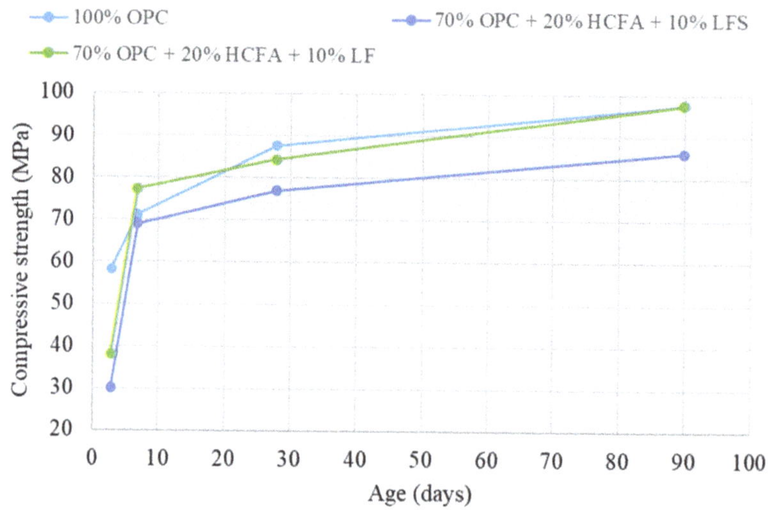

Figure 10. Compressive strength development at optimum water demand for ternary pastes.

SEM photos taken from samples at 28 days were used to explore the microstructure of the reference and ternary pastes (Figures 11–13). The pores observed were of similar size, while micro cracks were visible in all the cement pastes. Portlandite ($Ca(OH)_2$) crystals were identified in the pores of the reference paste, as a result of cement hydration (Figure 11). Ettringite needle-shaped crystals, on the other hand, were visible in the pores of the ternary paste with 70% OPC + 20% HCFA + 10% LF, confirming enhanced reactivity in the pore solution.

Figure 11. Portlandite crystal into a pore of 100% OPC paste at 28 days.

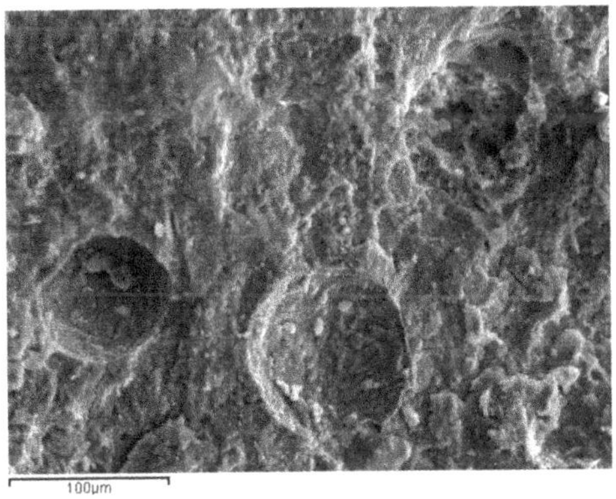

Figure 12. Microstructure of 70% OPC + 20% HCFA + 10% LFS paste at 28 days.

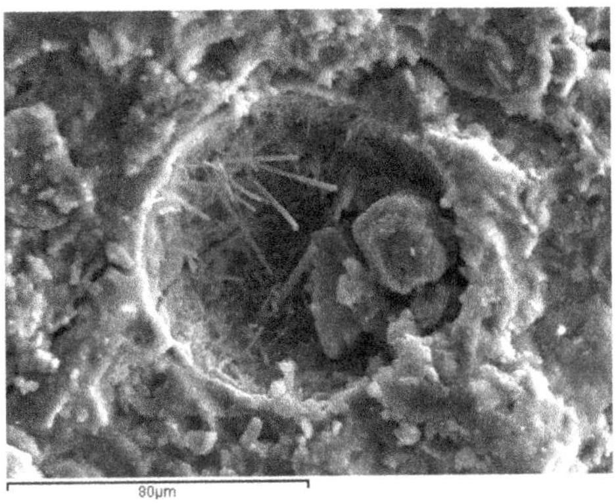

Figure 13. Effect of OPC replacement with LF on fresh paste consistency.

4. Discussion

The results showed that the design of binders with SCMs can be optimized by considering their physical and mechanical characteristics. HCFA had reduced fineness and increased water demand compared to LFS and LF, which rendered pastes with OPC substituted with HCFA less workable. Combining OPC with HCFA and LFS or LF in ternary pastes showed a solution to this problem. Furthermore, considering packing density can aid the design process and help design pastes with either the lowest w/cm ratio at maximum solid concentration for strength optimization or with the w/cm ratio for the required consistency. Fineness has been shown to influence the packing density and consistency of pastes greatly, but reactivity and synergy between binders also seems to be very important for compressive strength development. The reactivity of HCFA seemed to contribute more to strength development compared to the finer LFS and LF, despite the increased water demand. Regarding fineness, the use of SCMs coarser than cement can still lead to equal packing densities but grinding the SCMs to higher fineness could lead to further improvements. The ternary mixtures with OPC, HCFA, and LFS or LF, however, showed packing densities similar to that of 100% OPC paste. HCFA with LF in ternary pastes with OPC showed good synergy in cement-based binders, reaching the strength development of the reference paste, despite having a higher w/cm ratio at the 30% replacement rate, which was not the case when LFS was used as the third constituent. Overall, it seems that designing ternary binders with suitable SCMs by considering particle packing could compensate for strength loss and result in equal performance with reduced cement content.

Funding: The present research was funded by the Research Committee of the Aristotle University of Thessaloniki, Greece.

Institutional Review Board Statement: Not applicable.

Informed Consent Statement: Not applicable.

Data Availability Statement: The data presented in this study are available on request from the corresponding author.

Acknowledgments: The author would like to thank Eirini-Chrysanthi Tsardaka for her help in preparing the pastes and Eleni Pavlidou for her help with SEM.

Conflicts of Interest: The authors declare no conflict of interest.

References

1. Scrivener, K.L.; Juilland, P.; Monteiro, P.J.M. Advances in understanding hydration of Portland cement. *Cem. Concr. Res.* **2015**, *78*, 38–56. [CrossRef]
2. De Weerdt, K.; Haha, M.B.; Le Saout, G.; Kjellsen, K.O.; Justnes, H.; Lothenbach, B. Hydration mechanisms of ternary Portland cements containing limestone powder and fly ash. *Cem. Concr. Res.* **2011**, *41*, 279–291. [CrossRef]
3. Wang, C.; Yang, C.; Liu, F.; Wan, C.; Pu, X. Cement & Concrete Composites Preparation of Ultra-High Performance Concrete with common technology and materials. *Cem. Concr. Compos.* **2012**, *34*, 538–544. [CrossRef]
4. Snellings, R.; Salze, A.; Scrivener, K.L. Use of X-ray diffraction to quantify amorphous supplementary cementitious materials in anhydrous and hydrated blended cements. *Cem. Concr. Res.* **2014**, *64*, 89–98. [CrossRef]
5. Lothenbach, B.; Scrivener, K.; Hooton, R.D. Supplementary cementitious materials. *Cem. Concr. Res.* **2011**, *41*, 1244–1256. [CrossRef]
6. Long, G.; Wang, X.; Xie, Y. Very-high-performance concrete with ultrafine powders. *Cem. Concr. Res.* **2002**, *32*, 601–605. [CrossRef]
7. Proske, T.; Hainer, S.; Rezvani, M.; Graubner, C.A. Eco-friendly concretes with reduced water and cement contents-Mix design principles and laboratory tests. *Cem. Concr. Res.* **2013**, *51*, 38–46. [CrossRef]
8. Anastasiou, E.K.; Liapis, A.; Papayianni, I. Comparative life cycle assessment of concrete road pavements using industrial by-products as alternative materials. *Resour. Conserv. Recycl.* **2015**, *101*. [CrossRef]
9. Mueller, F.V.; Wallevik, O.H.; Khayat, K.H. Linking solid particle packing of Eco-SCC to material performance. *Cem. Concr. Compos.* **2014**, *54*, 117–125. [CrossRef]
10. Bentz, D.P.; Sato, T.; De La Varga, I.; Weiss, W.J. Fine limestone additions to regulate setting in high volume fly ash mixtures. *Cem. Concr. Compos.* **2012**, *34*, 11–17. [CrossRef]
11. Juenger, M.C.G.; Siddique, R. Recent advances in understanding the role of supplementary cementitious materials in concrete. *Cem. Concr. Res.* **2015**, *78*, 71–80. [CrossRef]
12. Mehdipour, I.; Khayat, K.H. Effect of supplementary cementitious material content and binder dispersion on packing density and compressive strength of sustainable cement paste. *ACI Mater. J.* **2016**, *113*, 361–372. [CrossRef]
13. Bentz, D.P.; Hansen, A.S.; Guynn, J.M. Optimization of cement and fly ash particle sizes to produce sustainable concretes. *Cem. Concr. Compos.* **2011**, *33*, 824–831. [CrossRef]
14. Soliman, N.A.; Tagnit-Hamou, A. Using particle packing and statistical approach to optimize eco-efficient ultra-high-performance concrete. *ACI Mater. J.* **2017**, *114*, 847–858. [CrossRef]
15. Wang, X.; Yu, R.; Song, Q.; Shui, Z.; Liu, Z.; Wu, S.; Hou, D. Optimized design of ultra-high performance concrete (UHPC) with a high wet packing density. *Cem. Concr. Res.* **2019**, *126*, 105921. [CrossRef]
16. Arora, A.; Aguayo, M.; Hansen, H.; Castro, C.; Federspiel, E.; Mobasher, B.; Neithalath, N. Microstructural packing- and rheology-based binder selection and characterization for Ultra-high Performance Concrete (UHPC). *Cem. Concr. Res.* **2018**, *103*, 179–190. [CrossRef]
17. Jones, M.R.; Zheng, L.; Newlands, M.D. Comparison of particle packing models for proportioning concrete constituents for minimum voids ratio. *Mater. Struct. Constr.* **2002**, *34*, 301–309. [CrossRef]
18. Fennis, S.A.A.M.; Walraven, J.C. Using particle packing technology for sustainable concrete mixture design. *Heron* **2012**, *57*, 73–101.
19. Yu, R.; Spiesz, P.; Brouwers, H.J.H. Mix design and properties assessment of Ultra-High Performance Fibre Reinforced Concrete (UHPFRC). *Cem. Concr. Res.* **2014**, *56*, 29–39. [CrossRef]
20. De Larrard, F. *Concrete Mixture Proportioning: A Scientific Approach*; E & FN Spon: London, UK, 1999.
21. Fennis, S.A.A.M.; Walraven, J.C.; Den Uijl, J.A. Compaction-interaction packing mode.Regarding the effect of fillers in concrete mixture design. *Mater. Struct. Constr.* **2013**, *46*, 463–478. [CrossRef]
22. Zhang, T.; Yu, Q.; Wei, J.; Zhang, P. Efficient utilization of cementitious materials to produce sustainable blended cement. *Cem. Concr. Compos.* **2012**, *34*, 692–699. [CrossRef]
23. De Larrard, F.; Sedran, T. Mixture-proportioning of high-performance concrete. *Cem. Concr. Res.* **2002**, *32*, 1699–1704. [CrossRef]

24. Mehdipour, I.; Khayat, K.H. Understanding the role of particle packing characteristics in rheo-physical properties of cementitious suspensions: A literature review. *Constr. Build. Mater.* **2018**, *161*, 340–353. [CrossRef]
25. De Belie, N.; Soutsos, M.; Gruyaert, E. *Properties of Fresh and Hardened Concrete Containing Supplementary Cementitious Materials: State-Of-The-Art Report of the RILEM Technical Committee 238-SCM, Working Group 4*; Springer: Ghent, Belgium, 2018; ISBN 978-3-319-70606-1.
26. Felekoğlu, B.; Türkel, S.; Kalyoncu, H. Optimization of fineness to maximize the strength activity of high-calcium ground fly ash-Portland cement composites. *Constr. Build. Mater.* **2009**, *23*, 2053–2061. [CrossRef]
27. Papadakis, V.G. Effect of fly ash on Portland cement systems. Part II. High-calcium fly ash. *Cem. Concr. Res.* **2000**, *30*, 1647–1654. [CrossRef]
28. Sideris, K.K.; Tassos, C.; Chatzopoulos, A. Production of durable self-compacting concrete using ladle furnace slag (LFS) as filler material. *Procedia Eng.* **2015**, *108*, 592–597. [CrossRef]
29. Setién, J.; Hernández, D.; González, J.J. Characterization of ladle furnace basic slag for use as a construction material. *Constr. Build. Mater.* **2009**, *23*, 1788–1794. [CrossRef]
30. Huang, W.; Kazemi-Kamyab, H.; Sun, W.; Scrivener, K. Effect of cement substitution by limestone on the hydration and microstructural development of ultra-high performance concrete (UHPC). *Cem. Concr. Compos.* **2017**, *77*, 86–101. [CrossRef]
31. Lothenbach, B.; Le Saout, G.; Gallucci, E.; Scrivener, K. Influence of limestone on the hydration of Portland cements. *Cem. Concr. Res.* **2008**, *38*, 848–860. [CrossRef]
32. Scrivener, K.L.; Nonat, A. Hydration of cementitious materials, present and future. *Cem. Concr. Res.* **2011**, *41*, 651–665. [CrossRef]
33. Kwan, A.K.H.; Wong, H.H.C. Packing density of cementitious materials: Part 2-packing and flow of OPC + PFA + CSF. *Mater. Struct. Constr.* **2008**, *41*, 773–784. [CrossRef]
34. Marchetti, G.; Rahhal, V.F.; Irassar, E.F. Influence of packing density and water film thickness on early-age properties of cement pasteswith limestone filler and metakaolin. *Mater. Struct. Constr.* **2017**, *50*, 1–11. [CrossRef]
35. Wong, H.H.C.; Kwan, A.K.H. Packing density of cementitious materials: Part 1—Measurement using a wet packing method. *Mater. Struct.* **2008**, *41*, 689–701. [CrossRef]
36. *EN 197-1:2011: Cement. Composition, Specifications and Conformity Criteria for Common Cements*; CEN: Brussels, Belgium, 2011.
37. *ASTM C188-14: Standard Test Method for Density of Hydraulic Cement*; ASTM International: West Conshohocken, PA, USA, 2014.
38. *EN 196-3:2016; Methods of Testing Cement. Determination of Setting Times and Soundness*; CEN: Brussels, Belgium, 2016.
39. Lecomte, A.; Mechling, J.M.; Diliberto, C. Compaction index of cement paste of normal consistency. *Constr. Build. Mater.* **2009**, *23*, 3279–3286. [CrossRef]
40. Anastasiou, E.K.; Papayianni, I.; Papachristoforou, M. Behavior of self compacting concrete containing ladle furnace slag and steel fiber reinforcement. *Mater. Des.* **2014**, *59*. [CrossRef]
41. Papayianni, J. Use of a high-calcium fly ash in blended type cement production. *Cem. Concr. Compos.* **1993**, *15*, 231–235. [CrossRef]
42. Knop, Y.; Peled, A. Sustainable blended cements-Influences of packing density on cement paste chemical efficiency. *Materials* **2018**, *11*, 625. [CrossRef]
43. Sakai, E.; Masuda, K.; Kakinuma, Y.; Aikawa, Y. Effects of shape and packing density of powder particles on the fluidity of cement pastes with limestone powder. *J. Adv. Concr. Technol.* **2009**, *7*, 347–354. [CrossRef]
44. Nawaz, A.; Julnipitawong, P.; Krammart, P.; Tangtermsirikul, S. Effect and limitation of free lime content in cement-fly ash mixtures. *Constr. Build. Mater.* **2016**, *102*, 515–530. [CrossRef]
45. Kwan, A.K.H.; Chen, J.J. Roles of Packing Density and Water Film Thickness in Rheology and Strength of Cement Paste. *J. Adv. Concr. Technol.* **2012**, *10*, 332–344. [CrossRef]
46. Fennis, S.A.A.M.; Walraven, J.C.; Nijland, T.G. Measuring the packing density to lower the cement content in concrete. In *Tailor Made Concrete Structures*; Taylor & Francis Group: London, UK, 2008; Volume 108. [CrossRef]
47. De Weerdt, K.; Kjellsen, K.O.; Sellevold, E.; Justnes, H. Synergy between fly ash and limestone powder in ternary cements. *Cem. Concr. Compos.* **2011**, *33*, 30–38. [CrossRef]
48. Thongsanitgarn, P.; Wongkeo, W.; Chaipanich, A.; Poon, C.S. Heat of hydration of Portland high-calcium fly ash cement incorporating limestone powder: Effect of limestone particle size. *Constr. Build. Mater.* **2014**, *66*, 410–417. [CrossRef]
49. Isaia, G.C.; Gastaldini, A.L.G.; Moraes, R. Physical and pozzolanic action of mineral additions on the mechanical strength of high-performance concrete. *Cem. Concr. Compos.* **2003**, *25*, 69–76. [CrossRef]
50. Kakali, G.; Tsivilis, S.; Aggeli, E.; Bati, M. Hydration products of C3A, C3S and Portland cement in the presence of CaCO3. *Cem. Concr. Res.* **2000**, *30*, 1073–1077. [CrossRef]

materials MDPI

Article

Strength Parameters of Foamed Geopolymer Reinforced with GFRP Mesh

Rafał Krzywoń * and Szymon Dawczyński

Faculty of Civil Engineering, Silesian University of Technology, 44-100 Gliwice, Poland; szymon.dawczynski@polsl.pl
* Correspondence: rafal.krzywon@polsl.pl; Tel.: +48-32-237-22-62

Abstract: The foaming of geopolymers lowers their density, thus opening up new environmental benefits, including acoustic and thermal insulation. At the same time, foaming disturbs the homogeneity of the material, which worsens the strength parameters, and particularly those related to tension, which can be improved by introducing reinforcement. This paper presents the results of research on foamed geopolymers reinforced with glass fiber meshes, a type of reinforcement that provides an adequate bond. The samples tested here were based on three types of coal fly ash, and were foamed with varying doses of hydrogen peroxide. Samples were cured at 40 °C and were tested after 28 days of maturing at ambient temperature. The strength parameters of the synthesized geopolymers were determined via laboratory testing, and were used to evaluate load-bearing capacity models of the tested samples reinforced with glass fiber mesh. The results showed the importance of the type of ash on the strength properties and efficiency of reinforcement. At the same time, a slight deterioration in the glass fibers was noticed; this was caused by the presence of sodium hydroxide solution, which was used as an activator.

Keywords: foamed geopolymer; reinforcing lightweight concrete; glass fiber composites

Citation: Krzywoń, R.; Dawczyński, S. Strength Parameters of Foamed Geopolymer Reinforced with GFRP Mesh. *Materials* **2021**, *14*, 689. https://doi.org/10.3390/ma14030689

Academic Editor: Hubert Rahier
Received: 29 December 2020
Accepted: 29 January 2021
Published: 2 February 2021

1. Introduction

Coal fly ashes are valuable raw materials for the building materials industry, and especially for the manufacturing of cement and concrete. As a result of the transformation of the energy industry, new types of ashes are now being produced; these are mixtures of products from simultaneous coal combustion and gas desulphurization processes (ashes from fluidized bed boilers). They often contain high amounts of SO_3 and CaO from unburned coal. For these reasons, they are sometimes treated as waste that is unsuitable for use in traditional cement production technologies. It is therefore necessary to look for other applications for these materials, one of which may be the synthesis of geopolymers.

Due to its significant content of silicon and aluminum, fly ash is an attractive material for use in the synthesis of geopolymers [1,2]. The traditional precursor, based on metakaolin, can be effectively replaced by fly ash of type F. The usefulness of fly ash in the synthesis of geopolymers is determined by its content of silicon and aluminum, since the Si/Al ratio determines the formation of a desirable type of zeolite. According to Tanaka et al. [3], a ratio of 0.9 gives a material that can be identified as a single-phase Na–A zeolite. This material is also produced at a lower rate at a ratio of around 1.7, and its crystallinity increases sharply at a ratio of 4.3.

Many studies have proved the superb strength properties of fly ash-based GPC (geopolymer cement) [4–6], which are comparable to those of OPC (ordinary Portland cement). They also provide good durability, and typically have better sulfate [7] and acid resistance [4,8], and excellent fire resistance [9]. Since it is cured at high temperatures, GPC has acceptably low shrinkage [10].

Another factor that may affect the properties of this material is the presence of calcium. In early studies, a higher amount of calcium (above 20%) was recognized as a contaminant, which caused a decrease in strength [11] and reduced the rate of geopolymer synthesis [12], and which produced various hydrates [13,14]. Recent studies, however, have highlighted other features with a beneficial effect on the properties of fresh mixture and hardened geopolymer [15]. The simultaneous formation of calcium silicate hydrate compounds in addition to aluminosilicate products under certain conditions improves the strength properties [16]; one example would be synthesis at ambient temperature [17,18]. Although thermal curing can still improve the properties of lignite coal fly ash [19,20], its properties are still worse than those of low-calcium Class F geopolymer [18].

The entrapment of air in the structure of a foamed geopolymer improves its thermal and sound insulating properties [21–23]. According to Cui et al. [24], foamed geopolymers may have better thermal and mechanical properties than foamed OPC. The lower density means that the weight of the structure is reduced, which allows for an additional reserve of load capacity. Another important benefit is the increase in compressive strength after exposure to elevated temperatures, a phenomenon that can be explained as an effect of geopolymer polycondensation and sintering at high temperatures [25]. Furthermore, the pores provide space to counteract the damage by heat [26]. Geopolymer foams can resist temperatures of up to 1000 °C without decomposition, and their shapes do not suffer dimensional changes within a temperature range of 400–800 °C [27]. These features mean that foamed geopolymers are very suitable for industrial applications as refractories, including furnace insulation and chimney cladding [28].

The main negative effect of foaming geopolymers is the reduction in the strength parameters. This problem affects most lightweight materials, and results from the disturbance in the homogeneous structure due to the presence of pores [29,30]. The relationship between the abovementioned advantages of foaming and strength reduction always involves a compromise, and the optimum parameters can be found by the proper addition of a foaming agent. Among the most common foaming techniques are the entrapment of air during mixing with surfactants (organic and inorganic technical foams) [31], and chemical foaming by the addition of peroxides such as hydrogen peroxide [32], sodium perborate [23], sodium hypochlorite [33] and the alkaline oxidation of metals, such as zinc, metallic silicon or aluminum [22,34,35].

Foamed geopolymers are brittle, and the ratio of the tensile to compressive strength, although better than for OPC concretes, is still very unfavorable. One remedy for this disadvantage may be the use of reinforcement. The most frequently studied method of reinforcing a geopolymer is the introduction of dispersed fibers to its structure, and various types of fibers have been investigated, including natural fibers (e.g., hemp [36], abaca [37]), and both organic (e.g., PVA [38], PP [39]) and inorganic high-strength fibers (e.g., glass [40], basalt [35,41] and carbon [42]). The most important limitation on the efficiency of reinforcement of foamed materials is the bond. In solid concrete, a suitable bond is ensured by chemical adhesion between the cement paste and steel, as well as the mechanical interlocking between the ribbed surface of the rebar and aggregate particles. The contact surface in porous materials is limited by the presence of voids, and flat fiber reinforcement does not usually provide interlocking conditions. The low thickness of the walls separating the pores and the brittleness of cellular concrete causes crushing of the local contact with the rebar rib. Although there has been no research on the reinforcement of foamed geopolymers with steel rebars, a study of this type of reinforcement in foamed OPC concrete shows that at a density of 1200 kg/m^3, the strength of the bond is reduced by a factor of eight [43]. According to the authors of this study, the effectiveness of the bond in foamed concrete can be successfully improved by expanding the contact zone. In practice, this could be realized through the application of textile [44] or mesh reinforcements [45,46], in which a perpendicular thread provides anchorage for the fibers in the direction of the internal forces. This concept has been applied for many years to thin-walled concrete structures; it is called ferrocement, and was developed five years before the reinforced concrete.

In this paper, we develop the concept of reinforcing discussed above. Tests were carried out on 18 sets of prisms made of foamed fly-ash-based-geopolymer, reinforced with glass fiber mesh. The aim was to assess the impact of the bulk density on the mechanical properties of the foamed geopolymer and the flexural behavior of beams strengthened with GFRP (glass-fiber-reinforced polymer). The parameters investigated here included changes in the content of foaming agent (1%, 2%, and 3%), the origin of the fly ash, and localization of the reinforcement (none, external, and internal).

2. Experimental Program

Our research program was divided into two parts. The first involved material research, and the goal was to determine the mechanical properties of the geopolymers by carrying out testing on cylinders and beams. To assess the compressive and flexural strength, prismatic beams (40 × 40 × 160 mm^3) were cast according to EN 196-1:2016 [47], and for each mixture, three cylinders (60 × 120 mm^2) were simultaneously made to test the modulus of elasticity and cylindrical compressive strength (Figure 1a,b).

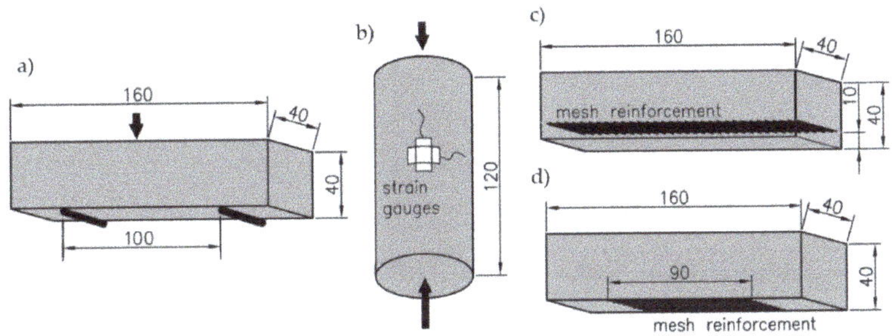

Figure 1. Dimensions of the sample and location of reinforcement: (**a**) beam with internal reinforcement; (**b**) beam with bottom reinforcement; (**c**) cylinder with strain sensors; (**d**) beam with bottom reinforcement. Unit: mm.

Using the same mixture, glass-mesh-reinforced beams were made for bending resistance tests. Figure 1c,d show the dimensions of the beams and the location of the reinforcing mesh. The meshes were arranged so that each beam contained eight bundles of fibers.

The applications of glass fiber meshes in geopolymers have not yet been studied. Based on an analogy to OPC, there may be a danger that the penetration of alkaline solutions can severely damage glass fiber in terms of a loss of toughness and strength, and embrittlement [48] (p. 160). On the other hand, fiberglass has a beneficial chemical composition [49], and most studies show that dispersed glass fibers are effective in increasing the strength of a geopolymer without giving rise to embrittlement problems [50,51]. To ensure resistance to alkalis, a fiberglass lathing mesh was used. This was originally designed for cement plaster reinforcement, and is based on C-glass fiber impregnated with an alkali-resistant dispersion. The mesh size was 4.5 × 5 mm^2, and the weight was 160 g/m^2. The breaking strength was 25 kN/m.

2.1. Materials and Methods

The foamed geopolymers tested here were synthesized from fly ash, drawn from the three largest coal-fired power plants in Poland: the Jaworzno power plant, which is fired with anthracite coal, and the Belchatow and Turow power plants, which are fired with lignite coal. The chemical compositions of the fly ash materials were identified by X-ray photoelectron spectroscopy (Thermo Scientific™, Waltham, MA, USA) (using the K-Alpha™ X-ray Photoelectron Spectrometer XPS System), and are shown in Table 1.

Table 1. Chemical composition of fly ashes (mass %).

Power Plant	C	SiO_2	Al_2O_3	CaO	SO_3	MgO	Fe_3O_3	K_2O
Jaworzno	6.87	45.04	30.88	8.13	0	1.86	4.29	2.1
Belchatow	9.18	16.65	11.15	21.40	35.58	1.87	4.16	0
Turow	8.73	18.15	13.30	22.27	29.25	2.79	4.33	0

The densities of the coal fly ashes were determined based on the EN 1097-6 [52] standard, and are presented in Table 2. The results indicate that fly ashes from the incineration of lignite coal have a higher density than those from anthracite coal.

Table 2. Densities of the fly ashes (kg/m^3).

Fly Ash Origin	Jaworzno	Belchatow	Turow
Density kg/m^3	2200	2600	2740

An activator based on a sodium silicate solution (Na_2O 8.6%, SiO_2 27.8%, water 63.2%) and 10 M sodium hydroxide was used. The components of the activator were mixed before use. The mass ratio of fly ash/sodium silicate/sodium hydroxide was constant for all mixtures, and was equal to 3/1.5/1 (Table 3). The mixture proportions were optimized in the strength tests of samples based on a not-foamed geopolymers.

Table 3. Composition of foamed geopolymer mixtures.

Mixture	Fly Ash Origin	Fly Ash kg/m^3	Sodium Silicate kg/m^3	Sodium Hydroxide kg/m^3	Hydrogen Peroxide wt %
J_1%	Jaworzno	540.0	270	180.0	1%
J_2%		403.6	201.8	134.6	2%
J_3%		327.8	163.9	109.3	3%
B_1%	Bełchatow	601.6	300.8	200.1	1%
B_2%		434.2	217.1	144.7	2%
B_3%		351.3	175.6	117.1	3%
T_1%	Turow	748.4	374.2	249.5	1%
T_2%		458.7	229.4	152.9	2%
T_3%		372.5	186.3	124.2	3%

The geopolymers were foamed by the addition of a 30% solution of hydrogen peroxide. To vary the densities of the geopolymers, three different concentrations of hydrogen peroxide with respect to the precursor were used: 1%, 2%, and 3% of the total weight ratio. In an alkaline environment, hydrogen peroxide decomposes into water and oxygen [53]; the introduction of oxygen into the mixture causes it to grow in volume. The advantage of using hydrogen peroxide for foaming is the generally homogeneous distribution of the macro-pores [29] and the increase in viscosity of the geopolymer paste [54]. However, there are also disadvantages, such as a reduction in the rate of the geopolymerization process [29]. The growing process is restricted by the hardening of the sample, which makes it relatively difficult to predict the final volume.

The same production conditions were used for all samples. The mixing procedure was as follows: the alkaline activator components were mixed for two minutes, while the precursor was simultaneously poured into the mixing bowl. After adding the activator, the components were mixed for three minutes at a constant stirrer speed of 100 RPM. Finally, hydrogen peroxide was added, and the materials were mixed for a further minute. Immediately after mixing, the samples were poured into molds containing fixed reinforcing meshes. After 10 min of growing, the excess geopolymer paste was removed, and the molds were sealed and cured in a heat chamber at 40 °C for 24 h. All the specimens were then de-molded and kept at room conditions until the test.

A total of 27 types of beams were cast, each set consisting of six identical beams made from the same mix. In addition, three cylinders were made for each mix, in order to determine the strength properties.

The following system is used here to describe the types of samples. The first letter represents the origin of the ash (J-Jaworzno, T-Turow and B-Belchatow), and the content of the foaming agent is then specified (1%, 2%, and 3%). In the case of the prismatic samples, the last two letters indicate the type of reinforcement (no-pure geopolymer (Figure 1a), in–internal mesh (Figure 1c), and bt-mesh at the bottom (Figure 1d)). For example, J_2%_in represents a beam sample made of fly ash obtained from the Jaworzno Power Plant and reinforced with an internal mesh.

2.2. Testing

All the specimens were tested after 28 days of curing. Before testing was carried out, the bulk density of each sample was measured.

The compressive tests of the cylinders were performed using a FORM + TEST Prüff-systeme MEGA3 Testing Machine for compression testing at a loading speed of 0.1 kN/s. Strains were measured with two groups of strain gauges of length 10 mm, orthogonally adhered to the side surface of the cylinder (Figure 2a). Measurements were recorded using a Z-TECH 64-channel Wheatstone bridge.

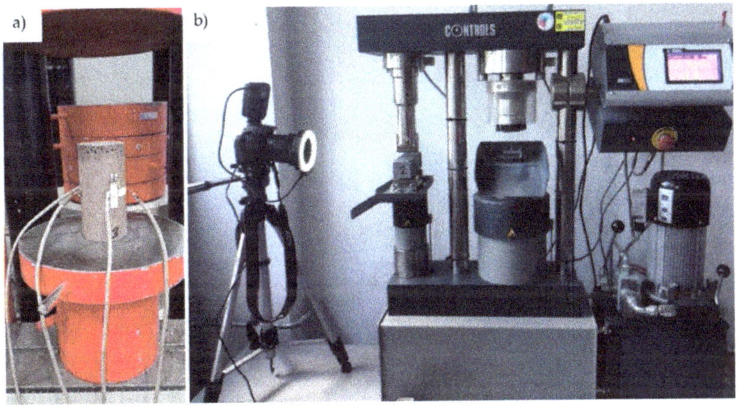

Figure 2. Samples during the test: (a) flexural test of the beam; (b) cylinder compressive test.

The tests of the beams were carried out according to EN 196-1:2016, and the flexural strength was determined for each type of beam. In addition, for the pure geopolymer, compressive strength tests were done using the cube samples that remained after the bending test. The tests were performed with a Controls 65-L27C12 Universal Testing Machine at a loading speed of 0.05 kN/s. During each test, the deformation of the reinforced samples was measured using the DIC (digital image correlation) method. Pictures were taken at one-second intervals using a Canon EOS100D camera (Figure 2b). Strains and displacements were computed with GOM Correlate software.

3. Experimental Results and Analysis

3.1. Mechanical Properties of Foamed Geopolymers

The results of the cylinder tests are shown in Table 4, which gives the mean values for the three cylinders. The modulus of elasticity and Poisson's ratio are defined as secant values within 40% of the strength.

Table 4. Average results of cylinder tests.

	Density kN/m^3	Strength MPa	SD	Young Modulus GPa	SD	Poisson's Ratio
J_1%	10.03	9.14	±0.72	2.75	±0.14	0.18
J_2%	7.78	6.36	±0.12	1.49	±0.12	0.19
J_3%	6.22	2.42	±0.28	1.48	±0.14	0.18
T_1%	13.44	4.27	±1.04	0.98	±0.036	0.18
T_2%	9.17	2.44	±0.46	0.63	±0.037	0.19
T_3%	7.17	1.55	±0.32	0.32	±0.052	0.19
B_1%	11.01	4.33	±0.43	1.26	±0.16	0.18
B_2%	8.78	2.37	±0.35	0.68	±0.053	0.18
B_3%	6.52	0.68	±0.05	0.31	±0.024	0.17

Geopolymers synthesized from fly ash obtained from anthracite coal combustion (the Jaworzno power plant) had the best strength properties. The cylinder (uniaxial) compressive strength of geopolymer made of fly ash from Jaworzno was on average 114% higher than the values obtained for geopolymers synthesized using suspensions from lignite coal-fired power plants (Turow and Belchatow). An even greater difference was seen in the modulus of elasticity, which was on average 213% higher, and in the case of geopolymer foamed with 3% H_2O_2 content, up to 370% higher. These results were gained for geopolymer made from Jaworzno fly ash, despite the average lowest density in the comparison group for the same quantity of foaming agent.

The modulus of elasticity for the foamed geopolymers based on fly ash from the Jaworzno power plant was comparable to those of foamed OPC concretes. Figure 3 shows a comparison with test results for foamed concrete of different densities published by Kozłowski [55] and Drusa [56].

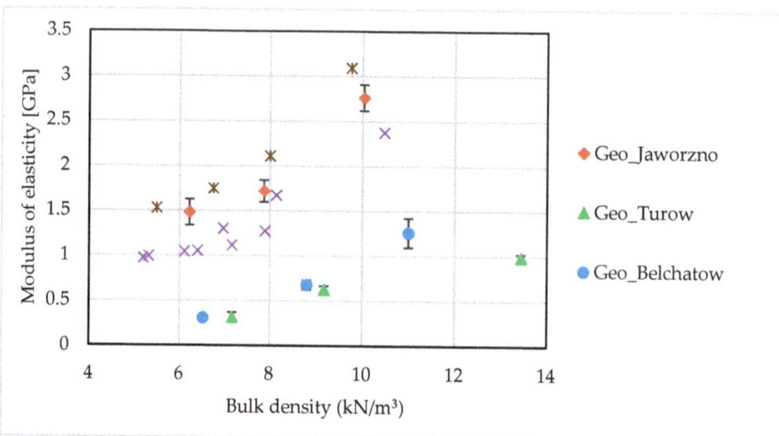

Figure 3. Secant modulus of elasticity vs. bulk density of tested samples: comparison with the results of foamed ordinary Portland cement (OPC) concrete tests performed by Kozlowski and Kadela [55] and Drusa et al. [56].

Figures 4–6 show the stress-strain relationships for geopolymers synthesized from the three different types of fly ash.

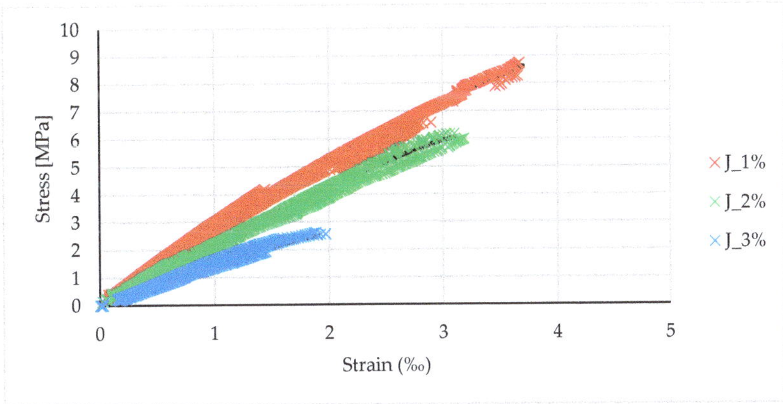

Figure 4. Stress-strain relationship for samples based on Jaworzno fly ash.

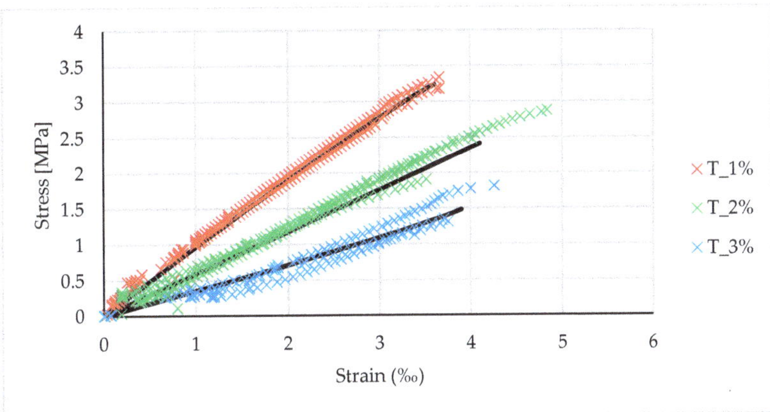

Figure 5. Stress-strain relationship for samples based on Turow fly ash.

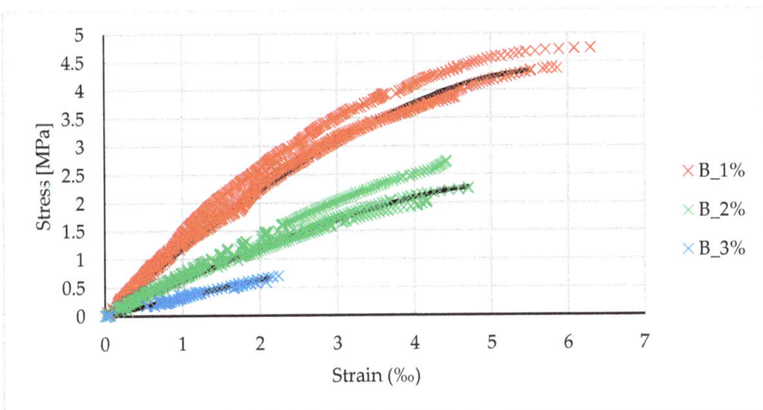

Figure 6. The stress-strain relationship for samples based on Belchatow fly ash.

By analogy with OPC concrete, the stress-strain relationship can be described by the following function:

$$\sigma = \sigma_u \left[1 - \left(1 - \frac{\varepsilon}{\varepsilon_u} \right)^n \right], \tag{1}$$

where σ_u and ε_u are the average strength and the corresponding ultimate strain obtained during the tests. The function in (1) was adjusted to the test results by determining the value of the power exponent n. The results of these analyses are summarized in Table 5. A value of $n = 2$ gives the parabolic relationship recommended for concrete, while $n = 1$ gives a linear relationship. Figures 4–6 show the curves corresponding to the functions described by the parameters listed in Table 5. The fit of the functions was verified by maximizing the determination coefficient. The results obtained for R-squared are presented in the last column of Table 5.

Table 5. Parameters for the function in (1) describing the stress-strain relationship for the foamed geopolymer.

	Strength MPa	Ultimate Strain ‰	Exponent n	Correlation with Test Results (R^2)
J_1%	9.14	3.93	1.2	0.993
J_2%	6.36	3.14	1.2	0.994
J_3%	2.42	1.98	1.24	0.986
T_1%	4.27	3.64	1.09	0.996
T_2%	2.44	4.12	1	0.990
T_3%	1.55	3.96	0.9	0.964
B_1%	4.33	5.56	1.6	0.985
B_2%	2.37	4.71	1.35	0.973
B_3%	0.68	2.15	1.08	0.993

The strengths obtained in the cylinder tests were confirmed by the beam tests, and these are summarized in Table 6, which shows the density, cube compressive strength, and tensile flexural strength. In additional, to enable further comparisons with the reinforced beams, it shows the failure force for which the flexural strength was calculated. The results were similar to those of the cylinder tests, as the average cube compressive strength was 73% higher and flexural strength 29% higher for the fly ash from Jaworzno than for those from the other two power plants.

Table 6. Average results of beam tests.

	Density kN/m^3	Compressive Strength GPa	Flexural Strength MPa	Ratio Flexural to Compressive	Failure Force (Bending) kN
J_1%_no	9.90	8.13	1.27	0.23	0.54
J_2%_no	7.40	3.97	0.83	0.21	0.35
J_3%_no	6.01	1.66	0.61	0.36	0.26
T_1%_no	13.72	5.26	1.17	0.22	0.50
T_2%_no	8.41	2.20	0.94	0.42	0.41
T_3%_no	6.84	1.30	0.40	0.31	0.18
B_1%_no	11.02	4.32	0.71	0.16	0.30
B_2%_no	7.96	2.25	0.85	0.37	0.37
B_3%_no	6.44	0.61	0.36	0.59	0.15

The foaming caused a severe drop in strength. For example, the compressive strength of a plain (not foamed) geopolymer, based on fly ashes from Jaworzno, is on average 40 MPa. Strength decreases with decrease of geopolymer density. Similar relationships of strength and density were obtained in studies published by other authors [23,27]. Moreover, to a similar extent, foaming affects the strength of concretes based on Portland cement [46,55,56].

3.2. Efficiency of Mesh Reinforcement

Undoubtedly, the structural suitability of materials with compressive strength below 8MPa is limited to structures of secondary importance such as partition walls, leveling layers, etc. The authors' experience shows that foamed concretes with such densities are increasingly used in load-bearing elements such as foundations, structural walls or ceilings. Usually these are composite structures with a thin concrete slab or reinforced with cores. The introduction of reinforcement should simplify such structures. The second effect of the reinforcement may be the reduction of susceptibility of usually fragile foamed geopolymer to accidental cracking.

The main objective of this part of research was to evaluate the effectiveness of reinforcement with glass fiber mesh. The results of these tests are presented in Tables 7 and 8, which show the magnitude of the failure force in the bending test and the corresponding deflection, with a description of the failure mode. Three typical modes of failure were observed: rupturing of mesh fibers, delamination of the mesh, and crushing of concrete in the compressed zone, which was usually preceded by the development of flexural cracks. Examples of these are shown in Figure 7.

Table 7. Test results for beams reinforced at the bottom.

	Density kN/m^3	Failure Force kN	Deflection at Failure mm	Failure Mode
J_1%_bt	9.93	0.74	0.41	Rupture
J_2%_bt	7.49	0.46	0.48	Rupture/delamination
J_3%_bt	6.64	0.47	0.51	Delamination
T_1%_bt	13.38	0.71	0.14	Rupture
T_2%_bt	8.42	0.50	0.36	Delamination
T_3%_bt	6.26	0.28	0.27	Delamination
B_1%_bt	11.02	0.35	0.23	Rupture/delamination
B_2%_bt	7.78	0.45	0.68	Delamination
B_3%_bt	6.08	0.30	0.48	Delamination

Table 8. Test results for internally reinforced beams.

	Density kN/m^3	Failure Force kN	Deflection at Failure mm	Failure Mode
J_1%_in	10.01	0.68	0.58	Rupture
J_2%_in	7.76	0.56	1.06	Rupture
J_3%_in	6.45	0.66	1.18	Rupture/crushing
T_1%_in	13.64	0.73	0.52	Rupture
T_2%_in	9.05	0.35	0.15	Rupture
T_3%_in	6.16	0.29	0.56	Crushing/rupture
B_1%_in	11.31	0.46	0.57	Rupture
B_2%_in	8.15	0.28	0.60	Rupture
B_3%_in	6.13	0.29	0.68	Crushing/rupture

The use of reinforcement should increase the load-bearing capacity. It is clear that the reinforcement fulfilled this requirement and that the carrying capacity of the reinforced samples was greater than for models without reinforcement. The best results were obtained for the samples with the highest density, which is in line with expectations. For example, a 43% increase in flexural capacity was obtained for the T_1%_in model, and a 90% increase for the B_1%_bt model. It should be noted that all of the samples were reinforced with the same fiberglass mesh, so the greatest strengthening effects could therefore be expected for models made of the weakest material, i.e., those with the lowest density. A graphical comparison of the failure forces in the bending test is shown in the diagram in Figure 8.

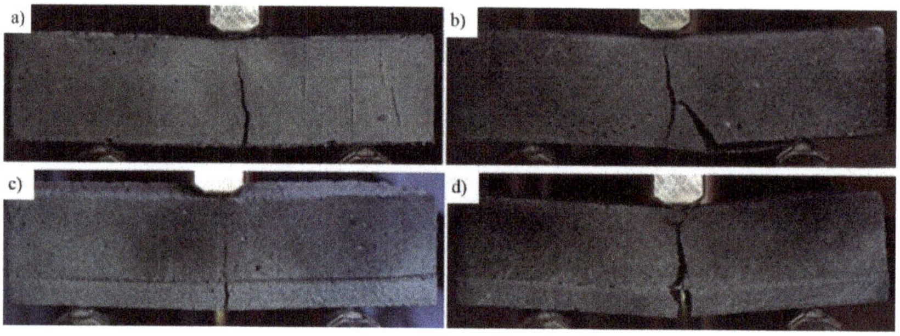

Figure 7. Observed failure modes: (**a**) rupture of the bottom fiber mesh; (**b**) delamination of the bottom fiber mesh; (**c**) rupture of the fiber mesh; (**d**) simultaneous crushing of concrete and fiber rupture.

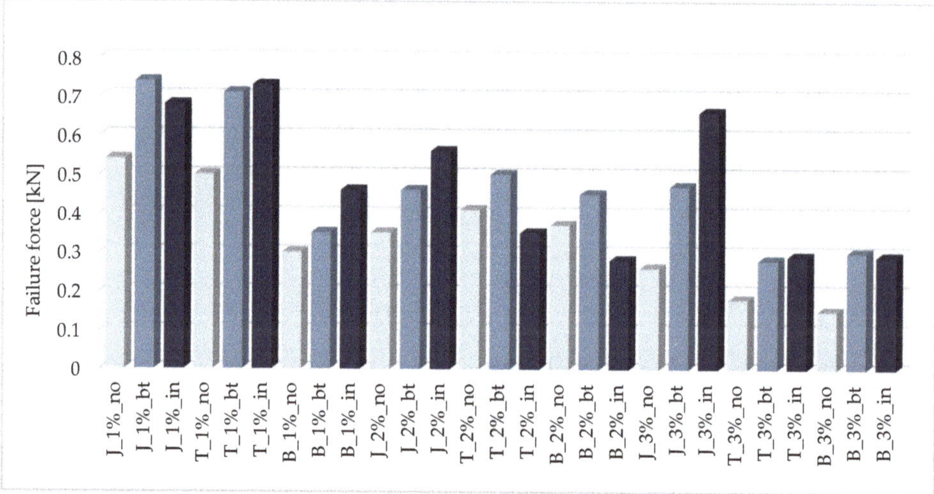

Figure 8. Comparison of failure forces in the bending test.

Theoretically, the flexural capacity of a concrete section is most strongly influenced by the reinforcement strength and the arm of the internal forces, and the mechanical properties of the concrete itself are of secondary importance. For the models tested here, the change in load capacity after using a more foamed geopolymer was not expected to drop by more than about 12%. This hypothesis was confirmed only for the internally reinforced samples based on fly ashes from the Jaworzno power plant (J_1%_in, J_2%_in, and J_3%_in). Of the models made of C-type ash, only the T_1%_in and T_1%_bt models had a load capacity that was comparable to those based on Jaworzno fly ash.

Despite the larger arm of the internal force, the load capacity for most samples reinforced with mesh attached along the bottom surface was lower. This phenomenon can be easily explained. An analysis of the failure modes (summarized in Table 7 and illustrated in Figure 7) shows that only the densest geopolymers (foamed with 1% H_2O_2) provided a sufficient bond. Models foamed with additions of 2% and 3% of the foaming agent mostly broke as a result of debonding of the fiberglass mesh. The presence of pores reduces the adhesion surface, creating a poorer bond to the composite fibers, and this led to premature failure before the strength of glass fibers was exceeded. The cause of the early failure of the internally reinforced samples was different; in general, these failed after the rupture of composite fibers (Figure 7c). An inspection of the broken sample showed increased brittleness of the composite (Figure 9). The most probable reason for this is corrosion of the fiberglass caused by the alkalinity of the geopolymer. Although the tests of the samples showed a similar pH of below 11.5 for all samples, this may exceed 13.5 for a fresh mix [57]. This effect was magnified in a fresh mixture by the presence of an unreacted sodium hydroxide activator [58,59]. In the studies carried out here, all the geopolymers were fabricated using the same ratio of activator to precursor. In lignite coal fly ashes, which are poorer in aluminum and silicon, the consumption of sodium hydroxide in the reaction mechanism (geopolymer synthesis) may be slower, thus extending the exposure time of the glass mesh to the alkaline solution and causing increased brittleness of the fibers, as shown by the samples based on the Belchatow and Turow fly ashes.

Figure 9. Brittle rupture of fiberglass in which the visible breakthrough does not contain free fibers.

Most of the internally reinforced beams broke at deflections of between 0.5 and 1 mm, with the exceptions being the T_1%_bt and T_2%_in beams, which failed at a deflection of around 0.15 mm (Tables 7 and 8). The reason for this specific behavior can be found by analyzing the load-deflection relationship, as shown in Figure 10. For most beams, there is an almost linear increase in the deflection in the first phase until cracking occurs (Figure 10a). There is then a rapid increase in the deflection, associated with the development of the crack, until the composite takes all the internal tensile force. We then see a further increase in the transferred load, until the fibers break. The cracking in both beams shown in Figure 10 occurred at a load of approximately 0.34 kN, and the sample shown in Figure 10b failed shortly after cracking. Premature rupture of the brittle glass fibers of the T_2%_in sample did not allow for the increase in the bearing capacity characterizing the post-crack phase, as described above. A similar effect was caused by composite delamination in the T_1%_bt sample. An increase in the brittleness of the fiberglass was noticed in all beams made of geopolymer synthesized from C-type ashes (Turow and Belchatow); however, strongly foamed material allowed for greater deflection due to crushing of the compressed zone (Figure 7d).

3.3. Analytical Model

Determination of the ultimate moment of resistance of composite-strengthened geopolymer concrete is based on similar assumptions to the calculations for steel-reinforced concrete elements. The most important are: (i) the cross-section remains a plane after deformation; (ii) the strains in composite fiber and surrounding concrete are compatible; and (iii) the tensile strength of the concrete can be ignored. Of course, due to its specific material properties, an appropriate strain limit should be assumed for geopolymers. Likewise, the stress-strain relationship must be based on a suitable function. Figure 11 illustrates the assumptions made in the model described here.

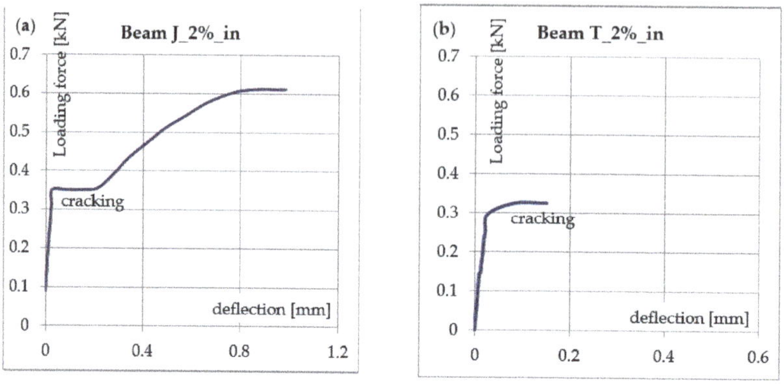

Figure 10. Loading force-deflection curves: (**a**) typical behavior of an internally reinforced beam; (**b**) premature failure caused by brittle fiber rupture or fiberglass debonding.

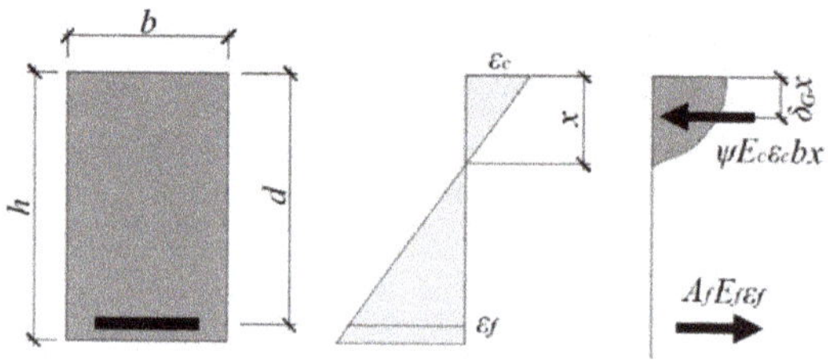

Figure 11. Model of a fiberglass reinforced beam under flexure.

The bending resistance of a cross-section can be determined based on the equilibrium of internal forces, according to the following Equations (2) and (3):

$$M_E = 0.25F_E \cdot l_b = F_f(h - \delta_G x), \tag{2}$$

$$\psi\varepsilon_c E_c bx = F_f, \tag{3}$$

where:

F_f—fiber breaking force,
l_b = 100 mm (the span of beam),
ε_f, ε_c—strains of the composite fiber and the outmost geopolymer fiber, respectively,
E_f, E_c—elasticity moduli for the composite fiber and concrete [MPa], respectively.

The lower limit of the cross-section bearing capacity is the cracking moment:

$$M_{cr} = f_{ct}bh^2/6. \tag{4}$$

where f_{ct} is the tensile strength of foamed geopolymer.

The plain section remains plain, therefore:

$$\varepsilon_c = \varepsilon_f \frac{x}{d-x} = \frac{F_f}{E_f A_f}\frac{x}{d-x}. \tag{5}$$

As shown in Figures 4–6, the stress-strain relationship can be expressed by the parabolic function in (1). In this case, the area of the compressive zone is characterized by the parameter ψ and the location of its center of gravity by the parameter δ_G (Figure 11). The values of these parameters change with an increase in the strain, reaching a maximum at the ultimate strain. For a linear stress-strain relationship (triangular), $\psi = 0.5$, and $\delta_G = \frac{1}{3}$. Table 9 shows the values of the coefficients ψ and δ_G for the geopolymers tested here, calculated for the parabolic relationship in (1) based on the assumption that the ultimate deformation of the compressed zone is reached, as shown in Table 5. The dependence of ψ and δ_G on the deformation significantly complicates the calculations. The parameters presented in Table 9 refer to the situation in which the ultimate strain is achieved in the compressed concrete; for lower strains, these parameters will be lower, aiming at the abovementioned values characterizing the linear relationship. The adoption of a simplified, triangular model over the entire range of deformation leads to a slight underestimate of the load capacity, which can be neglected. This is evidenced in the last column of Table 9, which shows the ratio between the capacity calculated for the triangular relationship and the capacity calculated for the parabolic Relationship (1).

Table 9. Parameters for the compressed zone of the foamed geopolymer and the capacity reduction ratio for a simplified stress-strain model.

	ψ	δ_G	**Capacity Reduction Ratio for a Triangular Model**
J_1%	0.545	0.344	0.953
J_2%	0.545	0.344	0.952
J_3%	0.553	0.345	0.943
T_1%	0.522	0.338	0.974
T_2%	0.500	0.333	1
T_3%	0.474	0.327	1.03
B_1%	0.615	0.361	0.887
B_2%	0.574	0.351	0.919
B_3%	0.519	0.338	0.976

For the "bt" specimens, i.e., those strengthened along the bottom surface, the force that breaks the composite fibers should not exceed the bond resistance F_b. To reflect this phenomenon, the end anchorage model was adopted (given in the *fib* bulletin 90 [60]). When the distance to the first crack l_b is shorter than the required anchorage length, the debonding force should be limited using the formula proposed by Chen and Teng [61]:

$$F_b(l_b) = F_b\beta_l = F_b \sin\left(\frac{\pi}{2}\frac{l_b}{l_{b,\lim}}\right). \tag{6}$$

The bond force for a fully anchored composite is equal to:

$$F_b = 0.25b\sqrt{2E_f t_f f_c^{2/3}}, \tag{7}$$

and the maximum bond length, according to [60], is:

$$l_{b,\lim} = 0.9\pi\sqrt{\frac{E_f t_f}{8 f_c^{2/3}}}, \tag{8}$$

where:

t_f—thickness of composite (mm),
f_c—compressive strength of foamed geopolymer (MPa).

The results of a theoretical analysis of the tested beams are presented in Table 10. Figure 12 shows a comparison of these theoretical results with the values obtained from laboratory testing, expressed as a ratio of the empirical laboratory value to the result predicted using the above formulas.

Table 10. Comparison of laboratory determined and theoretically predicted failure forces.

	Laboratory Determined Moment Nm	Maximum Bond Length $l_{b,lim}$ mm	Delamination Force kN	Fiber Breaking Force kN	Predicted Failure Moment Nm
J_1%_bt	18.5	96.2	4.15	0.72	27.1
J_2%_bt	11.5	105	3.30	0.72	26.6
J_3%_bt	11.7	134	1.77	0.72	19.8
T_1%_bt	17.7	116	2.55	0.72	26.1
T_2%_bt	12.5	134	1.77	0.72	24.2
T_3%_bt	7.0	150	1.32	0.72	14.8
B_1%_bt	8.75	115	2.57	0.72	26.4
B_2%_bt	11.3	134	1.75	0.72	25.7
B_3%_bt	7.5	184	0.76	0.72	7.86
J_1%_in	17.0			0.72	20.2
J_2%_in	14.0			0.72	19.7
J_3%_in	16.5			0.72	12.5
T_1%_in	18.2			0.72	17.7
T_2%_in	8.75			0.72	15.1
T_3%_in	7.25			0.72	9.06
B_1%_in	11.5			0.72	19.6
B_2%_in	7.00			0.72	18.1
B_3%_in	7.25			0.72	4.81

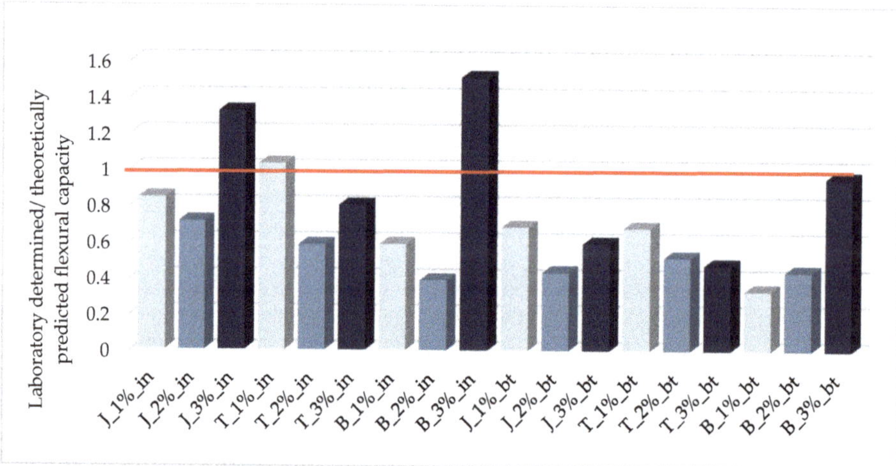

Figure 12. Ratios of laboratory determined/predicted flexural capacities of tested beams.

An analysis of the values listed in Table 10 shows that the delamination force is greater than the fiber breaking force for the GFRP composite used here, for all models, a result that is inconsistent with the observed modes of failure (listed in Table 7). All models foamed by the addition of hydrogen peroxide >1% failed due to delamination, indicating that the delamination model adopted here significantly overestimates the results. In future, it will be necessary to develop a model that is suitable for porous materials. To the best of the authors' knowledge, such a model has not yet been developed for externally reinforced expanded concretes.

The second reason for overestimating the load capacity in the case of internally reinforced models is the increased brittleness of the glass fibers as a result of accelerated corrosion in an alkaline environment, as described above.

The probability of premature delamination increases with the degree of foaming. The decreasing strength of the foamed geopolymer may be identified as the main factor promoting debonding. When analyzing most of the methods of predicting the debonding strength, a strong correlation between fracture energy and substrate strength can be seen [62]. For this reason, the Formula (6) can be corrected by a factor that depends on the strength of the tested geopolymer. Figure 13 shows the effect of the geopolymer strength on the relative error of the debonding force estimation (expressed as the ratio of the tensile force of the composite fiber corresponding to the laboratory-obtained bending resistance and the debonding force calculated according to Expression (6)).

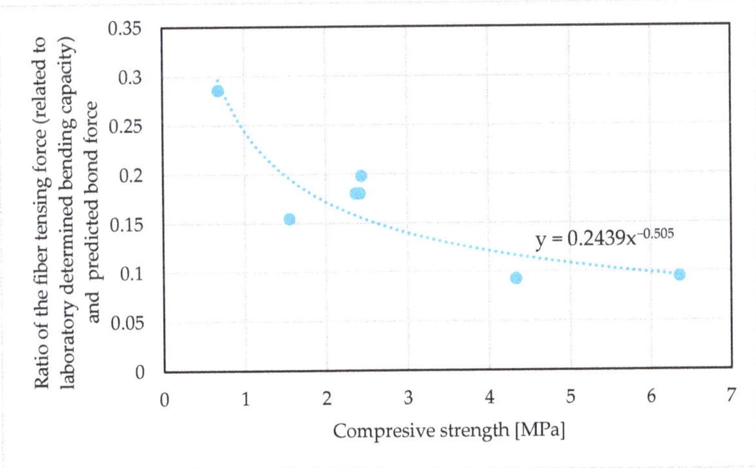

Figure 13. Effect of the compressive strength on the relative error of the adhesion force estimation.

The analysis of the error in estimating the debonding force enclosed the samples for which this form of failure was observed (Table 7). Relative error can be estimated with the use of the trend function shown in Figure 13. The value of this function expresses the correction coefficient of the bond force. Introducing it into (7), the modified formula for the bond to the foamed geopolymer can be obtained:

$$F_b = 0.2439 f_c^{-0.505} 0.25 b\sqrt{2E_f t_f f_c^{2/3}} \approx 0.0625 b\sqrt{2E_f t_f f_c^{-1/3}}, \quad (9)$$

To find the correction factor for the corrosion aging effect, the laboratory determined and predicted flexural capacities were compared. The research showed faster corrosion in geopolymers based on lignite coal fly ash, therefore the analysis was carried out separately for samples made of ashes from Jaworzno and samples made of ashes from Turow and Belchatow. Figure 14 shows the effect of content of foaming agent on the aging factor expressed as a ratio of flexural capacity determined in laboratory and theoretically predicted for ultimate fiber breaking force (listed in Table 10).

Given in Figure 14 trend functions express the estimation error and at the same time the correction factor for to the corrosion aging effect. For the anthracite-coal-fly-ash based foamed geopolymers the fiber breaking force F_f given in the Formulas (1) and (2) can be reduced according to the Equation (10):

$$F_{ffg} = 0.8F_f \quad (10)$$

Consequently, for the lignite-coal-fly-ash based foamed geopolymers, the reduced fiber breaking force F_{ffg} is equal:

$$F_{ffg} = (0.8 - 0.15D_{fa})F_f \quad (11)$$

where D_{fa} is the content of foaming agent in the geopolymer mixture (given in %).

To check the correctness of proposed empirically modified method, the theoretical flexural capacity of the tested samples was recalculated using the presented formulas. The results of these analyzes are shown in Figure 15. The comparison of the graphs in

Figures 12 and 15 shows a significant improvement in the accuracy of the load-bearing capacity estimation. The correlation coefficient for modified model is 0.835.

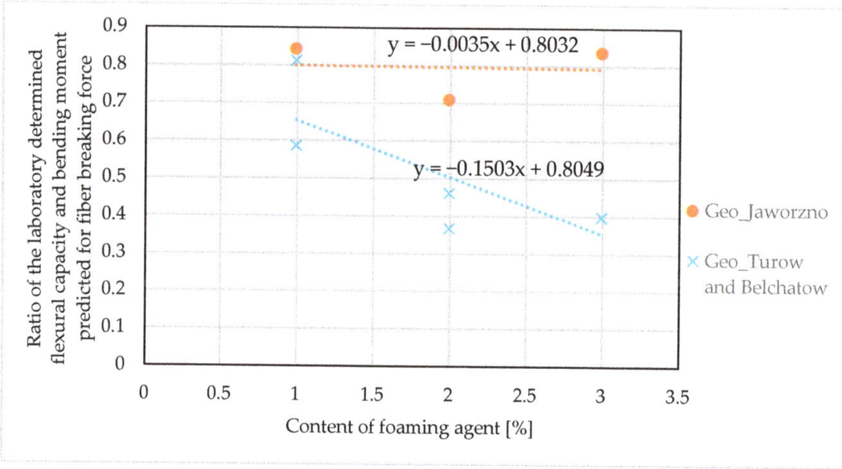

Figure 14. Effect of the content of foaming agent on the aging effect of glass fibers.

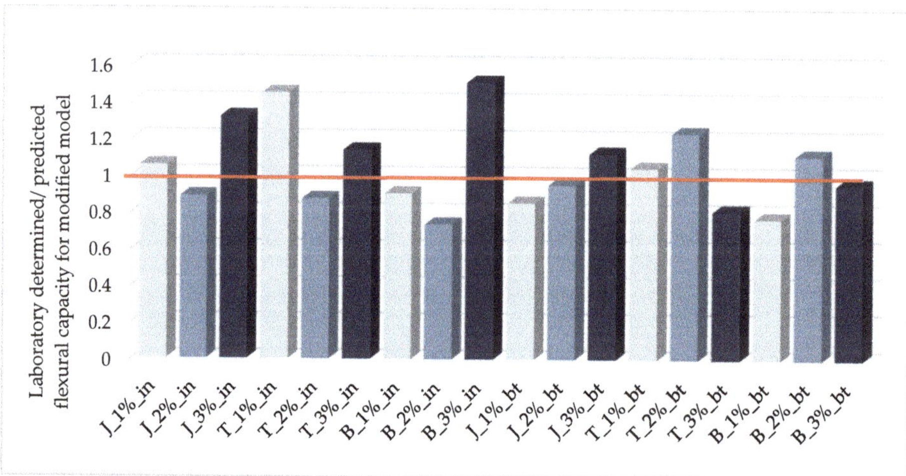

Figure 15. Ratios of laboratory determined/predicted flexural capacities of tested beams for empirically modified model.

4. Conclusions

This study investigated a series of foamed geopolymer samples reinforced with fiberglass mesh. The precursors used in the synthesis of the geopolymer were fly ashes obtained from the combustion of two types of coal: anthracite and lignite. The effects of reinforcement with GFRP mesh and the type of fly ash on the strength properties of geopolymer were explored, and the results of this research and theoretical analyses give rise to the following conclusions:

(1) The origin of fly ash is of great importance for the mechanical properties of the geopolymer concrete. The compressive strengths of geopolymers based on anthracite coal fly ash are at least 73% higher than those based on lignite coal, which is rich

in calcium. Similar differences were obtained in tests of the tensile strength and the modulus of elasticity. The worst mechanical properties were shown by geopolymer made from ashes from the Belchatow power plant.

(2) The deficiencies in the tensile strength can be effectively improved by reinforcement of the composite mesh. The effectiveness of this reinforcement depends on its location and the density of the foamed geopolymer. For lower densities, insufficient compressive strength leads to premature failure, which may be caused by the brittle breaking of the glass fibers or delamination.

(3) Of the beams reinforced along the bottom surface, failure due to debonding was most commonly observed. Only the samples made of fly ashes from Jaworzno and Turow and foamed with 1% H_2O_2 content were all broken due to mesh rupture. The proposed load capacity prediction model, which was based on existing debonding formulas originally developed for plain concrete, significantly overestimated the bond capacity.

(4) The effect of the chemical integrity of the fiberglass (Si content) [49] on improving the bond to the geopolymer was lower than expected. The suitability of glass fibers for geopolymer reinforcement, as reported by other researchers [49–51], was not confirmed. Chemical decomposition of the glass fibers was observed during testing, especially in the case of geopolymers based on type-C fly ashes. An increase in the brittleness of the fibers in samples inspected after testing was particularly noticeable.

(5) The proposed empirical model allows for a more precise determination of the load capacity. It takes into account the lower bond of the mesh composite to the foamed geopolymer and the aging effect of the glass fibers during the geopolymerization reaction.

Future research should focus on using reinforcing meshes that are more resistant to alkaline solutions, based on carbon or basalt fibers, and on developing a bonding model that is suitable for porous materials.

Author Contributions: Conceptualization, R.K. and S.D.; methodology, R.K. and S.D.; investigation, R.K. and S.D.; data curation, R.K.; writing—original draft preparation, R.K.; writing—review and editing, R.K. and S.D.; visualization, R.K. All authors have read and agreed to the published version of the manuscript.

Funding: This research was funded by the European Union for financing the project No 645696 RICON and the Silesian University of Technology (Grant no BK-296/RB6/2020).

Institutional Review Board Statement: Not applicable.

Informed Consent Statement: Not applicable.

Data Availability Statement: The data presented in this study are available on request from the corresponding author.

Conflicts of Interest: The authors declare no conflict of interest.

References

1. Shigemoto, N.; Hayashi, H.; Miyaura, K. Selective formation of Na-X zeolite from coal fly ash by fusion with sodium hydroxide prior to hydrothermal reaction. *J. Mater. Sci.* **1993**, *28*, 4781–4786. [CrossRef]
2. Inada, M.; Yukari, E.; Enomoto, N.; Hojo, J. Synthesis of zeolite from coal fly ashes with different silica-alumina composition. *Fuel* **2004**, *84*, 299–304. [CrossRef]
3. Tanaka, H.; Eguchi, H.; Fujimoto, S.; Hino, R. Two-step process for synthesis of a single phase Na–A zeolite from coal fly ash by dialysis. *Fuel* **2006**, *85*, 1329–1334. [CrossRef]
4. Rangan, B.V. Fly Ash-Based Geopolymer Concrete. In Proceedings of the International Workshop on Geopolymer Cement and Concrete, Mumbai, India, 7 December 2010; Allied Publishers Private Limited: Mumbai, India, 2010; pp. 68–106.
5. Kong, D.L.; Sanjayan, J.G.; Sagoe-Crentsil, K. Comparative performance of geopolymers made with metakaolin and fly ash after exposure to elevated temperatures. *Cem. Concr. Res.* **2007**, *37*, 1583–1589. [CrossRef]
6. Hung, T.V.; Dong, V.D.; Long, N.N.; Hien, T.D. Study on the Mechanical Properties of the Fly Ash Geopolymer Concrete. *Int. J. Civ. Eng. Technol.* **2017**, *8*, 950–957.
7. Naghizadeh, A.; Ekolu, S.O. Behaviour of fly ash geopolymer binders under exposure to alkaline media. *Asian J. Civ. Eng.* **2019**, *20*, 785–798. [CrossRef]

8. Bakharev, T. Resistance of geopolymer materials to acid attack. *Cem. Concr. Res.* **2005**, *35*, 658–670. [CrossRef]
9. Sarker, P.K.; Kelly, S.; Yao, Z. Effect of fire exposure on cracking, spalling and residual strength of fly ash geopolymer concrete. *Mater. Des.* **2014**, *63*, 584–592. [CrossRef]
10. Ridtirud, C.; Chindaprasirt, P.; Pimraksa, K. Factors affecting the shrinkage of fly ash geopolymers. *Int. J. Miner. Metall. Mater.* **2011**, *18*, 100–104. [CrossRef]
11. Yip, C.K.; van Deventer, J.S.J. Microanalysis of calcium silicate hydrate gel formed within a geopolymer binder. *J. Mater. Sci.* **2003**, *38*, 3851–3860. [CrossRef]
12. Wang, S.D.; Scrivener, K.L.; Pratt, P.L. Factors affecting the strength of alkali activated slag. *Cem. Concr. Res.* **1994**, *24*, 1033–1043. [CrossRef]
13. Luz Granizo, M.; Alonso, S.; Blanco-Varela, M.T.; Palomo, A. Alkaline activation of metakaolin: Effect of calcium hydroxide in the products of reaction. *J. Am. Ceram. Soc.* **2002**, *85*, 225–231. [CrossRef]
14. Sciubidło, A.; Majchrzak-Kuceba, I.; Nowak, W. Influence of the chemical composition of the fly ashes on the efficiency of the synthesis zeolites Na-X. In *Polska Inżynieria Środowiska Pięć Lat po Wstąpieniu do Unii Europejskiej*; Ozonek, J., Pawłowski, A., Eds.; Komitet Inżynierii Środowiska PAN: Lublin, Poland, 2009; Volume 59, pp. 225–237.
15. Dawczyński, S.; Kajzer, A.; Górski, M. Investigation on strength development in geopolymer made of power plant fly ash suspension. In Proceedings of the RILEM Proceedings PRO 121 SynerCrete'18—Interdisciplinary Approaches for Cement-Based Materials and Structural Concrete: Synergizing Expertise and Bridging Scales of Space and Time, Funchal, Portugal, 24–26 October 2018.
16. Diaz, E.I.; Allouche, E.N.; Eklund, S. Factors affecting the suitability of fly ash as source material for geopolymers. *Fuel* **2010**, *89*, 992–996. [CrossRef]
17. Guo, X.L.; Shi, H.S.; Dick, W.A. Compressive strength and microstructural characteristics of class C fly ash geopolymer. *Cem. Concr. Compos.* **2010**, *32*, 142–147. [CrossRef]
18. Wardhono, A. Comparison Study of Class F and Class C Fly Ashes as Cement Replacement Material on Strength Development of Non-Cement Mortar. *IOP Conf. Ser. Mater. Sci. Eng.* **2018**, *288*, 012019. [CrossRef]
19. Gunasekara, C.; Dirgantara, R.; Law, D.W.; Setunge, S. Effect of Curing Conditions on Microstructure and Pore-Structure of Brown Coal Fly Ash Geopolymers. *Appl. Sci.* **2019**, *9*, 3138. [CrossRef]
20. Dawczyński, S.; Soczyński, M.; Górski, M. Feasibility and strength properties of the geopolymeric binder made of fly ash suspension. *MATEC Web Conf.* **2019**, *262*, 06001. [CrossRef]
21. Henon, J.; Alzina, A.; Absi, J.; Smith, D.S.; Rossignol, S. Potassium geopolymer foams made with silica fume pore forming agent for thermal insulation. *J. Porous Mater.* **2012**, *20*, 37–46. [CrossRef]
22. Sanjayan, J.G.; Nazari, A.; Chen, L.; Nguyen, G.H. Physical and mechanical properties of lightweight aerated geopolymer. *Constr. Build. Mater.* **2015**, *79*, 236–244. [CrossRef]
23. Abdollahnejad, Z.; Pacheco-Torgal, F.; Felix, T.; Tahri, W.; Barroso Aguiar, J. Mix design, properties and cost analysis of fly ash-based geopolymer foam. *Constr. Build. Mater.* **2015**, *80*, 18–30. [CrossRef]
24. Cui, Y.; Wang, D. Effects of Water on Pore Structure and Thermal Conductivity of Fly Ash-Based Foam Geopolymers. *Adv. Mater. Sci. Eng.* **2019**. [CrossRef]
25. Łach, M.; Mierzwiński, D.; Korniejenko, K.; Mikuła, J. Geopolymer foam as a passive fire protection. *MATEC Web Conf.* **2018**, *247*, 00031. [CrossRef]
26. Cheng-Yong, H.; Yun-Ming, L.; Al Bakri Abdullah, M.M.; Hussin, K. Thermal Resistance Variations of Fly Ash Geopolymers: Foaming Responses. *Sci. Rep.* **2017**, *7*, 45355. [CrossRef] [PubMed]
27. Skvara, F.; Sulc, R.; Tisler, Z.; Skricík, P.; Smilauer, V.; Zlamalova Cílova, Z. Preparation and properties of fly ash-based geopolymer foams. *Ceram. Silik.* **2014**, *58*, 188–197.
28. Lahoti, M.; Tan, K.H.; Yang, E. A critical review of geopolymer properties for structural fire-resistance applications. *Constr. Build. Mater.* **2019**, *221*, 514–526. [CrossRef]
29. Masi, G.; Rickard, W.; Vickers, L.; Bignozzi, M.; Van Riessen, A. A comparison between different foaming methods for the synthesis of light weight geopolymers. *Ceram. Int.* **2014**, *40*, 13891–13902. [CrossRef]
30. Zhang, Z.; Wang, H. The Pore Characteristics of Geopolymer Foam Concrete and Their Impact on the Compressive Strength and Modulus. *Front. Mater.* **2016**, *3*. [CrossRef]
31. Kim, H.; Lee, S.; Han, Y.; Park, J. Control of pore size in ceramic foams: Influence of surfactant concentration. *Mater. Chem. Phys.* **2009**, *113*, 441–444. [CrossRef]
32. Huiskes, D.M.A.; Keulen, A.; Yu, Q.L.; Brouwers, H.J.H. Design and performance evaluation of ultra-lightweight geopolymer concrete. *Mater. Des.* **2016**, *89*, 516–526. [CrossRef]
33. Boke, N.; Birch, G.D.; Nyale, S.M.; Petrik, L.F. New synthesis method for the production of coal fly ash-based foamed geopolymers. *Constr. Build. Mater.* **2015**, *75*, 189–199. [CrossRef]
34. Anggarini, U.; Pratapa, S.; Purnomo, V.; Sukmana, N.C. A comparative study of the utilization of synthetic foaming agent and aluminum powder as poreforming agents in lightweight geopolymer synthesis. *Open Chem.* **2019**, *17*, 629–638. [CrossRef]
35. Le, V.S.; Hájková, P.; Kovacic, V.; Bakalova, T.; Voleský, L.; Le, C.H.; Cecoon Seifert, K.; Pereira Peres, A.; Louda, P. Thermal Conductivity of Reinforced Geopolymer Foams. *Ceram. Silikáty* **2019**, *63*, 365–373. [CrossRef]

36. Galzerano, B.; Formisano, A.; Durante, M.; Iucolano, F.; Caputo, D.; Liguori, B. Hemp reinforcement in lightweight geopolymers. *J. Compos. Mater.* **2018**, *52*, 2313–2320. [CrossRef]
37. Ngo, J.P.; Promentilla, M.A. Development of Fiber-reinforced Foamed Fly Ash Geopolymer. *MATEC Web Conf.* **2018**, *156*, 05018.
38. Masi, G.; Rickard, W.; Bignozzi, M.; Van Riessen, A. The influence of short fibres and foaming agents on the physical and thermal behaviour of geopolymer composites. *Adv. Sci. Technol.* **2014**, *92*, 56–61. [CrossRef]
39. Abdollahnejad, Z.; Zhang, Z.; Wang, H.; Mastali, M. Comparative Study on the Drying Shrinkage and Mechanical Properties of Geopolymer Foam Concrete Incorporating Different Dosages of Fiber, Sand and Foam Agents. In *High Tech Concrete: Where Technology and Engineering Meet, Proceedings of the 2017 Fib Symposium, Maastricht, The Netherlands, 12–14 June 2017*; Hordijk, D., Luković, M., Eds.; Springer: Cham, Switzerland, 2018.
40. Natali, A.; Manzi, S.; Bignozzia, M.C. Novel fiber-reinforced composite materials based on sustainable geopolymer matrix. *Procedia Eng.* **2011**, *21*, 1124–1131. [CrossRef]
41. Masi, G.; Rickard, W.; Bignozzi, M.; Van Riessen, A. The effect of organic and inorganic fibres on the mechanical and thermal properties of aluminate activated geopolymers. *Compos. Part B* **2015**, *76*, 218–228. [CrossRef]
42. Lee, J.H.; Wattanasiriwech, S.; Wattanasiriwech, D. Preparation of Carbon Fiber Reinforced Metakaolin Based-Geopolymer Foams. *Key Eng. Mater.* **2018**, *766*, 19–27. [CrossRef]
43. De Villers, J.P.; van Zijl, G.; van Royen, A.S. Bond of deformed steel reinforcement in lightweight foamed concrete. *Struct. Concr.* **2017**, *18*, 496–506. [CrossRef]
44. Falliano, D.; De Domenico, D.; Ricciardi, G.; Gugliandolo, E. Improving the flexural capacity of extrudable foamed concrete with glass-fiber bi-directional grid reinforcement: An experimental study. *Compos. Struct.* **2019**, *209*, 45–59. [CrossRef]
45. Liu, J.; Wua, C.; Li, C.; Dong, W.; Sua, Y.; Li, J.; Cui, N.; Zeng, F.; Dai, L.; Meng, Q.; et al. Blast testing of high performance geopolymer composite walls reinforced with steel wire mesh and aluminium foam. *Constr. Build. Mater.* **2019**, *197*, 533–547. [CrossRef]
46. Hulimka, J.; Krzywoń, R.; Jędrzejewska, A. Laboratory Tests of Foam Concrete Slabs Reinforced with Composite Grid. *Procedia Eng.* **2017**, *193*, 337–344. [CrossRef]
47. European Committee for Standardization. *EN 196-1: 2016. Methods of Testing Cement—Part 1: Determination of Strength*; CEN-CENELEC Management Centre: Brussels, Belgium, 2016.
48. Pacheco-Torgal, F.; Melchers, R.; de Belie, N.; Shi, X.; Van Tittelboom, K.; Saez Perez, A. *Eco-Efficient Repair and Rehabilitation of Concrete Infrastructures*; Woodhead Publishing: Cambridge, UK, 2017.
49. Steinerova, M.; Matulova, L.; Vermach, P.; Kotas, J. The Brittleness and Chemical Stability of Optimized Geopolymer Composites. *Materials* **2017**, *10*, 396. [CrossRef] [PubMed]
50. Nematollahi, B.; Sanjayan, J.; Chai, J.X.H.; Lu, T.M. Properties of Fresh and Hardened Glass Fiber Reinforced Fly Ash Based Geopolymer Concrete. *Key Eng. Mater.* **2014**, *594–595*, 629–633. [CrossRef]
51. Alekhya, A.; Mahesh, Y. Experimental Investigations on Properties of Glass Fiber Reinforced Geopolymer Concrete Composites. *IOSR J. Mech. Civ. Eng.* **2017**, *14*, 1–10.
52. European Committee for Standardization. *EN 1097-6: 2013. Tests for Mechanical and Physical Properties of Aggregates. Determination of Particle Density and Water Absorption*; CEN-CENELEC Management Centre: Brussels, Belgium, 2013.
53. Ducman, V.; Korat, L. Characterization of geopolymer fly-ash based foams obtained with the addition of Al powder or H_2O_2 as foaming agents. *Mater. Charact.* **2016**, *113*, 207–213. [CrossRef]
54. Davidovits, J. *Geopolymer: Chemistry and Applications*; Geopolymer Institute: Saint-Quentin, France, 2008.
55. Kozłowski, M.; Kadela, M. Mechanical Characterization of Lightweight Foamed Concrete. *Adv. Mater. Sci. Eng.* **2018**. [CrossRef]
56. Drusa, M.; Vlcek, J.; Scherfel, W.; Sedlar, B. Testing of Foam Concrete for Definition of Layer Interacting with Subsoil in Geotechnical Applications. *Int. J. Geomate* **2019**, *17*, 115–120. [CrossRef]
57. Koenig, A.; Mahmoud, H.; Baehre, O.; Dehn, F. Alkalinity and Its Consequences for the Performance of Steel-Reinforced Geopolymer Materials. *Molecules* **2020**, *25*, 2359. [CrossRef]
58. Coricciati, A.; Corvaglia, P.; Mosheyev, G. Durability of Fibers in Aggressive Alkaline Environment. In Proceedings of the ICCM-17—17th International Conference on Composite Materials, Edinburgh, UK, 27–31 July 2009.
59. Amaro, A.M.; Reis, P.N.B.; Neto, M.A.; Louro, C. Effects of Alkaline and Acid Solutions on Glass/Epoxy Composites. *Polym. Degrad. Stab.* **2013**, *98*, 853–862. [CrossRef]
60. FIB Bulletin 90. *Externally Applied FRP Reinforcement for Concrete Structures*; Fédération Internationale du Béton (FIB): Lausanne, Switzerland, 2019; p. 229.
61. Chen, J.F.; Teng, J.G. Anchorage strength models for FRP and steel plates bonded to concrete. *J. Struct. Eng.* **2001**, *127*, 784–791. [CrossRef]
62. Krzywon, R. Assessment of existing bond models for externally 3 bonded SRP composites. *Appl. Sci.* **2020**, *10*, 8593. [CrossRef]

materials MDPI

Article

Research on the Thermal Properties of Fireplace Concrete Materials Containing Various Mineral Aggregates Enriched by Organic and Inorganic Fibers

Agata Stempkowska [1,*], Joanna Mastalska-Popławska [2], Piotr Izak [2], Łukasz Wójcik [2], Tomasz Gawenda [1] and Marzena Karbowy [3]

1 Faculty of Mining and Geoengineering, AGH University of Science and Technology, Mickiewicza 30 Av., 30-094 Krakow, Poland; gawenda@agh.edu.pl
2 Faculty of Materials Science and Ceramics, AGH University of Science and Technology, Mickiewicza 30 Av., 30-094 Krakow, Poland; jmast@agh.edu.pl (J.M.-P.); izak@agh.edu.pl (P.I.); lukwoj@agh.edu.pl (Ł.W.)
3 Northstar Poland, 27 Stycznia 47-48 St., 64-980 Trzcianka, Poland; marzena.karbowy@northstar.pl
* Correspondence: stemp@agh.edu.pl

Abstract: This work presents a summary of research on concrete fireplace materials made of various mineral aggregates and enriched with steel and organic fibers. To determine the optimal applications of such concretes, their ability to accumulate heat and their other physicochemical parameters were tested and analyzed. Studies on the behavior of concrete materials during cooling are reported, and the ability of such materials to accumulate heat is evaluated using calculations. In addition, tests were performed on the loss of mass during heating, as well as on the mechanical bending strength and microstructures of these materials. Studies have shown that the behavior of concrete materials at high temperatures can be modified and adapted for specific purposes. The addition of fibers to concrete improves both the mechanical properties of mortars and the heat flow in concrete materials.

Keywords: concrete; thermal properties; structural properties

Citation: Stempkowska, A.; Mastalska-Popławska, J.; Izak, P.; Wójcik, Ł.; Gawenda, T.; Karbowy, M. Research on the Thermal Properties of Fireplace Concrete Materials Containing Various Mineral Aggregates Enriched by Organic and Inorganic Fibers. *Materials* **2021**, *14*, 904. https://doi.org/10.3390/ma14040904

Academic Editor: Frank Collins
Received: 7 December 2020
Accepted: 9 February 2021
Published: 14 February 2021

1. Introduction

Over recent decades, the energy demands in residential buildings have increased, and there is now a great need to balance energy consumption. Energy-efficient buildings are becoming increasingly desirable due to rising energy costs and increasing awareness of global warming [1]. The thermal properties of concrete materials are attracting increasingly more attention, not only because of their influence on the energy efficiency of buildings but also due to the structural properties and functionality of such materials. Modern concrete materials containing various complementary cement materials, various types of aggregates (including light and recycled aggregates), and fibers are increasingly used in transport structures such as sidewalks and bridge decks, as well as in large foundations (ready-mixed concrete), where thermal behavior is important and sensitive to construction properties [2]. However, there are few publications on the heat-accumulation properties of concrete elements used in fireplaces. Usually, the attention is focused on the role of the thermal conductivity coefficient in the insulation of buildings, and information can also be found on the behavior of concrete at high temperatures above 500 °C, for which the strong influence of the aggregates used and effective reinforcement with steel and polypropylene fibers are highlighted [3]. The type of aggregate is also important in concretes operating at increased temperatures. Fine aggregates based on sand show a polymorphic transformation at 573 °C and have a high coefficient of thermal expansion, which causes micro-cracking in the material and weakening of the structure [4]. Interesting results were also obtained by examining the compressive strength and pore distribution in concrete working at above 500 °C [5,6]. The addition of expanded glass to concrete also significantly reduces thermal conductivity [7], so such materials have good heat accumulation capacities [8]. This publication focuses on concrete materials designated for heating devices. Heating furnaces are usually made of natural soapstone (a green–grey rock containing mainly magnesium aluminosilicates), which has a very good thermal accumulation capacity. In these furnaces, thanks to the accumulation capacity of soapstone and technical solutions (e.g., multiple circulation and swirling of flue gases), the temperature in the furnace reaches about 1200 °C (normally about 600 °C), which enables the combustion of flue gases (burning of soot). This results in high furnace efficiency and has a positive impact on the environment [9]. The temperatures in accumulation furnaces reach 400 °C and higher. Simple concrete is rarely used as a material in these furnaces due to its low resistance to such temperatures (the safe temperature limit is 450 °C). This is why ceramics are used in these applications, as ceramics are much more resistant to high temperatures, especially chamotte. The behavior of cement mortars at high temperatures can also be modified and adapted for specific purposes. For this purpose, it is worthwhile to trace the physicochemical processes that occur under the influence of high temperatures.

To control the process of thermal accumulation, it is necessary to determine the thermal phenomena and transformations taking place in the materials. The energy absorbed by a body during heating or lost during cooling is proportional to the product of the body weight m and the temperature difference of the body ΔT before and after the thermal transformation. The principle of thermal conversion ability ΔQ can be rendered as [10]:

$$\Delta Q = c_V \times m \times \Delta T \text{ (J)} \tag{1}$$

where the c_v capacity of a substance is considered the amount of heat needed to raise the temperature of the substance by one degree. The thermal capacity, which is measured per unit of mass of a substance, is called specific heat c_p (expressed in J/kgK). This quantity is not a constant value and depends primarily on temperature (Figure 1) [11]. In many amorphous, glassy, and crystalline substances, the specific heat increases simultaneously with a temperature increase, as well as at high temperatures.

Another value that characterizes materials in terms of their thermal properties is their thermal volume capacity. The value b is calculated as the product of the thermal capacity c_V and the density ρ of the material from which the body is made:

$$b = c_V \times \rho \text{ (J/(m}^3\text{K))} \tag{2}$$

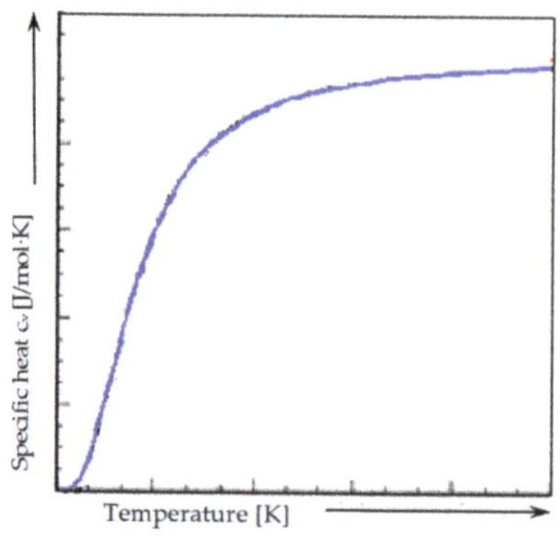

Figure 1. Thermal capacity dependence in function of temperature.

The values of thermal capacity b for building materials can be very different, ranging from about 25 kJ/m^{3}K for insulation materials to almost 3000 kJ/m^{3}K for heat accumulators. For example, for ordinary chamotte bricks, the thermal energy accumulated in a unit of volume is about 1000 kJ/m^{3}K, while for concretes resistant to high temperatures, with the same volume, much more energy can be accumulated (~2700 kJ/m^{3}K). In general, heat-resistant concretes can accumulate about three times more energy than ordinary chamotte coverings. The main factor influencing the value of concrete is its composition, as this capacity for concrete results from the volumetric thermal capacity of its ingredients, which are additive quantities [10,11].

Volume heat capacity is the amount of energy taken in during heating or released during cooling 1 m^3 of the material, changing the material's temperature by one degree. In other words, volume heat capacity is the energy that increases (or decreases) the temperature of a given material of a unit volume by a unit of temperature. Volume heat capacity is not sufficient, however, to describe the ability to accumulate heat. An additional parameter characterizing the efficiency of the material accumulation phenomenon is the energy that can be accumulated in a unit of the material's volume for a certain period of time Δt. The maximum energy that can be accumulated in a given $b_{\max}$ material can be described by the following formula [11]:

$$b_{\max} = b \times \Delta t \ (\mathrm{J/m^3}) \tag{3}$$

It follows that different building and mineral materials have different capacities to store energy, the value of which determines if the material is a good heat accumulator and can be used as such in space-heating systems. An important element for thermal properties is the use of natural [3], light [12,13], hybrid [14], or recycled [15,16] aggregates. The properties of concrete materials with rubber waste particles were also studied [17]. In the literature review, studies on the thermal properties of concretes and mortars containing recycled glass as fine aggregates [18] and reinforced concretes containing steel, plastic, and glass fibers [19] were found. Concrete is a multi-phase and complex system. Therefore, numerical methods based on multi-species modeling seem very promising for the analysis of concrete [20]. Numerical simulations have been widely used to analyze the influence of aggregates on the transport properties in concrete, including heat transport. These methods allow one to predict the important properties of concretes such as their resistance to chloride corrosion [21], which is significantly influenced by temperature [22], as well as the shapes of the aggregate grains [23,24].

Another parameter that determines the thermal properties of materials is the time needed to release (emission) the accumulated energy. With a given amount of accumulated energy, the emission time of that energy must not be too short (or too much heat will be released in a unit of time) or too long (or too little heat will be released in a unit of time and will be insufficient, e.g., for heating the room). In a thermodynamic system (e.g., in an isolated room) containing a body with a temperature higher than room temperature, entropy will strive to achieve the maximum value, i.e., the equilibrium state of the whole system, to equalize the temperature in the whole area. This process always occurs automatically and spontaneously. The phenomenon of heat transfer from the body to the system is called the emission of heat energy. The measure of thermal energy emission E is the thermal power P, which is determined by the following relation [25]:

$$P = E/t \ (\mathrm{J/s}). \tag{4}$$

Heat-accumulating materials used in practical applications should have a relatively high heat power. The heat that is delivered to the storage material then causes a proportional increase in temperature. Sensible heat is presently the most popular method of thermal energy storage. This change can be registered with the senses or via sensors [26]. Sensible heat magazines accumulate thermal energy by heating or cooling the stored material. These magazines then use the thermal capacity and temperature changes of the material during the charging and discharging processes. The amount of accumulated heat depends on the mass and medium of heat used for storage and the temperature difference between the initial and final states [27,28]. There have also been attempts to model the behavior of concrete materials under the influence of high temperature and to estimate their mechanical properties and heat accumulation capacities [29,30].

2. Research Significance

Materials used for the construction of fireplaces should have special thermal properties which can allow for the accumulation and release of heat. Heat accumulation refers to the ability of a material to accumulate and store a certain amount of thermal energy inside itself, which can then be released for a period of time by the material. That is why we are looking for ways to accumulate heat when there is excess and use it when there is a deficiency. Foundationally, this involves finding optimal solutions from the perspective of so-called thermal comfort, i.e., obtaining and maintaining the necessary temperatures inside objects for comfortable functioning in unfavorable external conditions. One example is seasonal accumulation, which involves storing heat energy in summer and using that energy in the autumn–winter season. Reverse processes are used to cool rooms in summer conditions. The purpose of this work is to try to determine whether concretes made with mineral aggregates and various organic additives have different heat accumulation capacities. This would allow the concrete to be used not only for the manufacturing of elements, including prefabricated elements, at increased temperatures but also where such elements could accumulate and release heat (e.g., fireplace covers). Therefore, the basic parameters characterizing the usefulness of such materials, i.e., volume and mass thermal capacity, maximum heat accumulation capacity, and high temperature behavior (differential thermal analysis(DTA) tests), were assessed.

3. Materials and Methods

3.1. Materials and Mix Proportions

Samples for testing were supplied from Northstar and were used as fireplace insulation inserts. The composition of the individual concrete materials is shown in Table 1.

Table 1. Concrete mix compositions used in research.

Sample	Ingredient	% wt
1 Concrete with sandstone aggregate enriched with glass fibers	Portland cement CEM I 52.5 R	28.528
	Sand aggregate 0–2 mm	45.645
	Quartz meal 0.75	13.313
	Glass fibers 12 mm	0.266
	Pigment	0.799
	Superplasticizer + additives	1.654
2 Concrete with silicate and refractory aggregate enriched with polypropylene fibers	High alumina cement 70	15.1
	Portland cement CEM I 52.5 R	4.9
	Fine silicate aggregate 0–2 mm	12.0
	Refractory aggregate 0–5 mm	67.4
	Polypropylene fibers 3 mm	0.141
	Pigment	0.304
	Superplasticizer + additives	0.100
3 Concrete with silicate and carbonate aggregates enriched with glass and polypropylene fibers	Portland cement CEM I 52.5 R	31.343
	Carbonate aggregate 8–16 mm	47.062
	Silicate aggregate 0–4 mm	6.268
	Glass fibers 12 mm	0.209
	Polypropylene fibers 3 mm	0.020
	Pigment	1.567
	Superplasticizer + additives	2.037
4 Foam concrete with sand aggregate enriched with glass and steel fibers	Portland cement CEM I 52.5 R	26.725
	Sand 0–2 mm	61.824
	Glass fibres 12 mm	1.58
	Steel fibres 25 mm	0.677
	Superplasticizer + additives	1.242

Table 1. *Cont.*

Sample	Ingredient	% wt
5 Concrete with mafic agregate enriched with polypropylene and steel fibres	Portland cement CEM I 52.5 R	4.740
	High alumina cement 40	11.037
	Mafic fine aggregate 0.75–3 mm	32.298
	Mafic aggregate 2–8 mm	40.355
	Polypropylene fibres 6 and 20 mm	0.025
	Steel fibres 12 mm	0.677
	Pigment	0.406
	Superplasticizer + additives	0.304

3.2. *Research Methodology*

3.2.1. Thermal Properties

There are many methods for testing the thermal properties of concrete materials [31–33]. In this publication, studies were carried out with the use of a thermal imaging camera NEC ThermoGear G100, which enabled the recording and visualization of temperature distribution on object surfaces (mapping thermal images of the objects). This system works on the principle of processing the infrared radiation emitted or reflected by these objects into an electrical signal and then into the image viewed on the screen as a so-called thermogram. Thermal imaging enables the detection of many properties of plastics in a way that no other technology provides. Samples of 10 cm × 10 cm were cut out of the tested concrete. Then, the samples were placed in a laboratory dryer and heated to 160 °C for 2 h. Next, the samples were taken out, immediately placed on a pedestal (Figure 2), and the temperatures were measured based on the cooling time. These tests were carried out with the use of a thermal imaging camera that was placed 70 cm from the tested sample.

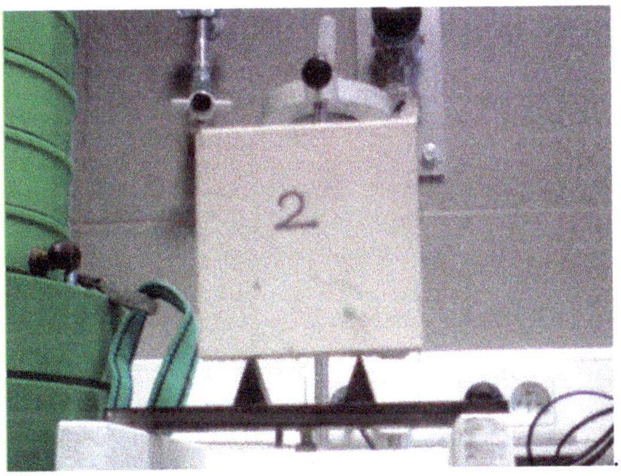

Figure 2. Sample during the thermal imaging test.

3.2.2. Differential Thermal Analysis

Concrete plastics and gas emissions were examined using an STA 449 F3 Jupiter Thermal Analyzer (Netzsch) and a coupled quadrupole mass spectrometer TA-QMS Coupling (Netzsch). The measurements were carried out in alumina crucibles at a heating rate of 10 °C/min in a temperature range of 30–600 °C under an air and argon atmosphere with a constant flow of 20 mL/min. This research focused on the analysis of concrete materials 2, 3, and 5 and the influence of additives modifying the strength properties in terms of gas emissions.

3.2.3. Mechanical Pproperties

Samples for mechanical strength testing were prepared in accordance with PN-EN 206 + A1:2016–12 [34], and bending strength measurement was performed after 1, 7, and 28 days of maturation. These measurements used bars with rectangular cross-sections and dimensions of 40 × 40 × 160 mm^3. The mechanical strength of the tested fireplace materials was determined using the three-point bending method. The specimen was placed in a special holder in a testing machine (Zwick/Roell 2.5, Ulm (Germany)) (Figure 3). A measurement was then performed three times.

Figure 3. Measurement of the bending strength via three-point bending.

3.2.4. Microstructure of the Fireplace Concrete

Samples of the delivered concrete materials were taken at random and analyzed on a scanning microscope equipped with an energy dispersive spectroscopy(EDS) analyzer (Ultra High Resolution Scanning Electron Microscope with Field Emission Gun (FEG)–NOVA NANO SEM 200 (manufacturer FEI EUROPE COMPANY), cooperating with an EDS analyzer by EDAX). Photographs at a magnification of about 100,000 (20,000) times are shown in Figures 5–10. Under each photograph, the EDS analysis is shown on the marked points.

4. Results

4.1. Thermal Properties

To standardize the data needed to calculate, among others, thermal parameters, the mass, volume, and density of the samples were determined (Table 2).

Table 2. Physical parameters of the tested samples.

Sample	Mass (g)	Volume (cm^3)	Density (g/cm^3)
1	447.24	210.05	2.13
2	351.52	199.31	1.76
3	372.98	176.7	2.11
4	282	188.6	1.5
5	589.33	272.65	2.16

Table 3 outlines the temperatures obtained depending on the cooling time. At the beginning of the measurement, the temperature between measurements was measured every few seconds, every 30 s, and then every 1 min. Using the thermal imaging camera, it was possible to implement a program for reading the temperature values on the surfaces of the samples in different places (Figure 4). To calculate the thermal capacity, the value a is used, i.e., the temperature measured in the middle of the sample.

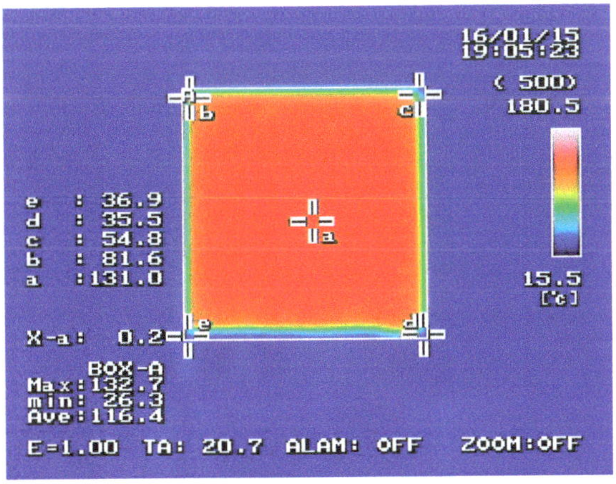

Figure 4. Example thermal image of the temperature distribution on the surface of the tested sample.

Table 3. Temperature of the tested samples depending on the cooling time.

Time (min)	Temperature (°C)				
	1	2	3	4	5
0	134.8	131.8	128.6	132.1	137.7
1	130.7	125.4	119	127.3	124.6
2	126.4	121.6	115.6	121.7	122.5
3	123.5	116	111.7	113.8	118.3
4	119.5	110.6	109.5	110.4	115.8
5	115	106.2	103.6	104.5	111.2
6	112.6	102.3	101	100.1	109.5
7	108.1	98.5	96.6	96.7	104.3
8	104.4	94.6	90	88	103.3
9	100.9	91.2	88.3	86.1	99.9
10	97.8	86.9	87.8	80.2	95.2
12	91.2	81.6	86.2	73.6	91.6
14	85.9	74.6	79.3	68.5	87.6
16	79.5	71.5	71.5	64.2	82.6
18	76.5	67.5	68.8	56.7	77.2
20	72.2	63	65.6	53.4	76.9
25	63.4	53.4	58.9	46.9	74
30	57.2	47.6	54.3	42	67.4
35	52.8	43.3	50.3	38.4	58.4
40	48.8	38.7	48.6	38	54.6
45	45.5	37.2	46.1	36.2	52.6
50	40.9	36.4	45.8	35.8	48.1

Figure 5 shows a diagram of temperature changes (value a) on the surfaces of the cooling samples.

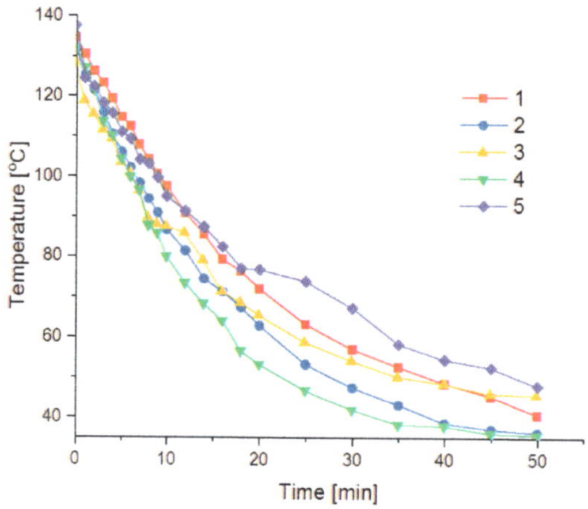

Figure 5. Combined temperature distribution diagram on the surface of cooling samples.

Calculation of Heat Capacity

The specific heat is an additive quantity, i.e., each degree of freedom in a given system contributes to the total specific heat of the system, and, as a result, the total specific heat is the sum of the different contributions [35]. On the basis of the percentage content of individual raw materials (Table 1) and their table values of c_V specific heat (Table 4) [36], the total specific heat of the examined materials was calculated:

$$c_V = \Sigma\%x \times c_{Vx} \quad (5)$$

where $\%x$ is the percentage of raw material in the sample [-], and c_{Vx} is the table-specific heat value (J/kgK).

The calculations are presented in Table 5.

Table 4. Specific heat of raw materials used to prepare samples.

Ingredient	c_V (J/kg·K)
Sand	800
Cement	504
Water	4190
Silicate aggregate	1006
Mafic aggregate	990
Carbonate aggregate	920
Refractory aggregate	1750
Glass fiber	840
Polypropylene fiber	1460
Steel fiber	440

Table 5. The total specific heat of the tested samples.

Sample	Σ c_V (J/kg·K)
1	1017.79
2	765.78
3	1170.03
4	990.58
5	1120.15

On the basis of the value of total specific heat Σc_V (Table 5), the mass of the tested samples m (Table 2) and ΔT (the difference in the temperature of a given body before and after thermal transformation; Table 6) and ΔQ, i.e., the change in the thermal energy of the tested materials, were calculated. The results of the calculations are presented in Table 6.

Table 6. The ΔT and ΔQ values of the materials tested.

Sample	ΔT (K)	ΔQ (kJ)
1	93.9	42.72
2	95.4	25.64
3	82.8	36.13
4	96.3	26.9
5	89.6	59.11

Another value that characterizes the thermal properties of materials is volumetric thermal capacity b, which describes the ability of amaterial to accumulate heat. The calculated values of the volumetric thermal capacity are shown in Table 7.

Table 7. Volume thermal capacity of the tested materials.

Sample	b (kJ/m^3·K)	b_{max} (MJ/m^3)
1	2167.89	203.57
2	1347.77	128.50
3	2468.76	204.43
4	1485.87	143.01
5	2419.52	216.83

When describing the thermal properties of fireplace materials, it is also important to know the thermal power of the tested materials P, i.e., the parameter determining the transfer of heat from the body to the system. The relevant calculations are presented in Table 8.

In the initial cooling phase (up to 10 min), all analyzed samples lost heat at a similar rate; then, the trend changed. The curves in Figure 5 and the temperature values in Table 3 indicate that sample no. 4 (with 60.477% weight of sand content) lost heat the fastest, while sample no. 5 (with 32.027% weight of sodium–calcium feldspar content) lost heat the slowest.

Table 8. Thermal power of the tested materials.

Sample	P (W)
1	14.24
2	8.55
3	12.04
4	8.97
5	19.70

Beyond the different compositions of the samples, this phenomenon could also be influenced by the fact that the samples differed significantly in their density. The density of sample no. 4 was 1.50 g/cm^3, while that of sample no. 5 was 2.16 g/cm^3. The values of their thermal parameters (thermal power, thermal volume capacity, and thermal energy), however, were found to be similar. The highest values were recorded for sample no. 5 consisting of magnesium silicates and sodium–calcium feldspar. The weakest thermal parameters were found for sample no. 4, as well as sample no. 2, which consisted mainly of cement, sand, and refractory aggregates. This is consistent with the literature data.

For example, materials with the highest specific density always have the highest thermal capacity [37]. The thermal capacity of metals with a density of 7000–9000 kg/m^3 is 1,500,000–3,500,000 J/(m^{3}K). Casting rocks used as aggregates in concretes have an even smaller thermal capacity than metals (granite is about 1,800,000 J/(m^{3}K)). Brick and sand have an even lower volumetric heat capacity at about 1,200,000 J/(m^{3}K). For normal and refractory concrete with a density of about 2400 kg/m^3, the volume heat capacity is about 2,770,000 J/(m^{3}K) [38,39].

4.2. Differential Thermal Analysis

The thermal curve differential thermal analysis/thermal gravimetric analysis (DTA/TG), differential scanning calorimetry (DSC) analysis, and total ion current (TIC) gas emission analysis are presented in Figures 6–8. Materials containing organic fibers were selected for these analyses (Table 1). Based on the performed tests, it can be concluded that the emission of the detected gases is related to the gases supplied to the sample during the sample's heating, i.e., O_2(32) and N_2(28), as well as Ar(39), introduced as an inert gas to protect the microbalance. The emission of water was observed in all of the samples, and the amount of water changed successively during the heating of the samples. The highest amount of water was observed at around 140 °C, and the lowest amount, related to the dehydroxylation of the cement products, was observed at 480 °C. In the tested samples, weight loss in the temperature range of 30–600 °C was also observed. The complex endothermic effect observed in the temperature range from about 50 to about 350 °C is related to the dehydration of silicates of type C–S–H, hydrated calcium aluminates and aluminosulphates, and the decomposition of gypsum. Another endothermic effect in the temperature range from about 400 to about 410 °C is attributed to C_2SH decomposition ($2CaO{\cdot}SiO_2{\cdot}H_2O$). The endothermic effect in the temperature range from about 490 to about 510 °C is attributed to $Ca(OH)_2$dehydroxylation. The total weight loss of the samples varied and was 10.2% for sample 2, almost 7% for sample 3, and almost 9% for sample 5. This is due to differences in the composition of the individual materials.

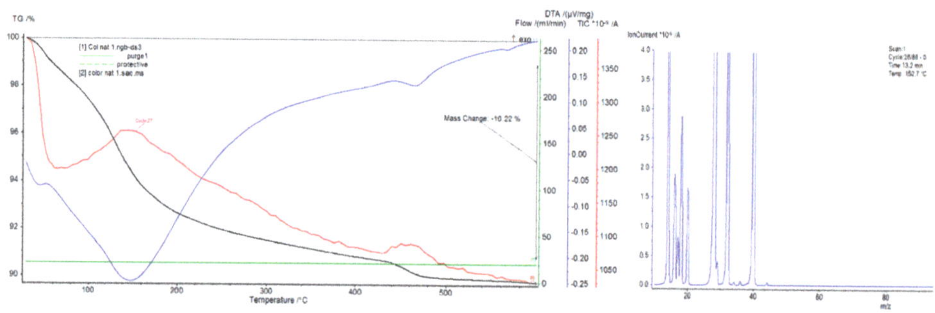

Figure 6. Thermal analysis differential thermal analysis/differential scanning calorimetry (DTA/DSC) of sample 2.

The observed mass loss is different in particular types of concretes. These differences most likely result from different aging times, i.e., the time from concrete preparation to measurement (different humidity values), and different amounts and types of aggregates. In general, it can be concluded that the addition of modifiers causes an increase in the temperature of the cycle in which the DTA/DSC analysis takes place and causes an increase in the weight loss. Organic additives likely facilitate the emission of water at about 140 °C.

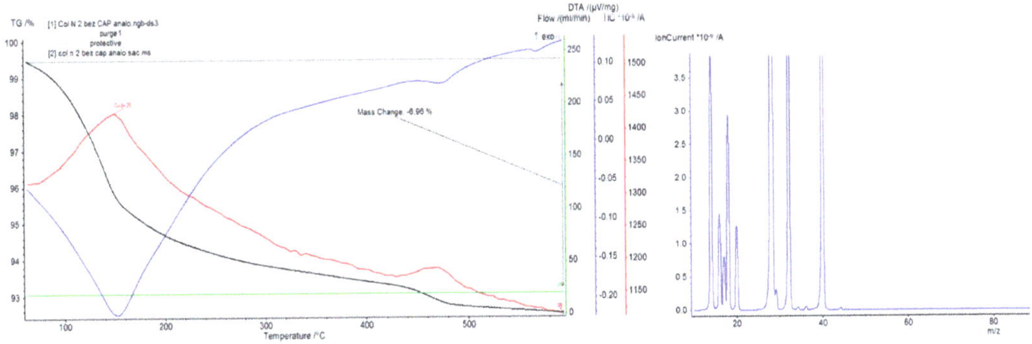

Figure 7. Thermal analysis DTA/DSC of sample 3.

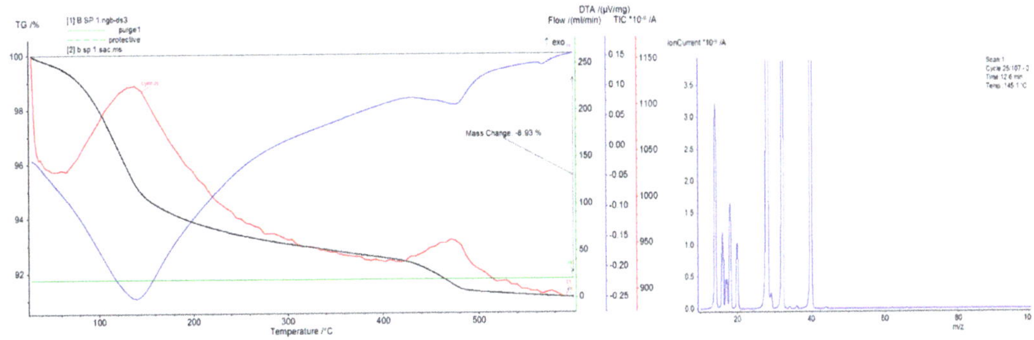

Figure 8. Thermal analysis DTA/DSC of sample 5.

The tests showed that the specific heat of the concrete material strongly depends on the amount of free water that is released in the temperature range of 50–90 °C. Values of the specific heat measured by this method significantly drop at higher temperatures, which is likely related to the absorption of heat during the dehydration reactions of concrete materials.

4.3. Mechanical Properties

Table 9 shows the maximum values of the mechanical strength under bending for the tested fireplace materials after the different aging periods.

Samples 1 and 3 showed the highest mechanical strength at more than twice that of the other tested materials. Their bending strength after 28 days of aging was 12.3 and 13.2 MPa, respectively. The main additives in both samples were cement and aggregate. As a reinforcing additive, sample 1 included glass fibers, while sample 3 used glass and polypropylene fibers. Samples 2, 4, and 5 included a minor addition of steel fibers, which are also used as a strengthening additive to maintain the concrete's structure during operations at high temperatures, causing significant stress in the system. Sample 5, despite the addition of reinforcing fibers, was characterized by low mechanical strength (4.8 MPa) due to the addition of a mafic aggregate containing more than 2% free biotite. Increased mica content causes a significant and constant decrease in compressive strength [40], which is especially important in the case of the fine aggregates (e.g., mafic sand) [41].

Table 9. Values of the mechanical strength under bending of the tested fireplace materials.

Sample Symbol	Maturation Time (days)		
	1	7	28
	Mechanical (MPa)		
1	8.5	9.9	12.3
2	3.5	4.3	5
3	7.4	10	13.2
4	2.9	4.1	4.7
5	1.8	3	4.8

4.4. Microstructure of Fireplace Concrete

Scanning electron microscope(SEM) microphotographs with the energy dispersive spectroscopy (EDS)system of the tested samples are shown in Figures 9–13. Based on the conducted research, it is possible to determine the compact microstructure of all of the concrete materials, except for sample 4, which is characterized by the presence of large spherical pores (Figure 12). A compact structure was also present in the concrete materials containing glass fibers (Figure 9) or polypropylene fibers (Figures 11–13). Generally, a homogeneous microstructure was found in the analyzed concrete materials; however, in the case of sample 5, the uneven distribution of fibers was noted (Figure 13). There were also visible pores (except for sample 4), which may have resulted from inappropriate selection of the aggregate graining (filler) of the concrete material or deaeration. This is an important factor because the presence of pores lowers the density and thermal properties of the concrete material. EDS measurements confirmed the presence of cement-derived aluminates and aluminosilicates, as well as the presence of fibers, aggregate grains, and pigments.

Quartz grains and zirconium were present in the microstructure of sample 1 (Figure 9). Sample 3 (Figure 11) contained increased amounts of calcium aluminosilicate and strengthening organic fibers. Sample 4 (Figure 12) is a foamed concrete featuring the addition of calcium carbonate. In the composition of sample 2 (Figure 10), increased amounts of magnesium, zirconium, and calcium can be seen. In particular, in micro-area b, dolomite crystals formed with the addition of titanium white reinforced with organic fibers were found. Concrete no. 5 (Figure 13) is a typical concrete reinforced with organic fibers and aggregate, where the presence of iron comes from the minerals of the aggregate.

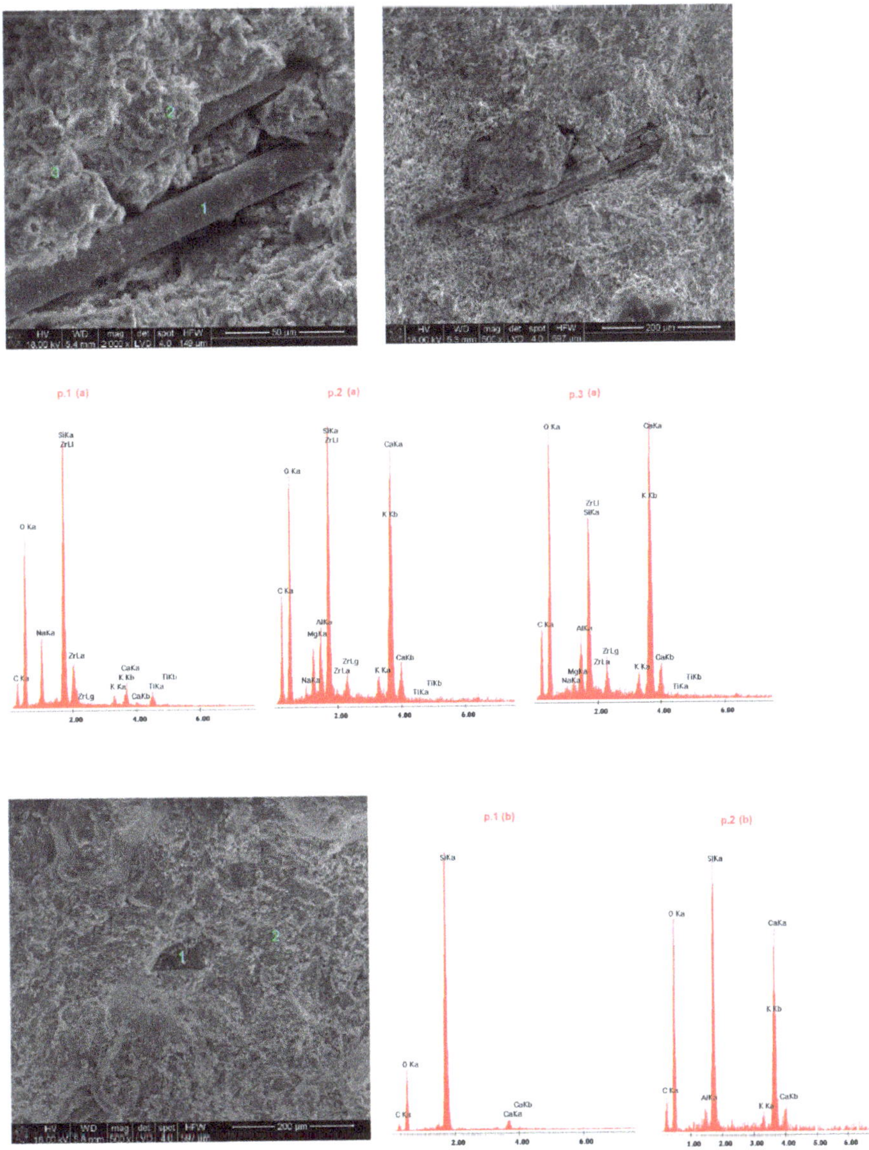

Figure 9. SEM microphotograph and energy dispersive spectroscopy (EDS) analysis of sample 1.

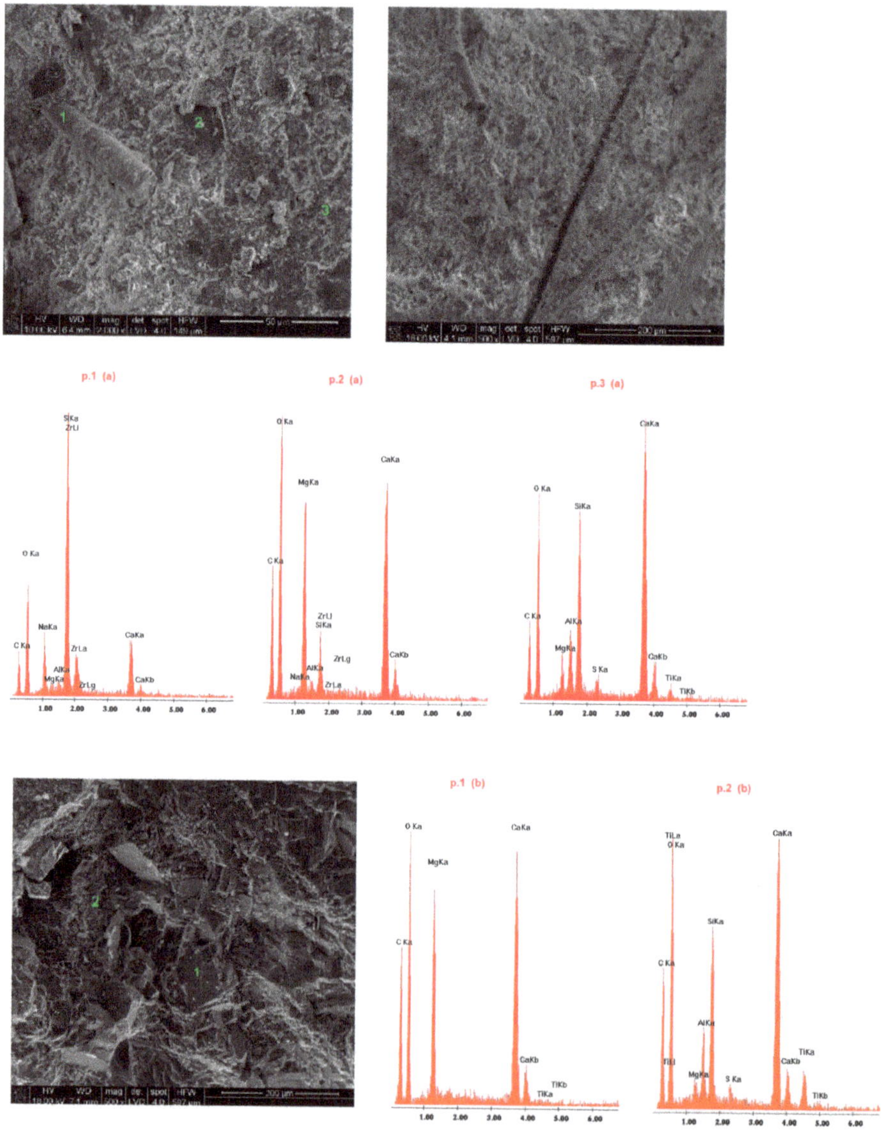

Figure 10. SEM microphotograph and EDS analysis of sample 2.

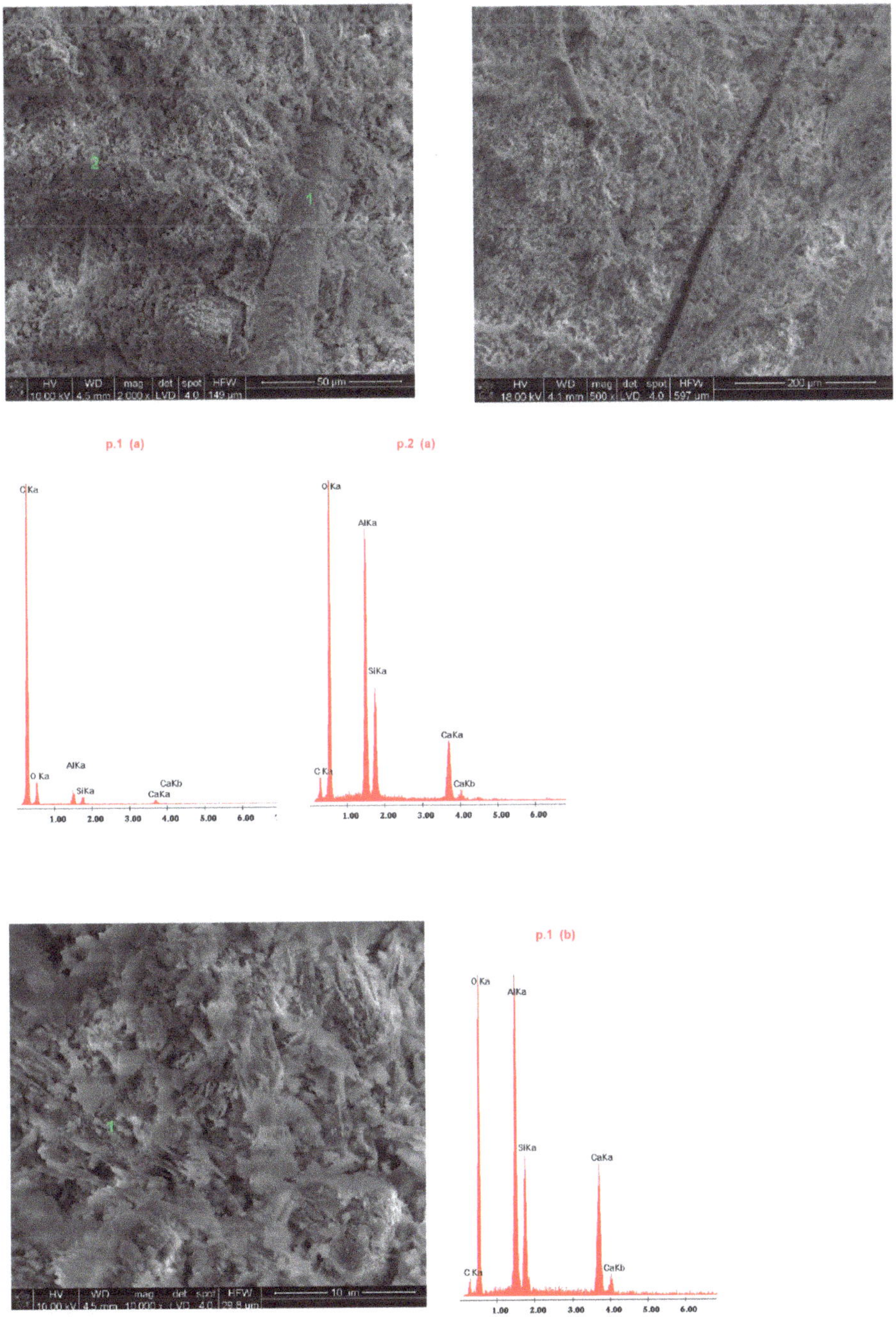

Figure 11. SEM microphotograph and EDS analysis of sample 3.

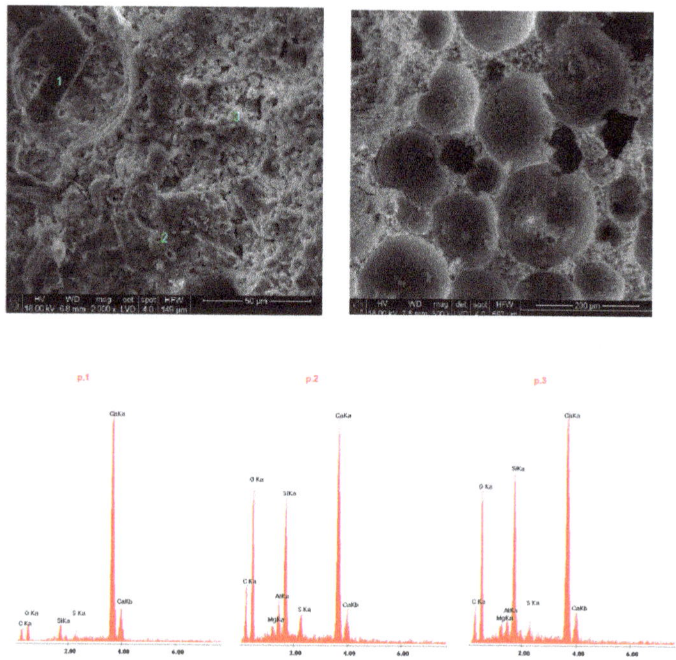

Figure 12. SEM microphotograph and EDS analysis of sample 4.

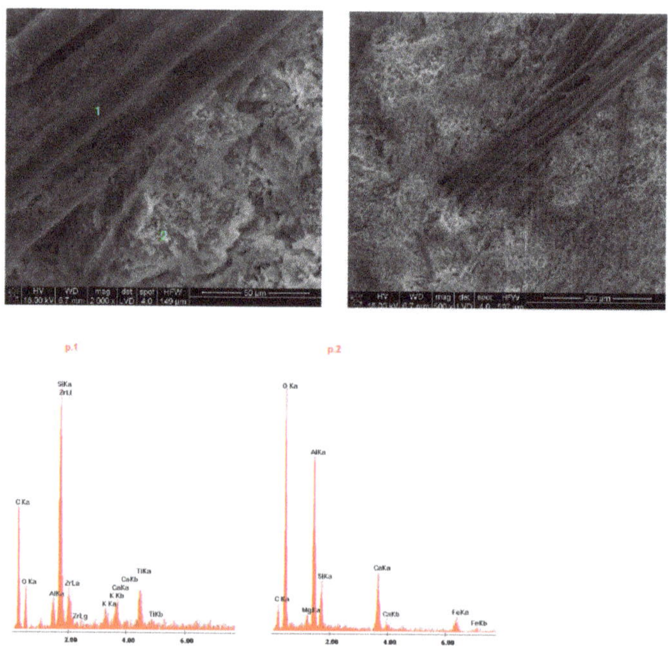

Figure 13. SEM microphotograph and EDS analysis of sample 5.

From an application perspective, concrete systems can be reduced to a mixture of three components: cement, fine aggregate, and water. Cement is a composite system in which aggregate grains are surrounded by a hardened cement slurry. Concrete formation is determined by the cement's hydration reactions. Cement is a mixture of solid phases and mainly contains various types of aluminates and calcium aluminosilicates: allite C_3S-($3CaO{\cdot}SiO_2$), bellite C_2S-($2CaO{\cdot}SiO_2$), tricalcium aluminateC_3A-($3CaO{\cdot}Al_2O_3$), brownmilleriteC_4AF-($4CaO{\cdot}Al_2O_3{\cdot}Fe_2O_3$), calcium oxide (free)-CaO, and gypsum-($CaSO_4{\cdot}0.5H_2O$).

Cement hydration is a multi-stage process. However, it can be assumed that the main phases of hydration are as follows:

$$\begin{array}{ll} \text{allite:} & 3CaO{\cdot}SiO_2 + 3H_2O \rightarrow C_{1.5}SH_{1.5} + 1.5{\cdot}H_2O \\ \text{bellite:} & 2CaO{\cdot}SiO_2 + 2H_2O \rightarrow C_{1.5}SH_{1.5} + 0.5\,H_2O \\ \text{gypsum:} & CaSO_4{\cdot}0.5H_2O + 1.5H_2O \rightarrow CaSO_4{\cdot}2H_2O \\ \text{calcium oxide:} & CaO + H_2O \rightarrow Ca(OH)_2 \end{array} \tag{6}$$

The hydration of tricalcium aluminate, however, is more complicated. There are three basic stages in this process. In stage I, under the initially high concentration of gypsum, ettringiteis formed according to the following reaction:

$$3CaO{\cdot}Al_2O_3 + 3CaSO_4{\cdot}2H_2O + 26H_2O \rightarrow 3CaO{\cdot}Al_2O_3{\cdot}3CaSO_4{\cdot}32H_2O. \tag{7}$$

In stage II, when gypsum is lacking, a transitional phase is formed:

$$3CaO{\cdot}Al_2O_3{\cdot}Ca(OH)_2{\cdot}12H_2O. \tag{8}$$

In stage III, the second stage phase reacts with ettringite, giving the low-sulfate form of aluminum sulphate according to the following reaction:

$$3CaO{\cdot}Al_2O_3{\cdot}3CaSO_4{\cdot}32H_2O + 3CaO{\cdot}Al_2O_3{\cdot}Ca(OH)_2{\cdot}12H_2O \rightarrow 3CaO{\cdot}Al_2O_3{\cdot}CaSO_4{\cdot}12H_2O + 2Ca(OH)_2 + 20H_2O. \tag{9}$$

The hydration of braunmillerite is similar to that of tricalcium aluminate. In the presence of gypsum and water, the following structures are formed:

$$4CaO{\cdot}Al_2O_3{\cdot}Fe_2O_3 + CaSO_4{\cdot}2H_2O + Ca(OH)_2 + aq \rightarrow 3CaO{\cdot}(Al_2O_3,\,Fe_2O_3){\cdot}3CaSO_4{\cdot}32H_2O. \tag{10}$$

After braunmillerite is exhausted, the structure becomes $3CaO{\cdot}(Al_2O_3,Fe_2O_3){\cdot}CaSO_4{\cdot}12H_2O$ [42,43].

The hardened cement slurry is largely dominated by two phases: the CSH gel phase and the portlandite ($Ca(OH)_2$) phase. There are also additional phases resulting from the hydration of tricalcium aluminate and braunmillerite. It should be noted that these phases contain large amounts of crystalline water, and this water is the main destructive factor for cement mortars at high temperatures, which accompanies the use of fireplaces and heating stoves. However, the content of elements such as magnesium, aluminum, calcium, and silicon significantly affects the thermal properties of the concrete [44].

5. Discussion

The addition of fibers to concrete improves the mechanical properties of mortars and may improve heat flow in concrete materials. At elevated temperatures, polypropylene melts and is absorbed by the surrounding cement matrix, creating a network of channels through which moisture and the resulting water vapor can flow [45]. Hardened cement mortar can also be treated as a fibrous composite in which the matrix is the hardened cement mortar. In the case of such a composite, the addition of a small amount of fibers, even with an elasticity modulus lower than that of the matrix, significantly improves the mechanical properties of the mortar. This is because during the hydration process of the paste components, a number of reactionsoccur, leading to the formation of mechanical stresses. The resulting stresses, which are related to internal contractions, are leveled until the structure stiffens. However, in the initial stage of hardening, the hardening paste has too low a mechanical strength to compensate for the stresses accompanying the hydration process (as confirmed by the results presented in Table 9).

The results of the mechanical strength after 28 days confirm this explanation. The highest bending strength was found for the samples with the highest amount of glass and polypropylene fibers, i.e., sample 1 (0.266 wt % of glass fibers) and sample 3 (0.209 wt % of glass fibers and 0.02 wt % of polypropylene fibers). The addition of fewer of these fibers or the addition of steel fibers (even in combination with glass fibers) caused more than a twofold decrease in mechanical strength in the described case, from 13.2 MPa for sample 3 to 4.7 MPa for sample 4.

When the proportion of organic fibers is so large that the fibers form a self-connected network of channels, it is known as a parallel transport model. That is, the transport of liquids and gases at temperatures above 400 °C takes place both through the network formed after melting the fibers and through the pores and cracks existing in the material. When there are fewer fibers, transport can be described by the so-called series-parallel model, where the incompletely connected network of channels is supplemented by the existing defects in the material [46,47].

The introduction of short polypropylene fibers, in addition to preventing detonation splashes during rapid heating or contact with fire, also has a very positive effect on the mechanical properties of mortars. Hardened cement mortar can be treated as a fibrous composite in which the matrix is hardened cement paste. In the case of such a composite, the addition of a small amount of fibers, even with a modulus of elasticity lower than that of the matrix, results in a significant improvement in the composite's mechanical properties. This is because several stress reactions take place during the hydration of the cement slurry components. In the subsequent period, the stresses arising from internal contraction are leveled until the structure is stiffened. However, in the initial stage of hardening, the hardening cement slurry still has too low a mechanical strength to compensate for the stresses accompanying the hydration. This yields the formation of a micro-fracture network, leading to material defects and a loss of continuity at the microscale. The addition of fibers causes bridging of the resulting cracks, which reduces their size and improves the material's resistance to brittle fractures [48].

However, the addition of short fibers to the cement matrix may regulate more than the matrix's hydration and hardening properties. Through the addition of fibers with an elasticity modulus higher than the matrix modulus, the strength properties of the material can be directly modulated. The addition of such fibers ensures that the material does not disintegrate when the cement matrix breaks, allowing the matrix to still transfer some loads.

The reactions taking place in the cement during heating can be divided into five stages:

I. At a temperature of about 100 °C, mechanically (capillary) bound moisture evaporates.
II. At a temperature of about 200 °C, physicochemically bound water is released due to adsorption.
III. At 400 °C, the release of chemically bound water in hydrated aluminosilicate compounds begins. This does not yet entail a significant decrease in mechanical strength.
IV. At a temperature of about 530 °C, due to the decomposition of calcium hydroxide, mechanical strength decreases rapidly.
V. At a temperature of about 1000 °C, the water is completely removed, and the structure disintegrates into a powder. The material sinters at even higher temperatures.

The effect of high temperature, i.e., that above 100 °C, on hardened cement material is very specific and causes the rapid desorption of moisture from the outer layers. The resulting water vapor flows towards the colder inner layers, where it is reabsorbed. As the temperature increases, the thickness of the heated layer increases. When the moisture-saturated layer does not move quickly enough, it is overtaken by the wandering temperature front, which leads to the evaporation of water at the front border and an increase in internal pressure. This creates tensile forces perpendicular to the temperature front, which in turn leads to rapid removal of the surface layers in the form of microcracks and spatters. This phenomenon mayoccur cyclically, destroying increasingly deeper layers of the material.

To counteract this phenomenon, it is necessary to improve the heat flow, which will entail a reduction in the pressure inside the heated element. The best way to do this is to create an interconnected network of channels in the material through which moisture and the resulting water vapor can flow, or toimprove the material's thermal conductivity.

This can be achieved via the introduction of a fibrous material that melts at an elevated temperature and is easily absorbed by the surrounding cement matrix. Polypropylene is perfect for this purpose. This material melts at 170 °C, which is below the temperature at which water vapor is rapidly released and does not damage the concrete matrix. At low temperatures (<100 °C), all forms of water in concrete (capillary, adsorptive, and crystalline) become very good heat accumulators (the heat of water evaporation is approximately 80 kcal/mol). Exterior cladding concrete shapes used in furnaces are a good choice for this purpose.

Another type of thermal accumulation involves using the excess heat energy generated inside the fireplace or the stove to heat rooms over a longer timeframe. Such furnaces are made of natural soapstone (a greenish-gray rock containing mainly magnesium aluminosilicates), which has a very good heat accumulation capacity. In these furnaces, due to the accumulation capacity of soapstone and various technical solutions (e.g., the multiple circulation and swirling of exhaust gases), the temperature in the furnace can reach about 1200 °C (normally about 600 °C), which allows the so-called afterburning of exhaust gases (soot combustion). This results in high efficiency of the furnace and, above all, has a positive impact on the environment.

The temperatures in storage furnaces can reach 400 °C and higher. Ordinary concrete is rarely used as a material for furnaces due to its low resistance to such temperatures (the safe limit temperature is 450 °C). This is why, as a rule, ceramics that are much more resistant to high temperatures are used for this application, most often fireclay. In our case, sample 5 showed the highest thermal power (Table 8) (19.70 W), whereas sample 2 showed the lowestthermal power (8.55 W). In addition to the significant effect of density (sample 5—2.16 g/cm^3; sample 2—1.76 g/m^3), composition itself plays an important role. Sample 2 consisted mainly of cement, sand, and refractory aggregate, whereas sample 5 consisted mainly of magnesium silicates and sodium–calcium feldspar. Therefore, it is worth analyzing, in each case, the physicochemical processes thatoccur during the formation of hydration materials and their destruction under the influence of high temperatures.

6. Conclusions

Concrete elements can be successfully used in the construction of domestic fireplaces, thus taking advantage of the good heat accumulation properties of these elements. Both thermal and strength properties can be modified significantly via the addition of various aggregates as well as propylene and steel fibers. This modification consists mainly in changing the mechanisms of the heat flow and the decomposition products of concrete elements during heating—mainly water, although the density of the materials also has a significant impact.

From the perspective of thermal energy use, our study showed the clear advantage of concrete materials containing magnesium and sodium–calcium aggregates (feldspars). The highest values were recorded for sample 5 (which contained over 70% of the listed components). In turn, from the perspective of mechanical properties, samples 1 and 3 showed the highest mechanical strength and were reinforced with glass and polypropylene fibers, respectively. The results, however, depend on the applied temperature range. Therefore, in the case of fireplaces, it is suggested to use different multi-layer systems of concrete materials to enable the long-term heating of rooms via heat accumulation.

Author Contributions: Conceptualization, A.S., J.M.-P. and P.I.; methodology, J.M.-P., Ł.W. and T.G.; software, A.S.; validation, A.S., P.I. and J.M.-P.; formal analysis, J.M.-P., Ł.W. and T.G.; investigation, A.S. and J.M.-P.; writing—original draft preparation, A.S.; writing—review and editing, J.M.-P.; visualization, A.S.; supervision, M.K. and P.I. All authors have read and agreed to the published version of the manuscript.

Funding: This research received no external funding.

Institutional Review Board Statement: Not applicable.

Informed Consent Statement: Not applicable.

Data Availability Statement: The data presented in this study are available on request from the corresponding author. The data are not publicly available due to the agreement with Northstar Poland.

Acknowledgments: We would like to thank Northstar Poland for its cooperation and for providing research materials.

Conflicts of Interest: The authors declare no conflict of interest.

References

1. Rao, V.V.; Parameshwaran, R.; Ram, V.V. PCM-mortar based construction materials for energy effcient buildings: A review on research trends. *Energy Build.* **2018**, *158*, 95–122. [CrossRef]
2. Sargam, Y.; Wang, K.; Alleman, J.E. Effects of Modern Concrete Materials on Thermal Conductivity. *Am. Soc. Civil Eng.* **2020**. [CrossRef]
3. Kakae, N.; Miyamoto, K.; Momma, T.; Sawada, S.; Kumagai, H.; Ohga, Y.; Hirai, H.; Abiru, T. Physical and thermal properties of concrete subjected to hightemperature. *J. Adv. Concr. Technol.* **2017**, *15*, 90–212. [CrossRef]
4. An, J.; Kim, S.S.; Nam, B.H.; Durham, S.A. Effect of Aggregate Mineralogy and Concrete Microstructure on Thermal Expansion and Strength Properties of Concrete. *Appl. Sci.* **2017**, *7*, 1307. [CrossRef]
5. Chan, Y.N.; Lou, X.; Sun, W. Compressive strength and pore structure of high-performance concrete after exposure to high temperature up to 800 °C. *Cem. Concr. Res.* **2000**, *30*, 247–251. [CrossRef]
6. Chan, Y.N.; Peng, G.F.; Anson, M. Residual strength and pore structure of high-strength concrete after exposure to high temperatures. *Cem. Concr. Compos.* **1999**, *34*, 1065–1069. [CrossRef]
7. Yousefi, A.; Tang, W.; Khavarian, M.; Fang, C.; Wang, S. Thermal and Mechanical Properties of Cement Mortar Composite Containing Recycled Expanded Glass Aggregate and Nano Titanium Dioxide. *Appl. Sci.* **2020**, *10*, 2246. [CrossRef]
8. Krishnamoorthy, R.R.; Zujip, J.A. Thermal conductivity and microstructure of concrete using recycle glass as a fine aggregate replacement. *Int. J. Emerg. Technol. Adv. Eng.* **2013**, *3*, 463–471.
9. Huhta, A.; Karki, A.; Hanski, E. A new method for testing thermal shock resistance properties of soapstone—Effects of microstructures and mineralogical variables. *Bull. Geol. Soc. Finl.* **2016**, *88*, 21–46. [CrossRef]
10. Garbalińska, H.; Bochenek, M. Thermal insulation and thermal accumulation of selected wall materials. In *Technical Transactions Architecture*; Publishing Cracow University of Technology: Krakow, Poland, 2011; Volume 11, pp. 89–96.
11. Szargut, J. *TermodynamikaTechniczna*; WydawnictwoPolitechnikiŚląskiej: Gliwice, Poland, 2000.
12. Cavalline, T.L.; Castrodale, R.W.; Freeman, C.; Wall, J. Impact of lightweight aggregate on concrete thermal properties. *ACI Mater. J.* **2017**, *114*, 945–956. [CrossRef]
13. Roberz, F.; Loonen, R.C.G.M.; Hoes, P.; Hensen, J.L.M. Ultra-lightweight concrete: Energy and Comfort performance evaluation in relation to buildings with low and high thermal mass. *Energy Build.* **2017**, *138*, 432–442. [CrossRef]
14. Marie, I. Thermal conductivity of hybrid recycled aggregate—Rubberized concrete. *Constr. Build. Mater.* **2017**, *133*, 516–524. [CrossRef]
15. Bravo, M.; Brito, J.D.; Evangelista, L. 2017 Thermal Performance of Concrete with Recycled Aggregates from CDW Plants. *Appl. Sci.* **2017**, *7*, 740. [CrossRef]
16. Zhu, L.; Dai, J.; Bai, G.; Zhang, F. Study on thermal properties of recycled aggregate concrete and recycled concrete blocks. *Constr. Build. Mater. Constr.* **2015**, *94*, 620–628. [CrossRef]
17. Benazzouk, A.; Douzane, O.; Mezreb, K.; Laidoudi, B.; Quéneudec, M. Thermal conductivity of cement composites containing rubber waste particles: Experimental study and modelling. *Constr. Build. Mater.* **2008**, *22*, 573–579. [CrossRef]
18. Chung, S.Y.; AbdElrahman, M.; Sikora, P.; Rucinska, T.; Horszczaruk, E.; Stephan, D. Evaluation of the effects of crushed and expanded waste glass aggregates on the material properties of lightweight concrete using image-based approaches. *Materials* **2017**, *10*, 1354. [CrossRef] [PubMed]
19. Nagy, B.; Nehme, S.G.; Szagri, D. Thermal Properties and Modeling of Fiber Reinforced Concrete. *Energy Procedia* **2015**, *78*, 2742–2747. [CrossRef]
20. Abyaneh, S.D.; Wong, H.S.; Buenfeld, N.R. Computational investigation of capillary absorption in concrete using a three-dimensional mesoscale approach. *Comput. Mater. Sci.* **2014**, *87*, 54–64. [CrossRef]

21. Liu, Q.F.; Feng, G.L.; Xia, J.; Yang, J.; Li, L. Ionic transport features in concrete composites containing various shaped aggregates: A numerical study. *Compos.Struct.* **2018**, *183*, 371–380. [CrossRef]
22. Mao, L.X.; Hu, Z.; Xia, J.; Feng, G.L.; Azim, I.; Yang, J.; Liu, Q.F. Multi-phase modelling of electrochemical rehabilitation for ASR and chloride affected concrete composites. *Compos. Struct.* **2019**, *207*, 176–189. [CrossRef]
23. Abyaneh, S.D.; Wong, H.S.; Buenfeld, N.R. Modelling the diffusivity of mortar and concrete using a three-dimensional mesostructure with several aggregate shapes. *Comput. Mater. Sci.* **2013**, *78*, 63–73. [CrossRef]
24. Zheng, J.J.; Zhou, X.Z.; Wu, Y.W.; Xin, J.Y. A numerical method for the chloride diffusivity in concrete with aggregate shape effect. *Constr. Build. Mater.* **2012**, *31*, 151–156. [CrossRef]
25. Zergadło, B.; Halicka, A. Analizawłaściwościcieplnychbetonu z kruszywem z odpadówceramikisanitarnej. *Bud. Archit.* **2011**, *9*, 39–49.
26. Jastrzębski, P.; Saługa, P. Innowacyjnemetodygromadzeniaciepła. *Zesz. Nauk. Inst. Gospod. Surowcami Miner. Energią PAN* **2018**, *105*, 225–232.
27. Kata, D. *SpiekanyAzotekGlinu o WysokimPrzewodnictwieCieplnym*; Wydawnictwa AGH: Kraków, Poland, 2015.
28. Avghad, S.N.; Keche, A.J.; Kousal, A. Thermal Energy Storage: A Review. *IOSR- J. Mech. Civ. Eng.* **2016**, *13*, 72–77.
29. Gencel, O.; Koksal, F.; Şahin, M.; Durgun, M.Y.; Lobland, H.E.H.; Brostow, W. Modeling of thermal conductivity of concrete with vermiculite using by artificial neural networks approaches. *Exp. Heat Transfer* **2013**, *26*, 360–383. [CrossRef]
30. Kurpinska, M.; Kułak, L. Predicting performance of lightweight concrete with granulated expanded Glass and Ash aggregate by means of using Artificial Neural Networks. *Materials* **2019**, *12*, 2002. [CrossRef] [PubMed]
31. Carlson, J.D.; Bhardwaj, R.; Phelan, P.E.; Kaloush, K.E.; Golden, J.S. Determining thermal conductivity of paving materials using cylindrical sample geometry. *J. Mater. Civ. Eng.* **2010**, *22*, 186–195. [CrossRef]
32. Devpura, A.; Phelan, P.E.; Prasher, R.S. Size effects on the thermal conductivity of polymers laden with highly conductive filler particles. *Microscale Themophysics Eng.* **2010**, *5*, 177–189. [CrossRef]
33. Patterson, N.; Yoon, S.; Macphee, D.E.; Imbabi, M.S. The cuboid method for measurement of thermal properties of cement-based materials using the guarded heat flow meter. *Constr. Build. Mater.* 2018. [CrossRef]
34. *PN-EN 206 + A1:2016-12; Concrete—Equirements, Properties, Production and Compliance*; Polish Committee for Standardization: Warsaw, Poland, 2016.
35. Grimvall, G. *Thermophysical Properties of Materials*; North-Holland: Amsterdam, The Netherlands, 1986.
36. Bailey, D.G. Heat Capacity of Minerals: A hands-On Introduction to Chemical Thermodynamics. In *Teaching Mineralogy, Mineralogical Society of America*; Mineralogical Society of America: Chantilly, VA, USA, 1997.
37. Stempkowska, A.; Izak, P.; Mastalska-Popławska, J.; Staszewska, M. The analysis ofthermal properties of selected rock materials by thermovision methods. *J. Pol. Miner. Eng. Soc.* **2018**, *20*, 337–344.
38. Stempkowska, A.; Gawenda, T.; Naziemiec, Z.; Adam Ostrowski, K.; Saramak, D.; Surowiak, A. Impact of the Geometrical Parameters of Dolomite Coarse Aggregate on the Thermal and Mechanic Properties of Preplaced Aggregate Concrete. *Materials* **2020**, *186*, 4358. [CrossRef] [PubMed]
39. Rutkowska, G. Betonzwykłyjakomateriał do akumulacjiciepła w budownictwie. *Architektura* **2011**, *10*, 33–41.
40. Mshali, M.R.; Visser, A.T. Influence of mica on unconfined compressive strength of a cement-treated weathered granite gravel. *J. South Afr. Inst. Civ. Eng.* **2012**, *54*, 71–77.
41. Tugrul, A.; Hasdemir, S.; Yılmaz, M. The Effect of Feldspar, Mica and Clay Minerals on Compressive Strength of Mortar. In *Engineering Geology for Society and Territory*; Springer: Cham, Switzerland, 2015; Volume 5. [CrossRef]
42. Taylor, H.F.W. *The Chemistry of Cements*; Academic Press Limited: London, UK, 1990.
43. Kurdowski, W. *ChemiaCementuiBetonu*; Wydawnictwa AGH: Kraków, Poland, 2010.
44. Gruner, M. *KorozjaiOchronaBetonu*; Arkady: Warszawa, Poland, 1983.
45. Królikowski, W. *TworzywaWzmocnioneiWłóknaWzmacniające*; WydawnictwoNaukowo—Techniczne: Warszawa, Polland, 1988.
46. Kalifa, P.; Chene, G.; Galle, C. High-temperature behavior of HPC with polypropylene fibers. From spalling to microstructure. *Cem. Concr. Res.* **2001**, *31*, 1487–1499. [CrossRef]
47. Kalifa, P.; Menneteau, F.D.; Quenard, D. Spalling and pore pressure in HPC at high temperatures. *Cem. Concr. Res.* **2000**, *30*, 1915–1927. [CrossRef]
48. Vitek, J.L. Fire resistance of concrete tunnel linings, Experimental research. In Proceedings of the 3rd Central European Congress on Concrete Engineering 2007—Innovative Materials and Technologies, Visegrad, Hungary, 17–18 September 2007.

materials

MDPI

Article

Effects of Amorphous Poly Alpha Olefin (APAO) and Polyphosphoric Acid (PPA) on the Rheological Properties, Compatibility and Stability of Asphalt Binder

Xiaoguang Pei and Weiyu Fan *

State Key Laboratory of Heavy Oil Processing, China University of Petroleum, Qingdao 266580, China; b15030089@s.upc.edu.cn
* Correspondence: 15853256892@139.com; Tel.: +86-158-5325-6892

Abstract: High production costs and poor storage stability have become important constraints in the manufacture of modified asphalt binder. To simplify the production process and reduce the production cost, amorphous poly alpha olefin (APAO) and polyphosphoric acid (PPA) were applied to prepare highly stable modified asphalt binder. The influence of APAO/PPA on the temperature sensitivity, rheological property, storage stability, compatibility and microstructure of neat binder were studied by rotational viscosity (RV), dynamic shear rheometer (DSR), bending beam rheometer (BBR) and Fourier transform infrared (FTIR) spectroscopy. The results show that the incorporation of APAO/PPA reduced the temperature sensitivity of neat binder. The combined effect of APAO/PPA contributed to the improvement in deformation resistance, which was evidenced by the increase in failure temperature and percent recovery. However, the compound modification of APAO/PPA decreased the binder's low-temperature performance. APAO strengthened the fatigue resistance of the binder, while PPA reduced the anti-fatigue performance. Composite modified asphalt binder with superior storage stability could be prepared, which was confirmed by the desired Cole–Cole plots and fluorescence imaging. Furthermore, chemical and physical reactions occurred during the APAO/PPA modification process. Overall, 2 wt.% (weight percentage) APAO and 1.5 wt.% PPA are recommended for the production of modified asphalt binder with remarkable rheological performance and storage stability.

Keywords: composite modified asphalt binder; rheological properties; storage stability; microstructure; amorphous poly alpha olefin; polyphosphoric acid

Citation: Pei, X.; Fan, W. Effects of Amorphous Poly Alpha Olefin (APAO) and Polyphosphoric Acid (PPA) on the Rheological Properties, Compatibility and Stability of Asphalt Binder. *Materials* **2021**, *14*, 2458. https://doi.org/10.3390/ma14092458

Academic Editor: Francesco Canestrari

Received: 8 April 2021
Accepted: 7 May 2021
Published: 10 May 2021

1. Introduction

Asphalt pavement has become the main expressway pavement structure in China due to its characteristics of superior mechanical strength, low noise, driving safety and ease of mechanized construction. However, a variety of pavement stresses such as ruts, cracks, loosening and water damage have emerged due to the rapid increase in heavy traffic and the fluctuations caused by extreme climate. To resolve these issues, researchers have adopted various methods to improve the performance and longevity of asphalt pavement. Polymer modification is the most commonly used method to provide asphalt binder with a wide operating temperature range.

Depending on the polymer modifier types, thermoplastic elastomers and plastomers are the two main categories [1]. Thermoplastic elastomers, specifically styrene–butadiene–styrene (SBS), represent the most successful asphalt modifiers to date. Such modifiers can effectively enhance the resistance of asphalt binder against rutting, fatigue cracking and low-temperature cracking. However, the main drawback of these polymers is their partial miscibility with asphalt binder, which leads to poor storage stability of polymer modified binder. Plastomers, such as polyethylene (PE) and ethylene–vinyl–acetate (EVA), can significantly improve the rutting resistance of binder, but the improvement of low-temperature anti-cracking properties is limited. Its incompatibility with asphalt binder also results in severe separation between polymer and asphalt binder, which is undesirable and restricts its application [2]. Moreover, these polymers must be blended with asphalt binder by a colloid mill or high-speed shearing equipment [3–5]. Equipment investment and high shearing temperatures increase the production cost of modified asphalt binder [6]. In addition, these modified asphalt binders can only be stored for a short period of time and therefore must be used as soon as possible to avoid undesirable phase separation. Although the application of a stabilizer can improve the storage stability of the modified binder, it also leads to an increase in production cost. Therefore, high production cost and poor storage stability are significant constraints currently faced in the manufacture of modified asphalt binder. To reduce the production cost without compromising on the properties of modified asphalt binder, the development of new types of modifiers to replace colloid mills and stabilizers and allow for lower processing temperatures and simplified production processes is of significant research interest in the field.

Amorphous poly alpha olefin (APAO) is a low molecular weight amorphous plastic material, which is highly compatible with asphalt binder [7,8]. It is formed by the copolymerization of α-olefin and has the characteristics of high fluidity, low crystallinity and high randomness. Previous studies have shown that APAO can be completely dissolved in asphalt binder when the APAO content is no more than 6 wt.% [7,9]. It has also been found that APAO modified binder with high stability can be processed without shearing and stabilizer. Kong et al. [10] showed that the processing temperature of APAO modified binder can be controlled at 165 °C, which is lower than the processing temperature of 180 °C for SBS modified binder. These results demonstrate that APAO is capable of producing modified asphalt binder at a lower processing temperature without shearing and stabilizer. Regarding the impact on asphalt binder performance, numerous studies on the mechanical properties of APAO modified binders have been conducted. Wei et al. [7,9] showed that APAO enhanced the elastic property and lowered the temperature susceptibility of binder, but had adverse effects on the creep stiffness and creep rate. Yan et al. [11] and Liu et al. [12] determined that APAO improved the high-temperature performances, storage stability and aging resistance of waste tire rubber (WTR) modified binder. However, its effect on low-temperature properties remains uncertain and is highly debated. Liu et al. [13] concluded that SBS/APAO modified binder had superior high- and intermediate-temperature properties and storage stability in comparison to SBS modified binder, while its low-temperature performance was equivalent to that of SBS modified binder. You et al. [14] found that APAO enhanced the high- and low-temperature properties as well as storage stability of terminal blend rubberized binder (TB). Yan et al. [15] showed that the incorporation of APAO strengthened the elastic recovery and aging resistance as well as low-temperature cracking resistance of EVA modified binder. Moreover, the results demonstrated that APAO was a good stabilizer to improve the stability of asphalt binder. Evidently, APAO is a modifier that satisfies the requirement due to its superior compatibility with binder, simple modification process and excellent modification effect. In addition, APAO has a comparable price to SBS. In order to further reduce the cost, it is important to choose another modifier that has good compatibility with asphalt binder and combine it with APAO for the modification of neat binder.

Polyphosphoric acid (PPA), a mineral acid, has good compatibility with asphalt binder and has been widely applied in asphalt binder modification [16,17]. The preparation process of PPA modified asphalt binder is relatively simple, which involves uniform mixing through stirring at 150–160 °C instead of high-speed shearing [18,19]. PPA does not separate from the asphalt phase when it modifies the asphalt binder in the absence of a stabilizer [20]. Previous research has shown that a small amount of PPA can markedly improve the Superpave performance grade (PG) of asphalt binder [18,21]. The useful temperature interval (UTI) is extended by PPA due to its characteristics of increasing the high-temperature stiffness without decreasing the low-temperature flexibility [22–25]. PPA was also adopted to modify asphalt binder with another polymer to strengthen its rheological behavior and reduce the cost [26–29]. Moreover, PPA can enhance the compatibility between polymer and asphalt binder during the composite modification process [27,30]. Overall, PPA is a desirable modifier that has similar characteristics to APAO in terms of good compatibility and a simple modification process.

The literature review shows that both APAO and PPA can produce highly stable modified asphalt binder at relatively low processing temperatures without shearing and stabilizer. The recent studies on APAO and PPA modification are summarized in Table 1. Thus, the combination of APAO and PPA can provide some enlightenment for improving the performance of modified asphalt binder and simplifying the process technology. To date, investigations regarding APAO/PPA modified binders have not been published, and the modification effect and mechanism of this unique modification are still unknown. Hence, systematic and comprehensive studies on the interaction between APAO/PPA and asphalt binder are still needed.

Table 1. The recent studies on APAO and PPA modified asphalt binder.

Reference	Major Objective	Major Key Findings
Behnood, A. et al. [1]	Commonly used asphalt modifier	The commonly used polymers for asphalt binder modification can be classified into thermoplastic elastomers and plastomers.
Vargas, M.A. et al. [2]	Compatibility of PE and asphalt binder	A substantial phase separation occurs between the PE and the asphalt binder.
Fang, C., Luo, W.-Q., Zhu, C. et al. [3–5]	Preparation process	Polymers, such as SBS, PE, EVA, must be blended with asphalt binder by high-speed shearing equipment.
Zhu, J. et al. [6]	Polymer modification	Equipment investment and high-shearing temperatures increase the production cost of the modified asphalt binder.
Wei, J., Liu, N. et al. [7,8]	New polymer	APAO is a low molecular weight amorphous plastic material, which is highly compatible with binder.
Wei, J. et al. [7,9]	Compatibility of APAO and asphalt binder	APAO can be completely dissolved in asphalt binder when the APAO content is no more than 6 wt.%.
Kong et al. [10]	Processing temperature	Processing temperature of APAO modified binder can be controlled at 165 °C.
Yan et al. [11] and Liu et al. [12]	Effect of APAO on asphalt binder properties	APAO improved the high-temperature performances, storage stability and aging resistance of WTR modified binder.
Liu et al. [13]	Effect of APAO on asphalt binder properties	SBS/APAO modified binder had superior high- and intermediate-temperature properties and storage stability compared to SBS modified binder.
You et al. [14]	Effect of APAO on asphalt binder properties	APAO enhanced the high- and low-temperature property and storage stability of TB.
Yan et al. [15]	Effect of APAO on asphalt binder properties	APAO strengthened the elastic recovery and aging resistance and low-temperature cracking resistance of EVA modified binder.
Polacco, G. [16] and Zhang, F. [17]	Compatibility of PPA and asphalt binder	PPA has good compatibility with asphalt binder and has been widely applied in asphalt binder modification.
Venkat Ramayya, V. [18] and Jiang, X. [19]	Preparation process	PPA modified binder can be prepared by stirring at 150–160 °C instead of high-speed shearing.
Arnold, T.S. et al. [20]	Compatibility of PPA and asphalt binder	PPA does not separate from the asphalt phase when it modifies the asphalt binder without a stabilizer.
Alam, S. et al. [21]	Effect of PPA on asphalt binder properties	A small amount of PPA can markedly improve the Superpave performance grade of asphalt binder.
Behnood, A., Cao, W.D., Baldino, N. et al. [22–25]	Effect of PPA on asphalt binder properties	The useful temperature interval is extended by PPA.
Zhang, F., Gama, D.A., Xiao, F. et al. [26–29]	Effect of PPA on asphalt binder properties	PPA is also adopted to modify asphalt binder with another polymer to strengthen its rheological behavior and reduce the cost.
Zhang, F. et al. [27,30]	Effect of PPA on asphalt binder properties	PPA can enhance the compatibility between polymer and asphalt binder.

This study aims to explore the compound modification effect of APAO and PPA on the base binder. The conventional properties, rheological behavior, storage stability, compatibility and microstructure of composite modified binders were evaluated. Furthermore, the modification effects of APAO/PPA compound modification were compared with those of APAO modification alone and PPA modification alone. The optimal proportions of APAO and PPA were also determined to provide basic parameters for industrial production. Finally, variations in functional groups of asphalt binder were revealed by FTIR spectroscopy measurements.

2. Experiment and Methods

2.1. Materials

Neat binder produced from Suizhong crude oil was adopted for modification, and its penetration grade was 80/100 according to the ASTM standard. Its basic specifications are presented in Table 2. PPA was supplied by Macklin in China. The phosphorous pentoxide (P_2O_5) concentration was greater than 85 wt.% and the density was 2.05 g/cm^3. The APAO-828 compound applied in this study was manufactured by Evonik Degussa of Germany. The key parameters of APAO-828 are summarized in Table 3.

Table 2. Physical properties of neat binder.

Item	Neat Binder	Specifications
Softening Point (°C)	45.8	ASTM D36 [31]
Penetration (25 °C, 0.1 mm)	90	ASTM D5 [32]
Ductility (10 °C, cm)	>100	ASTM D113 [33]
Viscosity (135 °C, Pa·s)	0.348	ASTM D4402 [34]

Table 3. Physical properties of APAO.

Item	APAO-828	Specifications
Softening Point (°C)	161	ASTM D36
Penetration (25 °C, 0.1 mm)	22	ASTM D5
Viscosity (190 °C, Pa·s)	25	ASTM D4402
Density (23 °C, g/cm^3)	0.87	ASTM D 1505 [35]
Glass Transition Temperature (°C)	−35	ASTM D3418 [36]
Molecular Weight (g/mol)	61,000	ASTM D 4001 [37]

2.2. Specimen Preparation

The compound modified binders were prepared by a high-speed stirring approach without a stabilizer. First, the neat binder was preheated to 165 °C in an iron cylinder and then APAO was slowly added. The temperature of the iron cylinder was maintained at 165 °C and the blend was stirred at 2000 rpm for 30 min. Then, PPA was poured into this cylinder and stirred for another 30 min. Subsequently, the prepared samples were subjected to analysis. Different contents of APAO and PPA were mixed together to prepare different modified binders. Four percentages of APAO (0 wt.%, 2 wt.%, 4 wt.% and 6 wt.%) and PPA (0 wt.%, 1.0 wt.%, 1.5 wt.% and 2.0 wt.%) (by the weight of neat binder) were employed in this research. For simplicity, the modified binders with various concentrations of APAO and PPA are labeled. The composition and corresponding abbreviations of all modified binders are listed in Table 4. The detailed technical flowchart of this study is shown in Figure 1.

Table 4. The composition and corresponding abbreviations of all modified binders.

Item	APAO (wt.%)	PPA (wt.%)	Item	APAO (wt.%)	PPA (wt.%)
Neat	0	0	A0P1.5	0	1.5
A2P0	2	0	A2P1.5	2	1.5
A4P0	4	0	A4P1.5	4	1.5
A6P0	6	0	A6P1.5	6	1.5
A0P1.0	0	1.0	A0P2.0	0	2.0
A2P1.0	2	1.0	A4P2.0	2	2.0
A4P1.0	4	1.0	A6P2.0	4	2.0
A6P1.0	6	1.0	A2P2.0	6	2.0

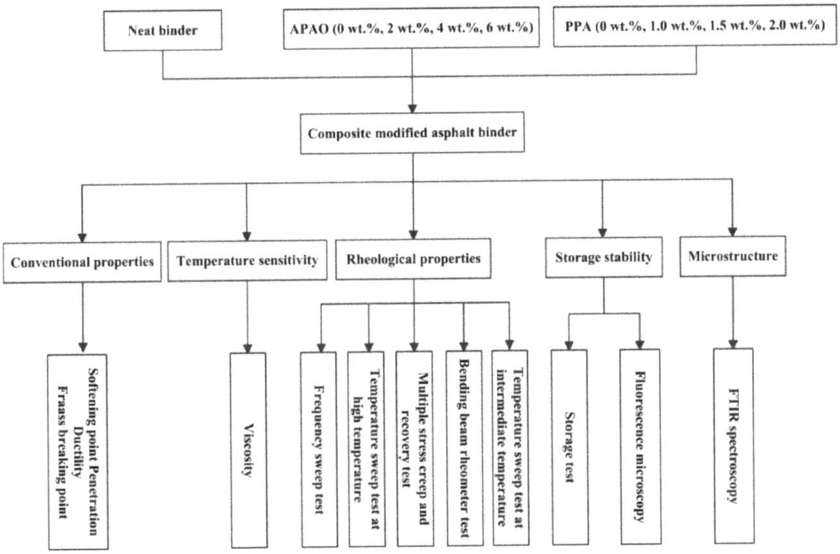

Figure 1. Flowchart of the experimental program.

2.3. Test Methods

2.3.1. Conventional Property Tests

The conventional property tests, including penetration at 25 °C (ASTM D 5), softening point (ASTM D 36), ductility at 10 °C (ASTM D113) and the Fraass breaking point (BS EN 12593) [38], were carried out.

The viscosity of the asphalt binder was tested with a Brookfield rotational viscometer according to ASTM D 4402. The measurements were conducted from 135 °C to 175 °C to assess its construction workability and temperature sensitivity.

2.3.2. Rheological Performance Tests

(1) DSR Test

A dynamic shear rheometer (DSR) with parallel plate geometry was applied to study the rheological behavior of asphalt binder at high and intermediate temperatures according to ASTM D 7175 [39]. Plates of a 25 mm diameter spaced 1 mm apart were selected for samples to detect the shear storage modulus G′ and shear loss modulus G″ at high temperatures. The linear viscoelastic range of each sample was identified by a strain sweep test. Each test was repeated twice to confirm a good repeatability of the data.

The frequency sweep test was carried out from 0.1 rad/s to 100 rad/s at 60 °C to obtain dynamic shear moduli. The temperature sweep test was performed at a fixed frequency of 10 rad/s and the temperature was raised from 30 °C to 90 °C with an interval of 5 °C to evaluate the high-temperature performance. Samples aged in the rolling thin film oven test (RTFOT) and pressure aging vessel (PAV) were subject to a temperature sweep test from 10 °C to 40 °C to investigate the fatigue performance.

(2) MSCR test

The multiple stress creep and recovery (MSCR) test, according to ASTM D 7405 [40], was performed to obtain the non-recoverable creep compliance and percent recovery, which could accurately reflect the mechanical response of the polymer modified asphalt binder. The test was conducted at 60 °C by applying shear stresses of 0.1 kPa and 3.2 kPa, respectively. The RTFOT aged samples were applied in this experiment.

(3) BBR test

The bending beam rheometer (BBR) test was employed to measure the low-temperature rheological performance of asphalt binder, according to ASTM D6648 [41]. Two main factors, creep stiffness (S) and creep rate (m-value), which indicate the low-temperature crack resistance, were obtained under a 60 s loading. The stiffness is the ratio of stress to the total strain under a certain duration and temperature. The m-value reflects the relaxation rate of asphalt binders. RTFOT-PAV aged samples were characterized at −6 °C, −12 °C and −18 °C, respectively.

2.3.3. Storage Test

The storage stability test was conducted to identify the phase separation degree between polymer and asphalt binder. According to ASTM D7173 [42], approximately 50 g of compound modified binder were poured into a cigar tube with a diameter and height of 2.5 cm and 14 cm, respectively. Then, the tube was sealed and placed vertically in an oven at 163 °C. After storage for 48 h, the tube was taken out of the oven and put into the refrigerator to cool down. The cooled tube was divided into three equal parts and the softening points of the upper and lower parts were tested.

2.3.4. Microstructure

(1) ATR-FTIR test

The functional groups of the asphalt binder were determined by attenuated total reflection Fourier transform infrared (ATR-FTIR) spectroscopy (Cary 630 FTIR Microscope, Agilent, Santa Clara, CA, USA). The infrared spectrum was recorded in the wavenumber range of 650 cm^{-1} to 4000 cm^{-1}.

(2) FM test

The dispersion of APAO and PPA in the binder was observed by a fluorescence microscope (FM) (Olympus, Tokyo, Japan). The magnification of the observed images was 400×.

3. Results and Discussion

3.1. *Conventional Properties*

The effects of APAO/PPA on conventional characteristics, such as the penetration, softening point, ductility and Fraass breaking point of the modified binder are shown in Figure 2a–d. It is evident from Figure 2a that the penetration of asphalt binder declines with increases in APAO content when the PPA concentration is constant. When APAO content increases from 0 wt.% to 6 wt.%, the reduction in penetration is 25.6%, 14.8%, 13.6% and 11.1% when the PPA content is 0 wt.%, 1.0 wt.%, 1.5 wt.% and 2.0 wt.%, respectively. This indicates that when APAO concentration increases by 6 wt.%, the reduction in penetration decreases with the increase in PPA content. Moreover, the reduced level of penetration through the incorporation of APAO is lower than that from PPA. The compound modification of APAO and PPA leads to a minimum penetration of 32 (0.1 mm), which is a reduction of 58 (0.1 mm) compared with the base binder. Figure 2b displays an ascending trend of softening point with increases in APAO or PPA concentration, indicating that both APAO and PPA could reinforce the high-temperature performances of asphalt binder. When the PPA dosage is 0 wt.%, 1.0 wt.%, 1.5 wt.% and 2.0 wt.%, a 6 wt.% increase in APAO leads to an 11.4%, 14.7%, 16.1% and 22.5% increase in softening point, respectively. This shows that the increment in the softening point increases with increases in PPA content when the APAO increases by 6 wt.%. Similarly, when the APAO dosage is 0 wt.%, 2 wt.%, 4 wt.% and 6 wt.%, the softening point increases by 39.7%, 40.8%, 49.5% and 53.7%, when the PPA content increases by 2.0 wt.%. It is noteworthy that the increase in softening point caused by PPA is greater than that by APAO.

Figure 2c,d present the variations of ductility at 10 °C and Fraass breaking point. Figure 2c shows that APAO significantly decreases the ductility of asphalt binder. When APAO increases by 6 wt.%, the reduction in ductility is 89.3%, 38.8%, 17.6% and 17.4%, when the PPA content is 0 wt.%, 1.0 wt.%, 1.5 wt.% and 2.0 wt.%, respectively. This indicates that when the APAO concentration increases by 6 wt.%, the reduction in ductility decreases with an increase in PPA content. After the incorporation of PPA, the ductility is further reduced on the basis of APAO. The compound modification of APAO and PPA results in a marked reduction in ductility, indicating that the low-temperature performance decreases. The literature [43,44] reveals that the incorporation of APAO is detrimental to the low-temperature properties of a mixture. The Fraass breaking point denotes the turning point of asphalt binder, where it transforms from a viscoelastic state to a brittle state. Figure 2d shows that APAO is harmful to the Fraass breaking point, while PPA is beneficial to the Fraass breaking point. When the PPA dosage is 0 wt.%, 1.0 wt.% and 2.0 wt.%, a 6 wt.% increase in APAO leads to a 15.4%, 20.0% and 22.2% increase in Fraass breaking point, respectively. Meanwhile, a 2.0 wt.% increase in PPA content contributes to a decrease by 38.5%, 30.8% and 27.3% in the Fraass breaking point, when the APAO content is 0 wt.%, 2 wt.% and 6 wt.%. The results show that PPA has a greater effect on the Fraass breaking point. The combined effect of APAO and PPA leads to a decrease in the Fraass breaking point, indicating that APAO/PPA could reduce the turning point of the binder's brittle state. In summary, the compound modification of APAO and PPA could enhance the high-temperature properties but decrease the low-temperature flexibility.

3.2. Temperature Sensitivity

The effects of APAO and PPA on changes in viscosity are shown in Figure 3. Notably, the viscosity increases as the content of APAO and PPA increases. The viscosity increases by 1.8 times when the APAO proportion increases from 0 wt.% to 6 wt.%, and the viscosity improves by 4.4 times as the PPA content increases by 2.0 wt.%. The results demonstrate that PPA has a greater effect on the increase in viscosity than APAO. Viscosity is a vital indicator to reveal the rheological behavior of asphalt binder, whereby SHRP requires that the 135 °C viscosity should be less than 3.0 Pa·s to guarantee construction workability. From Figure 3a,b, when the PPA proportion remains at 1.0 wt.% and 1.5 wt.%, and the APAO content reaches 6 wt.%, the viscosity meets the SHRP requirement. However, the viscosity exceeds 3.0 Pa·s when the PPA content increases to 2.0 wt.% and APAO content reaches 4 wt.%. Thus, to ensure the workability of asphalt binder, the APAO content could range from 0 wt.% to 6 wt.% when the PPA content is less than 2.0 wt.%, and the APAO proportion should not exceed 4 wt.% when the PPA content reaches 2.0 wt.%. As a result, two types of modified binders are chosen: one is 1.5 wt.% PPA combined with different contents of APAO, and the other is 2 wt.% APAO mixed with different concentrations of PPA.

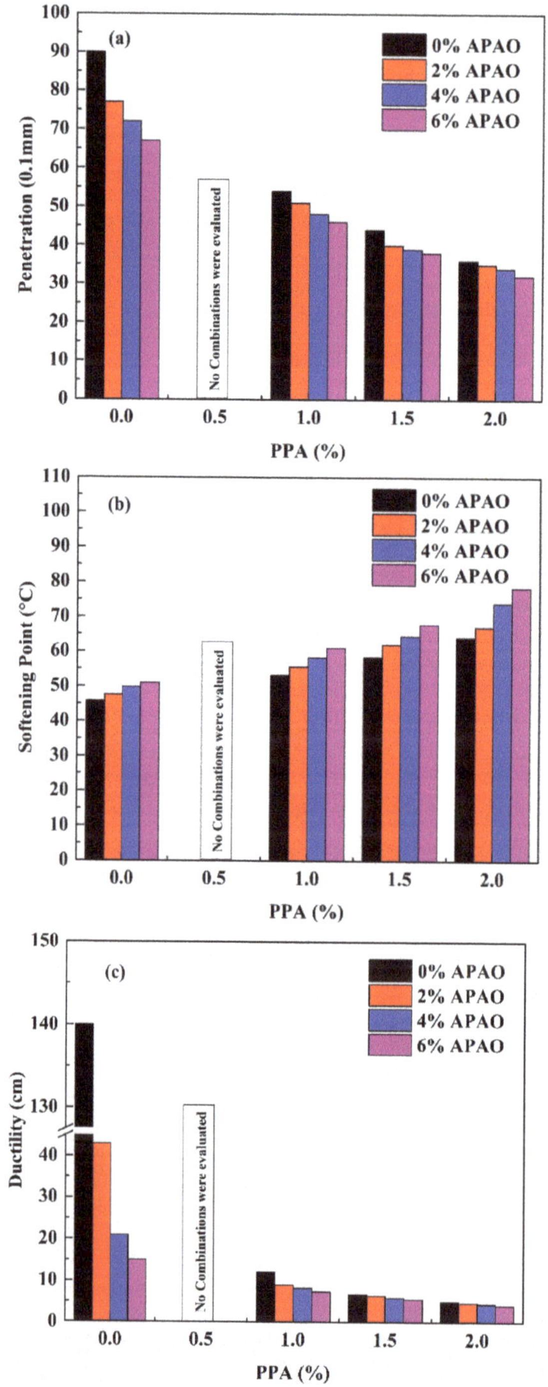

Figure 2. *Cont.*

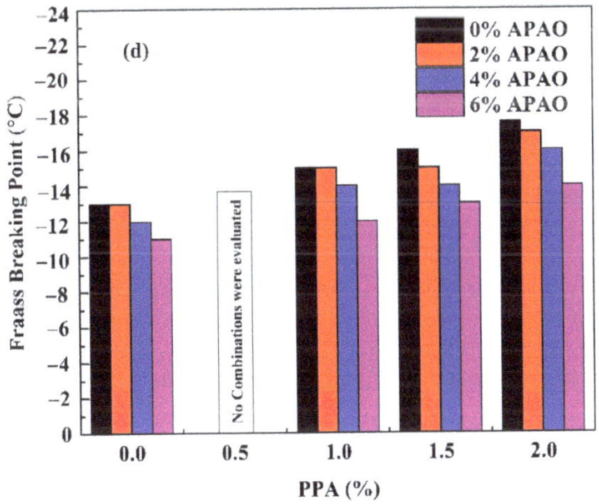

Figure 2. Conventional properties of all specimens: (**a**) penetration; (**b**) softening point; (**c**) ductility; (**d**) Fraass breaking point.

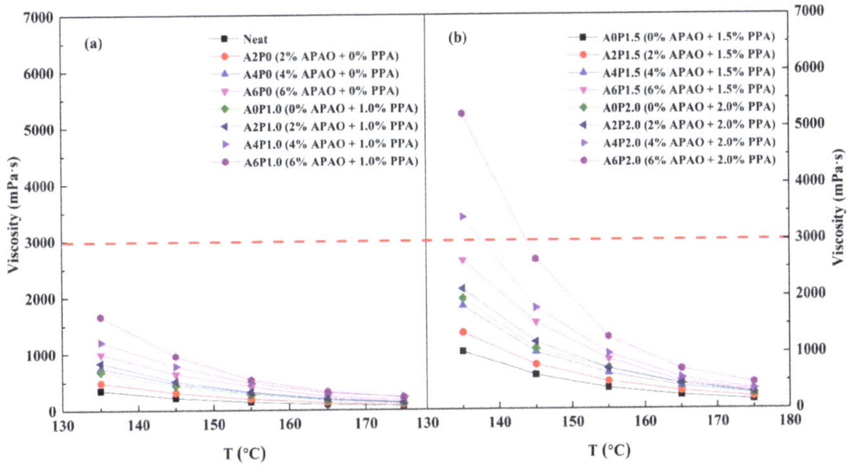

Figure 3. Variations in viscosity with temperature for all specimens: (**a**) composite modified binders with 0 wt.% and 1.0 wt.% PPA combined with 0-6 wt.% APAO; (**b**) composite modified binders with 1.5 wt.% and 2.0 wt.% PPA combined with 0-6 wt.% APAO.

The temperature sensitivity is an important rheological index for asphalt binder, which is usually determined by the viscosity–temperature curve. The asphalt binder's temperature sensitivity determines the characteristics of an asphalt pavement. Low temperature sensitivity of asphalt binder indicates better asphalt pavement performance. Figure 4a,b display the variations in viscosity of composite modified binder at different temperatures. Curve fitting is performed to show the correlation between viscosity and temperature and, from which, the slopes and coefficient of determinations (R^2) are obtained, as listed in Table 5. The R^2 value is greater than 0.99, which indicates that lg(lg(viscosity)) has a good linear correlation with lg(T). A lower slope value indicates that the asphalt binder is less susceptible to temperature changes. In comparison to neat binder, the modified asphalt binders A0P1.5 and A2P0 have lower slope values. Notably, the slope value of A2P0 is higher than that of A0P1.5, which indicates that the enhancement against temperature sensitivity of PPA is greater than that of APAO. It can be seen that the slope value of the fitted curve becomes smaller with increases in APAO or PPA content, indicating that APAO and PPA contribute to the reduction in temperature sensitivity. Furthermore, the mixing and compaction temperatures of all samples are determined according to AASHTO T312 viscosity requirements of (0.17 ± 0.02) Pa·s and (0.28 ± 0.02) Pa·s [45]. Table 6 shows the mixing and compaction temperatures for all tested samples. It is clear that the mixing temperature and compaction temperature of composite modified binder increase with increases in APAO and PPA content. The difference between the mixing temperature and the compaction temperature of neat binder is 11.21 °C, and the temperature difference increases with increasing content of APAO and PPA. The temperature difference is 13.40 °C for A6P1.5 and 13.17 °C for A2P2.0. This indicates that a higher temperature is needed to reach the same viscosity, which demonstrates that APAO and PPA reduce the temperature sensitivity of asphalt binder. Therefore, the incorporation of APAO and PPA has the potential to strengthen the asphalt pavement performance.

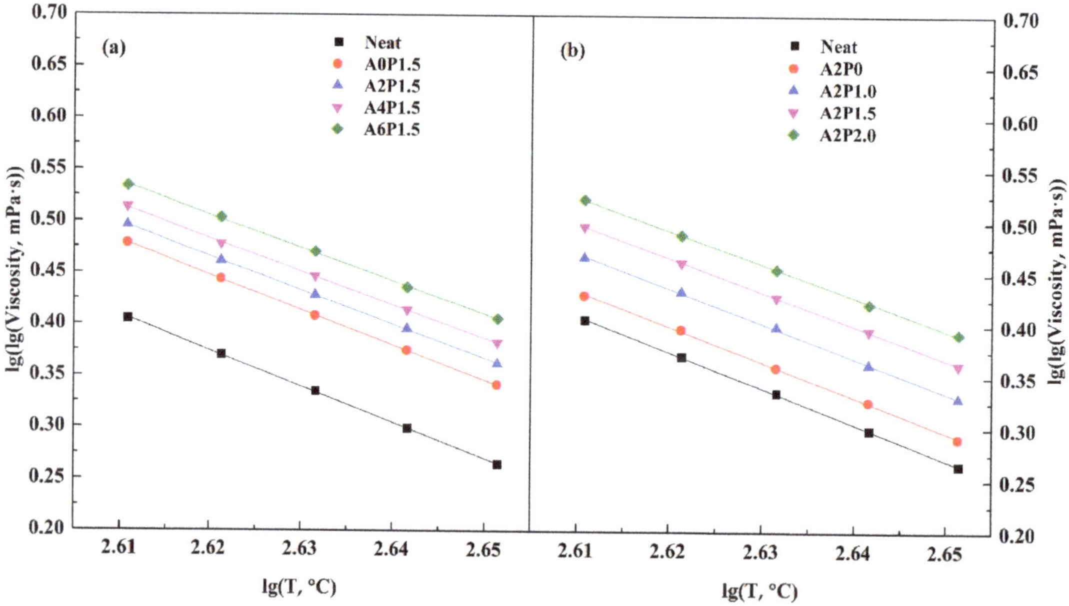

Figure 4. Viscosity–temperature relationship for composite modified binders: (**a**) composite modified binders with various contents of APAO; (**b**) composite modified binders with various contents of PPA.

Table 5. The slope and intercept values of fitting curves.

Item	Slope	Intercept	R^2
Neat	−3.4620	9.4445	0.9998
A0P1.5	−3.3734	9.2860	0.9999
A2P1.5	−3.2720	9.0385	0.9999
A4P1.5	−3.2116	8.8976	0.9998
A6P1.5	−3.1872	8.8566	0.9994
A2P0	−3.3957	9.2961	0.9994
A2P1.0	−3.3599	9.2391	0.9994
A2P1.5	−3.2720	9.0385	0.9999
A2P2.0	−3.2081	8.8975	0.9996

Table 6. The mixing and compaction temperatures for all samples.

Samples	Mixing Temperature (°C)	Compaction Temperature (°C)
Neat	150.90	139.69
A0P1.5	172.95	160.86
A2P1.5	179.64	166.99
A4P1.5	186.02	172.96
A6P1.5	194.04	180.64
A2P0	158.38	146.76
A2P1.0	169.56	157.51
A2P1.5	179.64	166.99
A2P2.0	189.07	175.90

3.3. *Rheological Properties*

3.3.1. Frequency Sweep Tests

Asphalt binder is a viscoelastic material whose stress–strain characteristics depend on temperature and time. The mechanical and viscoelastic behavior of asphalt binder varies with loading frequency. Thus, the effects of compound modification of APAO and PPA on the rheological behavior of base binder were investigated by frequency sweep tests.

Figure 5 shows the master curves of complex modulus (G*) and phase angle (δ) for the composite modified binders. Increases in the G* values and decreases in the δ values can be observed for all the tested binders. The complex modulus curve is elevated over the whole frequency range and the phase angle curve is lowered correspondingly. The increasing content of APAO and PPA causes an increase in G* values and a decrease in δ values, which indicates an increase in the elastic behavior of the asphalt binder. Moreover, G* values increase significantly at lower frequencies (higher temperatures) and increase slightly at higher frequencies (lower temperatures). This indicates that the incorporation of APAO and PPA improves the asphalt binder's high-temperature performance.

It can be noted that the values of G' and G'' vary with the concentration of APAO and PPA. As can be seen, both G' and G'' increase significantly as the contents of APAO or PPA increase. Remarkably, the increment in G' is much greater than that of G'', showing that G' has a strong dependence on the APAO or PPA concentration. As mentioned above, APAO or PPA are contributing factors to the enhancement of the elastic component, which further strengthens the elastic behavior. Moreover, the value of G' increases significantly when the APAO concentration is 2 wt.% and the PPA concentration exceeds 1.0 wt.%. Therefore, the PPA proportion should be optimized to enhance the synergistic modification effect of APAO.

From Figure 6, it is obvious that the shear storage modulus G' and shear loss modulus G'' increase as the frequency increases. This phenomenon signifies that asphalt binder exhibits more stiff and elastic characteristics with increases in vehicle travel speed. The increases in frequency accelerate the rate of increase of G', indicating that G' is highly sensitive to frequency. The compound modification of APAO and PPA markedly enhances G' and G'' in comparison to that of virgin binder.

3.3.2. Temperature Sweep Tests

Temperature is the main factor that affects the rheological behavior of asphalt binders [46]. The temperature dependence of G' and G'' of the compound modified binders is shown in Figure 7. Clearly, the parameters G' and G'' show a decreasing trend with increasing temperature. Moreover, the reduced level of G' is higher than that in G'' with increasing temperature, indicating that higher temperatures promote the transition of asphalt binder from an elastic to viscoelastic state. G'' always has a greater value than G', revealing that the viscous component is more dominant. Thus, the elastic component in asphalt binder should be enriched to enhance its high-temperature resistance.

The increasing concentration of APAO and PPA contributes to a marked enhancement of G' and G''. When the PPA content remains at 1.5 wt.%, increasing the concentration of

APAO leads to an increment in G′ and G″, with the former experiencing a higher growth rate. This suggests that APAO strengthens the elastic behavior which represents a stronger high-temperature deformation resistance. Similarly, when the APAO proportion is 2 wt.%, an increase in PPA concentration also enhances the G′ and G″ values. The G′ value increases significantly when the content of PPA exceeds 1.0 wt.%, providing evidence that an appropriate amount of PPA benefits the reinforcement of anti-rutting performance of composite modified asphalt binder. Therefore, the enhancement of viscoelastic performances of asphalt binder is based on the interaction of APAO and PPA.

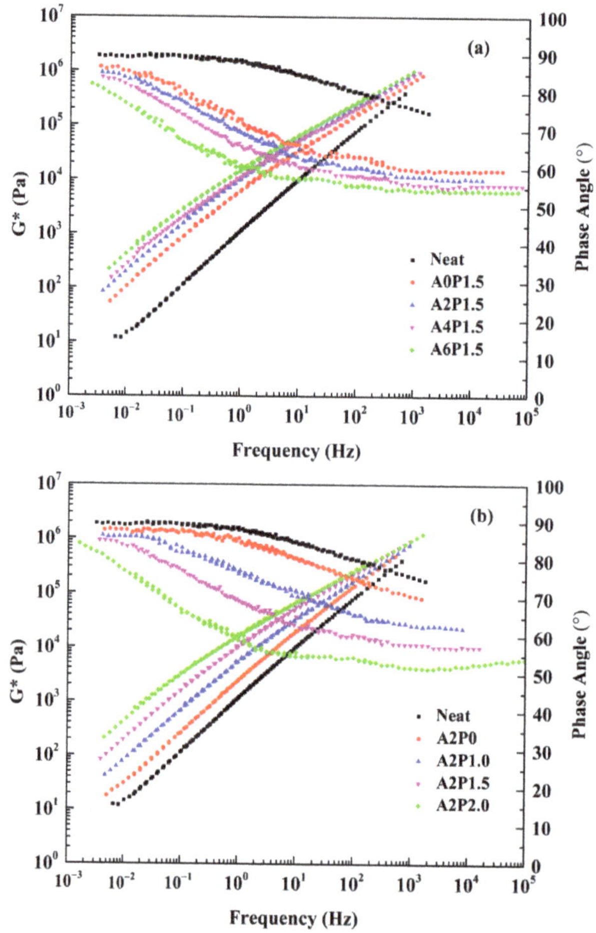

Figure 5. Complex modulus and phase angle master curves of composite modified binders: (**a**) composite modified binders with various contents of APAO; (**b**) composite modified binders with various contents of PPA.

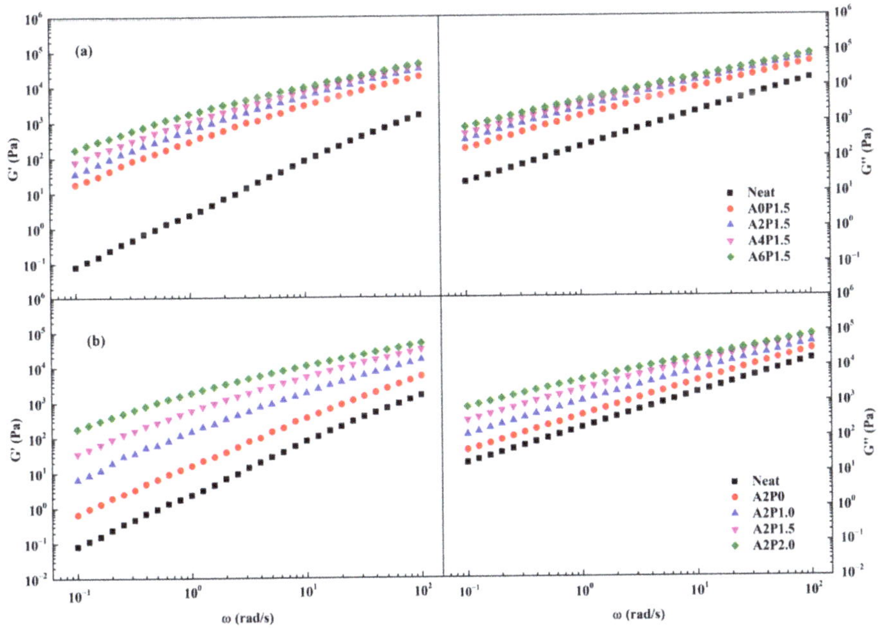

Figure 6. G′ and G″ variations with frequency for composite modified binders: (**a**) composite modified binders with various contents of APAO; (**b**) composite modified binders with various contents of PPA.

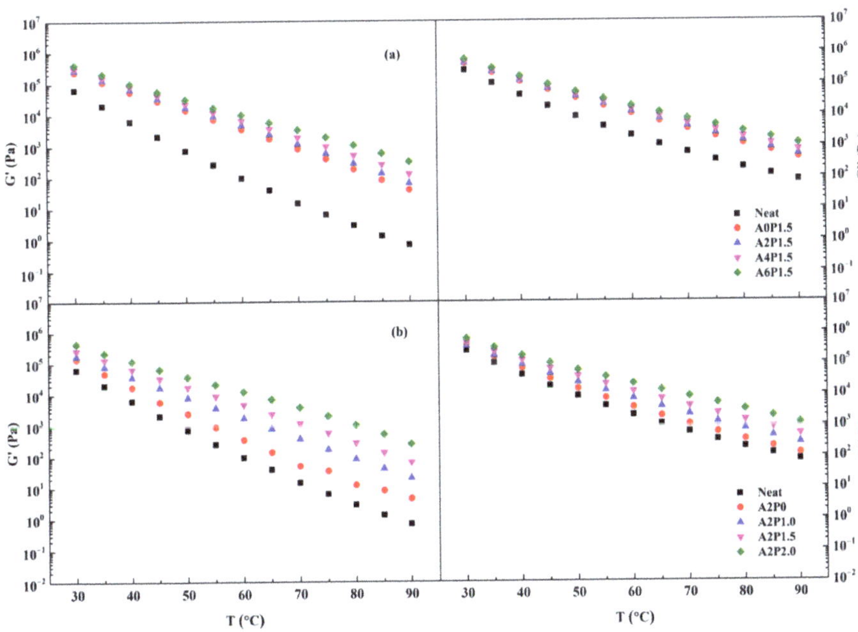

Figure 7. G′ and G″ variations with temperature for composite modified binders: (**a**) composite modified binders with various contents of APAO; (**b**) composite modified binders with various contents of PPA.

The rutting resistance index G*/sinδ is utilized to characterize the anti-rutting property of asphalt binders. Higher G*/sinδ signifies better resistance to permanent deformation. The calculated rutting index G*/sinδ values of all samples are displayed in Figure 8. As shown, G*/sinδ increases with increasing content of APAO, indicating that APAO benefits the anti-rutting property of the binder. Similarly, the rising concentration of PPA also leads to an increasing trend of G*/sinδ. It is worth noting that PPA has a greater contribution to the increment in G*/sinδ. Moreover, the failure temperatures of asphalt binders are acquired through calculation when the G*/sinδ value is 1.0 kPa. Failure temperatures of all tested samples are given in Table 7. Obviously, the failure temperatures increase with the addition of APAO and PPA. The incorporation of APAO markedly improves the failure temperature, with an increase of 9.38 °C for A6P1.5 compared to A0P1.5. However, PPA has a greater contribution to the increase in failure temperature. This result can be demonstrated by the fact that the failure temperature of A2P2.0 is 19.66 °C higher than that of A2P0. The failure temperature of modified binder A2P2.0 is 91.15 °C, which is 23.67 °C higher than that of virgin binder. Therefore, the combined effect of APAO and PPA can significantly improve the high-temperature properties of composite modified binder, with PPA contributing more to this improvement.

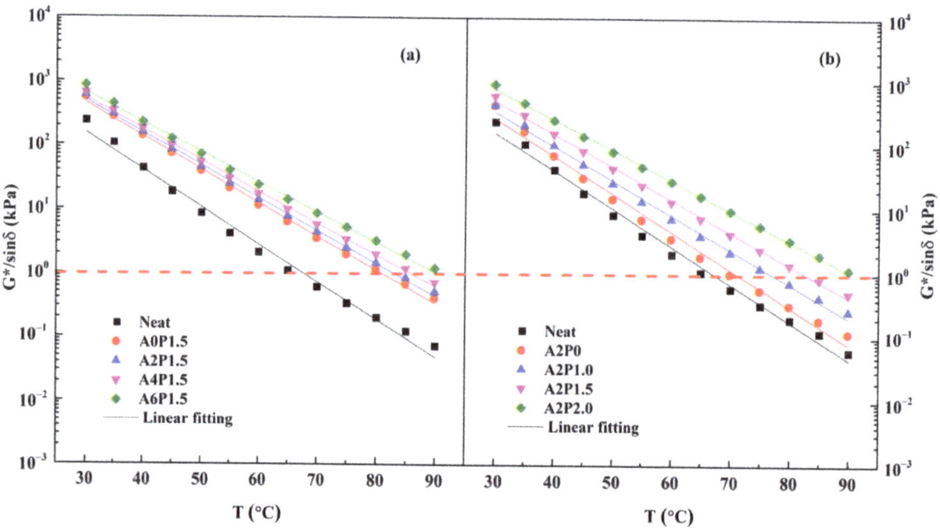

Figure 8. Variations in G*/sinδ with temperature for composite modified binders: (**a**) composite modified binders with various contents of APAO; (**b**) composite modified binders with various contents of PPA.

Table 7. Failure temperatures of all tested specimens.

Samples	Failure Temperature (°C)
Neat	67.48
A0P1.5	81.00
A2P1.5	83.03
A4P1.5	85.84
A6P1.5	90.38
A2P0	71.49
A2P1.0	77.35
A2P1.5	83.03
A2P2.0	91.15

3.3.3. Creep and Recovery Behavior

The rutting index G*/sinδ has a disadvantage in that it cannot reflect the delayed elastic deformation and recoverability of asphalt binder and thus fails to simulate real loading conditions [47]. The Federal Highway Administration (FHWA) proposed the MSCR test to more precisely evaluate the rheological behavior of modified binders at high temperatures [48]. A good correlation has been found between the non-recoverable creep compliance of the MSCR test and the anti-rutting property of asphalt mixtures [49–51]. Therefore, the MSCR test is employed to explore the influence of APAO and PPA on the high-temperature properties of the neat binder.

The samples aged by RTFOT were subjected to the MSCR test at 60 °C and shear stress levels of 0.1 kPa and 3.2 kPa, respectively. For instance, Figure 9 presents the variation of strain with time for all samples at 0.1 kPa. As can be seen, one cycle includes a one-second creep stage and a nine-second recovery stage. The strain level increases with increases in loading time in the creep stage. In the stage of recovery, the strain recovers instantly when the loading stress is removed, and the recovery rate decreases with the loading time. A residual strain still exists after each cycle of creep and recovery. The above results reflect the viscoelastic–plastic performance of asphalt binder. The elastic strain recovers rapidly, while the viscous strain recovers slowly after the loading stress is eliminated. The residual strain represents the permanent deformation. It is notable that the neat binder has the maximum residual strain while the modified binder A2P2.0 has the minimum residual strain. This phenomenon demonstrates that the combined effect of APAO and PPA can lower the permanent deformation of neat binder.

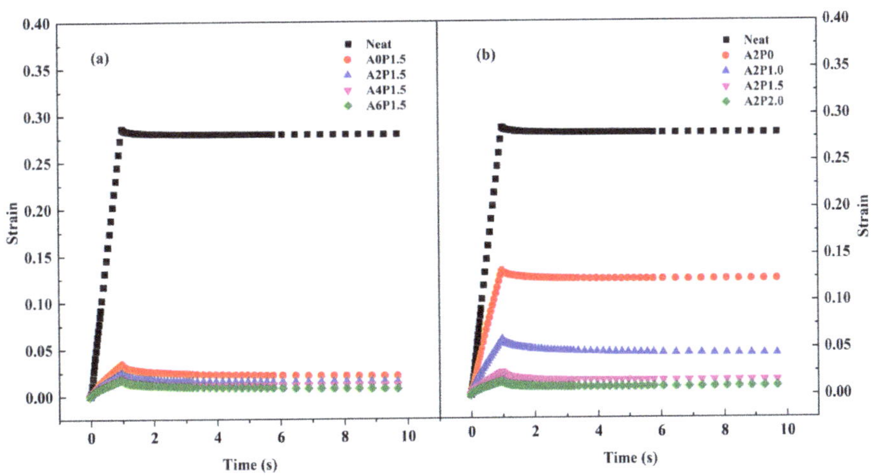

Figure 9. A creep–recovery cycle of the composite modified binder at 0.1 kPa and 60 °C: (**a**) composite modified binders with various contents of APAO; (**b**) composite modified binders with various contents of PPA.

Two main parameters of average non-recoverable creep compliance (J_{nr}) and average percent recovery (R) were proposed by the FHWA to evaluate the high-temperature viscoelastic characteristics of the modified binder. The J_{nr} and R are defined as the permanent deformation and elasticity indexes of asphalt binder, respectively. The higher the R value is, the lower the J_{nr} value is, indicating that the anti-rutting property of asphalt binder is better. Figures 10 and 11 show the J_{nr} and R values of composite modified asphalt binders at different stress levels. As observed from Figures 10a and 11a, the addition of 1.5 wt.% PPA alone decreases the J_{nr} and increases the R in comparison to neat binder, indicating that PPA reduces the permanent deformation of neat binder. Similarly, 2 wt.% APAO alone is also capable of enhancing the anti-rutting performance of neat binder, as shown in

Figures 10b and 11b. When the concentration of PPA remains constant, increasing content of APAO leads to a marked increase in the R value and a significant reduction in the J_{nr} value. Asphalt binders modified with a certain amount of APAO and increasing content of PPA also exhibit a similar increasing trend of the R value and decreasing trend of the J_{nr} value. Moreover, the R value increases by 51.50% and the J_{nr} value decreases by 68.48% as the APAO increases from 0 wt.% to 6 wt.% under 0.1 kPa stress. The R value increases by 6.89 times and the J_{nr} value decreases by 96.06% as the PPA increases from 0 wt.% to 2 wt.%. Relative to APAO, PPA contributes more to the decrease in J_{nr} and the increase in R. Therefore, this suggests that PPA contributes more to the improvement of the elasticity of asphalt binder.

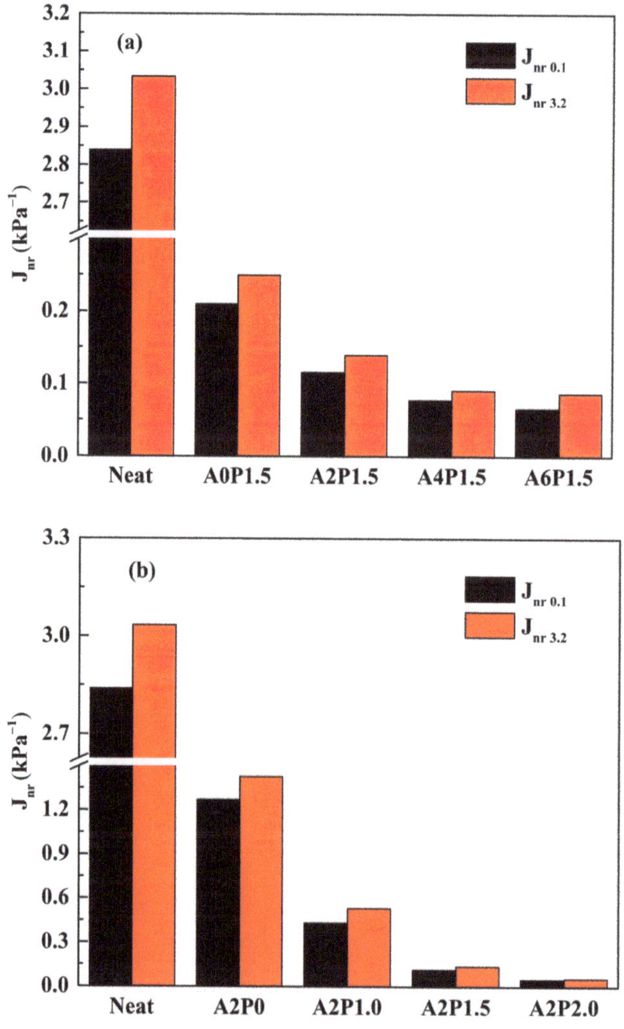

Figure 10. The J_{nr} of composite modified binders under 0.1 kPa and 3.2 kPa at 60 °C: (**a**) composite modified binders with various contents of APAO; (**b**) composite modified binders with various contents of PPA.

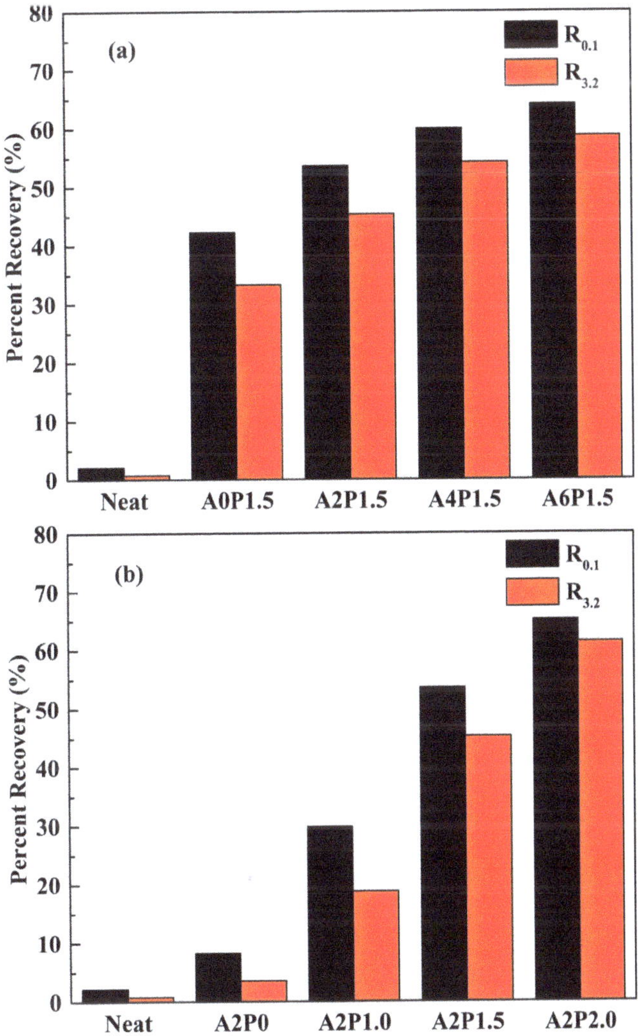

Figure 11. The R of composite modified binders under 0.1 kPa and 3.2 kPa at 60 °C: (**a**) composite modified binders with various contents of APAO; (**b**) composite modified binders with various contents of PPA.

The variations in J_{nr} and R are also affected by the applied stress level. The J_{nr} exhibits an increasing trend and the R presents a decreasing trend when the stress increases from 0.1 kPa to 3.2 kPa. This suggests that a higher stress level reduces the anti-rutting performance of the asphalt binder. Thus, an asphalt pavement with heavier traffic is prone to experiencing rutting stress. Furthermore, the R and J_{nr} values vary with APAO and PPA concentration. Taking the R value as an example, the R values of A0P1.5, A2P1.5, A4P1.5 and A6P1.5 decrease by 21.26%, 15.49%, 9.63% and 8.36%, respectively, due to the increased stress level. Similarly, the R values of A2P0, A2P1.0, A2P1.5 and A2P2.0 drop by 56.81%, 36.98%, 15.49% and 5.81%, respectively. The results of the data analysis reveal that the combined effect of APAO and PPA is helpful to mitigate the adverse impact of the increased stress level.

3.3.4. Low-Temperature Creep Behavior

An asphalt pavement should be flexible enough to avoid cracking at low temperatures, which is dependent upon the characteristics of the asphalt binder. The Strategic Highway Research Program (SHRP) recommends the BBR test to assess the binder's low-temperature properties. In the SHRP specification, the creep stiffness (S) and rate of creep (m-value) obtained under a 60 s loading are the two main parameters to identify the low-temperature properties of the binder. The creep stiffness represents the shrinkage stress of an asphalt pavement caused by temperature change, and the creep rate reflects the degree of stress reduction. Lower creep stiffness means less shrinkage stress due to a temperature drop, indicating better low-temperature crack resistance. Higher m-values represent a better stress dissipation property. It is well known that asphalt pavements with lower stiffness and higher m-value have better low-temperature cracking resistance performance. To investigate the influence of APAO and PPA on the low-temperature properties of neat binder, variations in stiffness and m-value are presented in Figures 12 and 13.

As can be seen, the stiffness value increases and the m-value declines as the temperature drops from −6 °C to −18 °C. The increase in stiffness indicates that the binder becomes stiffer and has greater shrinkage stress, and the decrease in m-value suggests that the stress dissipation rate becomes smaller. The incorporation of APAO and PPA has a marked influence on the variations in stiffness and m-value. As observed, the stiffness of virgin binder presents an increasing trend and the m-value exhibits a decreasing trend as the concentration of APAO increases. This result indicates that APAO lowers the binder's low-temperature performance. However, the increasing concentration of PPA further improves the stiffness and reduces the m-value, indicating that PPA decreases the low-temperature property on the basis of APAO modified asphalt binder. The stiffness increases by 23.47% as the APAO content rises from 0 wt.% to 6 wt.% at −6 °C, and the stiffness increases by 58.76% as the PPA content increases from 0 wt.% to 2.0 wt.%. The results show that the deterioration in the low-temperature performance caused by APAO is less than that by PPA.

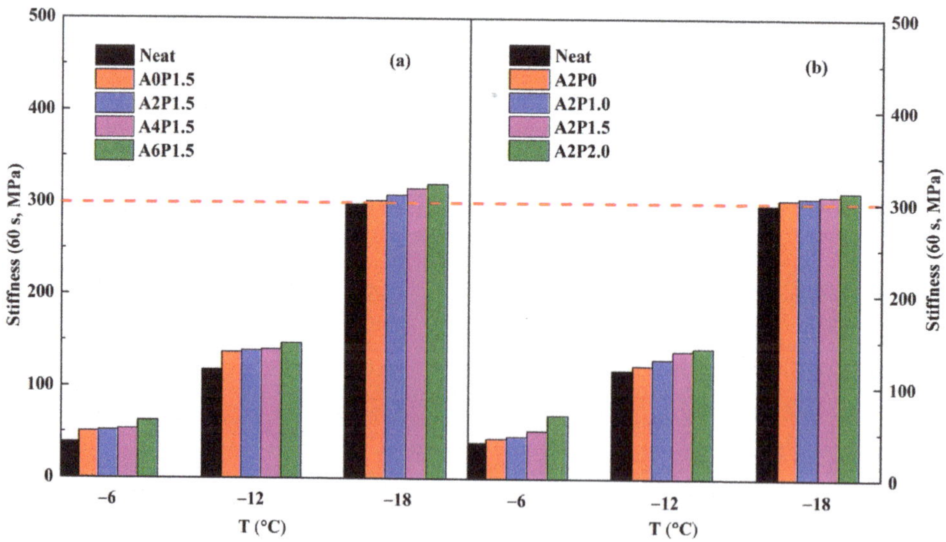

Figure 12. Creep stiffness variations with temperature for composite modified binders: (**a**) composite modified binders with various contents of APAO; (**b**) composite modified binders with various contents of PPA.

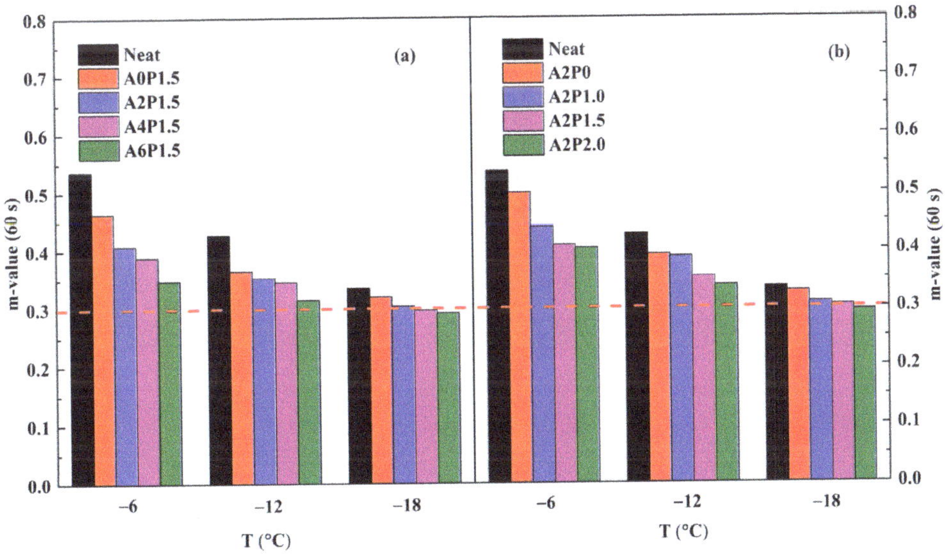

Figure 13. m-value variations with temperature for composite modified binders: (**a**) composite modified binders with various contents of APAO; (**b**) composite modified binders with various contents of PPA.

The stiffness of asphalt binder is less than 300 MPa and the m-value is higher than 0.3, which meet the requirements of the SHRP specification [52]. The neat binder can meet the requirement at −12 °C, but fails at −18 °C. After adding APAO and PPA, all composite modified asphalt binders fulfill the specifications at −12 °C. As expected, the composite modified binders fail to meet the requirement at −18 °C. It is clear that the composite modified binder has comparable low-temperature performance to the virgin binder. The aforementioned results indicate that the low-temperature grade is not reduced by the compound modification of APAO and PPA. Thus, the low-temperature PG grade of APAO/PPA modified binders can reach −22 °C, which meets the requirements of most areas in China. Therefore, it is clear that the compound modification of APAO/PPA is detrimental to the stiffness and m-value of asphalt binder, although it does not reduce the low-temperature PG grade.

3.3.5. Fatigue Behavior

Fatigue cracking of asphalt pavement has become an important stress factor influencing the long-term service performance of pavements. There are many factors contributing to fatigue cracking, such as climate change, binder properties, pavement design and structure and traffic volume. Among these influencing factors, it is well established that the anti-fatigue performance of asphalt binder plays a significant role in controlling the fatigue properties of asphalt mixtures [53]. The Superpave fatigue index G *sinδ, corresponding to the dissipated energy, is utilized to characterize the fatigue resistance of asphalt binder. Thus, the RTFOT-PAV aged specimens were subjected to a temperature sweep test to assess the binder's fatigue resistance, the G *sinδ values of which are displayed in Figure 14.

As observed, G *sinδ values decrease with increases in temperature. In the intermediate temperature range, a small G *sinδ value indicates a better anti-fatigue performance of the binder. As observed from Figure 14a, it is obvious that the neat binder has a lower G *sinδ value, which means the neat binder has better fatigue resistance. After adding 1.5 wt.% PPA, the G *sinδ value becomes larger, suggesting that the PPA reduces the anti-fatigue property of asphalt binder. On the other hand, the G *sinδ value declines with increasing APAO content, indicating that APAO strengthens the fatigue resistance of the asphalt binder. However, there is an intersection between the fitting curves of neat binder and modified binder A6P1.5, which makes it difficult to determine the fatigue temperature. Moreover, Figure 14b shows that the modified asphalt binder with 2 wt.% APAO has a better anti-fatigue resistance performance than neat binder, and the G *sinδ value increases with increases in PPA content. To quantitatively investigate the effects of compound modification of APAO and PPA on fatigue properties, the fatigue temperatures are determined.

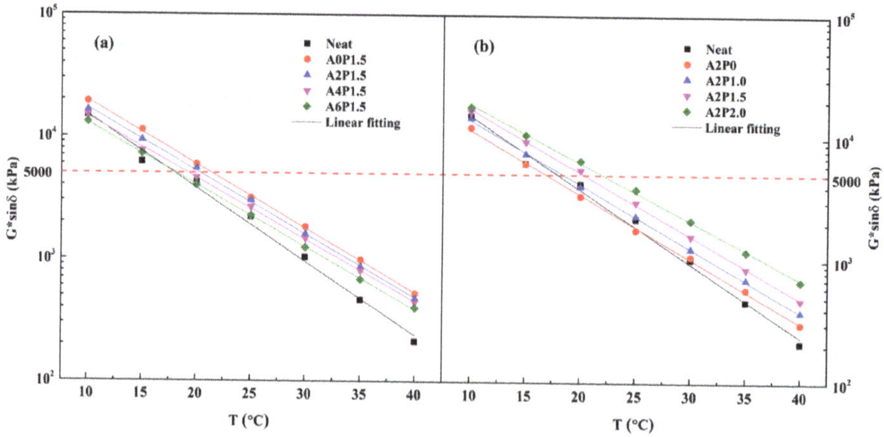

Figure 14. G *sinδ variations with temperature for composite modified binders: (**a**) composite modified binders with various contents of APAO; (**b**) composite modified binders with various contents of PPA.

In accordance with the SHRP specification, the fatigue temperatures of asphalt binder are calculated when the G *sinδ value reaches the criterion of 5000 kPa [54]. The specific values of fatigue temperature are shown in Table 8. Obviously, A2P0 has the lowest fatigue temperature, followed by the neat binder. The increasing content of APAO leads to reductions in fatigue temperature, indicating that APAO can promote the anti-fatigue property of the binder. In contrast, increasing content of PPA increases the fatigue temperature, which indicates that PPA reduces the anti-fatigue property of the asphalt binder. The modified binder A6P1.5 has a lower fatigue temperature than A2P2.0. The combined effects of APAO and PPA on the fatigue resistance of the binder depend on their respective concentrations. Therefore, the APAO and PPA content should be optimized to guarantee the fatigue resistance of composite modified binder.

Table 8. Fatigue temperatures of all tested specimens.

Samples	Fatigue Temperature (°C)
Neat	17.96
A0P1.5	21.39
A2P1.5	20.33
A4P1.5	19.22
A6P1.5	18.12
A2P0	16.89
A2P1.0	18.47
A2P1.5	20.33
A2P2.0	22.04

3.4. Storage Stability

Modified asphalt binder requires excellent storage stability during high-temperature storage, pumping, transportation and construction [55]. However, phase separation often occurs due to the incompatibility between polymers and asphalt binder. The principal reason for this is that the neat binder and polymer have great differences in density, molecular weight, polarity and solubility parameters. The phase separation leads to performance degradation in terms of anti-rutting, anti-cracking and fatigue resistance of modified binder, hindering the practical application of modified asphalt binder [56,57]. Therefore, the storage stability of composite modified binders containing various contents of APAO and PPA is evaluated by the storage test.

The temperature differences between the top and bottom parts in the storage test of modified binders are calculated, as shown in Table 9. To ensure the storage stability of the modified binder at a high temperature, a temperature difference of less than 2.5 °C is required. It is observed that the modified binder A0P1.5 is stable at elevated temperatures. After adding APAO, the temperature difference increases with increasing concentration of APAO, and phase separation becomes predominant for contents higher than 4 wt.%. Modified binder A2P0 has good storage stability due to the uniform dispersion of APAO in asphalt binder. After PPA is added, the softening point difference is less than 2.5 °C when PPA content reaches 2.0 wt.%. Such a phenomenon reveals that PPA has little influence on the storage stability of modified binders. In conclusion, the APAO proportion should be maintained at 2 wt.% to ensure storage stability.

Table 9. Softening point differences of composite modified binders.

Samples	Top Section (°C)	Bottom Section (°C)	Difference (°C)
A0P1.5	57.7	57.6	0.1
A2P1.5	61.8	61.5	0.3
A4P1.5	71.4	65.5	5.9
A6P1.5	74.1	67.6	6.5
A2P0	48.7	48.5	0.2
A2P1.0	54.7	54.4	0.3
A2P1.5	61.8	61.5	0.3
A2P2.0	67.8	66.5	1.3

It is generally believed that good compatibility between binder and polymer is essential to ensure the excellent stability of polymer modified binder. In view of the accuracy, the rheological method has a unique advantage in determining compatibility because it is sensitive to the differences between polymers and asphalt binder. At present, rheological methods such as master curves, black curves, Ham diagrams, and Cole–Cole diagrams are employed to characterize the compatibility of modified binders [54,55,58,59]. Among the above approaches, the Cole–Cole diagrams have proven to be the most effective method for determining compatibility. Thus, the compatibilities of APAO/PPA modified asphalt binder with various amounts of APAO and PPA were studied by a Cole–Cole diagram. The Cole–Cole plots comprise two variables, η' and η'', which are separated from the complex viscosity η^*. The compatibility is determined by the shape of the Cole–Cole diagram. A more symmetrical parabola represents better compatibility, while an asymmetrical parabola indicates poor compatibility.

The Cole–Cole diagrams of APAO/PPA modified binders at 60 °C are displayed in Figure 15. It is clear that the η'' values of samples initially increase and then decrease with increases in η', with a peak value appearing on the curve. The modified asphalt binders with various concentrations of APAO and PPA exhibit different curve shapes, which mainly depend on the effects of the modifier on the viscoelastic constituents of the asphalt binder. The left part and right part of the Cole–Cole diagram characterize the elastic and viscous performances of the binder, respectively. As shown in Figure 15a, the modified binder A0P1.5 exhibits a symmetrical shape, indicating better compatibility between PPA and neat binder. After APAO is added, the curves gradually transition from right to left with increasing content of APAO, showing that APAO can enhance the elastic components of modified asphalt binder. The curve is mostly concentrated on the left side when the APAO content reaches 6 wt.%, suggesting that the elastic component occupies a dominant position.

Moreover, the increase in PPA content also leads to improvement in elastic constituents, as demonstrated by the transition of curves from right to left. Remarkably, modified binder A2P1.5 shows the most symmetrical shape, followed by the modified binder A4P1.5. When the APAO concentration reaches 6 wt.% or the PPA content reaches 2.0 wt.%, the Cole–Cole curves show a tendency to deviate from the semicircle. This demonstrates that 2 wt.% APAO and 1.5 wt.% PPA are the optimal concentrations for the preparation of modified asphalt binder with superior compatibility.

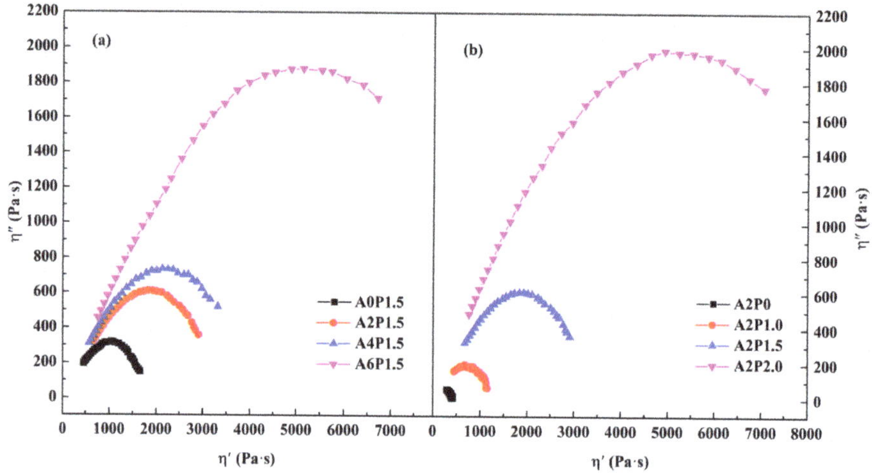

Figure 15. Cole–Cole diagrams of composite modified binders: (**a**) composite modified binders with various contents of APAO; (**b**) composite modified binders with various contents of PPA.

It is important to investigate the distribution of polymer in asphalt binder to explore the behavior of polymer modified binders. The morphologies of APAO/PPA modified binders were observed by a fluorescence microscope. Figure 16 presents the fluorescence images of tested samples where the APAO particles are bright yellow and the asphalt binder phase is black. As observed in Figure 16a, the APAO particles appear in the form of spherical particles and are sparsely dispersed in the binder. However, the dispersion state of APAO changed significantly after the addition of PPA. The particle size decreases markedly and the distribution becomes much more uniform with increasing content of PPA, as shown in Figure 16b–d. This indicates that PPA is conducive to dispersing the APAO particles, and proves that PPA strengthens the compatibility between APAO and binder. Furthermore, Figure 16c,e,f show that the particle size of APAO increases and the distribution tends to aggregate as the APAO concentration increases from 2 wt.% to 6 wt.%. When the APAO proportion exceeds 4 wt.%, the boundary between the APAO and the asphalt binder phase is obvious and the dispersion worsens, revealing that potential phase separation may occur. This phenomenon corroborates the results of the Cole–Cole plots.

3.5. FTIR Spectra

The analysis conducted above shows the variations of the conventional and rheological performances of modified binder after modification with APAO/PPA. More importantly, the microstructure of the modified binder primarily affects its macroscopic rheological behavior. FTIR has been widely used as an efficient method to evaluate the variations in chemical bonds and structures of modified binders. To ascertain the modification mechanism of the APAO/PPA modified binder, the infrared spectra of neat binder and APAO/PPA modified binder were acquired. Figure 17 displays the FTIR spectra of the modified binders.

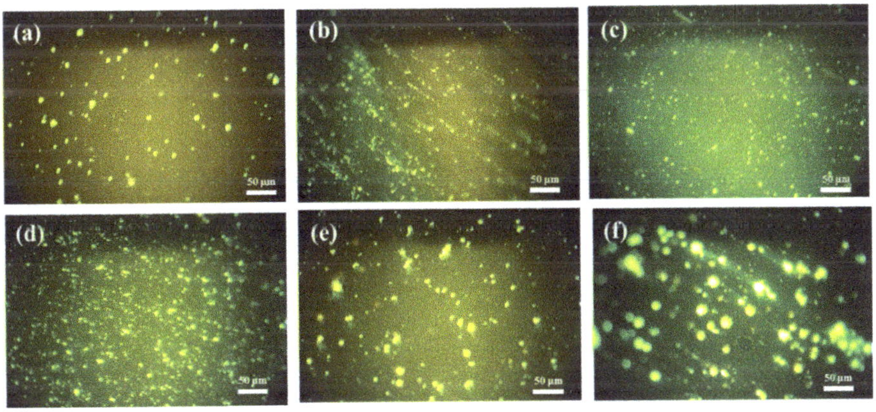

Figure 16. Morphology of tested samples (×400). (**a**) A2P0; (**b**) A2P1.0; (**c**) A2P1.5; (**d**) A2P2.0; (**e**) A4P1.5; and (**f**) A6P1.5.

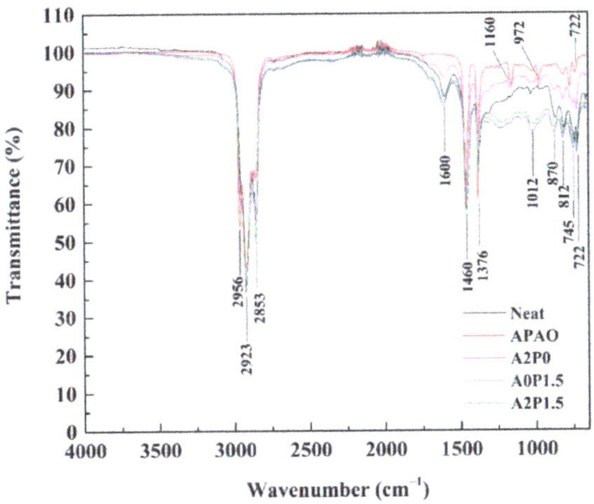

Figure 17. Infrared spectra of tested samples.

It is observed that the typical characteristic peaks at 2956 cm^{-1}, 2923 cm^{-1} and 2853 cm^{-1} denote the C-H stretching vibrations of aliphatic hydrogen. The absorption bands appearing at 1600 cm^{-1} correspond to the stretching vibrations of C=C. The C-H bending vibrations are located at 1460 cm^{-1} and 1376 cm^{-1}. The characteristic absorption peaks within the 700–900 cm^{-1} range are the C-H bending vibrations of the aromatic ring.

The APAO is formed through the polymerization reaction of α-olefines, and contains a polymerized methylene group. The 1160 cm^{-1} and 972 cm^{-1} peaks represent the wagging and rocking vibration of C-H, respectively. The methylene group corresponds to the absorption band at 722 cm^{-1}. The above peaks are considered to be typical characteristic peaks of APAO [7]. It is clear that there is no difference in absorption peaks between the neat binder and modified binder A2P0 in the infrared spectrum, indicating that APAO reacts with asphalt binder physically. However, after PPA is added, a new absorption band appears at 1012 cm^{-1}, representing the P-O stretching vibration [17,60]. Masson et al. [61–64] showed that a phosphorylated product was generated by the reaction of asphalt binder with PPA. Ge et al. [65] pointed out that PPA caused the generation and disappearance of some functional groups. These results reveal that PPA reacts with asphalt binder chemically. In conclusion, APAO and PPA modification involves both physical and chemical interactions.

4. Conclusions

To simplify the production process, APAO and PPA were blended to produce a composite modified binder at a lower processing temperature without shearing and stabilizer. The influence of APAO/PPA on the conventional performance, temperature sensitivity, rheological property, storage stability, compatibility and microstructure of neat binder was studied. According to the above results, the primary findings and conclusions can be summarized as follows:

(1) The addition of APAO and PPA improves the high-temperature performance of asphalt binder, in which the values of softening point and viscosity are increased. The low-temperature properties and the temperature sensitivity of asphalt binder can be reduced through the compound modification of APAO and PPA.
(2) The frequency sweep test shows that the APAO/PPA strengthens the elastic behavior of composite modified asphalt binder under different traffic speeds. The temperature sweep test indicates that APAO/PPA improves the rutting resistance of asphalt binder by increasing the failure temperature. In the MSCR test, the increase in percent recovery and the decrease in non-recoverable creep compliance demonstrate that APAO/PPA improves the permanent deformation resistance of asphalt binder.
(3) Both APAO and PPA are detrimental to the stiffness and m-value of asphalt binder, while they do not reduce the low-temperature PG grade. The low-temperature PG grade of APAO/PPA modified binder can reach −22 °C, which meets the requirements of most areas in China. APAO can strengthen the fatigue resistance of the binder, while PPA reduces the anti-fatigue performance.
(4) A highly stable composite modified binder could be prepared by optimizing the content of APAO and PPA, as was confirmed by Cole–Cole plots and fluorescence images. The FTIR test reveals that both physical and chemical reactions occur during the modification process.
(5) Considering the composite modified binder's rheological behavior and storage stability, the optimal dosage of APAO and PPA is 2 wt.% and 1.5 wt.%, respectively.

Author Contributions: Conceptualization, X.P. and W.F.; methodology, X.P.; formal analysis, X.P.; investigation, X.P.; writing—original draft preparation, X.P.; writing—review and editing, W.F.; supervision, W.F. Both authors have read and agreed to the published version of the manuscript.

Funding: This research received no external funding.

Institutional Review Board Statement: Not applicable.

Informed Consent Statement: Not applicable.

Data Availability Statement: The data presented in this study are available on request from the corresponding author.

Conflicts of Interest: The authors declare no conflict of interest.

References

1. Behnood, A.; Modiri Gharehveran, M. Morphology, rheology, and physical properties of polymer-modified asphalt binders. *Eur. Polym. J.* **2019**, *112*, 766–791. [CrossRef]
2. Vargas, M.A.; Vargas, M.A.; Sánchez-Sólis, A.; Manero, O. Asphalt/polyethylene blends: Rheological properties, microstructure and viscosity modeling. *Constr. Build. Mater.* **2013**, *45*, 243–250. [CrossRef]
3. Fang, C.; Liu, P.; Yu, R.; Liu, X. Preparation process to affect stability in waste polyethylene-modified bitumen. *Constr. Build. Mater.* **2014**, *54*, 320–325. [CrossRef]
4. Luo, W.Q.; Chen, J.C. Preparation and properties of bitumen modified by EVA graft copolymer. *Constr. Build. Mater.* **2011**, *25*, 1830–1835. [CrossRef]
5. Zhu, C.; Zhang, H.; Zhang, Y. Influence of layered silicate types on physical, rheological and aging properties of SBS modified asphalt with multi-dimensional nanomaterials. *Constr. Build. Mater.* **2019**, *228*, 116735. [CrossRef]
6. Zhu, J.; Birgisson, B.; Kringos, N. Polymer modification of bitumen: Advances and challenges. *Eur. Polym. J.* **2014**, *54*, 18–38. [CrossRef]
7. Wei, J.; Li, Y.; Dong, F.; Feng, H.; Zhang, Y. Study on the amorphous poly alpha olefin (APAO) modified asphalt binders. *Constr. Build. Mater.* **2014**, *66*, 105–112. [CrossRef]

8. Liu, N.; Yan, K.; You, L.; Chen, M. Laboratory testing on the anti aging performance of amorphous poly alpha olefin (APAO) modified asphalt binders. *Constr. Build. Mater.* **2018**, *189*, 460–469. [CrossRef]
9. Wei, J.; Liu, Z.; Zhang, Y. Rheological properties of amorphous poly alpha olefin (APAO) modified asphalt binders. *Constr. Build. Mater.* **2013**, *48*, 533–539. [CrossRef]
10. Kong, L.Y.; Zhou, J.C.; Li, W. Experimental research on APAO modified asphalt. *Technol. Hwy. Trans.* **2005**, *10*, 63–67.
11. Yan, K.; He, W.; Chen, M.; Liu, W. Laboratory investigation of waste tire rubber and amorphous poly alpha olefin modified asphalt. *Constr. Build. Mater.* **2016**, *129*, 256–265. [CrossRef]
12. Liu, W.; Yan, K.; Ge, D.; Chen, M. Effect of APAO on the aging properties of waste tire rubber modified asphalt binder. *Constr. Build. Mater.* **2018**, *175*, 333–341. [CrossRef]
13. Liu, J.; Yan, K.; Liu, W.; Zhao, X. Partially replacing Styrene-Butadiene-Styrene (SBS) with other asphalt binder modifier: Feasibility study. *Constr. Build. Mater.* **2020**, *249*, 118752. [CrossRef]
14. You, L.; Yan, K.; Wang, D.; Ge, D.; Song, X. Use of amorphous-poly-alpha-olefin as an additive to improve terminal blend rubberized asphalt. *Constr. Build. Mater.* **2019**, *228*, 116774. [CrossRef]
15. Yan, K.; Tian, S.; Chen, J.; Liu, J. High temperature rheological properties of APAO and EVA compound modified asphalt. *Constr. Build. Mater.* **2020**, *233*, 117246. [CrossRef]
16. Polacco, G.; Filippi, S.; Merusi, F.; Stastna, G. A review of the fundamentals of polymer-modified asphalts: Asphalt/polymer interactions and principles of compatibility. *Adv. Colloid Interface Sci.* **2015**, *224*, 72–112. [CrossRef] [PubMed]
17. Zhang, F.; Yu, J. The research for high-performance SBR compound modified asphalt. *Constr. Build. Mater.* **2010**, *24*, 410–418. [CrossRef]
18. Venkat Ramayya, V.; Vinayaka Ram, V.; Krishnaiah, S.; Sandra, A.K. Performance of VG30 paving grade bitumen modified with polyphosphoric acid at medium and high temperature regimes. *Constr. Build. Mater.* **2016**, *105*, 157–164. [CrossRef]
19. Jiang, X.; Li, P.; Ding, Z.; Yang, L.; Zhao, J. Investigations on viscosity and flow behavior of polyphosphoric acid (PPA) modified asphalt at high temperatures. *Constr. Build. Mater.* **2019**, *228*, 116610. [CrossRef]
20. Arnold, T.S.; Needham, S.P.; Youtcheff, J.S. Use of Phosphoric Acid as Modifier for Hot-Mix Asphalt. *Polyphosphoric Acid Modif. Asph. Bind.* **2009**, *7*, 40–51.
21. Alam, S.; Hossain, Z. Changes in fractional compositions of PPA and SBS modified asphalt binders. *Constr. Build. Mater.* **2017**, *152*, 386–393. [CrossRef]
22. Behnood, A.; Olek, J. Rheological properties of asphalt binders modified with styrene-butadiene-styrene (SBS), ground tire rubber (GTR), or polyphosphoric acid (PPA). *Constr. Build. Mater.* **2017**, *151*, 464–478. [CrossRef]
23. Cao, W.D.; Liu, S.T.; Mao, H.L. Experimental Study on Polyphosphoric Acid (PPA) Modified Asphalt Binders. *Adv. Mater. Res.* **2011**, *152*, 288–294. [CrossRef]
24. Baldino, N.; Gabriele, D.; Rossi, C.O.; Seta, L.; Lupi, F.R.; Caputo, P. Low temperature rheology of polyphosphoric acid (PPA) added bitumen. *Constr. Build. Mater.* **2012**, *36*, 592–596. [CrossRef]
25. Baldino, N.; Gabriele, D.; Lupi, F.R.; Oliviero Rossi, C.; Caputo, P.; Falvo, T. Rheological effects on bitumen of polyphosphoric acid (PPA) addition. *Constr. Build. Mater.* **2013**, *40*, 397–404. [CrossRef]
26. Zhang, F.; Hu, C.; Zhang, Y. The research for SIS compound modified asphalt. *Mater. Chem. Phys.* **2018**, *205*, 44–54. [CrossRef]
27. Zhang, F.; Hu, C.; Zhang, Y. Research for SEBS/PPA compound-modified asphalt. *J. Appl. Polym. Sci.* **2018**, *135*, 46085. [CrossRef]
28. Gama, D.A.; Yan, Y.; Rodrigues, J.K.G.; Roque, R. Optimizing the use of reactive terpolymer, polyphosphoric acid and high-density polyethylene to achieve asphalt binders with superior performance. *Constr. Build. Mater.* **2018**, *169*, 522–529. [CrossRef]
29. Xiao, F.; Amirkhanian, S.; Wang, H.; Hao, P. Rheological property investigations for polymer and polyphosphoric acid modified asphalt binders at high temperatures. *Constr. Build. Mater.* **2014**, *64*, 316–323. [CrossRef]
30. Zhang, F.; Hu, C. The research for SBS and SBR compound modified asphalts with polyphosphoric acid and sulfur. *Constr. Build. Mater.* **2013**, *43*, 461–468. [CrossRef]
31. American Society for Testing and Materials. *ASTM D36, Standard Test Method for Softening Point of Bitumen (Ring-and-Ball Apparatus)*; American Society for Testing and Materials: West Conshohocken, PA, USA, 2014.
32. American Society for Testing and Materials. *ASTM D5, Standard Test Method for Penetration of Bituminous Materials*; American Society for Testing and Materials: West Conshohocken, PA, USA, 2019.
33. American Society for Testing and Materials. *ASTM D113, Standard Test Method for Ductility of Asphalt Material*; American Society for Testing and Materials: West Conshohocken, PA, USA, 2017.
34. American Society for Testing and Materials. *ASTM D4402, Standard Test Method for Viscosity Determination of Asphalt at Elevated Temperatures using a Rotational Viscometer*; American Society for Testing and Materials: West Conshohocken, PA, USA, 2013.
35. American Society for Testing and Materials. *ASTM D1505, Standard Test Method for Density of Plastics by the Density-Gradient Technique*; American Society for Testing and Materials: West Conshohocken, PA, USA, 2003.
36. American Society for Testing and Materials. *ASTM D3418, Standard Test Method for Transition Temperatures and Enthalpies of Fusion and Crystallization of Polymers by Differential Scanning Calorimetry*; American Society for Testing and Materials: West Conshohocken, PA, USA, 2015.
37. American Society for Testing and Materials. *ASTM D4001, Standard Test Method for Determination of Weight-Average Molecular Weight of Polymers by Light Scattering*; American Society for Testing and Materials: West Conshohocken, PA, USA, 2020.
38. BS EN 12593. *Bitumen and Bituminous Binders-Determination of the Fraass Breaking Point*; BSI Standards Limited: London, UK, 2015.

39. American Society for Testing and Materials. *ASTM D7175, Standard Test Method for Determining the Rheological Properties of Asphalt Binder Using a Dynamic Shear Rheometer*; American Society for Testing and Materials: West Conshohocken, PA, USA, 2015.
40. American Society for Testing and Materials. *ASTM D7405, Standard Test Method for Multiple Stress Creep and Recovery (MSCR) of Asphalt Binder Using a Dynamic Shear Rheometer*; American Society for Testing and Materials: West Conshohocken, PA, USA, 2020.
41. American Society for Testing and Materials. *ASTM D6648, Standard Test Method for Determining the Flexural Creep Stiffness of Asphalt Binder Using the Bending Beam Rheometer (BBR)*; American Society for Testing and Materials: West Conshohocken, PA, USA, 2016.
42. American Society for Testing and Materials. *ASTM D7173, Standard Practice for Determining the Separation Tendency of Polymer from Polymer-Modified Asphalt*; American Society for Testing and Materials: West Conshohocken, PA, USA, 2020.
43. Yan, K.; Sun, H.; You, L.; Wu, S. Characteristics of waste tire rubber (WTR) and amorphous poly alpha olefin (APAO) compound modified porous asphalt mixtures. *Constr. Build. Mater.* **2020**, *253*, 119071. [CrossRef]
44. Wang, S.; Yan, K.; Ge, D.; Hong, Z. Laboratory research on the performance of stress-absorption interlayer (SAI) of waste tire rubber and amorphous ploy alpha olefin modified asphalt. *Constr. Build. Mater.* **2019**, *223*, 830–840. [CrossRef]
45. American Association of State Highway and Transportation Officials. *AASHTO Designation: T312-19, Standard Method of Test for Preparing and Determining the Density of Asphalt Mixture Specimens by Means of the Superpave Gyratory Compactor*; American Association of State Highway and Transportation Officials: Washington, DC, USA, 2019.
46. Lesueur, D. The colloidal structure of bitumen: Consequences on the rheology and on the mechanisms of bitumen modification. *Adv. Colloid Interface Sci.* **2009**, *145*, 42–82. [CrossRef]
47. Zhang, L.; Xing, C.; Gao, F.; Li, T.S.; Tan, Y.Q. Using DSR and MSCR tests to characterize high temperature performance of different rubber modified asphalt. *Constr. Build. Mater.* **2016**, *127*, 466–474. [CrossRef]
48. Subhy, A. Advanced analytical techniques in fatigue and rutting related characterisations of modified bitumen: Literature review. *Constr. Build. Mater.* **2017**, *156*, 28–45. [CrossRef]
49. Domingos, M.D.I.; Faxina, A.L.; Bernucci, L.L.B. Characterization of the rutting potential of modified asphalt binders and its correlation with the mixture's rut resistance. *Constr. Build. Mater.* **2017**, *144*, 207–213. [CrossRef]
50. Zhang, J.; Walubita, L.F.; Faruk, A.N.M.; Karki, P.; Simate, G.S. Use of the MSCR test to characterize the asphalt binder properties relative to HMA rutting performance—A laboratory study. *Constr. Build. Mater.* **2015**, *94*, 218–227. [CrossRef]
51. DuBois, E.; Mehta, D.Y.; Nolan, A. Correlation between multiple stress creep recovery (MSCR) results and polymer modification of binder. *Constr. Build. Mater.* **2014**, *65*, 184–190. [CrossRef]
52. Edwards, Y.; Tasdemir, Y.; Isacsson, U. Rheological effects of commercial waxes and polyphosphoric acid in bitumen 160/220—low temperature performance. *Fuel* **2006**, *85*, 989–997. [CrossRef]
53. Safaei, F.; Castorena, C. Material nonlinearity in asphalt binder fatigue testing and analysis. *Mater. Des.* **2017**, *133*, 376–389. [CrossRef]
54. Ren, S.; Liang, M.; Fan, W.; Zhang, Y.; Qian, C.; He, Y.; Shi, J. Investigating the effects of SBR on the properties of gilsonite modified asphalt. *Constr. Build. Mater.* **2018**, *190*, 1103–1116. [CrossRef]
55. Dong, F.; Yu, X.; Liu, S.; Wei, J. Rheological behaviors and microstructure of SBS/CR composite modified hard asphalt. *Constr. Build. Mater.* **2016**, *115*, 285–293. [CrossRef]
56. Karahrodi, M.H.; Jazani, O.M.; Paran, S.M.R.; Formela, K.; Saeb, M.R. Modification of thermal and rheological characteristics of bitumen by waste PET/GTR blends. *Constr. Build. Mater.* **2017**, *134*, 157–166. [CrossRef]
57. Liang, M.; Liang, P.; Fan, W.; Qian, C.; Xin, X.; Shi, J.; Nan, G. Thermo-rheological behavior and compatibility of modified asphalt with various styrene–butadiene structures in SBS copolymers. *Mater. Des.* **2015**, *88*, 177–185. [CrossRef]
58. Pei, X.; Fan, W.; Liu, Y.; Qian, C.; Wang, Y.; Jiang, Y.; Chen, S.; Wang, Z.; Han, L. The effect of oil sands de-oiled asphalt on rheological properties, compatibility and stability of asphalt binder. *Constr. Build. Mater.* **2020**, *263*, 120594. [CrossRef]
59. Liang, P.; Liang, M.; Fan, W.; Zhang, Y.; Qian, C.; Ren, S. Improving thermo-rheological behavior and compatibility of SBR modified asphalt by addition of polyphosphoric acid (PPA). *Constr. Build. Mater.* **2017**, *139*, 183–192. [CrossRef]
60. Olabemiwo, O.M.; Esan, A.O.; Adediran, G.O.; Bakare, H.O. The performance of Agbabu natural bitumen modified with polyphosphoric acid through fundamental and Fourier transform infrared spectroscopic investigations. *Case Stud. Constr. Mater.* **2016**, *5*, 39–45. [CrossRef]
61. Masson, J. Brief review of the chemistry of polyphosphoric acid (PPA) and bitumen. *Energy Fuels* **2008**, *22*, 2637–2640. [CrossRef]
62. Masson, J.; Gagné, M. Polyphosphoric acid (PPA)-modified bitumen: Disruption of the asphaltenes network based on the reaction of nonbasic nitrogen with PPA. *Energy Fuels* **2008**, *22*, 3402–3406. [CrossRef]
63. Masson, J.F. Ionic Pairs in Polyphosphoric Acid (PPA)-Modified Bitumen Insights from Model Compounds. *Energy Fuels* **2008**, *22*, 2637–2640. [CrossRef]
64. Masson, J.F.; Collins, P. Ftir study of the reaction of polyphosphoric acid and modelbitumen sulfur compounds. *Energy Fuels* **2009**, *23*, 440–442. [CrossRef]
65. Ge, D.; Yan, K.; You, L.; Wang, Z. Modification mechanism of asphalt modified with Sasobit and Polyphosphoric acid (PPA). *Constr. Build. Mater.* **2017**, *143*, 419–428. [CrossRef]

materials

MDPI

Article

The Influence of Hydrated Lime and Cellulose Ether Admixture on Water Retention, Rheology and Application Properties of Cement Plastering Mortars

Edyta Spychał * and Ryszard Dachowski

Faculty of Civil Engineering and Architecture, Kielce University of Technology, 25-314 Kielce, Poland; tobrd@tu.kielce.pl
* Correspondence: espychal@tu.kielce.pl

Abstract: In this article, the effect of hydrated lime and cellulose ether on the water retention, rheology, and application properties of plasters was studied. For mortars, the consistency, bulk density, and water retention were tested. Rheological measurements of pastes included yield stress and plastic viscosity. In addition to standard tests of mortars and examining the rheological properties of the pastes, a proprietary method for testing the application properties was proposed. The obtained research results made it possible to evaluate the performance of the tested plasters. An attempt was also made to correlate the rheological properties of pastes (plastic viscosity) to the water retention value. The influence of hydrated lime and cellulose ether on selected properties of pastes and plasters was also presented using the statistical Box–Behnken method. The subjective rating of an expert plasterer confirmed the necessity of the modification of plastering mortars with hydrated lime and cellulose ether. As shown, modification of cement plastering mortar with hydrated lime and cellulose ether at the same time allows obtaining a material with favorable technical and technological properties, especially mortars applied by machine.

Keywords: rheological properties; application properties; water retention; cement; hydrated lime; plaster; cellulose ether; viscosity

Citation: Spychał, E.; Dachowski, R. The Influence of Hydrated Lime and Cellulose Ether Admixture on Water Retention, Rheology and Application Properties of Cement Plastering Mortars. *Materials* **2021**, *14*, 5487. https://doi.org/10.3390/ma14195487

Academic Editor: A. Javier Sanchez-Herencia

Received: 31 August 2021
Accepted: 19 September 2021
Published: 23 September 2021

1. Introduction

The development of new technologies and construction chemicals has contributed to the popularization of dry mortars. The traditionally used cement, cement–lime, and lime plasters are being replaced by materials modified with chemical admixtures and additives. Currently produced mortars consist of several or even a dozen or so raw materials, which are selected so that the final product is characterized by optimal, previously assumed properties. Proper selection of the type and quantity of ingredients in mortars determines the fulfillment of technical and technological requirements. Due to many types and modifications of chemical admixtures, various types of masonry elements, and various substrates, this issue requires a lot of knowledge and practice. A wrong decision during the design of the mortar can lead to failure to meet standard requirements, loss of adhesion, difficulties in plaster application, and the appearance of shrinkage cracks on the material [1–3].

One of the most commonly used polymers in dry-mixed mortars is represented by cellulose ethers. They belong to the group of viscosity-modifying admixtures [4–7], which have been used as ingredients among others of masonry mortars, plasters, and ceramic tile adhesives [2,8–13]. Cellulose ethers are important admixtures in dry-mixed mortars because of their thickening effect and water retention ability [14–16]. Thanks to this admixture, mortars are easy to prepare, comfortable to use, and easy to handle [2,9].

The water retention and rheological properties of cement or lime mortars have been the subject of many studies. Patural et al. [17] found that the molecular weight is crucial to control water retention and consistency in cement mortars. It was noted that, as molecular weight increased, the yield stress diminished, the consistency increased, and the water retention improved. Betioli et al. [18] studied the effect of cellulose (HEMC) ether on the rheology of pastes and cement hydration. The correlation between the results from rheology and isothermal calorimetry was investigated. According to this study, the rheology of cement pastes changes during the induction period, when the reactivity of the system is low. The rheologic changes are due to the agglomeration of particles. In the first hours of hydration, cellulose ether increased the critical strain of the cement pastes measured using the strain sweep test when compared to the plain cement paste, probably by acting as a steric barrier, in addition to the retardation effect, according to the authors. A higher admixture content results in a higher and longer steric effect because the HEMC molecules adsorbed onto the cement particles and on C–S–H enhance the suspension stability due to steric repulsion. In [4], the effect of cellulose ether and guar gum on aerial lime-based mortars was explored. Standard and rheological investigations were supplemented with mercury intrusion porosimetry. The amount of mixing water was the greatest when cellulose ether was added, and this additive did not increase the water-retention capacity of the fresh mixtures. As assessed by the authors, high water absorption through capillarity, high permeability, and a long delay in setting time resulted in an undesirable general performance of this admixture in aerial lime-based mortars. Hydraulic lime-based mortars modified with cellulose ether were studied by the authors of [5,19,20]. In [5], the behavior of mortars with four different cellulose ethers was assessed. The results revealed an elevated air content and water retention in mortars with an increasing dose of polymer admixture, resulting in the decreased density of mortars in plastic state. The mechanical properties of modified mortars surpassed the reference ones at the age of 180 days despite the fact that these materials showed higher open porosity and water absorption. The enhanced porosity of mortars resulted in an improvement of their frost resistance and faster carbonation. Summarizing the research conducted, the authors stated that the effect of cellulose ether on hydraulic lime mortars is very similar to that on cement mortars, but it is different to that on air lime-based mortars.

In addition to the consistency, water retention, and rheological properties, the influence of cellulose ethers on cement hydration has also been assessed [6,21–23]. On the basis of the experimental results, it can be concluded that the addition of cellulose ether has a significant influence on the early hydration of cement, and this effect practically disappears after longer period. Cellulose ethers lead to a gradual slowing down of the C_3A hydration depending on admixture chemistry. According to observations, substitution groups (nature and content) seem to be more important controlling factors for C_3A hydration than molecular mass. The amount of the polymer admixture used is most important in the process of setting pastes and mortars. A much-reduced impact is visible when taking into account the viscosity of the admixture or the type of binder used.

The research on cement hydration was supplemented in [23] with a study of the microstructure observations under SEM (after 24 h of hydration). The microstructure of cement paste with cellulose ether had a high content of gypsum crystals. As shown by the EDS analyses of C–S–H in the paste with cellulose ether, this phase had a very low content of aluminum and sulfur. This confirmed the absence of monosulfate in this sample, as the content of aluminum and sulfur in the classical C–S–H phase increases when it forms a nano-mixture with monosulfate. As the author proved, the cellulose ether admixture inhibited the reaction of the C_3A phase with sulfate ions, which would have led to the formation of ettringite and monosulphate. In [24], the morphology of the hydrate crystals and the microstructure of the hardened cement matrix modified with polyvinyl alcohol–acetate and two cellulose ethers were studied. Polymer modification improved the cohesion of the bulk cement paste, while fewer microcracks were observed for the polymer-modified mortars.

Another analyzed issue in the literature is the influence of cellulose ethers on the shrinkage of mortars [25,26]. Messan et al. [25] presented a new testing method for investigating early-age shrinkage of cement-based material modified with glass fiber, cellulose ether, and redispersible powder. Results of this research showed the importance of the rheology of fresh mortar with regard to early-age shrinkage development. The authors proved that the admixture of cellulose ether decreases the early-age evaporation rate of cement-based materials. This has also been observed in the literature [26]. The use of a viscosity-modifying admixture showed a beneficial effect on mortar shrinkage. According to the authors, the action of cellulose ether reduces the surface tension of the pore solution, thus reducing capillary forces. Long-term water-binding capacity and high-level mortar humidity also contribute to reducing shrinkage deformation.

The nature and the scope of cellulose ether activities are not fully recognized and understood, especially in the case of cement–lime mortars. In addition, there is no information in the literature on how cellulose ether affects their application properties, which is an important issue in the case of plasters. Scientific research in the literature has focused on the assessment of the effect of this admixture on cement [1,9,22,23,27,28], lime [4,5,7,20], and gypsum [29] pastes and mortars. The research carried out by various authors has mainly covered the subject of the influence of this polymer admixture on the standard parameters, consistency, water retention, and rheological properties, among others.

The test results and their analysis presented in this article focus on the assessment of plasters using proprietary methods supplemented with standard methods of pastes and mortars. Importantly, the tests carried out can be applied to both cement-based mortars and cement–lime mortars.

2. Materials and Methods

2.1. Materials

A commercial Portland cement CEM I 42.5 R (Cemex, Chełm, Poland) [30], deeply separated hydrated lime (Alpol, ZSChiM "PIOTROWICE", Sitkówka, Poland) [31], two fractions of quartz sand 0.1–0.5 mm and 0.2–0.8 mm (Grudzeń Las, Grudzeń Las, Poland) [32], a polymer admixture of different viscosity and amount (WALOCEL, The Dow Chemical Company, Midland, MI, USA) [33], and tap water were used. Among the wide variety of cellulose ethers, the following were chosen in this research: hydroxypropyl methylcellulose (HPMC) with viscosity 3000 mPa·s and hydroxyethyl methylcellulose (HEMC) with viscosity 25,000 mPa·s and 45,000 mPa·s (for measurements using a Brookfield rheometer at 20 °C). These polymer admixtures had the form of a very fine white powder, with a grain size of less than 0.063 mm and a low degree of chemical modification. The chemical composition and physical properties of cement are given in Table 1. The chemical composition of hydrated lime is shown in Table 2.

Silicate blocks of N12 type with dimensions 0.25 m × 0.12 m × 0.22 m (H+H Company, Warsaw, Poland) [34] were used to test the application properties of plasters (for bricklaying the walls). The basic properties of these masonry elements are presented in Table 3.

Table 1. Chemical composition and physical properties of cement.

Chemical Composition (wt.%)		Physical Properties	
SiO_2	20.22	Water requirement of normal consistency (%)	28.8
Al_2O_3	4.43	Initial setting (min)	173
Fe_2O_3	3.65	Final setting (min)	237
CaO	64.06	Specific surface area (m^2/kg)	387.9
Na_2O	0.29	2 day compressive strength (MPa)	28.9
MgO	1.24	28 day compressive strength (MPa)	59.5
SO_3	3.31	2 day flexural strength (MPa)	5.1
Cl	0.093	28 day flexural strength (MPa)	8.4
K_2O	0.54	Loss on ignition (%)	3.81
Free CaO	1.83		

Table 2. Chemical composition of hydrated lime.

Chemical Composition (wt.%)	
CaO + MgO	95.17
MgO	0.80
CO_2	1.86
SO_3	0.41

Table 3. Chosen properties of silicate blocks.

Water absorption (%)	14 ± 2
Compressive strength (MPa)	15.0
Bond strength with PN-EN 998-2 standard, GPM [1] (MPa)	0.15
Bond strength with PN-EN 998-2 standard, TLM [2] (MPa)	0.30
Heat conductivity (W/m·K)	0.46
Average block weight (kg)	9.5

[1] For general-purpose mortars and light mortars; [2] for thin pointing mortars.

2.2. *Mix Proportions of Pastes and Mortars*

The Box–Behnken experiment plan was used to analyze the plasters [35,36]. Pastes and mortars were tested by taking into account the selected experimental plan. The analysis of the test results was performed using the STATISTICA computer program (StatSoft Company, Cracow, Poland) [37]. A solution was selected, as a result of which it was possible to conduct research for a three-factor, three-level model. A matrix with values of the coded factors is presented in Table 4, while a matrix with selected types of mortars and their composition is presented in Table 5. According to the selected experiment plan, it was necessary to carry out 15 elementary experiments each time—13 different experiments (13 mixtures of different compositions) and two repetitions (for the mixture corresponding to the central point of the Box–Behnken plan). Additionally, a base cement mortar was prepared and tested, which contained neither hydrated lime nor cellulose ether (paste and mortar marked with the symbol C0). The composition of the pastes was similar to the composition of the mortars, excluding fine aggregate.

Table 4. Box–Behnken experiment plan.

No.	Experiment Plan			Experiment Plan		
	X_1	X_2	X_3	X_1 Amount of Hydrated Lime [1] %	X_2 Amount of Cellulose Ether [2] %	X_3 Viscosity of Cellulose Ether mPa·s
1	−1	−1	−1	0	0.52	25,000
2	−1	1	0	0	3.12	25,000
3	−1	0	−1	0	1.82	3000
4	−1	0	1	0	1.82	45,000
5	1	−1	0	50	0.52	25,000
6	1	0	1	50	1.82	45,000
7	1	1	0	50	3.12	25,000
8	1	0	−1	50	1.82	3000
9	0	−1	−1	25	0.52	3000
10	0	1	−1	25	3.12	3000
11	0	−1	1	25	0.52	45,000
12	0	1	1	25	3.12	45,000
13	0	0	0	25	1.82	25,000
14	0	0	0	25	1.82	25,000
15	0	0	0	25	1.82	25,000

[1] The percentage of hydrated lime given in relation to the total amount of binder in the mortar; [2] the percentage of cellulose ether given in relation to the amount of binder in the mortar.

Table 5. Mortar and paste compositions defined according to the Box Behnken model.

Abbreviation of Mortar (Paste)	Amount of Binder (Cement + Lime) (g)	Amount of Sand 0.1–0.5 mm (g)	Amount of Sand 0.2–0.8 mm (g)	Amount of Cellulose Ether (g)
C0 [1]	96	437	467	-
C-0.52MV	96	437	467	0.50
C-3.12MV	96	437	467	3.00
C-1.82LV	96	437	467	1.75
C-1.82HV	96	437	467	1.75
C50L-0.52MV	48 + 48	437	467	0.50
C50L-1.82HV	48 + 48	437	467	1.75
C50L-3.12MV	48 + 48	437	467	3.00
C50L-1.82LV	48 + 48	437	467	1.75
C25L-0.52LV	72 + 24	437	467	0.50
C25L-3.12LV	72 + 24	437	467	3.00
C25L-0.52HV	72 + 24	437	467	0.50
C25L-3.12HV	72 + 24	437	467	3.00
C25L-1.82MV	72 + 24	437	467	1.75
C25L-1.82MV	72 + 24	437	467	1.75
C25L-1.82MV	72 + 24	437	467	1.75

[1] Base mortar (paste) without hydrated lime or cellulose ether.

Factors that were kept constant throughout this study were as follows:

- Total amount of binder in the mortar (96 g of binder per 1000 g of dry ingredients);
- Ratio of binder to fine aggregate (1:10);
- Consistency measured using a flow table (165 mm).

On the basis of the experiments carried out in accordance with the Box–Behnken model, it was possible to determine the quadratic formula (Equation (1)) for the tested property in each case, which was used to estimate the response area [36,38]. In Equation (1), Y_i is the dependent variable or output variable (tested parameter of pastes/mortars), b_0–b_{33} are factors, and X_1, X_2, and X_3 are dependent variables (input factors). For the tested properties of pastes and mortars, the value of the standard deviation was determined for the central point of the Box–Behnken experiment plan, i.e., sample numbers 13, 14, 15 (mortar abbreviated as C25-18.82MV).

$$Y_i = b_0 + b_1X_1 + b_2X_2 + b_3X_3 + b_{11}X_1^2 + b_{22}X_2^2 + b_{33}X_3^2 + b_{12}X_1X_2 + B_{12}X_1X_3 + b_{23}X_2X_3. \quad (1)$$

2.3. Methods

All pastes and mortars were prepared in an air-conditional laboratory at a temperature of 20 ± 2 °C and a relative humidity of 65% ± 5%. Binders, fine aggregate, and water were weighed with an accuracy of 0.1 g, and the chemical admixture was weighed with an accuracy of 0.0001 g. The mixing method was the same for all samples: 90 s of mechanical stirring at 45 rpm + 30 s pause + 90 s of mechanical stirring at 57 rpm.

Mortar consistency was assessed using two methods: a flow table according to the PN-EN 1015-3:2000 standard [39] and the fall cone method according to the PN-B-04500:1985 standard [40]. The consistency was measured using the method in [39] as the average of two measurements of the mortar flow on the flow table. The consistency measured using the method in [40] was defined as the depth of immersion of the cone in the tested mortar.

The bulk density of fresh mortars was calculated according to PN-EN 1015-6:2000 + A1:2007 standard [41].

The water retention value (WRV) was determined according to the procedure described in [42]. This parameter denotes the percentage water content remaining in the tested mortar after short-term contact with the filter paper. Each mortar was tested three times, determining its value after 10, 30, and 60 min, e.g., WRV10, WRV30, and WRV60. This nonstandard test allowed evaluating the behavior of the mortar in contact with a material (masonry element) with varied adsorption capacity.

Rheological measurements were carried out using the Viskomat NT rheometer (Schleibinger Testing Systems, Buchbach, Germany). The rheological behavior of pastes is described by the Bingham model [43–45], using g (N·mm) and h (N·mm·s) as parameters describing yield stress and plastic viscosity. These properties of the pastes were determined on the basis of the flow curves for decreasing shear rates, after 10 and 60 min. The temperature of pastes between individual measurements was constant (20 °C), set by an automatic thermostatic system. Each sample was well protected against water evaporation by being placed in a sealed container for the periods between studies.

The application properties of mortars were estimated on the basis of a subjective rating by an expert plasterer (Edyta Spychał's original method, which was used in her Ph.D. thesis) [46]. The first stage was to build a wall of silicate blocks (the masonry elements were seasoned for a period of 30 days in the research conditions). Then, the wall was primed. An image of the laid brick walls is shown in Figure 1, one with dimensions of 87.5 cm width and 154 cm height, and the other with dimensions of 137.5 cm width and 154 cm height.

Figure 1. View of the wall before plastering.

The assessment of mortars was carried out in two stages. On the first day, a specialist plastered the walls and evaluated the fresh materials in terms of consistency, viscosity, the ease of plastering, adhesion to the substrate, and potential difficulties during plastering. On the second day, the ease of surfacing (smoothing, surface correction) was determined.

The analysis of application properties consisted of the following elements:

1. Evaluation of application properties:
    - On a 0–10 point scale (10—material with the best application properties);
    - Based on a detailed descriptive assessment;
    - Based on the possible method of application: manual and/or machine-assisted).
2. Assessment of the ease and quality of surfacing:
    - On a 0–10 point scale (10—material with the best properties);
    - Based on a detailed descriptive assessment.

This article is a part of a wider analysis in which the influence of cellulose ether and hydrated lime on the selected properties of plasters was evaluated [46]. Some research in this direction was previously introduced in [2,6,10].

3. Results

3.1. Consistency and Bulk Density Measurements

In Table 6, the results of the amount of water in mortars, the consistency, and the bulk density for all samples are presented.

Table 6. Selected properties of fresh mortars.

Symbol of Mortar	Amount of Water (g)	Consistency (cm)	Bulk Density (kg/m^3)
C0	195	6.1	1950
C-0.52MV	148	6.8	1670
C-3.12MV	220	9.0	1210
C-1.82LV	142	8.8	1450
C-1.82HV	185	9.3	1310
C50L-0.52MV	176	8.5	1570
C50L-1.82HV	195	8.5	1390
C50L-3.12MV	225	9.1	1240
C50L-1.82LV	144	8.9	1470
C25L-0.52LV	165	6.8	1790
C25L-3.12LV	170	9.8	1210
C25L-0.52HV	165	9.0	1610
C25L-3.12HV	250	9.9	1250
C25L-1.82MV	175	8.8	1330
C25L-1.82MV	175	9.0	1330
C25L-1.82MV	175	8.8	1330

The consistency of the mortars measured using a flow table was selected experimentally, in accordance with the assumptions of the research, i.e., at a constant level of 165 mm. According to the values given in Table 6 and Figure 2, it can be seen that the amount of water needed to obtain the required consistency was highly variable and ranged from 142 g to 250 g. The amount of water needed for the reference mortar (C0) was 195 g. For the three mortars C-3.12MV, C50L-3.12MV, and C25L-3.12 HV, to obtain the required consistency, at least 220 g of water was required. This value was mainly due to the viscosity of cellulose ether used, which, in these cases, was 25,000 mPa·s and 45,000 mPa·s, as well as the maximum added amount of cellulose ether, which, in these cases, was 3.12%. Hence the conclusion, than when selecting the required amount of water, the amount and viscosity of this admixture first of all should be taken into account.

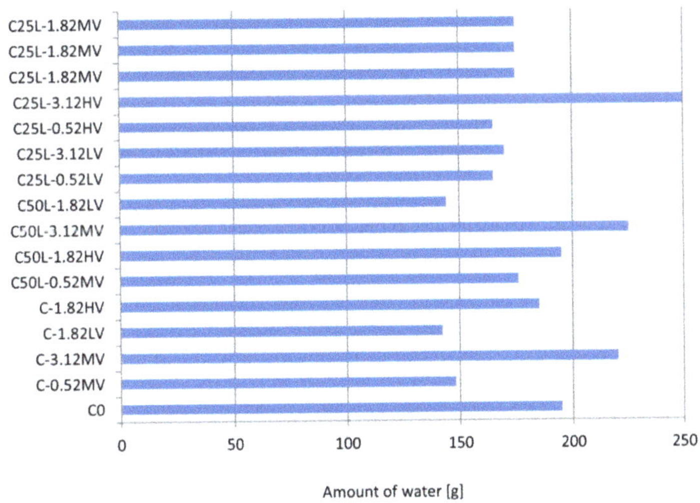

Figure 2. Amount of water in mortars.

Obtaining consistency at the same level for cement and cement–lime mortars modified with cellulose ether was associated with an increase in the amount of water when replacing a part of the cement binder with lime. This was due to the smaller grain size of lime compared to the grain size of cement, thus leading to its higher water demand (C-0.52MV and C50L-0.52MV mortars).

The use of cellulose ether in mortars is associated with a greater demand for water in order to obtain a consistency at a similar level compared to unmodified mortar. An increase in the admixture content is usually associated with an increase in the water demand of the mortar [3,12]. This research shows that this relationship was not always true, as observed for the examples of C-0.52MV, C-1.82LV, C50L-1.82LV, C25L-0.52LV, and C25L-0.52HV materials. These plasters required less water to achieve the assumed consistency, compared to the reference mortar C0. The reasons for this phenomenon are due to the properties of cellulose ether. When the viscosity of the liquid phase surrounding the grains of the binder and fine aggregate is too low, the dissolved polymer acts as a "lubricant", facilitating the movements of the mortar grains [29,47]. In this case, the action of the cellulose ether is similar to that of a plasticizing or fluidizing admixture.

Taking into account the test results from Table 6 and Figure 3, it can be concluded that the consistency of the mortars varied, ranging from 6.1 cm (C0) to 9.9 cm (C C25L-3.12HV). The standard deviation of the consistency test results [40] was equal to 0.094 cm. All tested mortars achieved a consistency similar to that of typical plasters used in practice [42,48]. The consistency of plastering mortars intended for manual application should be 6–9 cm, whereas the equivalent value for mechanical application should be 8–11 cm. The results obtained in this research fall within the range of consistency given in the literature [48].

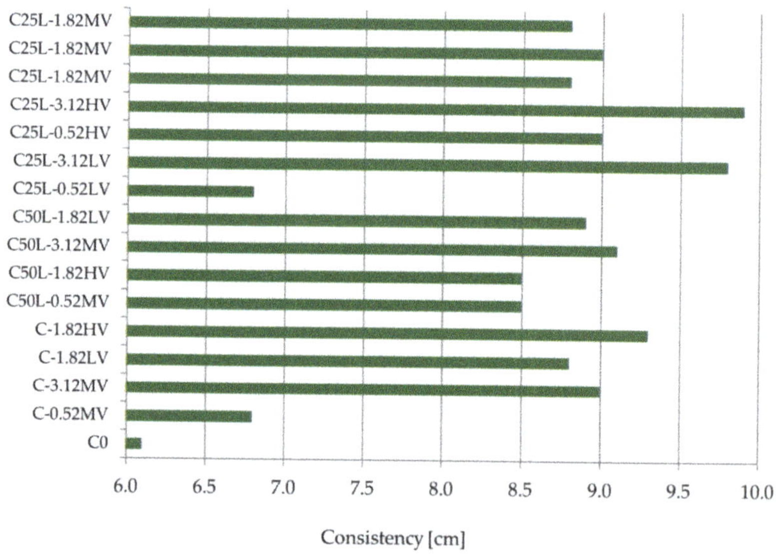

Figure 3. Consistency results of mortars according to PN-B-04500:1985.

Table 6 and Figure 4 present the range of bulk density values of fresh mortars. The standard deviation of the bulk density test results was equal to 0.000 kg/m^3. The lowest bulk density was achieved using the cement mortar modified with cellulose ether in the amount of 3.12% with a viscosity of 25,000 mPa·s (1210 kg/m^3—C3.12-MV), while the highest value was achieved by the unmodified mortar (C0)—1950 kg/m^3. This parameter can be used to indirectly judge the performance of the plasters (raw material efficiency). The material with the lowest bulk density is characterized by the best performance. Mortar efficiency is important not only from an economic but also from a technological point of view. All modified mortars had a lower bulk density than the base mortar, which reduced the consumption of materials. Thus, cellulose ether and hydrated lime had a positive effect on increasing the efficiency of the tested mortars, as well as reduced the bulk density. This result is in agreement (for cement mortar) with the literature [3,13].

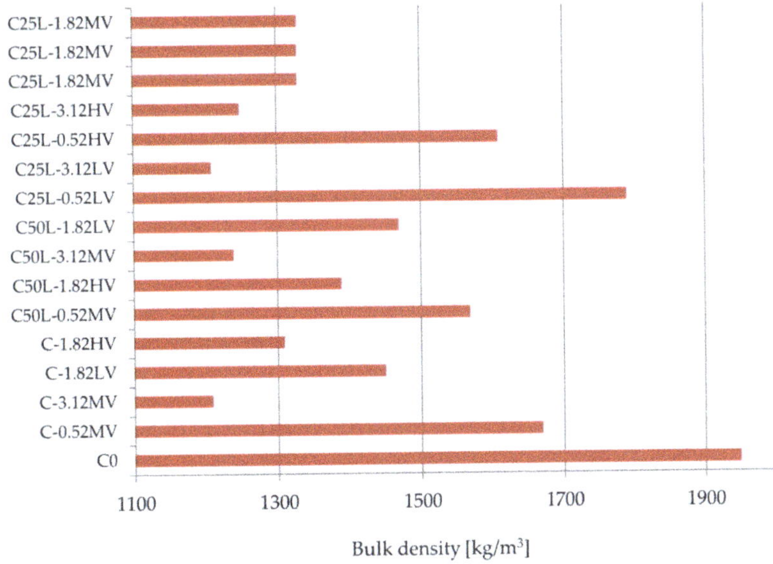

Figure 4. Results of bulk density of fresh mortars.

3.2. Water Retention of Mortars

Table 7 and Figure 5 present the results of the water retention value (WRV) after 10, 30, and 60 min. The standard deviation of the WRV test results was equal to 0.000% for WRV10 and WRV60 and equal to 0.047% for WRV30.

Table 7. Water retention results.

Symbol of Mortar	WRV10 (%)	WRV30 (%)	WRV60 (%)
C0	85.0	78.8	76.5
C-0.52MV	95.7	90.0	89.7
C-3.12MV	99.7	99.2	98.8
C-1.82LV	99.5	98.4	98.8
C-1.82HV	99.7	98.9	99.2
C50L-0.52MV	94.5	86.9	86.2
C50L-1.82HV	99.7	99.2	99.1
C50L-3.12MV	99.7	99.2	99.2
C50L-1.82LV	99.8	99.2	99.2
C25L-0.52LV	92.1	87.7	84.2
C25L-3.12LV	99.7	99.2	99.1
C25L-0.52HV	95.0	91.6	89.5
C25L-3.12HV	99.7	99.5	99.4
C25L-1.82MV	99.7	99.3	99.2
C25L-1.82MV	99.7	99.2	99.2
C25L-1.82MV	99.7	99.2	99.2

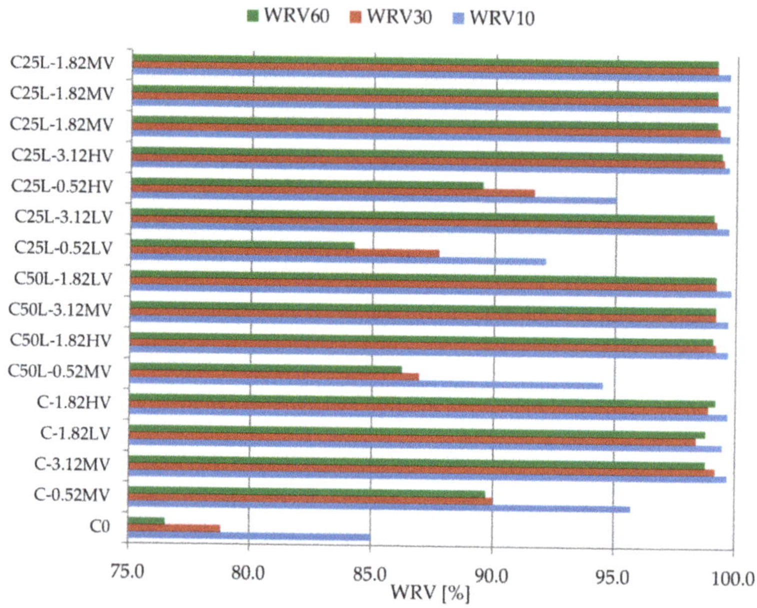

Figure 5. Change in water retention of tested mortars over time (WRV parameter).

Comparing the WRV parameter of the mortars, it can be seen that the reference mortar C0 had an initial value approximately 10–15% lower than that of the other mortars. After 15 min, is the water retention was 85.0%; however, after 60 min, it was 76.5%, which is a disadvantage, especially when the plaster is applied on a substrate with high absorbency or when plastering take place in conditions of changing temperature and humidity. In the case of materials C-0.52MV, C50L-0.52MV, C25L-0.52MV, and C25L-0.52MV WRV10, the water retention was greater than 90%, decreasing after 30 and 60 min due to the lowest cellulose ether content of 0.52% in each case. This proves the significant influence of the amount of admixture in terms of water retention in the mortar and its action over time. For the remaining materials, the water retention value remained at a level close to 100% throughout the entire study. Comparing C50L-1.82HV and C50L-1.82LV mortars with the same proportions of binder and amount of admixture, there was no visible influence of the polymer viscosity. The partial change from cement to lime binder was not as noticeable and important as the change in the amount of cellulose ether. According to Brumaud et al. [15], mortars can be classified as follows: C0 mortar would be a material with low water retention (WRV < 86%), C25L-0.52LV mortar would be characterized by medium water retention (86% < WRV < 94%), and the remining samples would be considered as having high water retention (WRV > 94%).

When assessing the test results of the water retention value in mortars after 10 min of testing, similar relationships were obtained in comparison with the literature data. The presence of cellulose ether in cement or lime mortars increases water retention [7,16,49]. Information about the effective action of the admixture in cement–lime mortars, which is not available in the literature, was confirmed in this article. In the case of mortars with a cement–lime binder, a strong influence of the polymer in terms of water retention was visible throughout the entire measurement. Figure 6 presents the response surface of the influence of X_1, X_2, and X_3 factors on WRV10 (Y_1). The fit factor of the R^2 model in this case was 0.962, which denotes that this model explained 96.2% of changes in the response value. As can be seen from Equations (2)–(4), the most important factor influencing WRV10 was the amount of cellulose ether (X_2 factor). For this factor, the significance level was $p < 0.05$, suggesting that the X_2 factor had a statistically significant influence on the examined feature.

Statistically, the amount of hydrated lime and cellulose ether viscosity had a lesser effect on the parameter tested. As the amount of admixture in the mortars increased, the WRV10 increased. Thus, modifying plasters with cellulose ether is a favorable solution from the point of view of water retention in the mortar. The utility level shown in the range from 0 to 1 in Figure 6 determines the influence of the defined factors on the tested parameter, suggesting the optimal arrangement of the factors X_1, X_2, and X_3. Adding as little as 1.2% cellulose ether enabled obtaining a high WRV10 (Figure 6).

$$Y_1 = 91.519 - 0.06X_1 + 0.0016{X_1}^2 + 6.002X_2 - 1.213{X_2}^2, \quad (2)$$

$$Y_1 = 91.519 - 0.06X_1 + 0.001{X_1}^2 + 0.0001X_3, \quad (3)$$

$$Y_1 = 91.519 + 6.002X_2 - 1.213{X_2}^2 + 0.0001X_3. \quad (4)$$

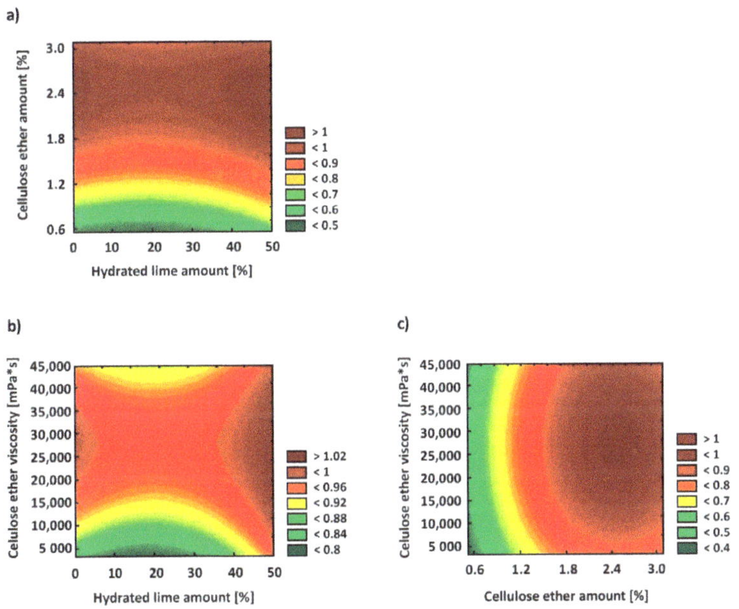

Figure 6. Utility function for the WRV10 test of mortars. (**a**) correlation between cellulose ether amount and hydrated lime amount; (**b**) correlation between cellulose ether viscosity and hydrated lime amount; (**c**) correlation between cellulose ether viscosity and cellulose ether amount.

3.3. Rheological Properties of Pastes

Table 8 and Figure 7 present the results of the rheological properties of pastes after 10 and 60 min. The standard deviation of the yield stress test results was equal to 0.354 N·mm for the measurements after 10 min and 0.514 N·mm for the measurements after 60 min, whereas the standard deviation of the plastic viscosity test results was equal to 0.052 N·mm·s for measurements after 10 min and 0.233 N·mm·s for the measurements after 60 min. When analyzing the above test results, it can be seen that the yield stress and plastic viscosity of the reference paste were the smallest. With an increase in the amount and viscosity of cellulose ether, both parameters increased after 10 and 60 min. Paste C50L-3.12MV with 50% hydrated lime content, cellulose ether content 3.12%, and viscosity 25,000 mPa·s had the biggest yield stress and plastic viscosity. The test results obtained for C50L-1.82HV, C50L-3.12MV, C25L-3.12HV, and C25L-1.82MV pastes, which translated into large values of rheological parameters, were associated with a higher percentage of cellulose ether and a high viscosity, in combination with the replacement of a part of the cement binder with hydrated lime binder. Information in addition to the measurement results of rheological parameters was also provided by the observations of the paste samples during the research. It is known that hydrated lime and cellulose ether affect the workability and plasticity of the material, as well as its cohesion, which was confirmed during the rheological tests. The appearance of C0 and C-0.52MV pastes after 60 min differed from that of the other samples, as shown in Figure 8a,b. In particular, visible differences were observed for the C0 paste. In the case of this sample (C0 paste), a clear separation of water from the rest of the ingredients could be noticed. The separation of water in the C-0.52MV paste was also visible, but to a smaller extent. Both samples (C0 and C-0.52MV) were characterized by the sedimentation of ingredients and a lack of homogeneity through the mass, confirming the successful modification of the pastes and mortars. Despite the fact that the C-0.52MV paste was modified with an admixture, the amount of cellulose ether seemed to be insufficient to prevent segregation of the ingredients of the paste. Comparing samples with the same amount and viscosity of cellulose ether, but differing in the type of binder (C-0.52MV and C50L-0.52MV), a beneficial effect of hydrated lime was observed. In sample C50L-0.52MV, there was no visible sedimentation of the components, while the leaven was coherent and homogeneous throughout the mass.

Table 8. Rheological properties.

Abbreviation of Paste	Yield Stress after 10 min, g (N·mm)	Yield Stress after 60 min, g (N·mm)	Plastic Viscosity after 10 min, h (N·mm·s)	Plastic Viscosity after 60 min, h (N·mm·s)
C0	0.09	0.08	0.43	0.43
C-0.52MV	0.48	0.35	2.17	2.13
C-3.12MV	11.78	14.38	10.26	13.58
C-1.82LV	1.43	0.62	6.32	5.89
C-1.82HV	2.02	4.22	3.66	5.73
C50L-0.52MV	1.31	1.07	2.60	2.43
C50L-1.82HV	15.06	16.33	9.19	12.07
C50L-3.12MV	31.81	26.89	20.39	19.21
C50L-1.82LV	3.48	2.43	10.54	9.40
C25L-0.52LV	0.32	0.46	1.32	1.30
C25L-3.12LV	5.47	6.29	11.31	13.78
C25L-0.52HV	1.61	0.97	3.82	3.31
C25L-3.12HV	22.86	22.98	13.42	12.79
C25L-1.82MV	12.13	10.97	11.98	12.18
C25L-1.82MV	12.88	11.15	11.86	12.75
C25L-1.82MV	12.13	12.14	11.95	12.43

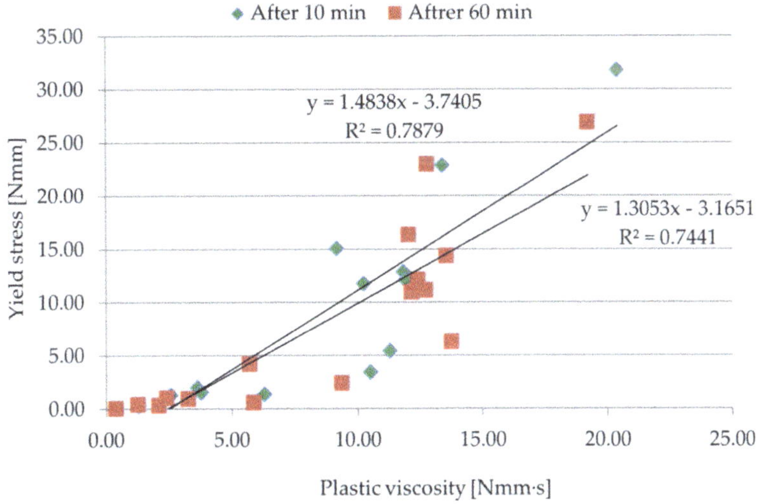

Figure 7. Correlation between yield stress and plastic viscosity of pastes.

(**a**) (**b**)

Figure 8. Photos showing sedimentation of pastes: (**a**) paste C0; (**b**) paste C-0.52MV.

A correlation between plastic viscosity and yield stress was established, as shown in Figure 7. The trend showed an increase in yield stress with the increase in plastic viscosity, which is consistent with the action of hydrated lime and cellulose ether. According to [50], there is a high correlation and significant interdependence of the examined parameters.

Figure 9 presents the response surface of the influence of X_1, X_2, and X_3 factors on the plastic viscosity after 10 min (Y_2). The fit factor of the R^2 model in this case was 0.902. As can be seen from Equations (5)–(7), the most important factor influencing the plastic viscosity was the amount of cellulose ether (X_2 factor). Statistically, the amount of hydrated lime and cellulose ether viscosity had a lesser effect on the parameter tested.

As the amount of admixture in the mortars increased, the plastic viscosity increased (similar to the statistical evaluation of the WRV).

$$Y_2 = -7.031 + 0.226X_1 - 0.002X_1^{2} + 7.641X_2 - 0.898X_2^{2}, \tag{5}$$

$$Y_2 = -7.031 + 0.226X_1 - 0.002X_1^{2} + 0.0003X_3, \tag{6}$$

$$Y_2 = -7.031 + 7.641X_2 - 0.898X_2^{2} + 0.0003X_3. \tag{7}$$

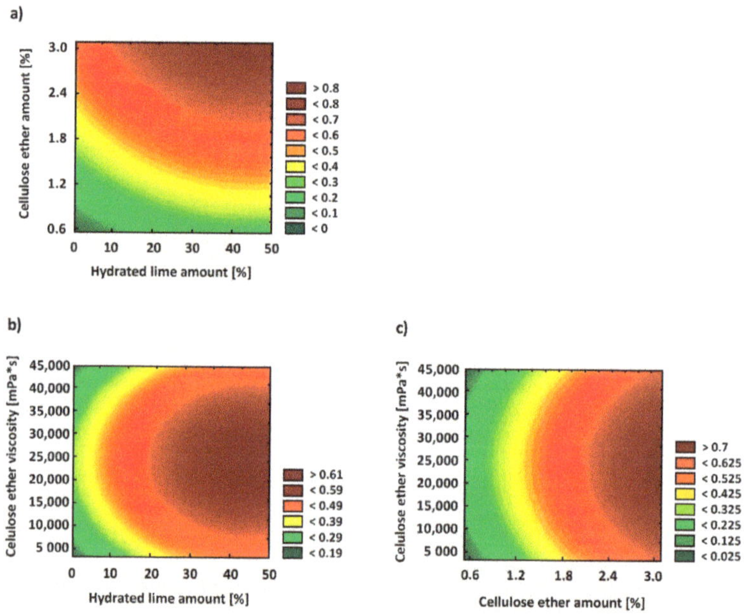

Figure 9. Utility function for the plastic viscosity after 10 min of pastes: (**a**) correlation between cellulose ether amount and hydrated lime amount; (**b**) correlation between cellulose ether viscosity and hydrated lime amount; (**c**) correlation between cellulose ether viscosity and cellulose ether amount.

3.4. Application Properties of Mortars

Table 9 shows the scoring of the quality of plastering mortars, the results of the assessment of the performance of the tested materials, and the methods of application of the plasters. The C-0.52MV, C-1.82HV, C50L-1.82HV, and C25L-1.82MV mortars received the highest score, followed by the C-1.82LV, C50L-0.52MV, and C50L-1.82LV mortars. The remaining plasters were of very poor quality. According to a specialist plasterer, mortars with the highest scores were those can be used as commercial products without any additional changes in their composition, whereby some of the tested mortars could be applied manually and others could be applied by machine. Mortar C50L-1.82HV was deemed a universal material, which could be applied manually or by machine. The methods of application were determined on the basis of the rheological properties. Mortars fit for a manual method of application had low plastic viscosity, whereas those fit for machine application had high viscosity. High plastic viscosity facilitates the application of the mortar, whereby the material clings to the float and does not run off the wall. Hydrated lime also improved the flexibility of the plasters and the ease of processing. Hydrated lime and cellulose ether were essential in the case of machine-applied plasters due to the method specificity. In addition, hydrated lime facilitated plaster processing, and cellulose ether enabled longer water retention in the mortar.

Table 9. Application properties of mortars.

Abbreviation of Mortar	The First Stage of the Assessment	The Second Stage of the Assessment	Manual Method of Application	Machine-Assisted Method of Application
C0	2	0	+	−
C-0.52MV	8.5	8	+	−
C-3.12MV	3	1	+	−
C-1.82LV	3	6	+	−
C-1.82HV	9	8	+	−
C50L-0.52MV	2	6	+	−
C50L-1.82HV	8	8	+	+
C50L-3.12MV	3.5	3.5	−	+
C50L-1.82LV	3.5	8	+	−
C25L-0.52LV	2	6	+	−
C25L-3.12LV	4	1	+	−
C25L-0.52HV	4	1	+	−
C25L-3.12HV	4	1	−	+
C25L-1.82MV	7	8	−	+
C25L-1.82MV	7	8	−	+
C25L-1.82MV	7	8	−	+

Figure 10 presents images of the walls during application (first-day assessment of application properties) and processing (second-day assessment).

(a) (b)

Figure 10. Images of walls (**a**) during application, and (**b**) during processing.

Due to the fact that the point assessment did not fully reflect the properties of each mortar, the results on the point scale were supplemented with a descriptive assessment. Table 10 shows a description of the application properties of the selected mortars. Materials were also visually assessed after 28 days of setting and hardening processes. Mortars containing hydrated lime in their composition had a much lighter color, which is advantageous from a technological and economic point of view. This allows a reduction in paint consumption for coating.

Table 10. Descriptive assessment of the workability of selected mortars.

Symbol of Mortar	Evaluation of Workability	Evaluation of Processing Quality
C0	Wet sand consistency; low viscosity	Lack of processing capabilities
C-0.52MV	Consistency good for manual coating; low efficiency of plaster; good workability properties	Plaster binds evenly, with good processing time
C-3.12MV	The consistency is very fluid, sticks too much to the float, and runs off the wall after a long time	Plaster is dry on top, wet inside, and characterized by a false bonding, along with difficult processing
C50L-1.82HV	Material with the best application properties and easy application; plaster clings to the float; consistency is very good; plaster is efficient	Plaster does not need water for final treatment

Figure 11 presents the response surface of the influence of X_1, X_2, and X_3 factors on the subjective rating by a specialist (Y_3). The fit factor of the R^2 model in this case was 0.749. Statistically, as can be seen from Equations (8)–(10), the most important factor was the amount of cellulose ether (X_2 factor). Adding as little as 1.82% cellulose ether led to favorable application properties (Figure 11).

$$Y_3 = 0.685 - 0.11X_1 - 0.0018{X_1}^2 + 13.932X_2 - 4.29{X_2}^2, \quad (8)$$

$$Y_3 = 0.685 - 0.11X_1 - 0.0018{X_1}^2 + 0.0003X_3, \quad (9)$$

$$Y_3 = 0.685 + 13.932X_2 - 4.29{X_2}^2 + 0.0003X_3. \quad (10)$$

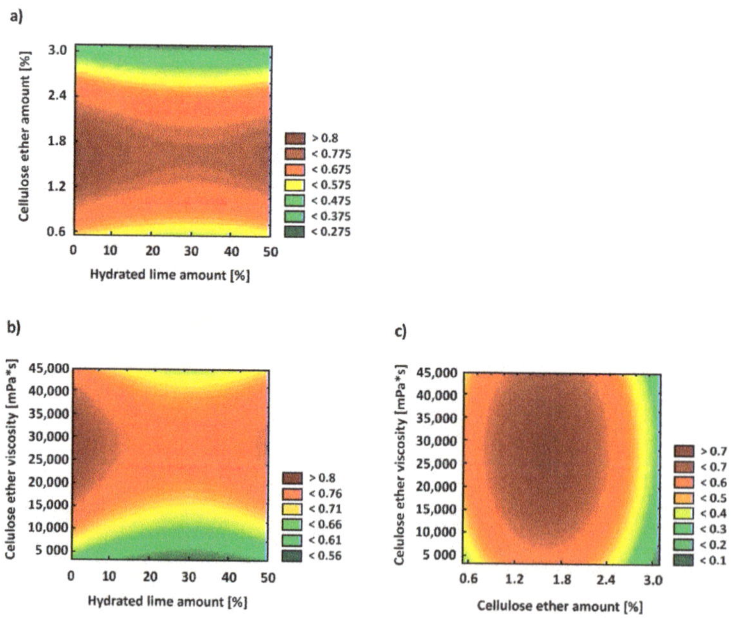

Figure 11. Utility function for subjective rating of mortars by a specialist: (**a**) correlation between cellulose ether amount and hydrated lime amount; (**b**) correlation between cellulose ether viscosity and hydrated lime amount; (**c**) correlation between cellulose ether viscosity and cellulose ether amount.

3.5. Correlation between Rheological Properties of Pastes and Chosen Properties of Mortars

A correlation between the rheological properties (plastic viscosity) and WRV was established, as shown in Figure 12 (correlation experimental results after 10 min) and in Figure 13 (correlation experimental results after 60 min). Comparing the individual test results, one can notice a high correlation and interdependence of the plastic viscosity and WRV, especially for materials based on a cement–lime binder (Figures 12b and 13b). The trend showed an increase in WRV with the increase in plastic viscosity. For pastes with a plastic viscosity over 8 N·mm·s, WRV10 and WRV60 were all within the limits. Similar correlations between the rheological properties and water retention value were obtained in the case of cement and lime pastes and mortars in [49,51].

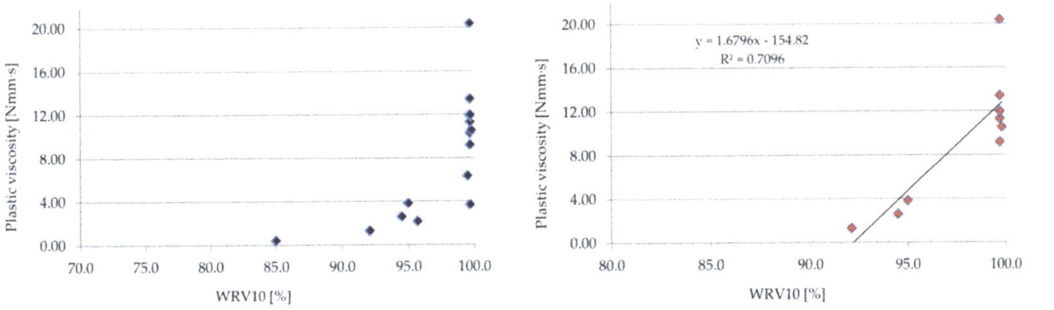

(a) (b)

Figure 12. Correlation between plastic viscosity and WRV10 (**a**) for all samples, and (**b**) only for samples with cement–lime binder.

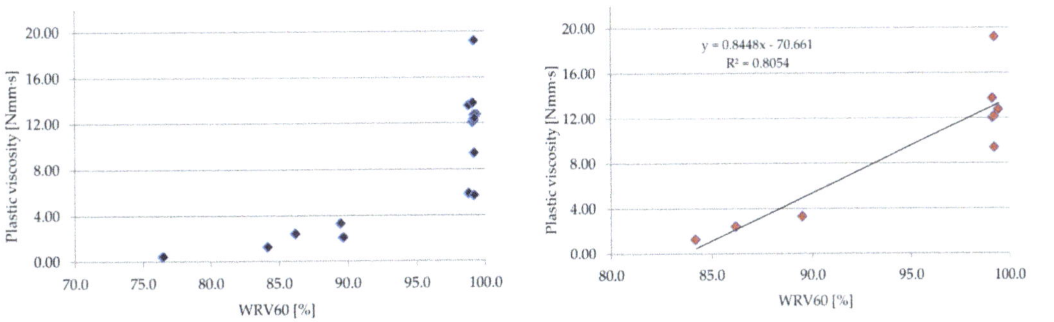

(a) (b)

Figure 13. Correlation between plastic viscosity and WRV60 (**a**) for all samples, and (**b**) only for samples with cement–lime binder.

4. Conclusions

On the basis of the results presented in this a article, the following conclusions can be derived:

- Modification of cement plastering mortar with both hydrated lime and cellulose ether allows obtaining a material with favorable technical and technological properties, especially for mechanical application.
- Correctly selected proportions of hydrated lime and polymer admixture with a specific viscosity ensure appropriate application properties: high efficiency, workability, and ease of plastering.

- An increase in the plastic viscosity of pastes influences the increase in WRV of the mortars.
- An increase in the amount and viscosity of cellulose ether increases the plastic viscosity and the yield stress.
- The subjective assessment of a specialist plasterer and the rheological parameters are valuable sources of information for determining the application properties of mortars. Indirectly, the results of these studies allow determining the method of mortar application.
- Providing the right consistency or obtaining a high WRV does not guarantee a high-quality material in terms of meeting the standard requirements and workability.
- The influence of cellulose ether on the tested properties of pastes and mortars (WRV, as well as rheological and application properties) is statistically more significant than the influence of hydrated lime.
- After concluding the research presented in the article, the authors propose that, related to the water retention mechanisms, the rheological properties of the mortar are inluential properties, but they should not be the main determinant in assessing the application properties of plasters or in selecting the method of mortar application.

Author Contributions: Conceptualization, E.S.; methodology, E.S. and R.D.; validation, E.S. and R.D.; formal analysis, E.S.; investigation, E.S.; resources, E.S.; data curation, E.S.; writing—original draft preparation, E.S.; writing—review and editing, E.S.; visualization, E.S. and R.D.; supervision, E.S.; project administration, E.S.; funding acquisition, E.S. and R.D. Both authors have read and agreed to the published version of the manuscript.

Funding: This research was funded by the Kielce University of Technology research program number 02.0.11.00/2.01.01.00.0000 SUBB.BKOB.21.001.

Institutional Review Board Statement: Not applicable.

Informed Consent Statement: Not applicable.

Data Availability Statement: No new data were created or analyzed in this study. Data sharing is not applicable to this article.

Conflicts of Interest: The authors declare no conflict of interest.

References

1. Brumaud, C.; Baumann, R.; Schmitz, M.; Radler, M.; Roussel, N. Cellulose ethers and yield stress of cement pastes. *Cem. Concr. Res.* **2014**, *55*, 14–21. [CrossRef]
2. Spychał, E. The rheology of cement pastes with the addition of hydrated lime and cellulose ether in comparison with selected properties of plastering mortars. *Cement Wapno Beton* **2020**, *25*, 21–30. [CrossRef]
3. Paiva, H.; Esteves, L.; Cachim, P.; Ferreira, V. Rheology and hardened properties of single-coat render mortars with different types of water retaining agents. *Constr. Build. Mater.* **2009**, *23*, 1141–1146. [CrossRef]
4. Izaguirre, A.; Lanas, J.; Alvarez, J.I. Characterization of aerial lime-based mortars modified by the addition of two different water-retaining agents. *Cem. Concr. Compos.* **2011**, *33*, 309–318. [CrossRef]
5. Vyšvařil, M.; Bayer, P. Cellulose ethers as water-retaining agents in natural hydraulic lime mortars. In Proceedings of the 13th International Conference Modern Building Materials, Structures and Techniques (MBMST 2019), Vilnius, Lithuania, 16–17 May 2019; pp. 194–200.
6. Spychał, E.; Czapik, P. The influence of HEMC on cement and cement-lime composites setting processes. *Materials* **2020**, *13*, 5814. [CrossRef]
7. Vyšvařil, M.; Hegrová, M.; Žižlavský, T. Influence of cellulose ethers on fresh state properties of lime mortars. *Solid State Phenom.* **2018**, *276*, 69–74. [CrossRef]
8. Brachaczek, W. Influence of cellulose ethers on the consistency, water retention and adhesion of renovating plasters. In *IOP Conference Series: Materials Science and Engineering*; IOP Publishing: Bristol, UK, 2019; Volume 471, p. 032020.
9. Pichniarczyk, P.; Niziurska, M. Properties of ceramic tile adhesives modified by different viscosity hydroxypropyl methylcellulose. *Constr. Build. Mater.* **2015**, *77*, 227–232. [CrossRef]
10. Spychał, E. The effect of lime and cellulose ether on selected properties of plastering mortar. *Procedia Eng.* **2015**, *108*, 324–331. [CrossRef]
11. Niziurska, M.; Pichniarczyk, P. Effect of hydroxypropyl methylcellulose viscosity on the properties of adhesive mortars. In Proceedings of the VI Konferencja Naukowo-Techniczna Zagadnienia Materiałowe w Inżynierii Lądowej MATBUD 2011, Cracow, Poland, 20–22 June 2011; pp. 261–270.

12. Chłądzyński, S.; Malata, G. Components of adhesive mortars. Part II Methylcellulose. *Izolacje* **2008**, *13*, 76–79.
13. Petit, J.-Y.; Wirquin, E. Evaluation of various cellulose ethers performance in ceramic tile adhesive mortars. *Int. J. Adhes. Adhes.* **2013**, *40*, 202–209. [CrossRef]
14. Kotwa, A.; Spychał, E. The influence of cellulose ethers on the chosen properties of cement mortar in the plastic state. *Struct. Environ.* **2016**, *8*, 153–159.
15. Brumaud, C.; Bessaies-Bey, H.; Mohler, C.; Baumann, R.; Schmitz, M.; Radler, M.; Roussel, N. Cellulose ether and water retention. *Cem. Concr. Res.* **2013**, *53*, 176–184. [CrossRef]
16. Bülichen, D.; Kainz, J.; Plank, J. Working mechanism of methyl hydroxyethyl cellulose (MHEC) as water retention agent. *Cem. Concr. Res.* **2012**, *42*, 953–959. [CrossRef]
17. Patural, L.; Marchal, P.; Govin, A.; Grosseau, P.; Ruot, B.; Devès, O. Cellulose ethers influence on water retention and consistency in cement-based mortars. *Cem. Concr. Res.* **2011**, *41*, 46–55. [CrossRef]
18. Betioli, A.M.; Gleize, P.J.P.; Silva, D.A.; John, V.M.; Pileggi, R.G. Effect of HMEC on the consolidation of cement pastes: Isothermal calorimetry versus oscillatory rheometry. *Cem. Concr. Res.* **2009**, *39*, 440–445. [CrossRef]
19. Žižlavský, T.; Bayer, P.; Vyšvařil, M. Bond properties of NHL-based mortars with viscosity-modifying water-retentive admixtures. *Minerals* **2021**, *11*, 685. [CrossRef]
20. Seabra, M.P.; Labrincha, J.A.; Ferreira, V.M. Rheological behaviour of hydraulic lime-based mortars. *J. Eur. Ceram. Soc.* **2007**, *27*, 1735–1741. [CrossRef]
21. Pourchez, J.; Grosseau, P.; Ruot, B. Current understanding of cellulose ethers impact on the hydration of C_3A and C_3A-sulphate systems. *Cem. Concr. Res.* **2009**, *39*, 664–669. [CrossRef]
22. Pourchez, J.; Govin, A.; Grosseau, P.; Guyonnet, R.; Guilhot, B.; Ruot, B. Alkaline stability of cellulose ethers and impact of their degradation products on cement hydration. *Cem. Concr. Res.* **2006**, *36*, 1252–1256. [CrossRef]
23. Pichniarczyk, P. Influence of methylcellulose on the hydration of cement. *Cement Wapno Beton.* **2015**, *3*, 186–192.
24. Knapen, E.; Van Gemert, D. Cement hydration and microstructure formation in the presence of water-soluble polymers. *Cem. Concr. Res.* **2009**, *39*, 6–13. [CrossRef]
25. Messan, A.; Ienny, P.; Nectoux, D. Free and restrained early-age shrinkage of mortar: Influence of glass fiber, cellulose ether and EVA (ethylene-vinyl acetate). *Cem. Concr. Compos.* **2011**, *33*, 402–410. [CrossRef]
26. Gołaszewski, J.; Cygan, G. Influence of VMA admixtures on early shrinkage of the cement mortars. *Civ. Environ. Eng.* **2011**, *2*, 263–266.
27. Kovalenko, Y.; Tokarchuk, V.; Poliuha, V. The effect of methyl hydroxyethyl cellulose on the cement matrix properties. *East.-Eur. J. Enterp. Technol.* **2020**, *3*, 28–33. [CrossRef]
28. Zhou, S.C.; Lv, H.L.; Li, N.; Zhang, J. Effect of chemical admixtures on the working and mechanical properties of ordinary dry-mixed mortar. *Adv. Mater. Sci. Eng.* **2019**, *2019*, 5978089. [CrossRef]
29. Pichniarczyk, P.; Niziurska, M.; Nosal, K. Influence of methylcellulose viscosity on properties of gypsum mortars. *Mater. Ceram.* **2012**, *64*, 558–562.
30. Available online: https://www.cemex.pl/cem-i-42-5-r (accessed on 26 August 2021).
31. Available online: http://www.alpol.pl/pl/katalog_produktow/go:28:209/ (accessed on 26 August 2021).
32. Available online: http://www.grudzenlas.pl/index.php?cat=oferta&page=piaski-techniczne (accessed on 26 August 2021).
33. Available online: https://www.dow.com/en-us/search.html?x11=category&q11=assets%7Cpages%7Cproducts&q=walocel&tab=all (accessed on 26 August 2021).
34. Available online: https://www.hplush.pl/h-h-silikaty-produkty-podstawowe (accessed on 26 August 2021).
35. Chmielewski, K.; Berczyński, S. *Matematic Statistics. Laboratory Excercises Using the Statistica PL Package*; Wydawnictwo Uczelniane Politechniki Szczecińskiej: Szczecin, Poland, 2002.
36. Available online: http://www.itl.nist.gov/div898/handbook/pri/section3/pri3362.htm (accessed on 26 August 2021).
37. Available online: https://www.statsoft.pl/Pelna-lista-programow-Statistica/?goto=statistica_enterprise#sc (accessed on 26 August 2021).
38. Zając, T.; Kulig, B. Estimation of the effect of selected agro technical factors on linseed croppingin $3^{4\text{-}1}$ experiment. *Rośliny Oleiste-Oilseed Crop.* **2001**, *22*, 597–608.
39. *PN-EN 1015-3:2000+A2:2007 Test Methods of Mortars for Walls. Determining the Consistency of Fresh Mortar (Using a Flow Table)*; Polish Committee for Standardization: Warszawa, Poland, 2007. (In Polish)
40. *PN-B-04500:1985 Mortars—Testing Physical and Strength Characteristics*; Polish Committee for Standardization: Warszawa, Poland, 1985. (In Polish)
41. *PN-EN 1015-6:2000+A1:2007 Test Methods of Mortars for Walls. Determination of the Bulk Density of the Fresh Mortar*; Polish Committee for Standardization: Warszawa, Poland, 2007. (In Polish)
42. Gawlicki, M.; Pichór, W.; Brylska, E.; Brylicki, W.; Łagosz, A.; Nocuń-Wczelik, W.; Petri, M.; Pytel, Z.; Roszczynialski, W.; Stolecki, J.; et al. *Podstawy Technologii Materiałów Budowlanych i Metody Badań*; Małolepszy, J., Ed.; Wydawnictwo AGH: Kraków, Poland, 2013.
43. Ponikiewski, T. The pasticity mortar with fibres and her rheological properties. *Zesz. Nauk. Politech. Śląskiej Ser. Bud.* **2002**, *95*, 465–474.

44. Gołaszewski, J.; Szwabowski, J. Influence of superpalsticizers on rheological behaviour of fresh cement mortars. *Cem. Concr. Res.* **2004**, *34*, 235–248. [CrossRef]
45. Grzeszczyk, S. *Reologia Zawiesin Cementowych*; Polska Akademia Nauk: Warsaw, Poland, 1999.
46. Spychał, E. Evaluation of Selected Properties of Plastering Mortar Modified with Hydrated Lime and Cellulose Ether. Ph.D. Thesis, Kielce University of Technology, Kielce, Poland, 2016.
47. Nocuń-Wczelik, W.; Trybalska, B. Effect of admixtures on the rate of hydration and microstructure of cement paste. *Cement Wapno Beton* **2007**, *6*, 284–289.
48. Gantner, E.; Chojczak, W. *Materiały Budowlane. Ćwiczenia Laboratoryjne. Spoiwa, Kruszywa, Zaprawy, Betony*; Oficyna Wydawnicza Politechniki Warszawskiej: Warsaw, Poland, 2013.
49. Chen, N.; Wang, P.; Zhao, L.; Zhang, G. Water retention mechanism of HPMC in cement mortar. *Materials* **2020**, *13*, 2918. [CrossRef] [PubMed]
50. Paszkowski, L. Wpływ Proszkowej Fazy Nf-Fe-B w Formowanych Wtryskowo Kompozytach Magnetycznych o Osnowie Polimerowej na ich Właściwości Magnetyczne i Mechaniczne. Ph.D. Thesis, Warsaw University of Technology, Warsaw, Poland, 2008.
51. Sébaibi, Y.; Dheilly, R.M.; Quéneudec, M. A study of the viscosity of lime—Cement paste: Influence of the physic-chemical characteristics of lime. *Constr. Build. Mater.* **2004**, *18*, 653–660. [CrossRef]

materials

MDPI

Article

The Optimization of Mix Proportion Design for SCC: Experimental Study and Grey Relational Analysis

Xinxin Ding [1,2,*], Mingshuang Zhao [1,2], Xue Qiu [1], Yupu Wang [1] and Yijie Ru [1]

1 International Joint Research Lab for Eco-Building Materials and Engineering of Henan, School of Civil Engineering and Communications, North China University of Water Resources and Electric Power, Zhengzhou 450045, China; zhaomingshuang@ncwu.edu.cn (M.Z.); xueqiu@stu.ncwu.edu.cn (X.Q.); yupuwang@stu.ncwu.edu.cn (Y.W.); yijieru@stu.ncwu.edu.cn (Y.R.)

2 Collaborative Innovation Center for Efficient Utilization of Water Resources, North China University of Water Resources and Electric Power, Zhengzhou 450045, China

* Correspondence: dingxinxin@ncwu.edu.cn

Abstract: The optimization of mix proportions based on the targeted fresh and hardened performances of self-compacting concrete (SCC) is a foundation for its transition from laboratory research to industrial production. In this paper, the mix proportions of various SCC mixtures were designed by the absolute volume method with changes in the content of river sand and manufactured sand, the content of fly ash and granulated ground blast furnace slag (GGBS) and the maximum particle sizes of coarse aggregates. This experimental study was carried out to verify the workability, density and cubic compressive strength of SCC. The results show that SCC demonstrated good performance with appropriate mix proportions of manufactured sand and river sand. A hybrid effect of fly ash and GGBS appeared on the fresh performance of SCC with a constant strength, and the coarse aggregate with a smaller maximum particle size was beneficial to the workability but detrimental to the compressive strength of SCC. Finally, the optimization of the mix proportion of SCC was evaluated by grey relational analysis, in which the weight of the indicators was determined by the entropy method to improve the evaluation credibility. As a result, the optimal mix proportions of SCC were selected.

Keywords: self-compacting concrete (SCC); mix proportion design; optimization; experimental verification; the grey relational analysis

Citation: Ding, X.; Zhao, M.; Qiu, X.; Wang, Y.; Ru, Y. The Optimization of Mix Proportion Design for SCC: Experimental Study and Grey Relational Analysis. *Materials* **2022**, *15*, 1305. https://doi.org/10.3390/ma15041305

Academic Editor: Dario De Domenico

Received: 20 December 2021
Accepted: 3 February 2022
Published: 10 February 2022

1. Introduction

Self-compacting concrete (SCC) is a kind of high fresh-performance concrete that fills and compacts in a given framework by its own weight without other assistance [1]. As opposed to other conventional vibrated concretes, the defining feature of SCC is its superior workability. Besides the typical requirements of setting time, cohesiveness and water retention, fresh SCC should also satisfy stricter requirements of filling ability, passing ability and segregation resistance. Furthermore, there is a closed relationship between the workability of fresh SCC and the basic mechanical properties of hardened SCC [2,3]. The correct proportioning of self-compacting concrete mixes based on the target fresh and hardened performances is a hot research point.

The mix proportion requires the segregation and settlement of fresh SCC, a large number of binder materials, a higher sand ratio and a smaller maximum particle size of coarse aggregate in order to maintain a sufficient yield value, ensure the viscosity of the fresh mixes and, ultimately, to reduce bleeding [4–7]. A traditional method to ensure the equal performance of SCC is to add mineral materials. The presence of fly ash in SCC reduces its water demand and improves its desired workability but also decreases its compressive strength [7–9]. The presence of granulated ground blast furnace slag (GGBS) in SCC improves its fresh properties as well as its continuing strength development [10,11]. A positive effect can be observed in SCC with both GGBS and fly ash added in an appropriate proportion and dosage [12]. The passing ability and the segregation resistance of fresh SCC reduce with increasing coarse aggregate size from 9.5 mm to 19 mm, while the compressive strength of hardened SCC increases [13]. A greater amount of binder material content is required to obtain self-compacting workability for an SCC mixed with manufactured sand compared with an SCC mixed with river sand. This is due to the special morphology features of manufactured sand, such as its rough surfaces, irregular shapes, angular edges and inevitable stone powder [14–16]. The workability of fresh SCC could change with hybrid sand mixed with different contents of river sand and manufactured sand due to the changes of particle grading of the hybrid sand [17].

Numerous methods have been used to study the properties of concrete, such as laboratory tests [2,18], field monitoring [19], mathematical statistics [20], numerical analysis [21] and software simulation [22]. As the workability and strength of SCC are affected by multiple factors, it is unclear which one is the major factor. Therefore, the optimal mix proportion of SCC is difficult to obtain from original tests. This encourages research that takes concrete as a system with random, gray and fuzzy uncertain messages and tries to predict and evaluate its properties by grey relational analysis [23–25]. The grey relational analysis theory was founded in 1982 by Deng in China; it is an uncertain systematic theory dealing with a "small sample" and "poor information" [26]. As a type of system analytical technology, the basic principle of grey relational analysis is to evaluate the relationships of connections between sequences according to the geometric shape of a sequence curve. The closer the geometric shapes, the greater the correlation between the corresponding sequences and vice versa.

Based on the above statement, the optimization of the mix proportion of SCC is important to ensure its transition from laboratory research to industrial production. In this paper, an experimental verification was conducted for the mix proportion design of SCC with multiple factors, including the type and content of mineral admixtures, the type and content of sand and the maximum particle size of the coarse aggregate. The test results were evaluated by grey relational analysis to clarify the main influencing factors for both the workability and strength of SCC.

2. Mix Proportion Design of SCC

2.1. Raw Materials

Ordinary Portland cement, with a strength grade of 52.5, fly ash of class II, and ground granulated blast furnace slag (GGBS) of class S95, was used as a binder material. The detailed physical and mechanical properties of this cement are listed in Table 1, where it can be seen that it met the specification of China code GB 175 [27]. The detailed physical and mechanical properties of the fly ash and GGBS are listed in Table 2, where it can be seen that they met the specification of China code GB/T 51003 [28]. Pictures of the binder materials are presented in Figure 1.

Table 1. Physical and mechanical properties of cement.

Fineness (45 μm, %)	Water Requirement of Normal Consistency (%)	Density (kg/m^3)	Setting Time (min)		Compressive Strength (MPa)		Flexural Strength (MPa)	
			Initial	Final	3 Days	28 Days	3 Days	28 Days
8.6	26.6	3195	176	222	34.9	57.8	6.23	8.83

Table 2. Physical and mechanical properties of fly ash and GGBS.

No.	Fineness (45 μm, %)	Density (kg/m^3)	Water Demand Ratio (%)	Specific Surface (cm^2/g)	Strength Activity Index (%)	Loss on Ignition (%)
Fly ash	6.9	2280	95	3590	84.3	5.9
GGBS	1.0	2998	-	4388	97.6	2.9

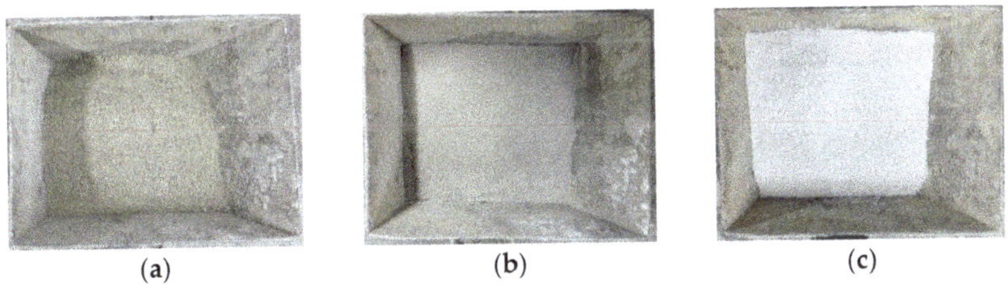

Figure 1. Pictures of binder materials: (a) cement; (b) fly ash; (c) GGBS.

Manufactured sand crushed from limestone and river sand were used as fine aggregates. Their physical and mechanical properties are listed in Table 3. Crushed limestones with maximum particle sizes of 20 mm, 16 mm, and 10 mm from 1 company were used as coarse aggregates. Their physical and mechanical properties are listed in Table 4. Their particle gradations are presented in Figure 2 and are adjusted according to the principle of maximum close-packing density. The gradations of fine and coarse aggregates met the specifications of China codes GB/T 14684 and GB/T 14685 [29,30]. Pictures of the fine and coarse aggregates are presented in Figure 3.

Table 3. Physical and mechanical properties of fine aggregates.

Type	Fineness Modulus	Apparent Density (kg/m^3)	Closed Packing Density (kg/m^3)	Bulk Density (kg/m^3)	Stone Powder Content (%)	Water Absorption (%)
Manufactured sand	2.78	2730	1698	1923	7.45	0.25
River sand	2.00	2597	1477	1641	-	0.15

Table 4. Physical and mechanical properties of coarse aggregates.

Maximum Particle Size (mm)	Apparent Density (kg/m^3)	Pile-Up Density (kg/m^3)		Crushed Index (%)	Content of Needle-Pieces (%)
		Loose	Close		
20	2770	1583	1682	7.2	4.70
16	2766	1575	1662	9.6	4.54
10	2711	1513	1575	16.8	1.32

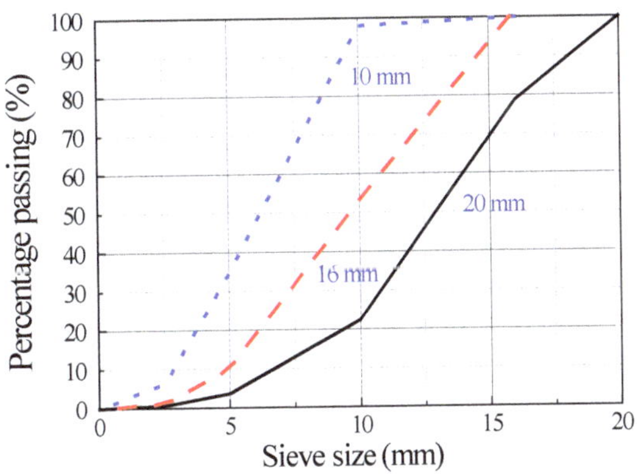

Figure 2. Particle size distribution of the coarse aggregate used.

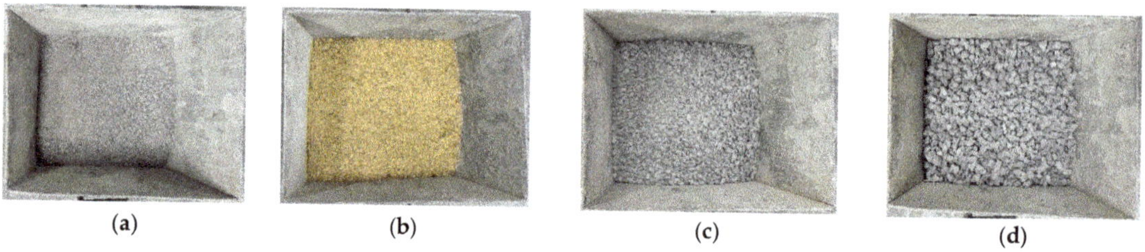

Figure 3. Pictures of aggregates: (**a**) manufactured sand; (**b**) river sand; (**c**) 5~10 mm aggregate; (**d**) 10~16 mm aggregate.

A polycarboxylate-based superplasticizer was used as a water reducer with a water-reducing rate of 29% and solid content of 23%. The mixing water was tap water.

2.2. Mix Proportion Design

In this study, the absolute volume method is used to design the mix proportion of the SCC [31,32]. The target compressive strength of SCC was 69 MPa, which is standard for concrete of strength grade C60 with a guarantee rate of 95%. The target workability of the fresh SCC was a slump larger than 260 mm, a slump flow of 700 mm, and a flow time T_{500} less than 5 s. The target pouring quality of the fresh SCC was a density greater than 2450 kg/m^3.

A total of 11 groups of mixtures for SCC were designed, with the influencing factors detailed in Table 5. The influencing factors included the ratio R_r of manufactured sand substituted by river sand, the ratios R_G and R_F of GGBS and fly ash in the total mass of the binders, and the maximum particle size of coarse aggregates (MPS). The water-to-binder ratio w/b was kept at 0.31, and the sand ratio β_s was kept constant at 50%.

Table 5. The influencing factors designed in the mix proportion of the SCCs.

No	w/b	β_s (%)	R_G (%)	R_F (%)	R_r (%)	MPS (mm)
R0	0.31	50	0	0	0	20
R27					27.5	
R35					35	
R45					45	
R100					100	
R35-F30	0.31	50	0	30	35	20
R35-G30			30	0		
R35-G20F10			20	10		
R35-G10F20			10	20		
R35-C16	0.31	50	0	0	35	16
R35-C10						10

The detailed mix proportions of the SCCs are presented in Table 6. The water content was 180 kg/m^3. The water reducer was kept constant at 1.6% of the total mass of the binders. This is to eliminate the influence of the water reducer content on the fresh performance of the SCC.

Table 6. The detailed mix proportions of the SCCs (kg/m^3).

No	Water	Cement	Mineral Admixture		Coarse Aggregate			Fine Aggregate		Water Reducer
			GGBS	Fly Ash	16~20	10~16	5~10	Manufactured Sand	River Sand	
R0	180	581	0	0	247	206	370	822	0	9.29
R27	180	581	0	0	247	206	370	596	226	9.29
R35	180	581	0	0	247	206	370	534	288	9.29
R45	180	581	0	0	247	206	370	452	370	9.29
R100	180	581	0	0	247	206	370	0	822	9.29
R35-F30	180	406	0	174	247	206	370	534	288	9.29
R35-G30	180	406	174	0	247	206	370	534	288	9.29
R35-G20F10	180	406	116	58	247	206	370	534	288	9.29
R35-G10F20	180	406	58	116	247	206	370	534	288	9.29
R35-C16	180	581	0	0	0	294	528	534	288	9.29
R35-C10	180	581	0	0	0	0	822	534	288	9.29

3. Experimental Verification

3.1. Fresh Performance

A horizontal shaft forced mixer with a maximum capacity of 100 L was used to mix the SCC. The slump, slump flow and slump flow time T_{500} of the fresh SCC were tested immediately after mixing according to the Chinese code JGJ/T 283 [1], which is identical to ASTM C1611/C1611M [33]. The slump flow time T_{500} is the time from the lifting of the slump cone to the diameter of the slump expansion surface reaching 500 mm [34]. Results with comparisons to target indices and specified filling ability are presented in Figure 4.

The reasonable mixing of machine-made sand and river sand can optimize the particle grading of sand and bring about good workability for fresh SCC [14,15]. In this test, the workability of SCC first decreased with the increasing ratio of river sand up to 27.5%, then increased with the ratio of river sand from 27.5% to 45%. This is similar to the study of pumped concrete with manufactured sand and river sand [17]. Compared with R0, R45 and R100, R27 has the worst fresh performance, with a smaller slump of 230 mm, a lower slump flow of 600 mm and a longer flow time T_{500} of 11 s, while R35 has good fresh performance, with a suitable slump of 245 mm, a higher slump flow of 700 mm and a shorter flow time T_{500} of 3.5 s.

Fly ash and GGBS are conducive to the slump and slump flow of fresh SCC R35-F30 and R35-G30 compared to R35. However, the addition of GGBS prolongs the T_{500}, while the addition of fly ash has no obvious influence on the T_{500}. The T_{500} of R35-G30 is 1.43 times that of R35, while the T_{500} of R35-G30 was equal to that of R35. Compared with R35-G30, R35-F30 had the same slump, a higher slump flow and a shorter T_{500}. This indicates that fly ash improved the working performance of the SCC. The reason is mainly due to the special morphological effects of fly ash, with microspheres featuring smooth surfaces, fine particle size and dense texture. Then, the fresh SCC with fly ash would adsorb a small amount of water in the mixing process [35], reducing the internal friction resistance and improving the flowability of the fresh concrete [8,36].

Compared with R35, R35-G10F20 and R35-G20F10 presented with a higher slump and slump flow. This indicated that the hybrid fly ash and GGBS contributed to the improvement of flowability of SCC, although no superposition effect of fly ash and GGBS was observed in these indexes. However, the negative superposition effect of hybrid fly ash and GGBS appeared on the T_{500}. The T_{500} of R35-G10F20 and R35-G20F10 were longer than both R35-F30 and R35-G30.

Compared with R35, R35-C16 had a better flowability, with a higher slump of 265 mm, a higher slump flow of 720 mm and a flow time T_{500} of 8.8 s, while R35-C10 had a similar slump, a slightly lower slump flow and the longest T_{500} of 11.3 s. This indicates that the gradation of the coarse aggregates had effects on the slump and slump flow, and especially on the flow time of the fresh SCC. The T_{500} of the SCC increased with the decreasing maximum particle size of the coarse aggregate.

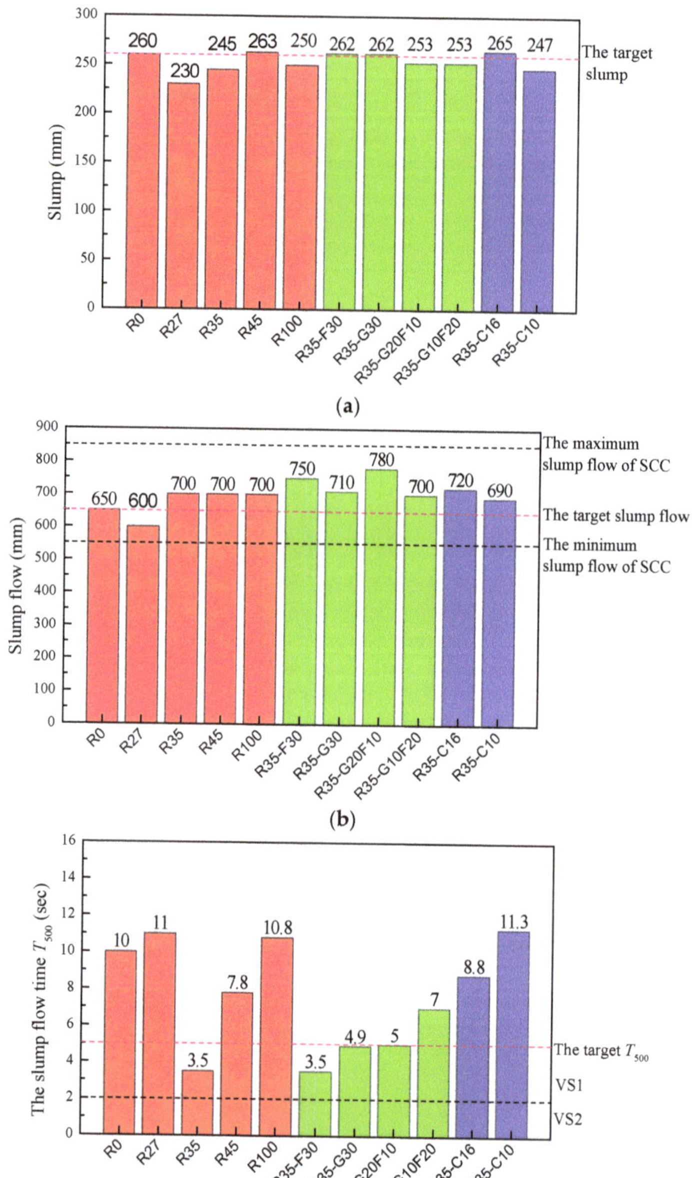

Figure 4. Workability of fresh SCC: (**a**) slump; (**b**) slump flow; (**c**) slump flow time T_{500}.

3.2. The Density

The density is an important index to present the self-compacting quality, which affects the performance of hardened SCC. Cubic specimens with dimensions of 150 mm were weighed after being cast for 24 h. The densities of the SCCs were obtained from the weight divided by the volume of the cubic specimen. Results with a comparison of the target values are presented in Figure 5.

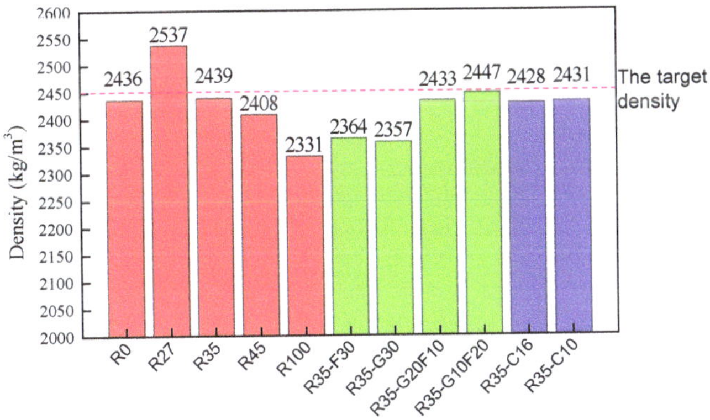

Figure 5. The densities of the SCCs.

Comparing R0 to R100, the SCC with manufactured sand had a density that was 4.5% higher than the SCC with river sand. This came from the lower density of river sand presented in Table 3. The mixing of manufactured sand and river sand in appropriate proportions could increase the density of SCC. The density of SCC first increased with the increasing ratio of river sand up to 27.5%, then decreased with the ratio of river sand from 27.5% to 100%. This indicates a hybrid effect of finer river sand to improve the particle grading of manufactured sand to promote the density of SCC. Compared with R0, R45 and R100, R27 had a maximum density of 2537 kg/m^3; R35 was next, with a density of 2439 kg/m^3.

Compared to R35, R35-F30 and R35-G30 had a density decrease of about 3%. R35-G10F20 and R35-G20F10, R35-C10 and R35-C16 had an almost equal density as R35. This indicates a slight decrease in the density of SCC with the single addition of fly ash and GGBS. The hybrid addition of GGBS and fly ash and the MPS of the coarse aggregate had no influence on the density of SCC.

3.3. The Strength

The cubic compressive strengths of the SCCs at a curing age of 28 days were tested in accordance with China code GB/T50081 [37], which is identical to British Standard BS EN 12390-3 [38]. Cubic specimens with dimensions of 150 mm were used. Three specimens were tested as a group. Results are presented in Figure 6.

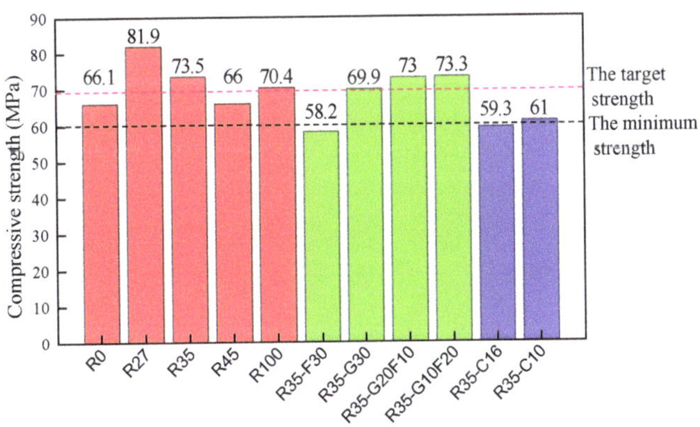

Figure 6. The cubic compressive strength of SCC.

The substitution of manufactured sand with river sand could improve the cubic compressive strength of SCC. Except for R45, other SCCs with ratios of river sand of 27.5% and 35% had higher compressive strength; specifically, their compressive strength was increased by 23.9% and 11.2% compared to R0. Meanwhile, R100 had a higher cubic compressive strength, which was about 6.5% greater than R0. This indicates the good relationship of cubic compressive strength with the density of SCCs with hybrid sand. Manufactured sand is characterized by its rough surface, irregular particle shape, angular edges and the inevitable presence of stone powder, while river sand is characterized by smooth, round and small particles [39]. This different morphology between manufactured sand and river sand could influence the microstructure of SCCs to present their corresponding cubic compressive strengths.

The compressive strength of R35-F30 was only 79.0% that of R35. This is due to the significant dilution effect of fly ash with a lower strength activity index of 84.3% [40–42]. Comparatively, due to the strength activity index of GGBS, which reached 97.6%, the compressive strength of R35-G30 was reduced only by 4.9% compared to R35. Meanwhile, R35-G10F20 and R35-G20F10 had similar compressive strengths to R35. This indicates that the hybrid of fly ash and GGBS had a positive superposition effect on the compressive strength of SCC [11,43].

The maximum particle size of the coarse aggregate had a significant effect on the compressive strength of the SCC. Compared to R35, with a maximum particle size of 20 mm, the compressive strengths of R35-C16 and R35-C10, with maximum respective particle sizes of 16 mm and 10 mm, were reduced by 19.3% and 17.0%. This is similar to the compressive strength observed in conventional vibrated concrete [41]. Under the same mix proportion, the skeleton effect of coarse aggregates in concrete is weakened with decreasing MPS due to the lack of rationally closed packing among coarse aggregates with a similar particle size. At the same time, the decreased compressive strength was also related to the increased crushing index of smaller coarse aggregates, as presented in Table 4.

4. Optimization by Grey Relational Analysis

Grey relational analysis is a modern mathematics method for factor analysis in a system [26]. It is always used to evaluate the significance of factors in a given system and to judge the factors in order of importance. Its fundamental principle is evaluating the level of similarity for a data sequence by relevancy calculations. A greater relevancy means a higher similarity. In this paper, grey relational analysis is conducted based on the test results as the sample data to select the optimal mixtures from these eleven groups of SCC mixtures.

Assuming that the original data is D in the sequence of slump, slump flow, T_{500}, density and cubic compressive strength, then,

$$D = \begin{bmatrix} d_1^1 & d_2^1 & \cdots & d_n^1 \\ d_1^2 & d_2^2 & \cdots & d_n^2 \\ \vdots & \vdots & & \vdots \\ d_1^m & d_2^m & \cdots & d_n^m \end{bmatrix} = \begin{bmatrix} 260 & 650 & 10.0 & 2436 & 66.1 \\ 230 & 600 & 11.0 & 2537 & 81.9 \\ 245 & 700 & 3.5 & 2439 & 73.5 \\ 263 & 700 & 7.8 & 2408 & 66.0 \\ 250 & 700 & 10.8 & 2331 & 70.4 \\ 262 & 750 & 3.5 & 2364 & 58.2 \\ 262 & 710 & 4.9 & 2357 & 69.9 \\ 253 & 780 & 5.0 & 2433 & 73.0 \\ 253 & 700 & 7.0 & 2447 & 73.3 \\ 265 & 720 & 8.8 & 2428 & 59.3 \\ 247 & 690 & 11.3 & 2431 & 61.0 \end{bmatrix} \quad (1)$$

where ${d_k}^i$ is the original data of the index k at the solution i. $m = 11$ and $n = 5$ in this paper.

In this study, the slump, slump flow, and slump flow time T_{500} are the dominant indicators for the workability of fresh SCC. The slump and slump flow are larger for better responses, while the slump flow time T_{500} is smaller for a better response. The density and the compressive strength are the dominant responses for the hardened SCC, and they are larger for better responses. Thus, the standard matrix C can be obtained with the standardizing Equations (2)–(4),

$$C_k^i = \frac{d_k^i - \min(d_k^i,\ i=1,2,\ldots m)}{\max(d_k^i,\ i=1,2,\ldots m) - \min(d_k^i,\ i=1,2,\ldots m)} \text{ (for the larger and better response)} \quad (2)$$

$$C_k^i = \frac{\max(d_k^i,\ i=1,2,\ldots m) - d_k^i}{\max(d_k^i,\ i=1,2,\ldots m) - \min(d_k^i,\ i=1,2,\ldots m)} \text{ (for the smaller and better response)} \quad (3)$$

$$C = \begin{bmatrix} C_1^1 & C_2^1 & \cdots & C_5^1 \\ C_1^2 & C_2^2 & \cdots & C_5^2 \\ \vdots & \vdots & & \vdots \\ C_1^{11} & C_2^{11} & \cdots & C_5^{11} \end{bmatrix} = \begin{bmatrix} 0.8571 & 0.2778 & 0.1667 & 0.5097 & 0.3333 \\ 0.0000 & 0.0000 & 0.0385 & 1.0000 & 1.0000 \\ 0.4286 & 0.5556 & 1.0000 & 0.5243 & 0.6456 \\ 0.9429 & 0.5556 & 0.4487 & 0.3738 & 0.3291 \\ 0.5714 & 0.5556 & 0.0641 & 0.0000 & 0.5148 \\ 0.9143 & 0.8333 & 1.0000 & 0.1602 & 0.0000 \\ 0.9143 & 0.6111 & 0.8205 & 0.1262 & 0.4937 \\ 0.6571 & 1.0000 & 0.8077 & 0.4951 & 0.6245 \\ 0.6571 & 0.5556 & 0.4943 & 0.5631 & 0.6371 \\ 1.0000 & 0.6667 & 0.3205 & 0.4709 & 0.0464 \\ 0.4857 & 0.5000 & 0.0000 & 0.4854 & 0.1181 \end{bmatrix} \quad (4)$$

where C^* is the referenced sequence.

$$C^* = [C_1^*, C_2^*, \ldots, C_5^*] = [1.0000 \quad 1.0000 \quad 1.0000 \quad 1.0000 \quad 1.0000] \quad (5)$$

The matrix C is the compared sequence, and the correlation coefficient $\xi_j(i)$ is obtained from Equation (6)

$$\xi_j(i) = \frac{\underset{j}{min}\underset{i}{min}\left|C_i^* - C_i^j\right| + \rho \underset{j}{max}\underset{i}{max}\left|C_i^* - C_i^j\right|}{\left|C_i^* - C_i^j\right| + \rho \underset{j}{max}\underset{i}{max}\left|C_i^* - C_i^j\right|}. \quad (6)$$

where the value range of ρ is [0, 1], and generally takes the form of $\rho = 0.5$.

Then, the evaluation matrix of indexes E is shown as follows.

$$E = \begin{bmatrix} \xi_1(1) & \xi_1(2) & \cdots & \xi_1(5) \\ \xi_2(1) & \xi_2(2) & \cdots & \xi_2(5) \\ \vdots & \vdots & & \vdots \\ \xi_{11}(1) & \xi_{11}(2) & \cdots & \xi_{11}(5) \end{bmatrix} = \begin{bmatrix} 0.7777 & 0.4091 & 0.3750 & 0.5049 & 0.4286 \\ 0.3333 & 0.3333 & 0.3421 & 1.0000 & 1.0000 \\ 0.4667 & 0.5294 & 1.0000 & 0.5125 & 0.5852 \\ 0.8975 & 0.5294 & 0.4756 & 0.4440 & 0.4270 \\ 0.5384 & 0.5294 & 0.3482 & 0.3333 & 0.5075 \\ 0.8537 & 0.7500 & 1.0000 & 0.3732 & 0.3333 \\ 0.8537 & 0.5625 & 0.7358 & 0.3640 & 0.4969 \\ 0.5932 & 1.0000 & 0.7222 & 0.4976 & 0.5711 \\ 0.5932 & 0.5294 & 0.4972 & 0.5337 & 0.5794 \\ 1.0000 & 0.6000 & 0.4239 & 0.4859 & 0.3440 \\ 0.4930 & 0.5000 & 0.3333 & 0.4928 & 0.3618 \end{bmatrix} \quad (7)$$

We can then determine the entropy weight for the index i by assuming the number of indexes and evaluation objects are $n = 5$ and $m = 11$, respectively:

$$f_{ij} = \frac{\xi_j(i)}{\sum_{j=1}^{m} \xi_j(i)} \quad (8)$$

Thus,

$$F = \begin{bmatrix} f_{11} & f_{12} & \cdots & f_{1n} \\ f_{21} & f_{22} & \cdots & f_{2n} \\ \vdots & \vdots & & \vdots \\ f_{m1} & f_{m2} & \cdots & f_{mn} \end{bmatrix} = \begin{bmatrix} 0.105 & 0.065 & 0.060 & 0.091 & 0.076 \\ 0.045 & 0.053 & 0.055 & 0.180 & 0.177 \\ 0.063 & 0.084 & 0.160 & 0.092 & 0.104 \\ 0.121 & 0.084 & 0.076 & 0.080 & 0.076 \\ 0.073 & 0.084 & 0.056 & 0.060 & 0.090 \\ 0.115 & 0.120 & 0.160 & 0.067 & 0.059 \\ 0.115 & 0.090 & 0.118 & 0.066 & 0.088 \\ 0.080 & 0.159 & 0.115 & 0.090 & 0.101 \\ 0.080 & 0.084 & 0.080 & 0.096 & 0.103 \\ 0.135 & 0.096 & 0.068 & 0.088 & 0.061 \\ 0.067 & 0.080 & 0.053 & 0.089 & 0.064 \end{bmatrix} \tag{9}$$

The entropy h_i of the index i is shown as

$$h_i = -k\sum_{j=1}^{11} f_{ij} ln f_{ij} \tag{10}$$

where $k = 1/ln\text{m} = 1/ln11 = 0.417$.

Thus,

$$\begin{aligned} \text{H} &= [h_1, h_2, \ldots, h_5] \\ &= [0.98044, 0.98323, 0.96412, 0.97988, 0.97778] \end{aligned} \tag{11}$$

The entropy weight for the index i is

$$\omega_i = \frac{1-h_i}{n-\sum_1^n h_i}\left(0 \le \omega_i \le 1, \sum_{i=1}^{n}\omega_i = 1\right) \tag{12}$$

The entropy weight matrix of various indexes is

$$\begin{aligned} \text{W} &= [\omega_1, \omega_2, \ldots, \omega_5] \\ &= [0.17076, 0.14638, 0.31319, 0.17567, 0.19400] \end{aligned} \tag{13}$$

Therefore, the comprehensive evaluation index is

$$\begin{aligned} R &= E \times W^T \\ &= [0.48197, 0.58253, 0.67393, 0.54055, 0.43551, 0.69897, 0.61890, 0.67207, \\ &\quad 0.54066, 0.54344, 0.41853] \end{aligned} \tag{14}$$

This means the sequence is R35-F30, R35, R35-G20F10, R35-G30, R27, R35-C16, R35-G10F20, R45, R0, R100, and R35-C10. The mixtures of R35-F30, R35 and R35-G20F10 are the relatively better groups of the eleven mixtures.

5. Discussion

Based on the test results of eleven SCCs in this study, their workability meets the requirements to be labeled as self-compacting. The SCCs marked as R27, R35, R100, R35-G30, R35-G20F10 and R35-G10F20 meet the target compressive strength of the SCC. Due to the complex influences of the properties and contents of raw materials, including fly ash, GGBS, manufactured sand, river sand and coarse aggregate, different optimal mixtures can be obtained by considering the performance. Considering the slump and the slump flow of fresh SCC, the mixtures R0, R45, R35-F30, R35-G30 and R35-C16 are optimal. Considering the flow time T_{500} of fresh SCC, the mixtures R35, R35-F30 and R35-G20F10 are optimal. Considering the density of SCC, the mixtures R0, R27, R35 and R35-G10F20 are optimal. Considering the compressive strength of SCC, the mixtures R27, R35, R100, R35-G30, R35-G20F10 and R35-G10F20 are optimal.

Therefore, it is difficult to determine the optimal mixture of SCC only by the test results to satisfy all aspects of performance. In this condition, grey relational analysis is used to obtain the optimal solution among these eleven SCC mixtures after comprehensively considering the workability, density and mechanical properties of the SCCs. As a result, R35-F30, R35 and R35-G20F10 are the better of the eleven mixtures. This agrees well with the experimental results. R35-F30 has a fresh performance with a high slump of 262 mm, a high slump flow of 750 mm and a short flow time T_{500} of 3.5 s, a density of 2364 kg/m^3, and a compressive strength of 58.2 MPa. R35 has a fresh performance with a suitable slump of 245 mm, a high slump flow of 700 mm and a short flow time T_{500} of 3.5 s, a density of 2439 kg/m^3, and a high compressive strength of 73.5 MPa. R35-G20F10 has a fresh performance with a high slump of 253 mm, a high slump flow of 780 mm, a short flow time T500 of 5 s, and a density of 2433 kg/m3 and a high compressive strength of 73.0 MPa.

Given all of the above, the experimental study is suitable for comparing the effects of the raw materials on one aspect of the performance of SCCs, such as workability, density or compressive strength. The grey relational analysis used in this study is applicable to further optimize the mix proportion of SCCs while considering the comprehensive performance of the SCC.

6. Conclusions

Based on the absolute volume method of mix proportion design of SCC, eleven groups of SCCs were designed with changes in the content of fly ash and GGBS, the content of manufactured sand and river sand, and the maximum particle sizes of coarse aggregates. The design was verified by experimental studies and evaluated by grey relational analysis. Conclusions can be drawn as follows.

The replacement of manufactured sand by a certain amount of river sand could improve the flowability of fresh SCC and increase the compressive strength of hardened SCC. Both the GGBS and fly ash are beneficial for the slump and slump flow of fresh SCC; however, they also increase the flow time. The hybrid effect of GGBS with fly ash would improve the workability of SCC with equal compressive strength. The maximum particle size of the coarse aggregate significantly influenced the performance of the SCC. Compared to an SCC with a maximum coarse aggregate particle size of 20 mm, an SCC with a maximum coarse aggregate particle size of 16mm demonstrated more favorable workability, but the cubic compressive strength of SCC decreased by nearly 20%.

Based on the test results, a grey correlation analysis model was built. Combined with the entropy evaluation method, the optimal mixtures of SCC were selected from these eleven group mixtures. The optimal mixture considering the workability of fresh SCC was found to be R35-F30, which demonstrated fresh performance with a slump of 265 mm, a slump flow of 750 mm and a flow time T_{500} of 3.5 s. The optimal mixture when comprehensively considering the workability and mechanical properties of SCC was found to be R35, which demonstrated fresh performance with a slump of 245 mm, a slump flow of 700 mm, a flow time T_{500} of 3.5 s, a density of 2439 kg/m^3 and a compressive strength of 73.5 MPa. This method has improved the objectivity of the evaluation and avoided the loss of information which can be used for the optimization of mix proportions of SCC.

Author Contributions: Conceptualization and methodology, X.D. and M.Z.; validation, X.D. and X.Q.; formal analysis, X.Q. and M.Z.; investigation, Y.W. and M.Z.; data curation, Y.R. and M.Z.; writing—original draft preparation, X.D. and Y.W; writing—review and editing, Y.R., M.Z. and X.D; funding acquisition, X.D. All authors have read and agreed to the published version of the manuscript.

Funding: This research was funded by the Natural Science Foundation of Henan, China, grant number "212300410192"; Key Scientific and Technological Research Project of University in Henan, China, grant number "20A560015".

Institutional Review Board Statement: Not applicable.

Informed Consent Statement: Not applicable.

Data Availability Statement: The data presented in this study are available in the submitted article.

Conflicts of Interest: The authors declare no conflict of interest.

References

1. Ministry of Housing and Urban-Rural Development of the People's Republic of China. *Technical Specification for Application of Self-Compacting Concrete*; JGJ/T 283-2012; China Building Industry Press: Beijing, China, 2012.
2. Ding, X.; Li, C.; Zhao, M.; Li, J.; Geng, H.; Lian, L. Tensile strength of self-compacting steel fiber reinforced concrete evaluated by different test methods. *Crystals* **2021**, *11*, 251. [CrossRef]
3. Siamardi, K. Optimization of fresh and hardened properties of structural light weight self-compacting concrete mix design using response surface methodology. *Constr. Build. Mater.* **2022**, *317*, 125928. [CrossRef]
4. Singh, N.; Kumar, P.; Goyal, P. Reviewing the behaviour of high volume fly ash based self compacting concrete. *J. Build. Eng.* **2019**, *26*, 100882. [CrossRef]
5. de la Rosa, Á.; Poveda, E.; Ruiz, G.; Cifuentes, H. Proportioning of self-compacting steel-fiber reinforced concrete mixes based on target plastic viscosity and compressive strength: Mix-design procedure & experimental validation. *Constr. Build. Mater.* **2018**, *189*, 409–419.
6. Beygi, M.H.; Kazemi, M.T.; Amiri, J.V.; Nikbin, I.M.; Rabbanifar, S.; Rahmani, E. Evaluation of the effect of maximum aggregate size on fracture behavior of self compacting concrete. *Constr. Build. Mater.* **2014**, *55*, 202–211. [CrossRef]
7. Sua-Iam, G.; Chatveera, B. A study on workability and mechanical properties of eco-sustainable self-compacting concrete incorporating PCB waste and fly ash. *J. Clean. Prod.* **2021**, *329*, 129523. [CrossRef]
8. Khan, M.; Ali, M. Improvement in concrete behavior with fly ash, silica-fume and coconut fibres. *Constr. Build. Mater.* **2019**, *203*, 174–187. [CrossRef]
9. Li, C.; Geng, H.; Zhou, S.; Dai, M.; Sun, B.; Li, F. Experimental Study on Preparation and Performance of Concrete with Large Content of Fly Ash. *Front. Mater.* **2022**, *8*, 764820. [CrossRef]
10. Al-Oran, A.A.A.; Safiee, N.A.; Nasir, N.A.M. Fresh and hardened properties of self-compacting concrete using metakaolin and GGBS as cement replacement. *Eur. J. Environ. Civ. Eng.* **2019**, *26*, 1–14. [CrossRef]
11. Dinakar, P.; Sethy, K.P.; Sahoo, U.C. Design of self-compacting concrete with ground granulated blast furnace slag. *Mater. Des.* **2013**, *43*, 161–169. [CrossRef]
12. Tadi, C.; Rao, T.C. Investigating the performance of self-compacting concrete pavement containing GGBS. *Mater. Today Proc.* **2022**, *49*, 2013–2018. [CrossRef]
13. Qiu, C.Y.; Yu, F. Research on plastic shrinkage cracking of self-compacting concrete based on orthogonal experiment. *Water Resour. Power* **2021**, *39*, 155–160.
14. Nikbin, I.; Beygi, M.; Kazemi, M.; Amiri, J.V.; Rahmani, E.; Rabbanifar, S.; Eslami, M. A comprehensive investigation into the effect of aging and coarse aggregate size and volume on mechanical properties of self-compacting concrete. *Mater. Des.* **2014**, *59*, 199–210. [CrossRef]
15. Ding, X.; Li, C.; Xu, Y.; Li, F.; Zhao, S. Experimental study on long-term compressive strength of concrete with manufactured sand. *Constr. Build. Mater.* **2016**, *108*, 67–73. [CrossRef]
16. Hou, W.-Q.; Yang, J.-J.; Zhang, Z.-X.; Yuan, X.-Q. Experimental study and application of manufactured sand self-compacting concrete in concrete-filled-steel-tube arch bridge: A case study. *Case Stud. Constr. Mater.* **2021**, *15*, e00718. [CrossRef]
17. Zhang, L. Experimental research of machine-made sand content on performance impact of mixing pump concrete. *Shanxi Sci. Technol. Commun.* **2016**, *2*, 84–86.
18. Sargam, Y.; Faytarouni, M.; Riding, K.; Wang, K.; Jahren, C.; Shen, J. Predicting thermal performance of a mass concrete foundation—A field monitoring case study. *Case. Stud. Constr. Mater.* **2019**, *11*, e00289. [CrossRef]
19. Ahangari, K.; Beygi, M.H.A.; Rezaei, Y. Applicability of fiber reinforced self-compacting concrete for tunnel lining. *Arab. J. Geosci.* **2013**, *6*, 3841–3846. [CrossRef]
20. Craeye, B.; Van Itterbeeck, P.; Desnerck, P.; Boel, V.; De Schutter, G. Modulus of elasticity and tensile strength of self-compacting concrete: Survey of experimental data and structural design codes. *Cem. Concr. Compos.* **2014**, *54*, 53–61. [CrossRef]
21. Asteris, P.G.; Kolovos, K. Self-compacting concrete strength prediction using surrogate models. *Neural Comput. Appl.* **2019**, *31*, 409–424. [CrossRef]
22. Deeb, R.; Kulasegaram, S.; Karihaloo, B.L. 3D modelling of the flow of self-compacting concrete with or without steel fibres. Part I: Slump flow test. *Comput. Part Mech.* **2014**, *1*, 373–389. [CrossRef]
23. Suji, D.; Adesina, A.; Mirdula, R. Optimization of self-compacting composite composition using Taguchi-Grey relational analysis. *Materialia* **2021**, *15*, 101027. [CrossRef]
24. Chang, C.Y.; Huang, R.; Lee, P.C.; Weng, T.L. Application of a weighted Grey-Taguchi method for optimizing recycled ag-gregate concrete mixtures. *Cem. Concr. Comp.* **2011**, *33*, 1038–1049. [CrossRef]
25. Zhu, L.; Zhao, C.; Dai, J. Prediction of compressive strength of recycled aggregate concrete based on gray correlation analysis. *Constr. Build. Mater.* **2021**, *273*, 121750. [CrossRef]
26. Deng, J. *Multidimensional Grey Planning*; Huazhong University Press: Wuhan, China, 1989.

27. The State Bureau of Quality and Technical Supervision of the People's Republic of China. *Common Portland Cement*; GB175-2007; China Standard Press: Beijing, China, 2008.
28. Ministry of Housing and Urban-Rural Development of the People's Republic of China. *Technical Code for Application of Mineral Admixture*; GB/T 51003-2014; China Building Industry Press: Beijing, China, 2014.
29. General Administration of Quality Supervision, Inspection and Quarantine of the People's Republic of China. *Sand for Construction*; GB/T 14684-2011; China Standard Press: Beijing, China, 2011.
30. General Administration of Quality Supervision, Inspection and Quarantine of the People's Republic of China. *Pebble and Crushed Stone for Construction*; GB/T 14685-2011; China Standard Press: Beijing, China, 2011.
31. Ding, X.; Li, C.; Han, B.; Lu, Y.; Zhao, S. Effects of different deformed steel-fibers on preparation and fundamental properties of self-compacting SFRC. *Constr. Build. Mater.* **2018**, *168*, 471–481. [CrossRef]
32. Zhao, M.; Ding, X.; Li, J.; Law, D. Numerical Analysis of Mix Proportion of Self-Compacting Concrete Compared to Ordinary Concrete. *Key Eng. Mater.* **2019**, *789*, 69–75. [CrossRef]
33. ASTM. *International, Standard Test Method for Slump Flow of Self-Consolidating Concrete*; ASTM C1611/C1611M; ASTM: West Conshohocken, PA, USA, 2009.
34. Ding, X.; Geng, H.; Shi, K.; Song, L.; Li, S.; Liu, G. Study on Adaptability of Test Methods for Workability of Fresh Self-Compacting SFRC. *Materials* **2021**, *14*, 5312. [CrossRef]
35. Kurda, R.; de Brito, J.; Silvestre, J.D. Influence of recycled aggregates and high contents of fly ash on concrete fresh properties. *Cem. Concr. Compos.* **2017**, *84*, 198–213. [CrossRef]
36. Akid, A.S.M.; Hossain, S.; Munshi, M.I.U.; Elahi, M.M.A.; Sobuz, M.H.R.; Tam, V.W.Y.; Islam, M.S. Assessing the influence of fly ash and polypropylene fiber on fresh, mechanical and durability properties of concrete. *J. King. Saud. Univ Eng. Sci.* 2021; *in press*. [CrossRef]
37. Ministry of Housing and Urban-Rural Development of the People's Republic of China. *Standard for Test Methods of Concrete Physical and Mechanical Properties*; GB/T50081-2019; China Building Industry Press: Beijing, China, 2019.
38. European Committee for Standardization. *Testing Hardened Concrete—Compressive Strength of Test Specimens*; BS EN 12390-3-2009; European Committee for Standardization: London, UK, 2009.
39. Zhao, S.; Ding, X.; Zhao, M.; Li, C.; Pei, S. Experimental study on tensile strength development of concrete with manufactured sand. *Constr. Build. Mater.* **2017**, *138*, 247–253. [CrossRef]
40. Ding, X.X.; Lu, Y.Z.; Zhao, M.S.; Chen, M.H.; Zhao, S.B. Experimental study on tensile properties of high-performance concrete with machine-made sand and fly-ash. *WIT Trans. Eng. Sci.* **2015**, 164–172.
41. Ding, X.X.; Lu, Y.Z.; Zhao, M.S.; Chen, M.H.; Zhao, S.B. Influence coefficient of fly-ash for determination of binder's compres-sive strength in mix proportion design of concrete. *WIT Trans. Eng. Sci.* **2015**, 115–123.
42. Liu, S.; Zhu, M.; Ding, X.; Ren, Z.; Zhao, S.; Zhao, M.; Dang, J. High-Durability Concrete with Supplementary Cementitious Admixtures Used in Corrosive Environments. *Crystals* **2021**, *11*, 196. [CrossRef]
43. Yang, F.L.; Ji, Y.H.; Liu, W.; Zhang, C.S.; Fang, D.Y. Experimental study on ultra-high-performance concrete with coarse aggregate. *Concrete* **2018**, *12*, 110–113.

materials

Article

Bond Behaviors of Steel Fiber in Mortar Affected by Inclination Angle and Fiber Spacing

Xinxin Ding [1], Mingshuang Zhao [1], Hang Li [1], Yuying Zhang [1], Yuanyuan Liu [1] and Shunbo Zhao [2,*]

1 International Joint Research Lab for Eco-Building Materials and Engineering of Henan, North China University of Water Resources and Electric Power, Zhengzhou 450045, China
2 Collaborative Innovation Center for Efficient Utilization of Water Resources, North China University of Water Resources and Electric Power, Zhengzhou 450045, China
* Correspondence: sbzhao@ncwu.edu.cn

Abstract: Considering the random orientation and distribution of steel fibers in concrete, the synergistic reinforcement of steel fibers on concrete is much complex than the bond of single fiber. It is meaningful to study the bond behavior of steel fiber during many actions. With the inclination angle of steel fiber to pullout direction and the fiber spacing as main factors, this paper carried out fifteen groups of pullout tests for hook-end steel fiber embedded in manufactured sand mortar. The inclination angle ranged from 0 to 60°, and the fiber spacing ranged from 3.5 mm to 21.2 mm. The characteristic pullout load-slip (*PL-S*) curve of steel fibers are given out after treating the original complete curves of each group test. The values of key points featured the debonding, peak and residual pullout loads and slips are determined from the characteristic *PL-S* curves. Based on a multi-index synthetical evaluation method, the nominal debonding strength, bond strength, residual bond strength and the debonding work, slipping work, and pullout work, as well as the debonding energy ratio, slipping energy ratio, and pullout energy ratio are analyzed. Results indicate that the bond performance represented by above indexes changes with the inclination angle and spacing of steel fibers. Except for the bond mechanism performing the same as aligned steel fibers by pullout test, the bond is dominated by the resistance of mortar to peeling off near pullout surface and scraping along pullout direction. When the inclination angle is over 15° or 30°, the bond performance is generally decreased, due to the peeling off of mortar on surface of transversal section with a certain depth. When the fiber spacing is over than 5 mm, the bond performance becomes worst due to the scraping out of mortar along with the slip of steel fibers.

Keywords: hook-end steel fiber; bond performance; pullout test; inclination angle; fiber spacing; pullout load-slip curve

Citation: Ding, X.; Zhao, M.; Li, H.; Zhang, Y.; Liu, Y.; Zhao, S. Bond Behaviors of Steel Fiber in Mortar Affected by Inclination Angle and Fiber Spacing. *Materials* **2022**, *15*, 6024. https://doi.org/10.3390/ma15176024

Academic Editor: Gwenn Le Saout

Received: 13 July 2022
Accepted: 29 August 2022
Published: 31 August 2022

1. Introduction

With short and random distributed steel fibers, a composite material is produced to be the steel fiber reinforced concrete (SFRC) [1–3]. This promotes the post-cracking performances of conventional concrete [4–7]. From the viewpoint of meso-mechanism, the bond of steel fiber is essential to the reinforcement effect, and the steel fibers bridging cracks of concrete matrix enhance the post-cracking performance of SFRC [2,4,7]. In other words, the reinforcement effect depends on the bond which may be affected by the orientation and distribution of steel fibers in concrete [8–10].

The bond mechanisms have attracted considerable attention in active research [11,12]. Many studies based on the pullout test of single or multi aligned fibers have been performed in order to understand the bond behaviors of steel fibers in concrete or mortar. Except for the geometry of steel fiber, rough surface of steel fiber [13] and high-performance cementitious matrix [10,14,15] benefit to the bond performance. By purposefully reinforcing the strength or the toughness of cementitious matrix, or both simultaneously, different deformed steel fibers can be selected with different bond behaviors [16–18].

Normally, the inclination of steel fibers avoids the direct pullout of steel fibers from cementitious matrix. These benefits to the steel fibers worked together with the cementitious matrix [13,19–21]; however, it leads a potential rupture of fibers and matrix with higher tensile stresses [22–28]. In this aspect, the influence of the inclination angle on the pullout behavior of steel fibers depends on the types, size and embedded length of steel fiber and the matrix strength [24,29]. As report by Chun [13], a higher bond strength with a lower energy absorption capacity was observed for the straight steel fibers with inclination angle to 45° in the ultra-high-performance concrete. The study of Huang [19] obtained that the ultimate pullout load and pullout energy increased with the increase in inclination angle from 0° to 45° for the brass-coated straight steel fiber embedded in reactive powder concrete. Similarly, other investigations also indicated that the pullout load reached the peak for straight steel fibers with inclination angle of 30° [20,21], 30° or 45° [22] and 45° [30], regardless of the fiber size and the matrix strength, although the consumption energy during pullout could keep increasing as angles to 60° [20].

Compared to that of the straight steel fibers, the pullout behavior of hook-end fibers presents a less sensitivity to the inclination angle. With the same length of 13 mm and diameter of 0.2 mm, when the inclination angle increased from 0° to 30° and 45°, the bond strengths of straight fiber increased by 19.2% and 52.9%, while those of hook-end fiber increased by 10.3% and 16.2%. Meanwhile, the bond strengths of hook-end fiber with a length of 25 mm and a diameter of 0.35 mm increased by 13.6% and 26.1%, respectively [27]. For both the hook-end steel fiber with a length of 60 mm and diameter of 0.75 mm and the straight fiber cut from the hook-end fiber, when the inclination angle increased from 0° to 30°, the peak pullout load of straight fiber with the embedded lengths of 20 mm and 30 mm increased by 124% and 31.2%, while those of hook-end fiber increased by 15% and 7%, the peak slip increased with the inclination angle [9]. The peak pullout load of hook-end steel fiber with a length of 30 mm and a diameter of 0.38 mm was similar with the inclination angels ranged from 0° to 30°, whereas it increased with the angle and reached the maximum around the inclination angle of 20° for hook-end steel fiber with a length of 60 mm and a diameter of 0.9 mm [24]. Wang [25] has carried out the pullout tests of hook-end steel fiber with a length of 35 mm and a diameter of 0.55 mm embedded in concrete with water to cement ratio of 0.49 at the inclination angles of 0°, 30°, 45° and 60°. The result showed that the bond strength decreased by 27.3% while the peak slip gradually increased with the inclination angle increased from 0° to 60°.

At present, few studies concerned the fiber group effect by pullout test using multi fibers. Feng [31] held that the group effect of the hook-end steel fiber weakened the bond between the fibers and magnesium phosphate cementitious matrix when the spacing of the fibers changed from 16 mm to 6 mm. Kim and Yoo [32,33] reported that approximately 30% lower bond strengths were obtained from the specimens with multiple fibers compared to those with a single straight fiber, a hook-end fiber or a twisted fiber. The average bond strengths of the hook-end and twisted steel fibers were improved by decreasing the fiber spacing up to 1 mm, corresponding to a volume fraction of 7% based on an assumption of perfect fiber distribution. Thus, the synergistic reinforcement mechanism of steel fibers on concrete needs to be enriched.

The pullout performances of different deformed steel fibers with hook-end, crimped, indentation, milled and large-end in the mortars with different strengths and aggregates have been systematically studied in previous works [14,16]. A multi-index synthetical evaluation method was built based on the key points of the characteristic pullout load-slip (*PL-S*) curve to quantitatively evaluate the bond strengths, energy dissipation abilities and toughness of steel fiber, which corresponded to the loading cases of cracking resistance, normal serviceability and ultimate bearing capacity of SFRC, respectively. Results showed that the whole bond performance after the slipping of steel fiber attributed from the increasing strength of mortar with manufactured sand, although distinct reinforcing and toughening effects presented on the bond of different fibers. Therefore, the synergistic working of steel fibers in the concrete matrix should be realized to only or simultaneously improve the strength and toughness of SFRC.

Based on above discussion of previous studies, three series of pullout tests for the hook-end steel fiber embedded in manufactured sand mortar were carried out in this study. The pullout behavior of steel fibers in groups with the implications of the inclination angle, the angle hybrid and the fiber spacing were examined. Two series were tested for steel fibers with the inclination angles and the angles hybrid of 0°, 15°, 30°, 45° and 60°. One series were tested with the number of aligned fibers of 1, 2, 9, 16 and 25. Subsequently, several important pullout parameters [14,16], including the debonding strength, bond strength, residual bond strength and the debonding work, slipping work and pullout work, as well as the debonding energy ratio, slipping energy ratio and pullout energy ratio, were analyzed.

2. Experimental

2.1. Preparation of Mortar

The raw materials used for mortar included ordinary silicate cement, fly ash, manufactured sand, high-performance polycarboxylate water reducer and tap water. The manufactured sand was made of limestone with a fineness modulus of 2.73 and a stone powder content of 7.3%. Properties of the raw materials are the same as previous reports [14,16]. The self-compacting workability of the fresh mortar was contributed from the addition of fly ash as mineral admixture, to ensure the accuracy of the fibers position in the mortar specimen. Details of mixture proportion and workability of mortar is summarized in Table 1.

Table 1. Mix proportion and workability of mortar.

Mix Proportion	Water to binder ratio *w/b*	0.31
	Water (kg/m^3)	277.9
	Cement (kg/m^3)	627.6
	Fly ash (kg/m^3)	269.0
	Manufactured sand (kg/m^3)	1110.4
	Water reducer (kg/m^3)	8.06
Micro slump flow (mm)		250

A planetary-type mortar mixer was used for the mixing of fresh mortar. The water and water reducer, the cement and fly ash were added in the mixing pot to mix for 30 s, then the manufactured sand was evenly added to continuously mix for 30 s. After that, the fresh mortar was mixed for another 30 s, and cast into the molds of specimens.

2.2. Details of Specimen

The steel fiber was hook-end shaped with a length of 29.8 mm, a diameter of 0.5 mm and a tensile strength f_{sf} of 1150 MPa.

A total of sixty dog-bone shape specimens were used in the study. Three series with fifteen group of pullout test were designed; each group had four specimens—refer to the pullout specimen design in the specification of China code CECS 13 [34]. Figure 1a presents the geometric details of specimen for Series IA with two pairs of inclined steel fibers. The specimens, marked as IA0, IA1, IA2, IA3 and IA4, respectively corresponded to the inclination angle of 0°, 15°, 30°, 45° and 60°. Figure 1b presents the geometric details of specimen for Series HIA with two pairs of steel fibers. One pair of steel fiber was arranged with inclination angle of 0°, 15°, 30°, 45° and 60°, respectively, while other pair aligned horizontally along the pullout direction. The specimens were respectively marked as HIA0, HIA1, HIA2, HIA3 and HIA4. Figure 1c presents the geometric details of specimen for Series NA with different number of aligned steel fibers. The area of 16 mm × 16 mm in the center of the cross section is designed as the area for fiber placement. The fiber number are designed as 1, 2, 9, 16 and 25, corresponding to the fiber center-to-center spacing L_{sf} of infinite, 21.2 mm, 7.5 mm, 5 mm and 3.5 mm. The specimens are marked as NA0, NA1,

NA2, NA3 and NA4, successively. Meanwhile, the specimens of Series IA and Series HIA with inclination angle of 0° were used as the ones with 4 aligned steel fibers in spacing of 15 mm.

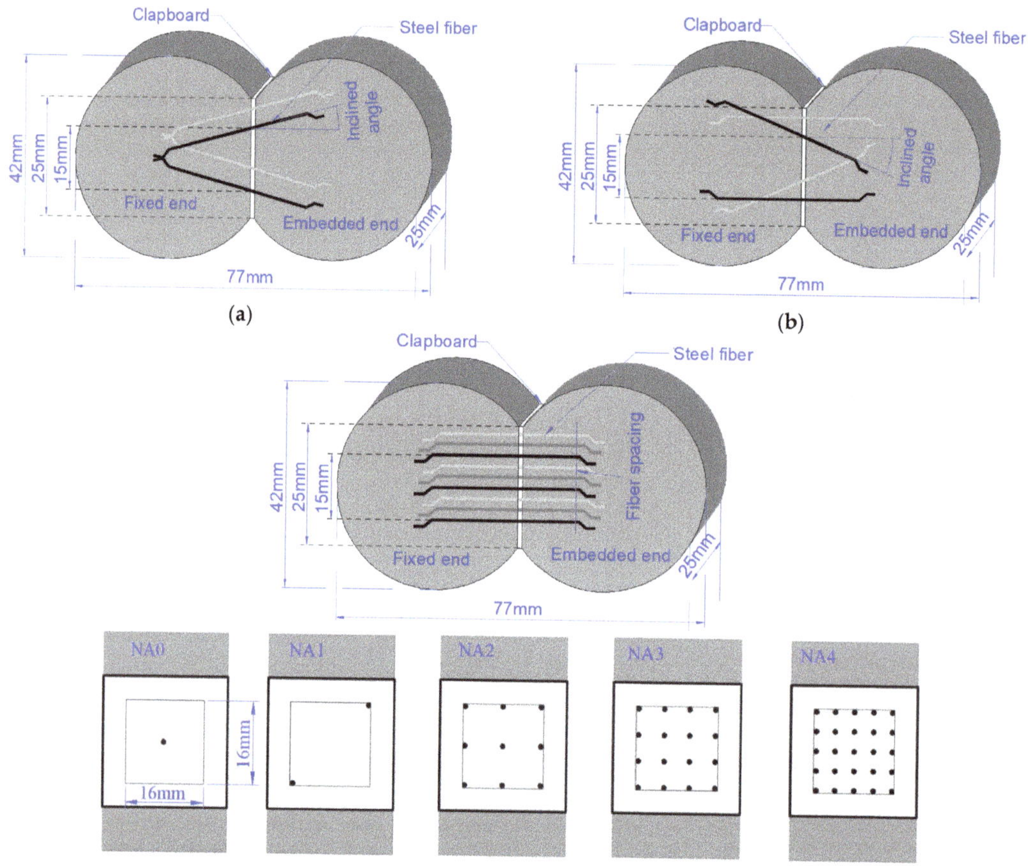

Figure 1. The geometric details of specimen for the pullout test. (**a**) Series IA (**b**) Series HIA (**c**) Series NA.

The pull-out test of steel fibers in mortar was realized by different bond length embedded in the two parts of specimen. One part was used to fix steel fiber with longer length in the mortar, while the steel fibers with shorter length in other part would be pulled out of the mortar. The embedded length and fixed length of steel fibers for Series IA, HIA and NA are given in Table 2.

Table 2. Details of the specimen designed for the pullout test.

Trials	Embedded Length (mm)	Fixed Length (mm)	Fiber Number	Inclination Angle (°)				Fiber Spacing (mm)	Influence Factor
				Fiber 1	Fiber 2	Fiber 3	Fiber 4		
IA0	10	18.8	4	0	0	0	0	15	Inclination angle for two pairs of steel fibers
IA1	10	18.8	4	15	15	15	15	/	
IA2	10	18.8	4	30	30	30	30	/	
IA3	10	18.8	4	45	45	45	45	/	
IA4	10	18.8	4	60	60	60	60	/	
HIA0	12	16.8	4	0	0	0	0	15	Inclination angle for a pair of steel fibers
HIA1	12	16.8	4	0	15	0	15	/	
HIA2	12	16.8	4	0	30	0	30	/	
HIA3	12	16.8	4	0	45	0	45	/	
HIA4	12	16.8	4	0	60	0	60	/	
NA0	11	17.8	1	0	0	0	0	/	Fiber spacing
NA1	11	17.8	2					21.2	
NA2	11	17.8	9					7.5	
NA3	11	17.8	16					5	
NA4	11	17.8	25					3.5	

2.3. Test Method

The specimen was formed in the dog-bone shape mold [14,16]. In order to minimize the disturbance of demolding on the bond between the fibers and the mortar matrix, the specimen was demolded after cast for 2 days, and placed into the standard curing box for next 26 days before testing. The standard curing box has a temperature of 20 ± 2 °C and humidity larger than 95%.

As in previous studies [14,16], the specimen was stretched using an electronic universal testing machine under a loading speed of 0.3 mm/min to obtain the completely tested pull-out load vs. slip curve (*PL-S* curve). Then, as shown in Figure 2, the four tested *PL-S* curves of each group specimens were processed to be a characteristic *PL-S* curve. The detailed processing method for the ascending portion and descending portion of the four tested *PL-S* curves are presented in previous study [16]. The feature values were extracted at the key points, which represent the slope changing in ascending portion, at peak and in descending portion of the characteristic *PL-S* curve.

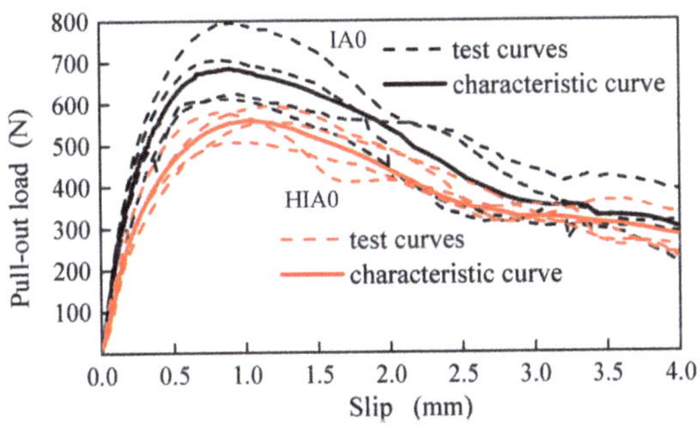

Figure 2. Two characteristic *PL-S* curves from the test curves of two groups of specimens.

Accompanied with each series of specimens, three mortar prisms with dimension of 40 mm × 40 mm × 160 mm were poured for determining its flexural and compressive strengths as per China code GB/T17671 [35]. Due to the difference of casting time and environment, a differential of strength exists in different series of specimens, even though they all have the same composition and test age. The compressive and flexural strengths of mortar used for Series IA, HIA and NA were 75.8 MPa and 11.13 MPa, 69.9 MPa and 8.35 MPa, 77.8 MPa and 12.38 MPa with standard variations of 6.6 MPa and 1.11 MPa, 5.8 MPa and 1.24 MPa, 3.89 MPa and 1.24 MPa, respectively.

3. Test Results and Analyses

3.1. Failure Modes

Two failure modes featured by the fiber pullout and the mortar spalling were observed in this study.

For series IA, the steel fibers of IA0 and IA1 were pullout with straightened hook-end; the steel fibers of IA2, IA3 and IA4 were pullout with straightened hook-end accompanied by the mortar spalling surrounded the fibers. As presented in Figure 3, the volume of mortar spalling significantly increases with the inclination angles. Although a similar peeling off area of the mortar presented on specimens with steel fibers inclined at angle 45° and 60°, a larger peeling off depth happened on the specimens with steel fibers at greater inclination angle. This phenomenon is also reported in the reference [23]. With the increase in inclination angle of steel fiber, the peeling off force perpendicular to transversal section increased during the pull-out of steel fibers, which would be much more increased with the process of steel fibers straightened. This results in the tensile stress that could be over the tensile strength of mortar. If that occurs, the mortar near surface of transversal section will peeling off.

Figure 3. Typical failure mode of Series IA.

Figure 4 presents the typical failure mode of Series HIA. The spalling areas of mortar near the straight and inclined fibers are highlighted as red and blue circles, respectively. All the steel fibers of HIA0, HIA1 and HIA2 were pullout with straightened hook-end. The aligned fibers of HIA3 and HIA4 were pullout with the straightened hook-end, while the inclined steel fibers were pullout with straightened hook-end accompanied by the mortar spalling surrounded the fibers. This also indicted the effect of inclination angle on the bond performance of steel fibers in mortar. With the increase in inclination angle, the straightening degree of the hook-end decreases, while the spalling volume of mortar significantly increases.

Figure 5 presents the typical failure mode of Series NA. All steel fibers of NA0, NA1, NA2 and NA3 were pullout with straightened hook-end, while the mortars of NA4 peeled off accompanied with slightly straightened hook-end of steel fibers. The tensile strength of the mortar near the fiber end is not enough to resist the stress transmitted by the fiber to the matrix. This indicates a rational spacing among steel fibers is necessary to ensure a sufficient surrounding mortar which can provides anchorage shear resistance of the interface between steel fiber and mortar.

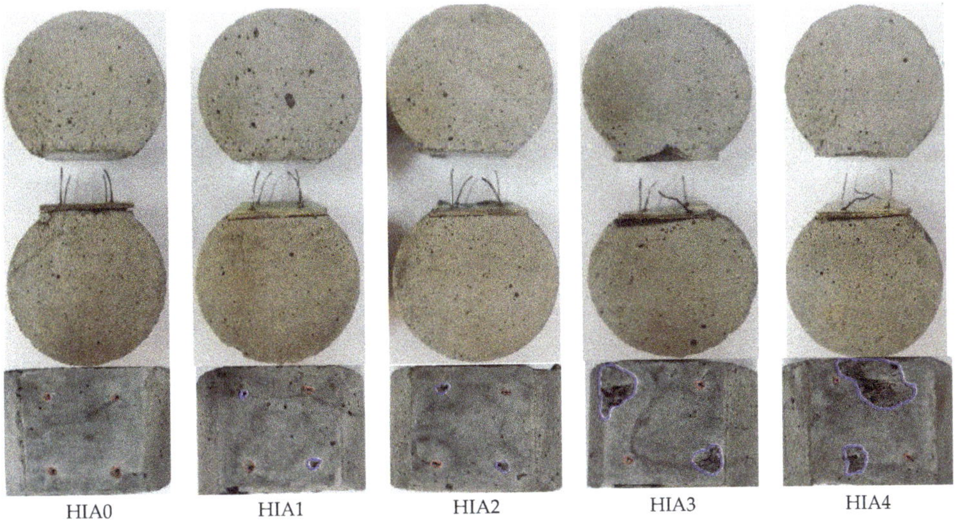

Figure 4. Typical failure mode of Series HIA.

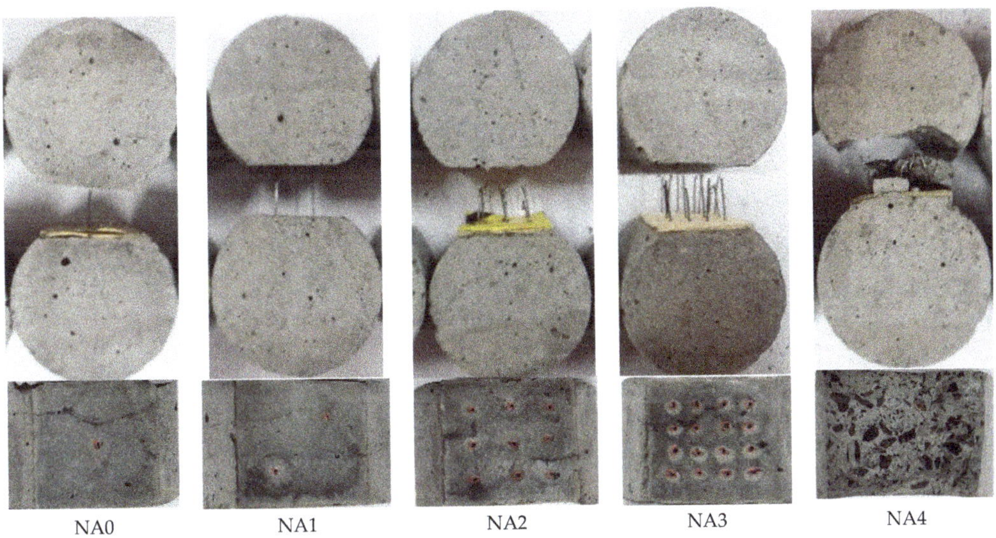

Figure 5. Typical failure mode of Series NA.

Based on the above description, in condition of the mortar with a compressive strength around 74.5 MPa, the steel fibers will be pullout with the straightened hook-end. If the inclination angle of steel fiber is over than 30°, the mortar will be peeled off from the transversal section. When the fiber spacing is 3.5 mm, the mortar surrounded the steel fibers can be scraped out during the pulling out of steel fibers. This indicates the bond performance will be affected by the orientation and distribution of steel fiber in concrete matrix.

3.2. The Characteristic Pullout Load-Slip Curve

Figure 6 presents the characteristic pullout load-slip (*PL-S*) curves of Series IA, HIA and NA. As explained in previous studies [14,16], the bond performance at debonding, peak and residual cases can be expressed by the corresponding point at the characteristic *PL-S* curve with a change of slope in ascending portion, at peak and in descending portion. They respectively represent the debonding load P_d and slip s_d, the peak load P_p and slip s_p, and the residual load P_r and slip s_r. Values of them are listed in Table 3.

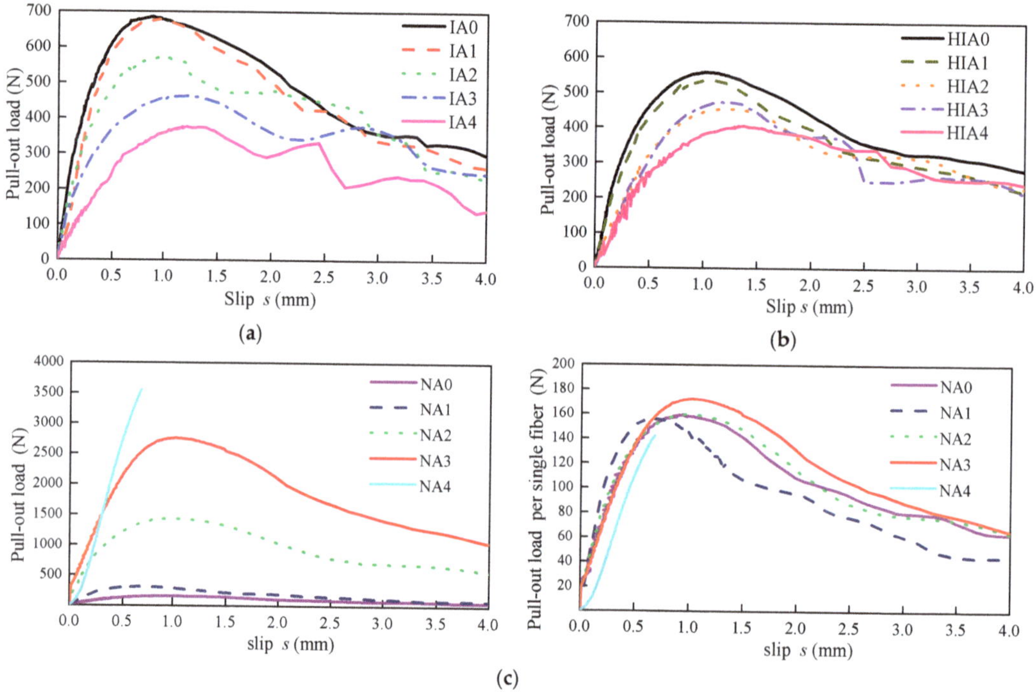

Figure 6. The characteristic PL-S curves. (**a**) Series IA, (**b**) Series HIA and (**c**) Series NA.

Table 3. Test values at key points of the characteristic *PL-S* curves.

Trials	Peak Point		Debonding Point		Residual Point	
	P_p (*N*)	s_p (mm)	P_d (*N*)	s_d (mm)	P_r (*N*)	s_r (mm)
IA0	685.2	0.904	424.8	0.243	363.9	2.893
IA1	678.8	0.967	481.8	0.348	413.8	2.321
IA2	572.8	1.013	286.4	0.196	471.3	1.925
IA3	464.8	1.169	125.5	0.080	338.3	2.221
IA4	375.6	1.220	90.2	0.153	289.9	1.952
HIA0	559	1.028	212.4	0.137	325.1	2.980
HIA1	535.7	1.055	225.0	0.175	331.6	2.321
HIA2	455.7	1.215	259.7	0.379	320.4	2.552
HIA3	474.5	1.135	251.5	0.346	373.3	1.930
HIA4	406.1	1.370	101.5	0.170	338.3	2.466
NA0	159.3	0.946	113.1	0.380	95.9	2.458
NA1	312.4	0.842	203.0	0.216	190.8	2.007
NA2	1437.3	0.984	1078.2	0.378	714.6	2.792
NA3	2755.2	1.020	2176.0	0.503	1569.6	2.672
NA4	3550.0	0.674	3301.5	0.607	-	-

In Series IA, the curve of IA1 with the inclination angle of 15° almost coincides with that of IA0. With the inclination angle increased from 15° to 60°, the slope of the ascending portion decreases, the P_p decreased about 45.2%, while the peak-slip s_p increased about 26.2%. The regularity is also reported by the reference [25]. The P_d increased 13.4% with the inclination angle increased to 15°, then obviously decreased 78.8% with the angle continuously increased to 60°. The P_r increased 29.5% with the inclination angle increased to 30°, then decreased 20.3% with the angle continuously increased to 60°. Both slips s_d and s_r presented the decrease trends with the increased inclination angle. The fluctuation of the descending portion of the curve gradually increases, and the area under curve gradually decreases. The variation of peak-slip is consistent with the reference [36]. Similar regularities are observed in the characteristic curves of Series HIA, the change degree of which almost reduced by half, due to a pair of steel fibers aligned to the pullout direction of series HIA.

In Series NA, the complete *PL-S* curves of NA0, NA1, NA2 and NA3 are obtained with failure mode of fibers pullout. The P_d, P_p and P_r of NA0, NA1 and NA2 are positively correlated with the number of fibers of 1, 2 and 9, successively. The similar slips s_d, s_p and s_r of NA0, NA1, NA2 were observed, while, the P_d and P_p of NA3 are significantly higher than the load values of NA0 multiplying the fiber number. NA4 only got the ascending portion of the curves, due to the failure mode of mortar peeled off.

3.3. Bond Strengths and Strength Ratio

The nominal debonding strength τ_d, bond strength τ_{max} and residual bond strength τ_{res} are calculated using Formulas (1)–(3), and the test values of them are present in Figure 6. To reflect the debonding resistance of steel fiber from mortar and the loss rate of bond strength in the descending portion, the nominal fiber utilization efficiency u_{sf}, the nominal strength ratios u_{de} and u_{res} are calculated using Formulas (4)–(6), and the test values of them are listed in the Table 4.

$$\tau_d = \frac{P_d}{n\pi d_f(l_{f,em} - s_d)} \tag{1}$$

$$\tau_{max} = \frac{P_p}{n\pi d_f(l_{f,em} - s_p)} \tag{2}$$

$$\tau_{res} = \frac{P_r}{n\pi d_f(l_{f,em} - s_r)} \tag{3}$$

$$u_{sf} = \frac{4P_p}{n\pi d_f f_{sf}} \tag{4}$$

$$u_{de} = \frac{\tau_d}{\tau_{max}} \tag{5}$$

$$u_{res} = \frac{\tau_{res}}{\tau_{max}} \tag{6}$$

Table 4. The nominal strength ratios.

Series	IA0	IA1	IA2	IA3	IA4
u_{sf} (%)	75.9	75.2	63.4	51.5	41.6
u_{de} (%)	58.5	67.2	46.5	24.5	21.9
u_{res} (%)	64.7	69.5	89.7	80.6	82.8
Series	**HIA0**	**HIA1**	**HIA2**	**HIA3**	**HIA4**
u_{sf} (%)	61.9	59.3	50.5	52.5	45.0
u_{de} (%)	35.1	38.9	52.9	49.4	22.5
u_{res} (%)	70.7	70.0	80.3	84.9	92.9
Series	**NA0**	**NA1**	**NA2**	**NA3**	**NA4**
u_{sf} (%)	70.6	69.2	70.8	76.3	-
u_{de} (%)	67.5	61.5	71.1	75.4	-
u_{res} (%)	69.7	68.2	59.5	67.1	-

Figure 7a shows the variations of τ_d, τ_{max} and τ_{res} of Series IA with the inclination angle. Slight inclination of steel fibers benefits to the resistance of debonding and the residual bond. This leads that τ_d and τ_{res} reach the maximum at the angle of 15° and the angle of 30° with an increment of 14% and 17%, respectively, compared to that of IA0. However, the τ_{max} trends to decrease with the increase in inclination angle from 0° to 15°, and decreases linearly with the inclination angle increased from 15° to 60°. The τ_{max} of IA4 is about 44% lower than that of IA0. This is similar to the pullout test result of single hook-end steel fiber [31].

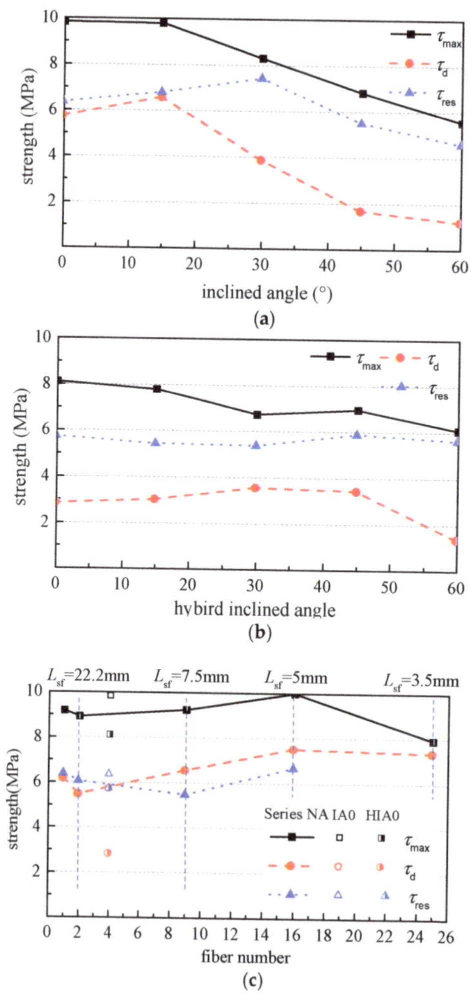

Figure 7. The bond strengths. (**a**) Series IA, (**b**) Series HIA and (**c**) Series NA.

With the increase in the inclination angle, the risk of mortar cracking increases and the fiber utilization rate decreases. The nominal fiber utilization efficiency u_{sf} decreases by 45.2% with the angle increased to 60°. The nominal strength ratio u_{de} increases by 15% with the inclination angle increased from 0° to 15°, and then decreases by 64% with the angle increased from 15° to 60°. This illustrates that the debonding resistance is sensitive to the inclination angle. The nominal strength ratio u_{res} increases by 38% with the inclination angle increased from 0° to 30°, and then changes a little with the continuous increase in the angle. This indicates that the loss rate of bond strength can be reduced with a larger inclination angle of steel fiber, due to the better bond retention by the compressive action of peeling off force perpendicularly on the steel fibers.

Figure 7b shows the variations of τ_d, τ_{max} and τ_{res} of Series HIA with the hybrid inclination angle. Compared with Series IA, half of steel fibers aligned to the pullout direction. This lightened the effect of inclination angle on the bond of steel fibers. With the hybrid action of aligned and inclined steel fibers, the τ_d of HIA2 and HIA3 are 24% and 21% higher than that of HIA0. Although the τ_{max} still trends to decrease with the increase in the inclination angle, the decrement becomes slower. The decrement is 25% with the inclination angle increased from 15° to 60°, which is 56.8% that of the IA4. Therefore, the reduction of τ_{max} comes from the decreased bond strength of the inclined steel fiber in Series HIA, no hybrid effect exists among inclined and aligned steel fibers. A slight increase in the τ_{res} appears with the increase in inclination angle. This is due to the better bond retention of inclined steel fiber during the pullout.

Except for the large inclination angle of 60°, the smaller angle benefits to the debonding resistance and the retention of residual bond. The strength ratio u_{de} increases 50.7% with the inclination angle from 0° to 30°, and then decreases 35.7% with the inclination angle continuously increased from 30° to 60°. The strength ratio u_{res} increases 31.4% with the inclination angle from 0° to 60°. The nominal fiber utilization efficiency u_{sf} decreased by 27.3% with the angle increased to 60°. It is almost half decrease rate compared with the regularity of u_{sf} for series IA. It also illustrates that no hybrid effect exists among inclined and aligned steel fibers.

Figure 7c shows the variations of τ_d, τ_{max} and τ_{res} of Series NA with the fiber number. The influence of fiber number on bond strength essentially relates to the influence of fiber spacing. There are slight decreases of the τ_d, τ_{max}, τ_{res}, u_{sf}, u_{de} and u_{res} of NA1 compared with those of NA0. This may be due to the eccentric pullout on the two steel fibers during the loading process. In addition, the reduction rate of the bond strengths of multiply fibers compared with single fiber is smaller than that with reported in the references [32,33]. This may attribute to the different cementitious matrix and pull-out test method. When the fiber number increased to 9 for NA2, the τ_d, τ_{max}, u_{sf} and u_{de} are basically equal to those of NA0, while the τ_{res} and u_{res} slightly decrease. When the fiber number reached to 16 for NA3, the τ_d, τ_{max}, u_{sf} and u_{de} increase by 20.2%, 8.1%, 8.1% and 11.7% compared with those of NA0, respectively. This indicates that a group effect of parallel fibers with fiber spacing no less than 5 mm benefits to the bond performances. Therefore, an interaction exists among steel fibers in concrete matrix if the steel fibers are uniformly distributed in parallel with a volume fraction over 0.78% (corresponding to L_{sf} of 5 mm).

The values of τ_d, τ_{max} and τ_{res} of IA0 and HIA0 are also presented in the Figure 6c. Ignoring the effect of embedded length, comparing the values of τ_d, τ_{max} and τ_{res} of IA0, HIA0, NA0, NA1 and NA2, all the bond strengths increase with the flexural strength of mortar. The result is consistent with the previous study [14].

In addition, to get the real bond performance of steel fiber without influenced by the eccentric loading, a reasonable number of fibers should be used in the pullout test [32,33,37,38]. Comprehensively considering the range and test accuracy of the test system in this study, four steel fibers symmetrically arranged to section centroid is a better of chose.

Considered the pullout process of inclined steel fibers, the bond of steel fiber to mortar not only comes from the shear stress on interface, but also from the perpendicular pressure on interface, as shown in Figure 8. The former is the same as that of aligned steel fiber, which comes from the chemical adhesion and the mechanical fraction on interface between fiber and mortar and the anchorage of hook-end [14,16]. The latter comes from the component force of pullout load, which directly relies on the pullout load and the peeling off resistance of mortar. If the component force of pullout load is over the peeling off resistance of mortar, the mortar will be peeled off. This leads to broken off mortar from the surface of transversal section to the inner along steel fiber, and results in a shortening of the bond length of steel fiber. The final presentation of the nominal bond strengths by Formulas (1) to (3) will be decreased. With the increase in inclination angle, the component force of pullout load increases to rise the possibility of the peeling off of mortar, as presented in Figures 3 and 4.

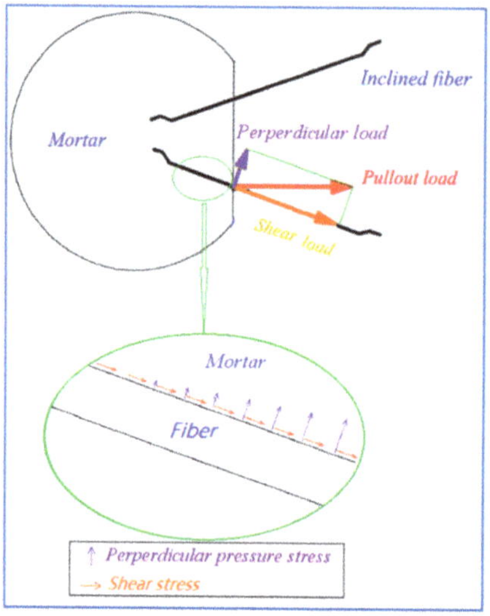

Figure 8. Actions on bond interface of inclined steel fiber.

3.4. Bond Works

The debonding work W_d, the slipping work W_p and the pullout work W_r are used for evaluating the energy dispersion during the bond-slip process. They are the areas under the characteristic *PL-S* curve with the slip from origin to the slips at debonding, peak and residual points, respectively. Formulas are listed as follow:

$$W_d = \int_0^{S_d} Pds \quad (7)$$

$$W_p = \int_0^{S_p} Pds \quad (8)$$

$$W_r = \int_0^{S_r} Pds \quad (9)$$

Table 5 presented the test values of W_d, W_p and W_r. In Series IA, the W_d increases by 30% with the inclination angle from 0 to 15°, and then decreases by 92% with the inclination angle continuously increased from 15° to 60°. The W_p has no change with the inclination angle from 0 to 15°, and then decreases with the increase in inclination angle. The W_p of IA4 is 34% lower than that of IA0. The W_r has a linear decrease with the increase in inclination angle. The W_r of IA4 is 64% lower than that of IA0. Therefore, when the inclination angle of steel fibers is larger than 15°, the bond energy will be sustainably decreased.

In Series HIA, the W_d obviously increases with the inclination angle of steel fibers, while the W_p and W_r trend to decrease with the increase in the inclination angle. This is consistent to the influence of the hybrid action of inclined to aligned steel fibers on the bond strength.

The influence of fiber spacing on the bond energy can be reflected by the bond works per single fiber. Therefore, the bond works W_d, W_p and W_r of Series NA divided the fiber number n are listed in Table 5. With the decrease in fiber spacing from 22.2 mm to 5 mm, the W_d/n, W_p/n and W_r/n increase 263%, 22% and 47%, respectively. This means an improving effect of reasonable fiber number on the bond energy. In addition, the lower bond works of NA1 indicate that the eccentric loading of pullout test for two steel fibers should be avoided to get real bond performance.

Table 5. Bond works.

Series	IA0	IA1	IA2	IA3	IA4
W_d (N·mm)	59	78	32	5	8
W_p (N·mm)	465	469	438	407	308
W_r (N·mm)	1543	1233	892	795	556
Series	**HIA0**	**HIA1**	**HIA2**	**HIA3**	**HIA4**
W_d (N·mm)	12	20	54	40	89
W_p (N·mm)	422	408	387	349	386
W_r (N·mm)	1276	967	893	691	794
Series	**NA0**	**NA1**	**NA2**	**NA3**	**NA4**
W_d/n (N·mm)	25	11	28	40	37
W_p/n (N·mm)	108	101	117	123	–
W_r/n (N·mm)	292	239	337	352	–

3.5. Energy Ratios

The debonding energy ratio R_d, the slipping energy ratio R_{dp} and the pullout energy ratio R_{pr} of hook-end steel fiber are used to represent the energy dissipation ability in different stages of the characteristic *PL-S* curve from the origin to the debonding point, from the debonding point to the peak point and from the peak point to the residual point [14]. Formulas can be written as follow,

$$R_d = \frac{W_d}{W_r} \tag{10}$$

$$R_{dp} = \frac{W_p - W_d}{W_r} \tag{11}$$

$$R_{pr} = \frac{W_r - W_p}{W_r} \tag{12}$$

As shown in Figure 9a, the R_d increases by 65% with the inclination angle from 0 to 15°, and then decreases by 84% with the inclination angle continuously increased to 60°. R_{dp} increases 110% and R_{pr} decreases 36% with the inclination angle increased from 0 to 60°. This indicates that the energy dissipation ability increases before peak-slip and decreases afterward. To be applied for the concrete structures, the inclined steel fiber is favorable to the crack control at the normal serviceability, while less toughness and lower energy dispersion ability at the ultimate bearing capacity.

As presented in Figure 9b, the R_d of Series HIA increases with the inclination angle of steel fiber. The R_d of HIA4 is 11.9 times that of HIA0. This indicates the hybrid effect of inclined and aligned steel fibers is favorable on the energy dissipation before cracking of concrete matrix. The R_{dp} and the R_{pr} of Series HIA present the similar changes to those of Series IA with the increase in inclination angle of steel fiber; however, the changes become slowly due to half number of steel fibers were inclined. The R_{dp} increases by 40% with the inclination angle from 0 to 45°, and then decreases by 16% afterward. The R_{pr} decreases by 23.2% with the inclination angle from 0 to 60°. This once again indicates that the energy dissipation ability increases before peak-slip and decreases afterward with the inclination angle of steel fibers.

As shown in Figure 9c, with the fiber spacing decreased from 22.5 mm to 5 mm, the R_d increases by 146%, the R_{dp} decreases by 37% and the R_{pr} increases by 13%. This indicated that the energy dissipation ability increases at the debonding process and the residual bond stage, which is consistent to the strengthening of tensile strength and toughness of steel fiber reinforced concrete. Meanwhile, it shows that steel fibers in concrete matrix should keep a reasonable spacing to develop their reinforcing contribution. Therefore, the content of steel fiber in concrete should be optimized. At the same time, the variation of energy ratios between NA0 and NA1 can be concluded to the validity of the pullout test with two steel fibers.

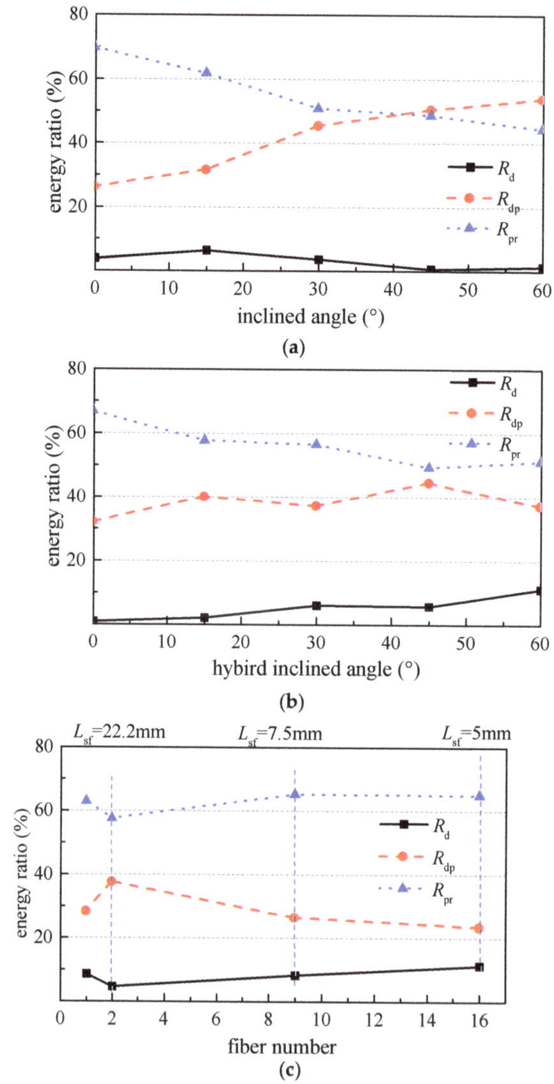

Figure 9. Influence of various factors on the bond energy ratios. (**a**) Series IA, (**b**) Series HIA and (**c**) Series NA.

4. Conclusions

The effect of inclination angle of steel fiber, hybrid inclined and aligned steel fibers and fiber spacing on the bond behaviors of steel fibers in manufactured sand mortar were experimentally studied by the pullout test, results are discussed based on the bond mechanisms of hook-end steel fiber. Conclusions can be drawn as follows:

The nominal debonding strength, debonding work and debonding energy ratio reached the maximum when the inclination angle was 15° and decrease afterward, while the nominal bond strength and the slipping work had slight changes at the inclination angle of 15° and decrease afterward. The nominal residual bond strength reached its maximum at the inclination angle of 30° and decrease afterward. The slipping energy ratio increases, and the pullout work and the pullout energy ratio decrease with the increase in inclination angle. This indicates that the inclination angle of steel fibers has obvious effect on the reinforcing effect of steel fiber on concrete matrix.

Compared with the bond behaviors of specimens with couples of inclined steel fibers, no hybrid effect among the inclined and aligned fibers are observed on the nominal debonding strength and the nominal bond strength. However, a favorable hybrid effect presented on the nominal residual bond strength and the debonding energy ratio. This indicates the different bond performance can be provided by steel fiber in different orientation.

The nominal debonding strength, bond strength and residual strength increases with the decrease in fiber spacing no less than 5 mm. The scraping failure of surrounding mortar appeared in condition of much smaller spacing of steel fibers, due to insufficient mortar resists the scraping force during pullout. Moreover, specimen with fiber number from 4 to 9 in this study are suitable for study the bond mechanism between the fiber and mortar matrix. Comprehensively considering the range and test accuracy of the test system in this study, four steel fibers symmetrically arranged to section centroid is a better choice.

Author Contributions: Conceptualization, X.D. and M.Z.; methodology, S.Z. and Y.Z.; validation, X.D. and S.Z.; formal analysis, M.Z., Y.L. and H.L.; investigation and data curation, H.L. and Y.Z.; writing—original draft preparation, X.D. and Y.L.; writing—review and editing, X.D. and S.Z.; supervision, S.Z.; funding acquisition, X.D. and S.Z. All authors have read and agreed to the published version of the manuscript.

Funding: This research was funded by National Natural Science Foundation of China, grant number 52108212, and Natural Science Foundation of Henan, grant number 212300410192.

Institutional Review Board Statement: Not applicable.

Informed Consent Statement: Not applicable.

Data Availability Statement: Data are available with the first author and can be shared with anyone upon reasonable request.

Conflicts of Interest: The authors declare no conflict of interest.

References

1. Ministry of Housing and Urban-Rural Development of the People's Republic of China. *Steel Fiber Reinforced Concrete*; JG/T 472-2015; China Architecture & Building Press: Beijing, China, 2015.
2. Zhao, M.L.; Li, J.; Law, D. Effects of flowability on SFRC fibre distribution and properties. *Magaz. Concr. Res.* **2017**, *69*, 1043–1054. [CrossRef]
3. Ding, X.X.; Li, C.Y.; Han, B.; Lu, Y.Z.; Zhao, S.B. Effects of different deformed steel-fibers on preparation and fundamental properties of self-Compacting SFRC. *Constr. Build. Mater.* **2018**, *168*, 471–481. [CrossRef]
4. Zhao, M.L.; Zhao, M.S.; Chen, M.H.; Li, J.; David, L. An experimental study on strength and toughness of steel fiber reinforced expanded-Shale lightweight concrete. *Constr. Build. Mater.* **2018**, *183*, 493–501. [CrossRef]
5. Ding, X.X.; Li, C.Y.; Zhao, M.L.; Li, J.; Geng, H.B.; Lian, L. Tensile strength of self-Compacting steel fiber reinforced concrete evaluated by different test methods. *Crystals* **2021**, *11*, 251. [CrossRef]
6. Li, C.Y.; Shang, P.R.; Li, F.L.; Feng, M.; Zhao, S.B. Shrinkage and mechanical properties of self-Compacting SFRC with calcium sulfoaluminate expansive agent. *Materials* **2020**, *13*, 588. [CrossRef] [PubMed]
7. Ding, X.X.; Li, C.Y.; Li, Y.Z.; Song, C.; Zhao, S.B. Experimental and numerical study on stress-Strain behavior of self-Compacting SFRC under uniaxial compression. *Constr. Build. Mater.* **2018**, *185*, 30–38. [CrossRef]
8. Zhao, M.S.; Zhang, X.Y.; Song, W.H.; Li, C.Y.; Zhao, S.B. Development of steel fiber reinforced expanded-Shale lightweight concrete with high freeze-Thaw resistance. *Adv. Mater. Sci. Eng.* **2018**, *2018*, 9573849. [CrossRef]
9. Cunha, V.M.C.F.; Barros, J.A.O.; Sena-Cruz, J.M. Pullout behavior of steel fibers in self-Compacting concrete. *J. Mater. Civ. Eng.* **2010**, *22*, 1–9. [CrossRef]

10. Zhao, M.L.; Li, J.; Xie, Y.M. Effect of vibration time on steel fibre distribution and flexural properties of steel fibre reinforced concrete with different flowability. *Case Stud. Constr. Mater.* **2022**, *16*, e01114.
11. Abdallah, S.; Fan, M.; Rees, D.W.A. Bonding mechanisms and strength of steel fiber–Reinforced cementitious composites: Overview. *J. Mater. Civ. Eng.* **2018**, *30*, 04018001. [CrossRef]
12. Isla, F.; Arganaraz, P.; Luccioni, B. Numerical modelling of steel fibers pull-Out from cementitious matrixes. *Constr. Build. Mater.* **2022**, *332*, 127373. [CrossRef]
13. Chun, B.; Kim, S.; Yoo, D.Y. Benefits of chemically treated steel fibers on enhancing the interfacial bond strength from ultra-high-Performance concrete. *Constr. Build. Mater.* **2021**, *294*, 123519. [CrossRef]
14. Ding, X.X.; Zhao, M.L.; Li, C.Y.; Li, J.; Zhao, X.S. A multi-Index synthetical evaluation of pull-Out behaviors of hooked-End steel fiber embedded in mortars. *Constr. Build. Mater.* **2021**, *276*, 122219. [CrossRef]
15. Yuan, M.; Liang, E.; Yan, D.H.; Wei, B.D.; Liu, Y.; Huang, L. Investigation on effect of mixture ratio on interfacial bonding properties of steel fiber-Matrix. *J. Chang. Univ. Nat. Sci. Ed.* **2020**, *40*, 57–66.
16. Ding, X.X.; Geng, H.B.; Zhao, M.L.; Chen, Z.; Li, J. Synergistic bond properties of different deformed steel fibers embedded in mortars wet-Sieved from self-Compacting SFRC. *Appl. Sci.* **2021**, *11*, 10144. [CrossRef]
17. Esmaeili, J.; Andalibi, K.; Gencel, O.; Maleki, F.K.; Maleki, V.A. Pull-Out and bond-Slip performance of steel fibers with various ends shapes embedded in polymer-Modified concrete. *Constr. Build. Mater.* **2021**, *271*, 121531. [CrossRef]
18. Zhan, Y.; Meschke, G. Analytical model for the pullout behavior of straight and hooked-End steel fibers. *J. Eng. Mech.* **2014**, *140*, 04014091. [CrossRef]
19. Huang, L.; Yuan, M.; Wei, B.D.; Yan, D.H.; Liu, Y. Experimental investigation on single fiber pullout behaviour on steel fiber-Matrix of reactive powder concrete (RPC). *Constr. Build. Mater.* **2022**, *318*, 125899. [CrossRef]
20. Chen, L.; Sun, W.W.; Chen, B.C.; Shi, Z.; Lai, J.Z.; Feng, J. Multiscale study of fibre orientation effect on pullout and tensile behavior of steel fibre reinforced concrete. *Constr.Build. Mater.* **2021**, *283*, 122506. [CrossRef]
21. Huo, L.Y.; Bi, J.H.; Zhao, Y.; Wang, Z.Y. Constitutive model of steel fiber reinforced concrete by coupling the fiber inclining and spacing effect. *Constr. Build. Mater.* **2021**, *280*, 122423. [CrossRef]
22. Lee, Y.; Kang, S.T.; Kim, J.K. Pullout behavior of inclined steel fiber in an ultra-High strength cementitious matrix. *Constr. Build. Mater.* **2010**, *24*, 2030–2041. [CrossRef]
23. Yoo, D.Y.; Choi, H.J.; Kim, S.H. Bond-Slip response of novel half-Hooked steel fibers in ultra-High-Performance concrete. *Constr. Build. Mater.* **2019**, *224*, 743–761. [CrossRef]
24. Cao, Y.Y.Y.; Yu, Q.L. Effect of inclination angle on hooked end steel fiber pullout behavior in ultra-High performance concrete. *Compos. Struct.* **2018**, *201*, 151–160. [CrossRef]
25. Wang, X.W.; Guo, R.; Tian, W.L.; Zhao, X.Y. Experimental investigation of the bonding and slipping performance of the deformed steel-Fiber. *J.-Hebei Univ. Technol.* **2007**, *36*, 110–114.
26. Sun, W.; Gao, J.M.; Qin, H.G. Studies on bond strength of interface between fiber and matrix in steel fiber reinforced concrete. *J. Chin. Ceram. Soci.* **1985**, *13*, 292–300.
27. Qi, J.; Wu, Z.; Ma, Z.J.; Wang, J. Pullout behavior of straight and hooked-End steel fibers in UHPC matrix with various embedded angles. *Constr. Build. Mater.* **2018**, *191*, 764–774. [CrossRef]
28. Zhang, J.; Li, V.C. Effect of inclination angle on fiber rupture load in fiber reinforced cementitious composites. *Compos. Sci. Technol.* **2002**, *62*, 775–781. [CrossRef]
29. Laranjeira, F.; Aguado, A.; Molins, C. Predicting the pullout response of inclined straight steel fibers. *Mater. Struct.* **2010**, *43*, 875–895. [CrossRef]
30. Krahl, P.A.; Gidrão, G.D.M.S.; Neto, R.B.; Carrazedo, R. Effect of curing age on pullout behavior of aligned and inclined steel fibers embedded in UHPFRC. *Constr. Build. Mater.* **2021**, *266*, 121188. [CrossRef]
31. Feng, H.; Sheikh, M.N.; Hadi, M.N.; Feng, L.; Gao, D.; Zhao, J. Pullout behaviour of different types of steel fibres embedded in magnesium phosphate cementitious matrix. *Int. J. Concr. Struct. Mater.* **2019**, *13*, 33. [CrossRef]
32. Kim, J.J.; Yoo, D.Y. Effects of fiber shape and distance on the pullout behavior of steel fibers embedded in ultra-High-Performance concrete. *Cem. Concr. Compos.* **2019**, *103*, 213–223. [CrossRef]
33. Yoo, D.Y.; Kim, J.J.; Park, J.J. Effect of fiber spacing on dynamic pullout behavior of multiple straight steel fibers in ultra-High-Performance concrete. *Constr. Build. Mater.* **2019**, *210*, 461–472. [CrossRef]
34. *CECS13:2009*; China Association for Engineering Construction Standardization. Standard Test Methods for Fiber Reinforced Concrete. China Planning Press: Beijing, China, 2009.
35. *GB/T17671-1999*; The State Bureau of Quality Technical Supervision. Method of Testing Cements-Determination of Strength. The State Bureau of Quality Technical Supervision: Beijing, China, 1999.
36. Breitenbücher, R.; Meschke, G.; Song, F.; Zhan, Y. Experimental, analytical and numerical analysis of the pullout behaviour of steel fibres considering different fibre types, inclinations and concrete strengths. *Struct. Concr.* **2014**, *15*, 126–135. [CrossRef]
37. Xu, J.E.; Sun, J.Y.; Luo, G.D. Experimental study on the bond between steel fiber and concrete by orthogonal design. In Proceedings of the Fourth China Academic Conference on Fiber Reinforced Cement and Concrete, Nanjing, China, 10 November 1992.
38. Yang, M. Strengthening and Toughening Mechanism of Steel Fiber Reinforced High-Strength Concrete and Design Method Based on Toughness. Ph.D. Thesis, Dalian University of Technology, Dalian, China, 2006.

MDPI
St. Alban-Anlage 66
4052 Basel
Switzerland
Tel. +41 61 683 77 34
Fax +41 61 302 89 18
www.mdpi.com

Materials Editorial Office
E-mail: materials@mdpi.com
www.mdpi.com/journal/materials

www.ingramcontent.com/pod-product-compliance
Lightning Source LLC
Chambersburg PA
CBHW061757210926
51850CB00006B/579
9783036567426